Lecture Notes in Civil Engineering

Volume 300

Lecture Notes in Civil Engineering (LNCE) publishes the latest developments in Civil Engineering—quickly, informally and in top quality. Though original research reported in proceedings and post-proceedings represents the core of LNCE, edited volumes of exceptionally high quality and interest may also be considered for publication. Volumes published in LNCE embrace all aspects and subfields of, as well as new challenges in, Civil Engineering. Topics in the series include:

- Construction and Structural Mechanics
- Building Materials
- Concrete, Steel and Timber Structures
- Geotechnical Engineering
- Earthquake Engineering
- Coastal Engineering
- Ocean and Offshore Engineering; Ships and Floating Structures
- Hydraulics, Hydrology and Water Resources Engineering
- Environmental Engineering and Sustainability
- Structural Health and Monitoring
- Surveying and Geographical Information Systems
- Indoor Environments
- Transportation and Traffic
- Risk Analysis
- Safety and Security

To submit a proposal or request further information, please contact the appropriate Springer Editor:

- Pierpaolo Riva at pierpaolo.riva@springer.com (Europe and Americas);
- Swati Meherishi at swati.meherishi@springer.com (Asia—except China, Australia, and New Zealand);
- Wayne Hu at wayne.hu@springer.com (China).

All books in the series now indexed by Scopus and EI Compendex database!

Kasinathan Muthukkumaran · R. Ayothiraman ·
Sreevalsa Kolathayar
Editors

Soil Dynamics, Earthquake and Computational Geotechnical Engineering

Proceedings of the Indian Geotechnical Conference 2021 Volume 5

Springer

Editors
Kasinathan Muthukkumaran
Department of Civil Engineering
National Institute of Technology
Tiruchirappalli
Tiruchirappalli, Tamil Nadu, India

R. Ayothiraman
Department of Civil Engineering
Indian Institute of Technology, Delhi
New Delhi, Delhi, India

Sreevalsa Kolathayar
Department of Civil Engineering
National Institute of Technology
Surathkal, Karnataka, India

ISSN 2366-2557 ISSN 2366-2565 (electronic)
Lecture Notes in Civil Engineering
ISBN 978-981-19-6997-3 ISBN 978-981-19-6998-0 (eBook)
https://doi.org/10.1007/978-981-19-6998-0

This Springer imprint is published by the registered company Springer Nature Singapore Pte Ltd.
The registered company address is: 152 Beach Road, #21-01/04 Gateway East, Singapore 189721, Singapore

Preface

The Indian Geotechnical Society, Trichy (IGS-Trichy) Chapter, and National Institute of Technology (NIT), Tiruchirappalli, India, organized the Indian Geotechnical Conference (IGC-2021) at Trichy during 16–18 December 2021. The main theme of the conference was **"GEO–INDIA"—GEO**TECHNICS FOR **IN**FRASTRUCTURE **D**EVELOPMENT AND **I**NNOVATIVE **A**PPLICATIONS.

The sub-themes of the conference included:

1. Soil Behaviour and Characterization of Geomaterials
2. Geotechnical, Geological and Geophysical Investigation
3. Foundation Engineering
4. Ground Improvement Techniques
5. Geo-Environmental Engineering
6. Soil Dynamics and Earthquake Geotechnical Engineering
7. Earth Retaining Structures, Dams and Embankments
8. Slope Stability and Landslides
9. Transportation Geotechnics
10. Geosynthetics Application
11. Computational, Analytical and Numerical Modelling
12. Rock Engineering, Tunnelling, Deep Excavations and Underground Constructions
13. Forensic Geotechnical Engineering and Case Studies
14. Others: Behaviour of Unsaturated Soils, Offshore & Marine Geotechnics, Remote Sensing & GIS, Instrumentation & Monitoring, Retrofitting of Geotechnical Structures, Reliability in Geotechnical Engineering, Geotechnical Education, Codes & Standards, & any other relevant topic.

The proceedings of this conference consists of selected papers presented at the conference. The proceedings are divided into six volumes. A special issue on IGC-2021 keynote and theme lecture presentations were published by Indian Geotechnical Journal.

We sincerely thank all the authors who have contributed their papers to the conference proceedings. We also thank all the theme editors and reviewers who have been instrumental in giving their valuable inputs for improving the quality of the final papers. We greatly appreciate and thank all the student volunteers for their unwavering support that was instrumental in preparation of this proceedings. Finally, thanks to the Springer team for their support and full cooperation for publishing six volumes of this IGC-2021 proceedings.

Trichy, India

Kasinathan Muthukkumaran
Chairman

Contents

About the Editors

Dr. Kasinathan Muthukkumaran is currently Professor in Civil Engineering at National Institute of Technology, Tiruchirappalli, India. He obtained Ph.D. in Soil-Structure Interaction and Marine Geotechnical Engineering from Indian Institute Technology Madras. He has published more than 150 papers in international and national journals and conferences. He has completed 5 R&D (including ISRO—**Chandrayaan-2 Mission project**) and 70 major consultancy projects in Geotechnical Engineering and published two patent including **"Moon Soil"** (A Method for Manufacture of Highland Lunar Soil Simulant). He has guided 10 Ph.D. scholars and 6 more are in progress, 5 MS (by research) and more than 40 M.Tech. students in Geotechnical and allied research areas. He is the **Founder Chairman** of Indian Geotechnical Society (IGS-Trichy) Trichy Chapter. He is a member of Technical Committee (TC-301 on **"Preservation of Historic Sites"**) of International Society for Soil Mechanics and Geotechnical Engineering. Prof. Muthukkumaran area of research is in geotechnical engineering, which includes pile foundation, soil-structure interaction, marine geotechnics and foundations, field instrumentation, geotechnical physical modeling, ground improvement and forensic geotechnical engineering. He has received DST Young Scientist Award, IGS-Smt. Indra Joshi Biennial Award and Keynote Paper Award—GEOMATE Conference 2015 at Osaka, Japan. He is an Associate Editor of *Australian Journal of Civil Engineering and Serving* editorial board member of several journals. Prof. Muthukkumaran has significant administrative contribution as Estate Officer, Associate Dean (Planning and Development), Member of Buildings and Works Committee and Member of **Board of Governors (BoG)** of National Institute of Technology, Tiruchirappalli and Member of Buildings and Works Committee, IIM Trichy. He has received **NIT Trichy Achiever Awards** for research publications, research projects, maximum citation and consultancy projects.

Dr. R. Ayothiraman is currently working as Professor in the Department of Civil Engineering, Indian Institute of Technology (IIT) Delhi. He completed his Graduation in Civil Engineering from Madurai Kamaraj University in April 1998 and Post-Graduation from College of Engineering Guindy in 2000 with Gold Medal. He obtained his Ph.D. in the area of Soil Dynamics from IIT Madras in 2005. His areas of research interest include soil dynamics, pile foundations, tunnelling in soils and rocks and ground improvement. He has published about 95 papers in reputed journals and conferences. He supervised nine Ph.D. theses and is currently supervising 13 Ph.D. research scholars, in addition to the supervision of about 85 M.Tech./B.Tech. students. He is a recipient of Young Engineer Award from IEI and DST and DAAD Fellowship. He has actively participated in organizing several short courses, workshops and national/international conferences. He has delivered invited lectures at many short courses, workshops, seminars and conferences. He is a member of several professional societies and reviewer of several reputed international and national journals. He is also serving as member of BIS for revising National Building Code and Codes related to machine foundation. He has been a consultant in providing technical advice for several metro projects including Delhi Metro, Kolkata Metro and Lucknow metro and several other projects of national importance.

Dr. Sreevalsa Kolathayar pursued M.Tech. from Indian Institute of Technology (IIT) Kanpur, Ph.D. from Indian Institute of Science (IISc) and served as International Research Staff at UPC Barcelona Tech Spain. He is presently Assistant Professor in the Department of Civil Engineering, National Institute of Technology (NIT) Karnataka, Surathkal, India. Dr. Sreevalsa has authored five books, edited 12 books and published over 80 research papers. He is associate editor of two international journals and acted as a reviewer for many. His research interests include ground motion attenuation and seismic hazard assessment, local site effects and liquefaction susceptibility, pseudo-dynamic approach for seismic loading, disaster risk reduction, geogrids and geocells and water geotechnics. He is currently the Secretary Indian chapter of International Association for Coastal Reservoir Research (IACRR) and Executive Committee Member of Indian Society of Earthquake Technology. In 2017, The New Indian Express honoured Dr. Sreevalsa with South India's Most Inspiring Young Teachers Award. He is the recipient of ISET D. K. Paul Research Award from Indian Society of Earthquake Technology in 2018. He is in the roster of two technical committees of ASCE Geo-Institute. He received "IEI Young Engineers Award" by The Institution of Engineers (India), in recognition of his contributions in the field of Civil Engineering. Recently, Dr. Sreevalsa was featured by the American Society of Civil Engineers in their Geostrata Magazine.

Chapter 1
Evaluation of Compactive Parameters of Soil Using Machine Learning

Jitendra Khatti and Kamaldeep Singh Grover

Introduction

The surface and subsurface of the earth are formed by soil and rocks. These soils are classified by engineering properties and index properties of soil. The compressibility, permeability, and strength properties are engineering properties of soil. The engineering properties is determined using laboratory and field methods. The maximum dry density and optimum moisture content are the most important engineering properties of soil, and these are determined using standard proctor test and modified proctor test [8]. The optimum moisture content and maximum dry density are the compaction parameters. The experimental procedure for determining the compaction parameters is a tedious and time-consuming task, and it also requires human resources.

Escaping from the tedious and time-consuming process, numerous researchers and scientists evolved different methods and methodologies to determine or predict soil parameters. The authors proposed MEP models to predict the OMC and MDD of soil, and authors predicted OMC and MDD with 0.9607 and 0.9263, respectively [10]. The investigators developed a regression model using SVM [6]. The investigators concluded that the proposed models predicted OMC and MDD with 0.9237 and 0.9487, respectively. The authors suggested that the AI approaches can be used to predict the OMC and MDD for modified and standard proctor tests [20]. The researchers evolved ANN models to predict the OMC and MDD of soil and reported that the performance of NN 10-5-1 OMC and NN 10-7-1 MDD model was 0.8855 and 0.9754, respectively [1]. The authors proposed a model to predict the dry density using the thermal conductivity of the soil [15]. The authors concluded that

J. Khatti (✉) · K. S. Grover
Department of Civil Engineering, RTU, Kota, Rajasthan 324010, India
e-mail: jitendrakhatti197@gmail.com

K. S. Grover
e-mail: ksgrover@rtu.ac.in

K. Muthukkumaran et al. (eds.), *Soil Dynamics, Earthquake and Computational Geotechnical Engineering*, Lecture Notes in Civil Engineering 300,
https://doi.org/10.1007/978-981-19-6998-0_1

the proposed model is able to predict the dry density with 0.9923 performance. The authors stated that the genetic algorithm approach is able to predict the MDD of soil [15]. The researchers developed ANN models to predict MDD and UCS [18] and reported that the BRNN, DENN, LMNN, SVM, FN, MARS, MLR models predicted MDD with 0.84, 0.88, 0.76, 0.93, 0.92, 0.94, and 0.45 performance. The authors evolved regression and ANN models. The performance of models was compared, and it was concluded that the CFB-10 predicted OMC and MDD with 0.9323 and 0.9263, respectively [2]. The authors evolved ANN models to predict the compaction parameters and reported that the evolved ANN models predicted OMC and MDD with 0.92 and 0.91 performance [12]. The investigators proposed a MARS model to predict the compaction parameters of coarse-grained soil [19]. The authors concluded that the performance of OMC and MDD models was 0.9381 and 0.9000 and that is more acceptable to previous models. The authors proposed MLR models to predict the OMC and MDD of soil [13]. The authors reported that the proposed MLR models have the potential to predict the OMC and MDD of soil. The author evolved the regression model to predict the OMC and MDD of soil using index properties [14]. The authors concluded that the MLR analysis provides reliable results. Many authors and researchers stated that the artificial intelligence approaches required fewer human resources and it is less time-consuming [3, 4, 7, 11, 16, 17]. It was also concluded that the AI approaches give reliable results of OMC and MDD of soil. From the literature study, it has been observed that the authors used different numbers of data for predicting OMC and MDD of soil. The MLR and ANN models have not been compared for same datasets. Hence, the optimum performance AI models is still doubtable.

Problem Statement

The present study has the following aims:

- Develop the artificial neural network models using different hyperparameters and determine the best performance ANN model.
- Draw the comparison between the best architecture ANN model and regression model to determine the optimum performance AI models.
- Map the comparison between the optimum performance AI models with published models in literature study.
- Predict the OMC and MDD of soil using the optimum performance AI models.

Methodology

Regression analysis and artificial neural network approaches have been used to predict the compaction parameters of the soil in this work. The detail of regression analysis and artificial neural network is given below.

Detail of Multilinear Regression Analysis

Regression analysis is a traditional method of prediction or forecasting of data. Regression analysis is two types; one is simple regression analysis, and the second is multiple regression analysis. The relationship is drawn between the one input and output variable in the simple regression analysis. The proposed relationship gives an equation to predict the output variable by input variable. In multiple regression analysis, the input and output variables are more than one, and the input variables with respect to output variables propose an equation. The simple regression analysis and multiple regression analysis are based on the linear or nonlinear approaches.

The multilinear regression (MLR) analysis has been performed in the present research work to evaluate the compaction parameters of soil. The fine content (FC), sand content (SC), specific gravity (SG), liquid limit (LL), and plasticity index (PI) have been used to develop the MLR models to predict the OMC and MDD of soil. The MLR_OMC and MLR_MDD models have been developed using the data analysis tool of Microsoft Excel 2019. An equation has been proposed by MLR_OMC and MLR_MDD model to predict the optimum moisture content and maximum dry density of soil, as written in Eqs. 1.1 and 1.2.

$$\begin{aligned} \mathrm{OMC}' = {} & 15.992 + 0.0243 * \mathrm{FC} - 0.083 * \mathrm{S} - 2.102 * \mathrm{SG} \\ & + 0.2508 * \mathrm{LL} - 0.071 * \mathrm{PI} \end{aligned} \quad (1.1)$$

$$\begin{aligned} \mathrm{MDD}' = {} & 1.725 - 0.002 * \mathrm{FC} + 0.0018 * \mathrm{S} + 0.0596 * \mathrm{SG} \\ & - 0.005 * \mathrm{LL} + 0.0058 * \mathrm{PI} \end{aligned} \quad (1.2)$$

where OMC′ and MDD′ are the predicted optimum moisture content and maximum dry density of 19 soil specimens.

Detail of Artificial Neural Network

An artificial neural network is an approach to machine learning techniques, and the machine learning techniques are a subset of artificial intelligence. The artificial neural network is nothing, but it is a network of different layers. The artificial neural network consists of an input layer, hidden layer(s), and output layer. Neurons connect these layers and transmit the information from one layer to another layer. The hidden layer(s) consists of activation function and bias. Similarly, the output layer consists of an activation function. These activation functions can be linear or nonlinear.

The multilayer perceptron (MLP) class-based artificial neural network models have been developed with different hidden layers and neurons in the present research work using MATLAB R2020a. The number of hidden layers has been selected in

Table 1.1 Hyperparameters of ANN models

Hyperparameters	Value
Hidden layers	One, two, three, four, five
Neurons	Two to eleven
Backpropagation algorithm(s)	Levenbergs–Marquardt
Normalizing function(s)	Min–max for input and log for output parameters
Activation function(s)	Sigmoid at hidden layers, Linear at the output layer
Train: Validation ratio	70: 30
Epochs	1000 (default)
Network type	Feed-forward backpropagation
Network class	Multilayer perceptron class (MLP)
Mu	0.001
Max fail	6
Min gradient	10e−7

Table 1.2 Architecture of ANN models

Model	Architecture	Remarks
Optimum moisture content	ANN_OMC_XHY	Where *X*, *Y* is the number of hidden layers and neurons
Maximum dry density	ANN_MDD_XHY	

the range of one to five with two to eleven neurons. The hyperparameters of artificial neural network models are given in Table 1.1.

A total of one hundred artificial neural network models have been developed to predict the OMC and MDD of soil. Fifty NN models of OMC and fifty NN models of MDD have been developed. The architecture of NN models of OMC and MDD is given in Table 1.2.

Data Analysis

In the present research work, 53 datasets have been used to develop the AI models and predict the compactive parameters of soil. The experiments were performed in the geotechnical laboratory to obtain these datasets. The tests of specific gravity, consistency limits, standard proctor test were performed as per IS 2720: 1980 (P-3), IS 2720: 1985 (P-5), and IS 2720: 1980 (P-7), respectively. The consistency limits of soil are liquid limit, plastic limit, and plasticity index, and standard proctor test is performed to determine the OMC and MDD of soil.

Dataset of Soil

A total of 53 datasets of soil have been used to carry out the present research work. The datasets consist of fine-grain content (FC), sand content (SC), specific gravity (SG), liquid limit (LL), plasticity index (PI), maximum dry density (MDD), and optimum moisture content (OMC). The datasets of soil are given in Table 1.3.

Statistics of Dataset of Soil

The statistics of datasets of soil have been mapped using the data analysis tool of Microsoft Excel 2019. The statistical parameters such as minima, maxima, mean, median, standard deviation, and confidence level (95%) of soil datasets have been calculated as shown in Table 1.4.

The sand content, fine content, LL, and PI affect the compaction parameters of soil. The correlation coefficient method has been used to map the relationship between input parameters (sand content, fine content, LL, and PI) with compaction parameters. The relationship between input parameters and compaction parameters of soil has been drawn in the form of Pearson matrix, as given in Table 1.5.

The coefficient of correlation of sand content, fine content, specific gravity, liquid limit, plasticity index with optimum moisture content is 0.96, 0.86, 0.31, 0.96, and 0.68. Similarly, the correlation coefficient of sand content, fine content, specific gravity, liquid limit, plasticity index with maximum dry density is 0.92, 0.92, 0.31, 0.91, and 0.54. The relationship between input and output parameters is shown in Fig. 1.1.

From Fig. 1.1, it has been observed that the fine content, sand content, liquid limit have strong correlation with OMC and MDD. The specific gravity and plasticity index have good correlation with OMC and MDD [9].

Training and Testing Datasets

Regression analysis and artificial neural network models have been developed to predict the compaction parameters of the soil. The linear approach has been used to develop regression models. Thirty-four datasets of soil have been used to train the proposed MLR and ANN models, and the rest of the datasets have been used to test the proposed models. These thirty-four soil datasets have been subdivided into 70 and 30% to train and validate the ANN models.

Table 1.3 Datasets of soil

S. No	FG (%)	S (%)	SG	LL (%)	PI (%)	OMC (%)	MDD (gm/cc)
1	87.00	13.00	2.62	46.80	18.43	22.79	1.58
2	4.00	96.00	2.71	26.40	9.29	8.09	1.97
3	0.00	82.00	2.70	26.86	11.59	8.75	2.01
4	37.00	25.00	2.62	38.57	15.68	19.35	1.74
5	18.00	82.00	2.69	26.07	12.07	10.12	1.93
6	5.00	95.00	2.71	26.26	9.18	8.21	1.96
7	16.00	84.00	2.70	25.74	11.10	9.89	1.94
8	17.20	70.40	2.69	28.45	17.86	10.90	1.95
9	35.17	60.88	2.53	28.61	10.87	11.94	1.85
10	13.04	68.84	2.69	28.62	18.09	11.20	1.94
11	8.39	72.87	2.69	28.26	15.25	10.22	1.96
12	0.80	97.46	2.70	26.46	9.80	7.61	1.98
13	72.00	28.00	2.64	38.29	15.12	16.73	1.72
14	28.00	63.00	2.53	28.84	13.19	11.85	1.89
15	42.00	55.00	2.65	28.23	10.62	12.28	1.81
16	15.00	85.00	2.70	25.63	10.63	9.76	1.94
17	27.00	34.00	2.58	32.41	15.92	15.80	1.81
18	60.00	40.00	2.73	32.98	12.46	15.21	1.76
19	8.00	92.00	2.71	25.88	9.02	8.65	1.96
20	9.00	78.00	2.70	27.23	13.32	9.34	1.99
21	100.00	0.00	2.65	54.18	22.08	24.72	1.53
22	52.00	30.00	2.80	35.51	13.91	17.35	1.73
23	85.00	15.00	2.62	45.81	18.04	21.99	1.60
24	39.00	35.00	2.64	32.05	12.94	16.02	1.80
25	6.00	20.00	2.73	42.04	30.90	15.02	1.94
26	21.00	31.00	2.65	34.39	21.84	15.67	1.79
27	44.00	50.00	2.71	28.59	10.93	12.46	1.80
28	2.00	98.00	2.71	26.71	9.56	7.91	1.97
29	14.00	86.00	2.71	25.57	10.21	9.62	1.94
30	15.00	84.00	2.70	25.77	11.11	9.66	1.95
31	0.00	100.00	2.70	27.06	9.85	7.81	1.97
32	10.00	88.00	2.71	25.66	9.69	8.83	1.97
33	21.00	35.00	2.64	32.38	19.15	15.34	1.81
34	3.00	94.00	2.70	26.09	9.64	7.74	1.99
35	97.00	3.00	2.65	52.42	21.36	24.63	1.54
36	15.00	64.00	2.69	28.95	19.79	12.49	1.91

(continued)

Table 1.3 (continued)

S. No	FG (%)	S (%)	SG	LL (%)	PI (%)	OMC (%)	MDD (gm/cc)
37	19.00	81.00	2.69	26.28	12.51	10.22	1.93
38	65.00	34.00	2.67	34.67	13.53	15.74	1.76
39	3.00	66.00	2.69	29.22	16.50	11.41	1.88
40	5.00	65.00	2.69	29.16	17.49	11.74	1.89
41	29.00	21.00	2.63	41.40	23.00	19.59	1.71
42	26.00	16.00	2.68	45.11	19.00	21.83	1.70
43	31.00	60.00	2.52	28.88	11.69	12.09	1.87
44	14.00	43.00	2.79	30.45	19.90	14.44	1.83
45	24.40	65.20	2.59	28.81	15.82	11.72	1.91
46	9.70	88.70	2.71	25.65	9.41	8.89	1.96
47	99.00	1.00	2.65	53.61	21.87	24.70	1.54
48	84.00	16.00	2.62	45.33	17.86	21.54	1.60
49	75.00	25.00	2.63	40.40	15.92	17.53	1.69
50	11.00	88.00	2.71	25.61	9.62	9.06	1.96
51	9.00	79.00	2.70	27.01	12.89	9.20	1.99
52	5.00	58.00	2.69	29.38	19.77	12.81	1.84
53	60.00	30.00	2.76	35.91	14.20	16.32	1.71

Table 1.4 Statistics of datasets of soil

Statistical parameters	FG (%)	S (%)	SG	LL (%)	PI (%)	OMC (%)	MDD (gm/cc)
Min	0.00	0.00	2.52	25.57	9.02	7.61	1.53
Max	100.00	100.00	2.80	54.18	30.90	24.72	2.01
Mean	30.11	56.46	2.67	32.39	14.74	13.49	1.84
Median	18.00	63.00	2.69	28.84	13.53	11.94	1.89
St. deviation	29.25	30.06	0.06	7.96	4.70	4.95	0.14
Confi. level (95%)	8.06	8.28	0.02	2.19	1.30	1.37	0.04

Table 1.5 Pearson matrix of datasets of soil

(*r*)	FG (%)	S (%)	SG	LL (%)	PI (%)	OMC (%)	MDD (gm/cc)
FG (%)	1.0000						
S (%)	0.6558	1.0000					
SG	0.3275	0.3024	1.0000				
LL (%)	0.8324	0.9075	0.2639	1.0000			
PI (%)	0.3513	0.7408	0.1222	0.7001	1.0000		
OMC (%)	0.8573	0.9643	0.3098	0.9598	0.6801	1.0000	
MDD (gm/cc)	0.9248	0.9208	0.3136	0.9161	0.5368	0.9666	1.0000

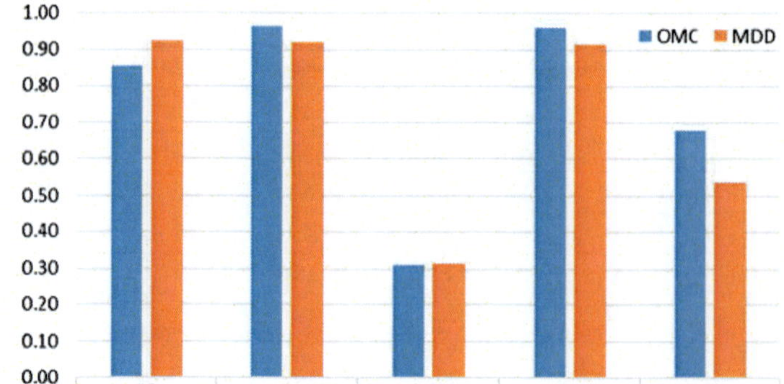

	FG	S	SG	LL	PI
OMC	0.86	0.96	0.31	0.96	0.68
MDD	0.92	0.92	0.31	0.92	0.54

Fig. 1.1 Relationship (r) between input and output parameters

Results and Discussion

The performance and results of multilinear regression analysis and artificial neural network models has been discussed in this section. The RMSE, R, and MAE have determined the performance of the proposed models using the following formulas:

$$\mathrm{RMSE} = \sqrt{\frac{1}{N}\sum_{i=1}^{N}(T_i - P_i)^2} \tag{1.3}$$

$$\mathrm{MAE} = \frac{1}{N}\left(\sum_{i=1}^{N}\mathrm{abs}(T_i - P)_i\right) \tag{1.4}$$

$$R = \frac{\sum_{i=1}^{N}(T_i - \overline{T})(P_i - \overline{P})}{\sqrt{\sum_{i=1}^{N}(T_i - \overline{T})^2\sum_{i=1}^{N}(P_i - \overline{P})^2}} \tag{1.5}$$

Multilinear Regression Analysis

The MLR_OMC and MLR_MDD models have been developed to predict the optimum moisture content and maximum dry density of soil specimens. Equations 1.1 and 1.2 have been proposed by MLR_OMC and MLR_MDD models and used for predicting OMC and MDD of soil. The training and testing performances of MLR_OMC and MLR_MDD models have been recorded and given in Table 1.6.

Table 1.6 Performance of MLR_OMC and MLR_MDD models

Model architecture	Training performance			Testing performance		
	RMSE	*R*	MAE	RMSE	*R*	MAE
MLR_OMC	0.7532	0.9888	0.6023	1.1132	0.9763	0.8384
MLR_MDD	0.0226	0.9866	0.0164	0.0500	0.9456	0.0376

Artificial Neural Network

The ANN_OMC_XHY and ANN_MDD_XHY models have been developed to predict the optimum moisture content and maximum dry density of soil specimens. The artificial neural network models have been developed with the different numbers of hidden layers and neurons. The ANN_OMC_1H8, ANN_OMC_2H6, ANN_OMC_3H8, ANN_OMC_4H9, and ANN_OMC_5H3 models of one, two, three, four, and five hidden layers for optimum moisture content have been identified as the best architecture neural network models. The performance of artificial neural network models for optimum moisture content is given in Table 1.7.

Similarly, the ANN_MDD_1H6, ANN_MDD_2H7, ANN_MDD_3H5, ANN_MDD_4H7, and ANN_MDD_5H1 models of one, two, three, four, and five hidden layers for maximum dry density have been identified as the best architecture neural network models. The performance of artificial neural network models for maximum dry density is given in Table 1.8.

Table 1.7 Performance of ANN_OMC models

Model architecture	Training performance			Testing performance		
	RMSE	*R*	MAE	RMSE	*R*	MAE
ANN_OMC_1H8	0.0144	0.9986	0.0170	0.0184	0.9992	0.0177
ANN_OMC_2H6	0.0049	0.9999	0.0060	0.0101	0.9990	0.0029
ANN_OMC_3H8	0.0064	0.9998	0.0401	0.0049	0.9998	0.0239
ANN_OMC_4H9	0.0188	0.9989	0.1011	0.0199	0.9983	0.1259
ANN_OMC_5H3	0.0259	0.9961	0.0071	0.0236	0.9988	0.0153

Table 1.8 Performance of ANN_MDD models

Model architecture	Training performance			Testing performance		
	RMSE	*R*	MAE	RMSE	*R*	MAE
ANN_MDD_1H6	0.0027	0.9985	0.0006	0.0037	0.9977	0.0006
ANN_MDD_2H7	0.0020	0.9992	0.0024	0.0021	0.9978	0.0020
ANN_MDD_3H5	0.0015	0.9994	0.0018	0.0037	0.9982	0.0026
ANN_MDD_4H7	0.0055	0.9952	0.0002	0.0037	0.9963	0.0005
ANN_MDD_5H1	0.0068	0.9889	0.0005	0.0015	0.9997	0.0008

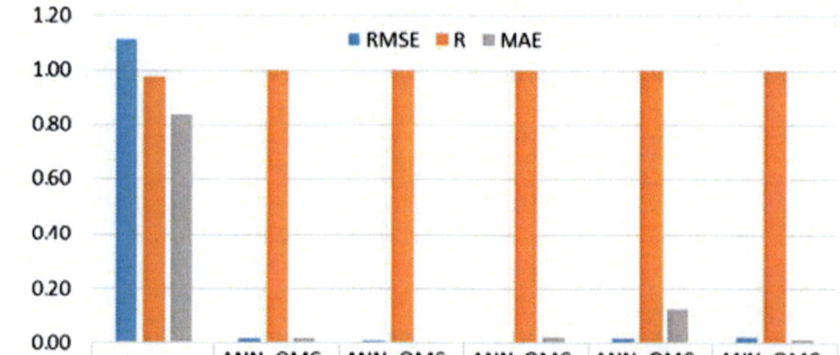

	MLR_OMC	ANN_OMC_1H8	ANN_OMC_2H6	ANN_OMC_3H8	ANN_OMC_4H9	ANN_OMC_5H3
RMSE	1.1132	0.0184	0.0101	0.0049	0.0199	0.0236
R	0.9763	0.9992	0.9990	0.9998	0.9983	0.9988
MAE	0.8384	0.0177	0.0029	0.0239	0.1259	0.0153

Fig. 1.2 Comparison of performance MLR_OMC and ANN_OMC models

Optimum Performance Model

The performance of the ANN and MLR models of optimum moisture content and maximum dry density has been compared to recognize the optimum performance AI models. The performance of the best architecture ANN_OMC_1H8, ANN_OMC_2H6, ANN_OMC_3H8, ANN_OMC_4H9, and ANN_OMC_5H3 models of optimum moisture content has been compared to MLR_OMC to recognize the optimum performance AI models. The comparison of the performance of MLR and ANN models is shown in Fig. 1.2.

From Fig. 1.2, it has been observed that the ANN_OMC_3H8 model has predicted OMC with 0.0049 RMSE. It has also been observed that the ANN_OMC_3H8 model has outperformed the MLR_OMC and other ANN_OMC with 0.9998 performance.

Similarly, the performance of the best architecture ANN_MDD_1H6, ANN_MDD_2H7, ANN_MDD_3H5, ANN_MDD_4H7, and ANN_MDD_5H1 models of maximum dry density has been compared to MLR_MDD to recognize the optimum performance AI models. The comparison of the performance of MLR and ANN models is shown in Fig. 1.3.

From Fig. 1.3, it has been observed that the ANN_MDD_5H1 model has predicted MDD with 0.0015 RMSE. It has also been observed that the ANN_MDD_5H1 model has outperformed the MLR_OMC and other ANN_OMC with 0.9997 performance.

The ANN_OMC_3H8 and ANN_MDD_5H1 models have been recognized as the optimum performance model to predict the OMC and MDD of soil. The optimum performance ANN_OMC_3H8 model has been used to predict the optimum moisture content of 19 soil specimens. From the results of predicted OMC, it has been observed that the predicted OMC is approximately equal to laboratory test results. The comparison of laboratory and predicted OMC is shown in Fig. 1.4.

Similarly, the optimum performance ANN_MDD_5H1 model has been used to predict the maximum dry density of 19 soil specimens. From the results of predicted

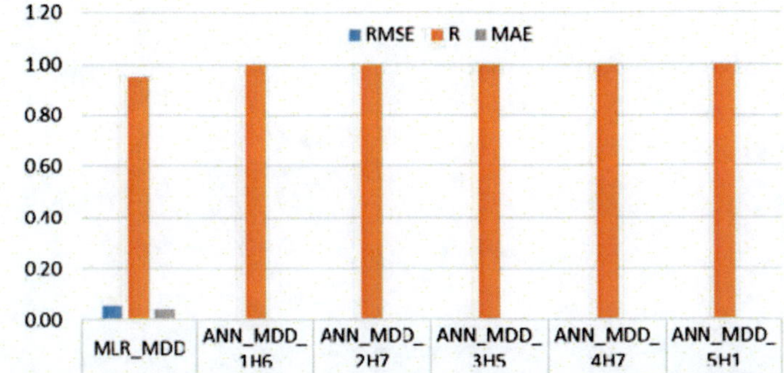

	MLR_MDD	ANN_MDD_1H6	ANN_MDD_2H7	ANN_MDD_3H5	ANN_MDD_4H7	ANN_MDD_5H1
RMSE	0.0500	0.0037	0.0021	0.0037	0.0037	0.0015
R	0.9456	0.9977	0.9978	0.9982	0.9963	0.9997
MAE	0.0376	0.0006	0.0020	0.0026	0.0005	0.0008

Fig. 1.3 Comparison of performance of MLR_MDD and ANN_MDD models

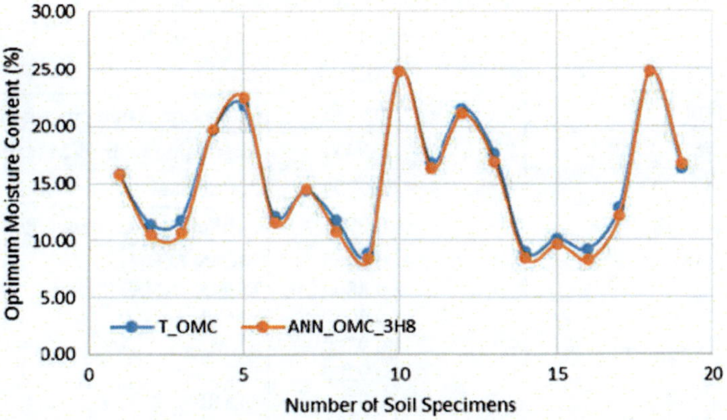

Fig. 1.4 Comparison of laboratory OMC (T_OMC) and predicted OMC

MDD, it has been observed that the predicted MDD is approximately equal to laboratory test results. The comparison of laboratory and predicted MDD is shown in Fig. 1.5.

Comparison with Literature Survey

The performance of the proposed ANN models has been compared with the literature survey, and it has been found that the proposed ANN model outperformed the models available in published research work. The comparison of the presently proposed model with the literature survey is given in Table 1.9.

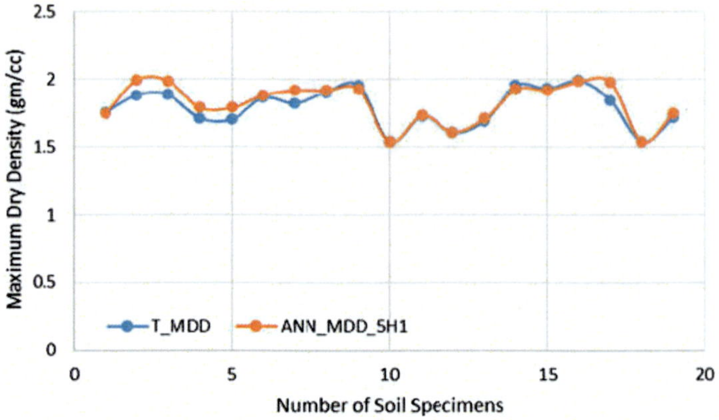

Fig. 1.5 Comparison of laboratory MDD (T_MDD) and predicted MDD

Table 1.9 Comparison of performance of models with the literature survey

Author(s)	Model performance		AI approach
	OMC	MDD	
Wang et al. [10]	0.9607	0.9263	Multi-expression programming
Hasnat et al. [6]	0.9237	0.9487	Regression with SVM
Salahudeen et al. [1]	0.8855	0.9754	Artificial neural networks
Sanuade et al. [15]	–	0.9923	Artificial neural networks
Suman et al. [18]	–	0.8400	BRNN (ANN)
	–	0.8800	DENN (ANN)
	–	0.7600	LMNN (ANN)
	–	0.9300	SVM
	–	0.9200	FN
	–	0.9400	MARS
	–	0.4500	MLR
Shrivastava et al. [2]	0.9323	0.9263	CFB-10 (ANN)
Jayan et al. [12]	0.9200	0.9100	Artificial neural network
Khuntia et al. [19]	0.9381	0.9000	MARS
Khatti et al. (present study)	0.9998	0.9997	Artificial neural network

Conclusion

Regression analysis and artificial neural network models were developed to predict the OMC and MDD of soil. The ANN_OMC_1H8, ANN_OMC_2H6, ANN_OMC_3H8, ANN_OMC_4H9, and ANN_OMC_5H3 models of optimum moisture content and ANN_MDD_1H6, ANN_MDD_2H7, ANN_MDD_3H5, ANN_MDD_4H7, and ANN_MDD_5H1 models of maximum dry density were identified as the best architecture ANN models. The performance of the best architecture ANN models of OMC and MDD was compared to MLR_OMC and MLR_MDD to identify the optimum performance model. The ANN_OMC_3H8 and ANN_MDD_5H1 ANN models were identified as the optimum performance models for predicting the OMC and MDD of soil. The comparison of the performance of the optimum performance model with presently available AI models in the literature survey was mapped. It was found that the optimum performance models have better performance and can be used to predict the OMC and MDD of soil.

References

1. Salahudeen AB, Ijimdiya TS, Eberemu AO, Osinubi KJ (2018) Artificial neural networks prediction of compaction characteristics of black cotton soil stabilized with cement kiln dust. J Soft Comput Civil Eng 2(3):50–71
2. Shrivastava AK, Jain PK (2016) Prediction of compaction parameters using regression and ANN tools. Int J Sci Res Dev 3(11):697–702
3. Majidi A, Lashgaripour G, Shoaie Z, Noruzi Nashlaji M, Firouzei Y (2014) Estimating compaction parameters of marl soils using multi-layer perceptron neural networks. J Balkan Tribological Assoc 20(2):170–198
4. Alavi AH, Gandomi AH, Mollahassani A, Heshmati AA, Rashed A (2010) Modeling of maximum dry density and optimum moisture content of stabilized soil using artificial neural networks. J Plant Nutr Soil Sci 173(3):368–379
5. Anjita NA, George CA, Krishnankutty SV (2017) Prediction of maximum dry density of soil using Genetic algorithm. Int J Eng Res Technol 6(3):550–552
6. Hasnat A, Hasan MdM, Islam MdR, Alim MdA (2019) Prediction of compaction parameters of soil using support vector machine. Curr Trends Civil Struct Eng 4(1):1–7
7. Patra CR, Sivakugan N, Das BM, Rout SK (2013) Correlations for relative density of clean sand with median grain size and compaction energy. Int J Geotech Eng 4(2):195–203
8. Arora KR (2004) Soil mechanics and foundation engineering. Standard Publisher Distributors, New Delhi, India
9. Smith GN (1986) Probability and statistics in civil engineering: an introduction. Collins, London
10. Wang H-L, Yin Z-Y (2020) High performance prediction of soil compaction parameters using multi expression programming. Eng Geol 279:105758
11. Mujtaba H, Farooq K, Sivakugan N, Das BM (2013) Correlation between gradational parameters and compaction characteristics of sandy soils. Int J Geotech Eng 7(4):395–401
12. Jayan J, Sankar N (2015) Prediction of compaction parameters of soils using artificial neural network. Asian J Eng Technol 3(4):368–375, special issue for ICETTAS'15
13. Farooq K, Khalid U, Mujtaba H (2016) Prediction of compaction characteristics of fine-grained soil using consistency limits. Arab J Sci Eng 41:1319–1928
14. Ng KS, Chew YM, Osman MH, Mohamad Ghazali SK (2015) Estimating maximum dry density and optimum moisture content of compacted soils. In: International conference on advances in civil and environmental engineering, Faculty of Civil Engineering, Universiti Teknologi MARA Pulau Pinang, July 2015
15. Sanuade OA, Adesina RB, Amosun JO, Fajana AO, Olaseeni OG (2017) Using artificial neural network to predict dry density of soil from thermal conductivity. Mater Geoenviron 64(3):169–180
16. Sivrikaya O, Kayadelen C, Cecen E (2013) Prediction of the compaction parameters for coarse-grained soil with fines content by MLR and GEP. Acta Geotechnica Slovenica 10(2):29–41
17. Al-saffar RZ, Khattab S, Yusif ST (2013) Prediction of soil's compaction parameters using artificial neural network. Al-Rafidain Eng 21(3):15–27

18. Suman S, Mahamaya M, Das SK (2016) Prediction of maximum dry density and unconfined compressive strength of cemented stabilized soil using artificial intelligence techniques. Int J Geosynth Ground Eng 2(11):1–11
19. Khuntia S, Mujtaba H, Patra C, Farooq K, Sivakugan N, Das BM (2015) Prediction of compaction parameters of coarse-grained soil using multivariate adaptive regression splines (MARS). Int J Geotech Eng 9(1):79–88
20. Khalid U, Ur Rehman Z (2018) Evaluation of compaction parameters of fine-grained soil using standard and modified efforts. Int J Geo-Eng 15(9):1–17

Chapter 2
Numerical Study of Tapered Pile Subjected to Cyclic Loading

Anuj Kumar Soni, Mohit Kumar, Sushil Kumar, Vishal Kumar, Muthukkumaran Kasinathan, and Prabu Thannasi

Introduction

With an increase in the depletion of non-renewable sources of energy and the alarming condition of the environment due to increased pollution, the new generation has shifted to renewable sources of energy so that our society can be more oriented towards sustainable development. Offshore wind energy, primarily offshore wind turbines, has become a large clean, renewable energy supplier. Onshore wind has been successfully harnessed in many countries. However, the land requirements and aesthetics of onshore turbines are often considered undesirable. Offshore wind farms have a number of clear advantages, namely (a) their limited aesthetic impact by locating them far from land; (b) high unrestricted wind speeds, which are generally more consistent than onshore; (c) higher power generation through the use of large-capacity turbines [1–3]. In general, the choice for an offshore wind support structure depends on water depth, soil conditions, and environmental conditions at the project location. Based on the structural configuration, the type of offshore wind support structures was divided into monopile structures, gravity structures, jacket structures, tripod structures, and floating structures (Fig. 2.1).

Pile is also being used to construct embankments on the side of bridges which are regularly subjected to the cyclic load as vehicles move over them. A pile support structure is exposed to both horizontal and vertical loads. The horizontal loads are transferred to the soil by mobilizing lateral resistance of the soil through bending, while the vertical loads are carried by the pile wall friction and tip resistance. The pile diameter shall be large enough to provide the required stiffness. Fabrication and installation of very large diameter piles can be difficult due to limitations on available

A. K. Soni (✉) · M. Kumar · S. Kumar · V. Kumar · M. Kasinathan · P. Thannasi
Department of Civil Engineering, National Institute of Technology, Tiruchirapalli, Tamilnadu, India
e-mail: kmk@nitt.edu

K. Muthukkumaran et al. (eds.), *Soil Dynamics, Earthquake and Computational Geotechnical Engineering*, Lecture Notes in Civil Engineering 300,
https://doi.org/10.1007/978-981-19-6998-0_2

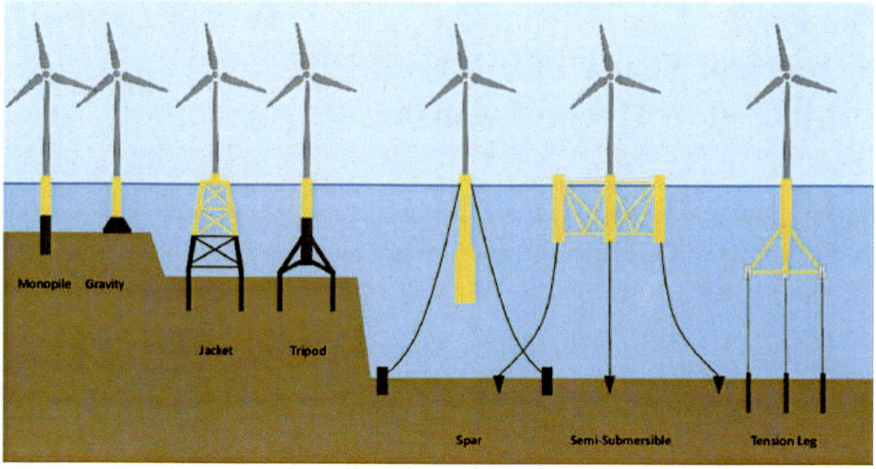

Fig. 2.1 Types of offshore wind support structures. *Source* Google image

steel plate sizes and pile driving capacity, respectively. In addition, conventional pile driving techniques generate noise that concerns surrounding marine life [4–6]. For a pile foundation, the self-weight generates the axial loads while wind and wave generate torsional loads, lateral loads, and bending moments in the structure. The main objective of the support structure design is to determine the dimensions of its components, taking into account operability, load resistance, and economics. This paper explains the numerical study on the tapered piles subjected to cyclic loading. The PLAXIS 3D was used for the analysis, and the results are validated and discussed. For analysis, two different materials, such as steel and aluminium piles, were taken and analysed for different loading conditions.

Validation in PLAXIS 3D

The HS small strain model has been validated, and the results obtained are discussed. Further, the results of the parametric study considering the three parameters have been discussed. A pile was tested in the field for its lateral capacity. The field situation has been simulated in the PLAXIS 3D by the use of HS small strain modelling [1]. HS small incorporates the loading history of soil and a strain-dependent stiffness. Therefore, it can, to some extent, be used to model cyclic loading. It accounts for increased stiffness of soil at small strain. The soil and pile properties for the PLAXIS analysis are given in Tables 2.1 and 2.2. The boundary of 36 m * 36 m * 15 m with a water table at 20 m below the ground surface was considered for the analysis. One-way cyclic load with a maximum amplitude of 960 kN and minimum amplitude of 480 kN for a duration of 40 s applied at a distance of 1 m from the top of the pile.

Table 2.1 Soil properties taken for PLAXIS analysis

S. No	Soil properties	Values
1	Oedometer stiffness, E_{oed}^{ref}	51,600 kN/m^2
2	Secant stiffness, E_{50}^{ref}	51,600 kN/m^2
3	Unload/reload stiffness, E_{ur}^{ref}	154,800 kN/m^2
4	Friction angle, ϕ_p	38.75°
5	Effective cohesion intercept, C	0.3
6	Angle of dilation, Ψ	8.75°
7	Shear modulus at very small strains, G_0^{ref}	118,480 kN/m^2
8	Reference shear strain (at $G_s = 0.722\ G_o$), $\gamma_{0.7}$	1.991×10^{-4}
9	Power of stress level dependency of stiffness, m	0.697
10	Unsaturated unit weight, γ_{unsat}	15.034 kN/m^3
11	Saturated unit weight, γ_{sat}	19.014 kN/m^3

Table 2.2 Pile properties are taken for PLAXIS analysis

S. No	Soil properties	Values
1	External diameter, ϕ_e	0.72 m
2	Internal diameter, ϕ_i	0.60 m
3	Total length of the pile	14.6 m
4	Embedded length	12 m
5	Young's modulus (E)	73.939×10^9 Pa
6	Unit weight (γ)	27.1 kN/m^3

Result in PLAXIS 3D

For the regular pile of the above-given dimension and one-way cyclic load applied, the following results were obtained from PLAXIS. From the result, it is concluded that the Hssmall strain model is giving the same result as that of the paper. Though there is some deviation in the initial phase, it might be due to the different software used; i.e. in the paper, ABAQUS has been used, and here, PLAXIS 3D has been used.

Parametric Study

The first parameter chosen is pile configuration. A regular pile has been replaced by a square tapered pile based on an equivalent volume approach, and all other conditions remained the same. The dimension of the squared tapered pile is the top square is 0.61 m × 0.61 m inscribed in a circle of radius 0.43 m, the bottom square is 0.41 m × 0.41 m inscribed in a circle of radius 0.29 m, and the length of the tapered pile is 14.6 m.

From the analysis, it is observed that the tapered pile's deflection is less than the regular pile by approximately 4 cm. The second parameter taken is the loading condition in which two-way cyclic loading has been applied on the tapered pile with a frequency of 0.3 Hz, and all other parameters are kept the same.

Results and Discussion

Analysis has been done by considering three broad parameters: pile configuration, loading conditions, and material properties. In pile configuration, a regular pile has been replaced by a tapered pile based on an equivalent volume approach. Volume has been kept the same to be convenient for the comparison and would give a realistic comparison. The pile has been subjected to one-way cyclic load in loading conditions and two-way cyclic load applied laterally. Soil properties have been varied by considering different relative densities. Pile material consisted of steel and aluminium.

Pile Configuration

From Fig. 2.2, the tapered pile subjected to cyclic load has undergone less deflection, i.e. regular pile experienced a deflection of 0.25 m. In contrast, the tapered pile experienced a deflection of 0.21 m.

Loading Conditions

Pile is under the critical condition when subjected to one-way cyclic load than pile under two-way cyclic load because deflection of tapered pile under two-way cyclic load is 0.12 m which is less than the corresponding values of one-way cyclic load (Fig. 2.3).

Materials Properties

When the steel pile is subjected to a two-way cyclic load, the resultant deflection of 0.08 m at the pile head, 0.07 m at the loading point, and 0.05 m at the ground position was observed for a relative density of 65% of soil (Fig. 2.4). Figure 2.5 shows the load versus deflection plot for different relative densities of soil. More deflection occurs at the top of the pile, followed by less deflection at the loading point and lesser deflection at the ground position (Fig. 2.6).

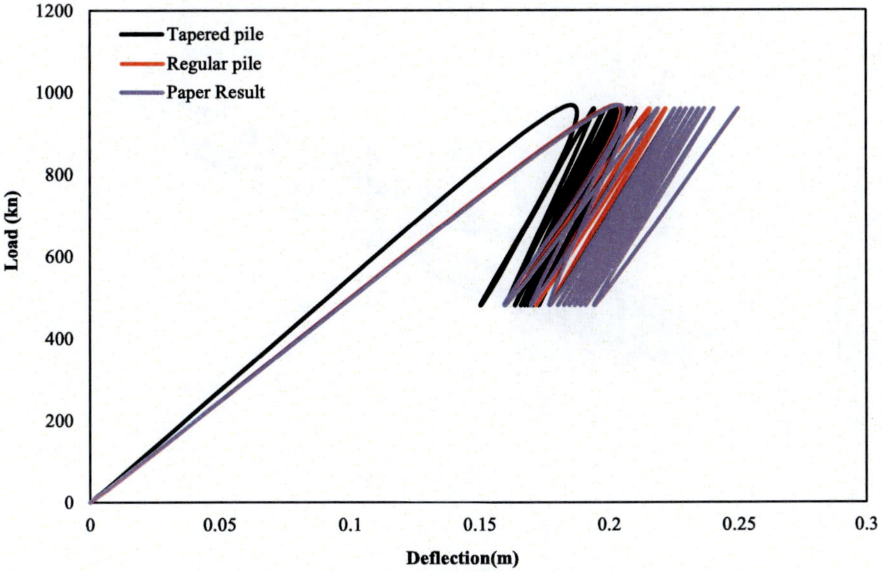

Fig. 2.2 Load versus deflection plot for different pile configuration

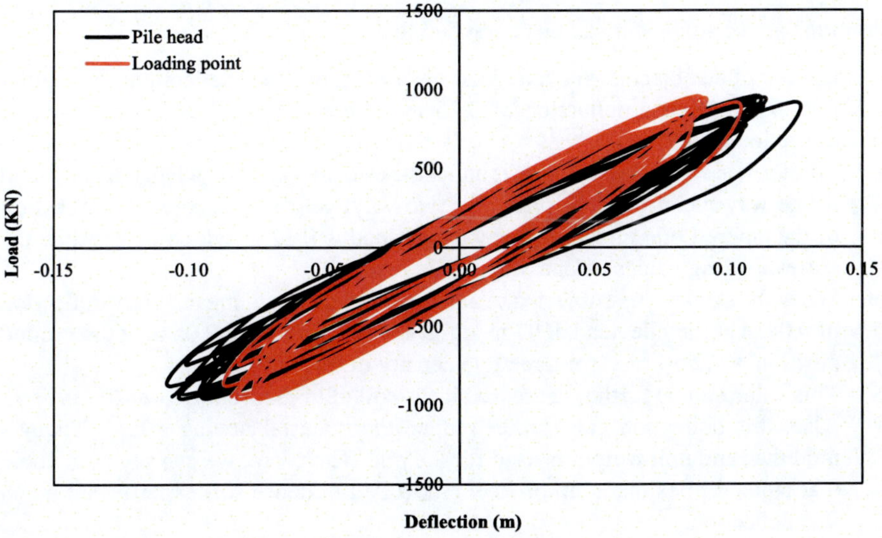

Fig. 2.3 Load versus deflection plot for two-way cyclic

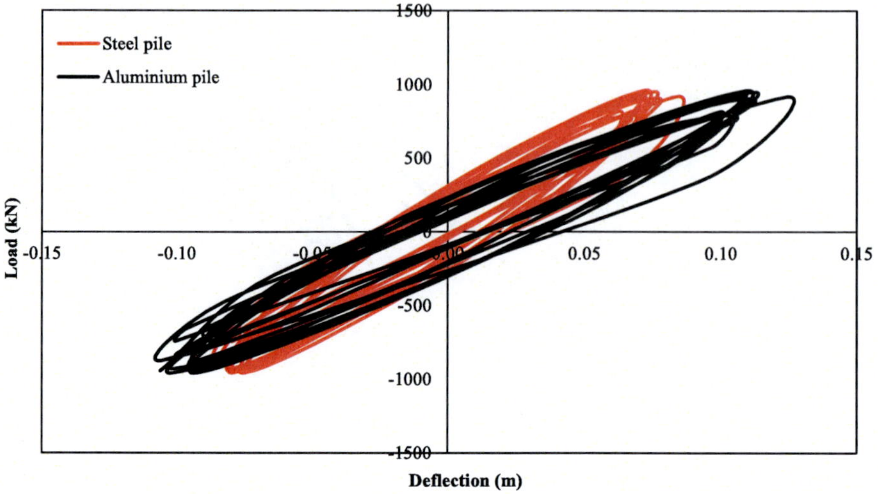

Fig. 2.4 Load versus deflection plot for different pile material

Conclusion

From the above studies, it can be concluded that

i. tapered pile subjected to cyclic load has undergone less deflection, i.e. regular pile experienced a deflection of 0.25 m, whereas tapered pile experienced a deflection of 0.21 m.
ii. Further, it is concluded that pile is under the critical condition when it is subjected to one-way cyclic load than pile under two-way cyclic load because the deflection of the tapered pile under two-way cyclic load is 0.12 m which is less than the corresponding values of one-way cyclic load.
iii. When the steel pile is subjected to a two-way cyclic load, the resultant deflection of 0.08 m at the pile head, 0.07 m at the loading point, and 0.05 m at the ground position was observed for a relative density of 65% of soil.
iv. This value can be further decreased if the soil of higher relative density is used. Also, this deflection can further reduce the material having a high Young's modulus, and unit weight is used for the pile. Therefore, we can say as a whole that tapered piles made up of steel embedded in dense soil experienced much less deflection.

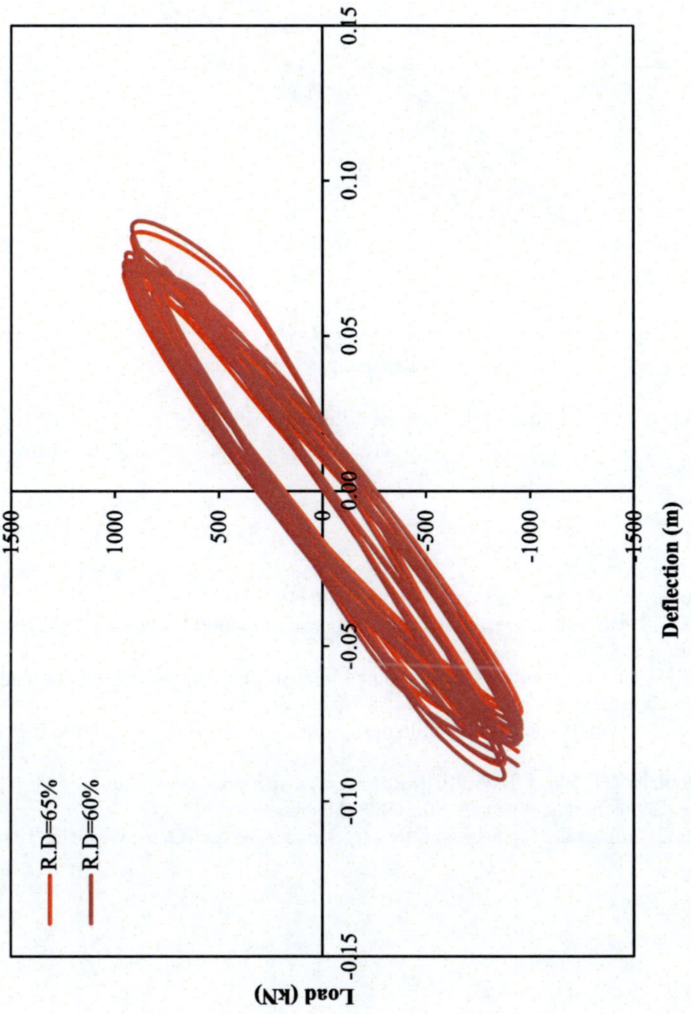

Fig. 2.5 Load versus deflection plot for different relative densities of soil

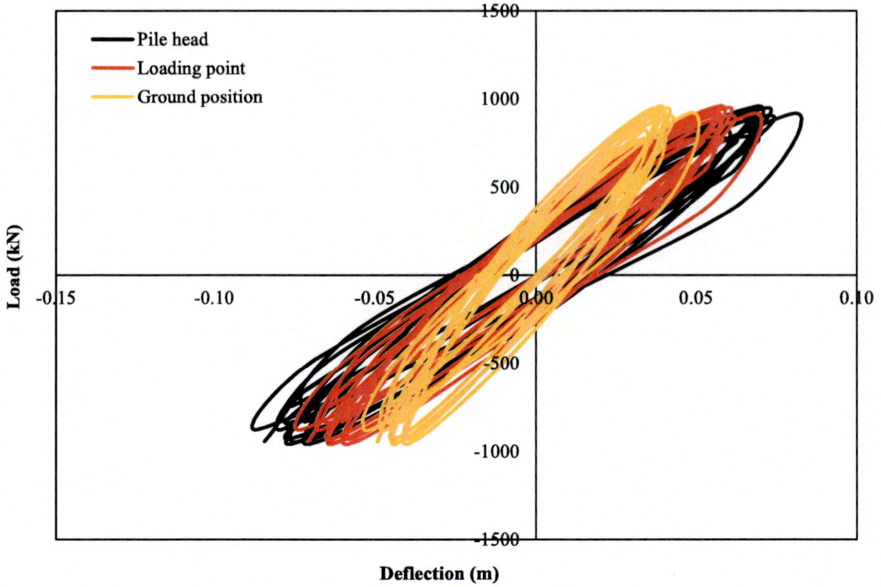

Fig. 2.6 Load versus deflection plot of steel pile at different positions

References

1. Giannakos S, Gerolymos N, Gazetas G (2012) Cyclic lateral response of piles in dry sand: finite element modeling and validation. Comput Geotech 44:116–131
2. Li IG (2015) Field tests to investigate the cyclic response of monopiles in sand. Proc Inst Civil Eng Geotech Eng 168(5)
3. Ismael NF (2010) Behaviour of step tapered bored piles in sand under static lateral loading. J Geotech Geoenviron Eng 6(5)
4. Doherty P (2011) Laterally loaded monopile design for offshore wind farms. Proc ICE Energ 165(1):7–17
5. Leblanc C, Houlsby GT, Byrne BW (2010) Response of stiff piles in sand to long-term cyclic lateral loading. Géotechnique 60(2):79–90. ICE Publishing
6. Poulos HG (1982) Single pile response to cyclic lateral load. J Geotech Geoenviron Eng 108:355–75

Chapter 3
Seepage Analysis of Resilient Rubble Mound Breakwater Under Tsunami Overflow: Numerical Analysis

Manu K. Sajan and Babloo Chaudhary

Introduction

Breakwaters are off shore manmade structures constructed to create a tranquil zone of water near the coast, mainly for the mooring and docking of vessels. Even though there are different types of breakwaters based on their functionality and construction methods, the RM type of breakwaters is the most commonly constructed. These breakwaters are constructed by placing quarry stones on seabed. Generally, the RM breakwaters have an armour layer with heavier rock blocks and core layer with comparatively lighter blocks.

A tsunami wave train is triggered due to the sudden vertical displacement of water in deep oceanic regions. This sudden displacement of water could be a result of an earthquake in the seabed, manmade explosions, impact of cosmic objects, landslides, etc. Even though the tsunamis have smaller amplitudes and larger wavelengths in deep oceans, as it approaches the shore, the amplitude increases tremendously. Thus, the impact height of tsunamis could be very high depending on the proximity of coastal lines to the origin of tsunami and energy released from the source. The existing RM breakwaters across different countries have not been necessarily designed to withstand the impact of tsunami waves. However, the coastal belts of these breakwaters could be subjected to a tsunami impact of height varying from 5 to 10 m with a mean return period of 500 years [1].

India has several RM breakwaters distributed along the western and eastern off shores of the peninsular. The longest breakwater in the country is the RM breakwater located at the north of the Ennore port, in Chennai [2]. The breakwater was subjected to the impact of 2004 Indian Ocean tsunami that had inundated into the coastal areas nearly 300 m with an impact height of 5 m [3]. It was observed from the complex

M. K. Sajan (✉) · B. Chaudhary
National Institute of Technology Karnataka, Surathkal, India
e-mail: manuksajan@gmail.com

K. Muthukkumaran et al. (eds.), *Soil Dynamics, Earthquake and Computational Geotechnical Engineering*, Lecture Notes in Civil Engineering 300,
https://doi.org/10.1007/978-981-19-6998-0_3

evaluation of satellite imageries that the longest breakwater of the country was moved around 1–3 m during the tsunami attack [4]. This clearly points out the necessity of developing counter measures to improve the resiliency of RM breakwaters against the tsunami impact. Even though the studies on improving the resiliency of composite breakwaters were conducted before [5–9], there were no similar studies reported on RM breakwaters.

The possible mechanism that could result in the failure of a RM breakwater under the impact of tsunami could be due to the seepage through the body and seabed beneath the breakwater [10]. In addition to this, the scouring of seabed and the mound due to tsunami overtopping could also result in the failure of the structure [11, 12]. The provision of impermeable faced gabions on the seaward side of the breakwater mound was found to be effective in preventing the seepage through foundation of composite breakwaters [5]. The addition of sheet piles beneath the composite breakwater, within the seabed, was observed to be effective in reducing the seepage through the soil [6]. Further, the provision of gabions on the harbour side slope of the composite breakwater foundation was effective in preventing the scouring due to tsunami overtopping [10].

The dynamic analysis of RM breakwater by finite element method using Plaxis software was done before [12, 13]. The method of modelling the breakwater core, amour layer, and concrete Accropode as soil models was justified by the reasonable results obtained in the numerical analysis. However, the behaviour of sand bed beneath the breakwater structure was not analysed in the study. Also the study focused on the seismic responses of the RM structure neglecting the possible hydrodynamic pressures.

In the present study, the numerical analysis has been conducted by using Plaxis software to evaluate the performance of the proposed countermeasures of incorporating sheet piles as well as specially designed gabions on RM breakwaters. The longest breakwater in India was chosen as the prototype. This breakwater was subjected to the impact of 2004 Indian ocean earthquake and tsunami in the past. The analysis was focused mainly on the effects of seepage through the mound and soil during a tsunami overflow condition. Therefore, the flow analysis in the Plaxis 2D software was utilized in the present study.

Method of Analysis

The analysis was conducted on the numerical model of the north breakwater located at the Ennore Port, Chennai, India. The RM breakwater has the core layer, armour layer, filter, and Accropodes. Rock blocks of different weights extracted from nearby quarry have been used for the construction of the breakwater as shown in Fig. 3.1. The breakwater span nearly 3 km in length in north–south direction, almost parallel to the Ennore coastal lines [14].

The armour and core layers were modelled as hardening soil model in Plaxis 2D with the properties as detailed in Table 3.1 [13], where the permeability parameters

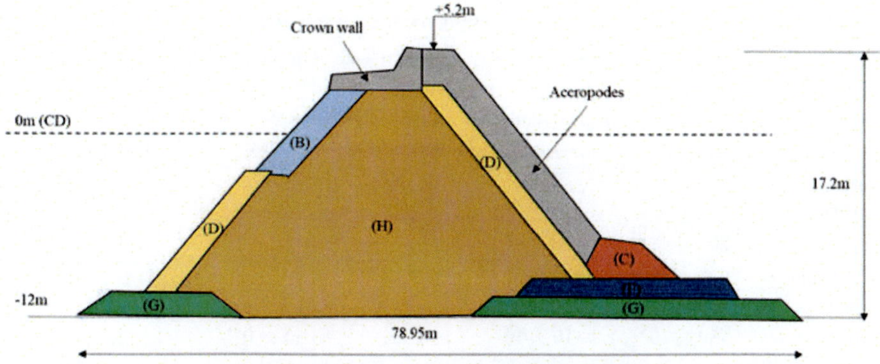

Grade of Rock	Weight (kg)
B	2500-7000
C	1000-5000
D	500-2000
F	100-500
Quarry Run (H)	1-1000
Filter G	1-50

Fig. 3.1 Sectional details of the north breakwater at Ennore port

of the armour and core layers have been defined sufficiently large enough to simulate the seepage through these layers [15]. Flow analysis was performed by simulating the tsunami overflow condition where the height of the water in the seaward side was 10 m above the mean sea level. The height difference between the seaward and harbour sides of the breakwater has triggered the seepage of water through the mound as well as the soil beneath. Two layers of seabed soil were considered with the bottom layer of higher relative density ($D_r = 90\%$) which was overlaid by the top layer of lower relative density ($D_r = 60\%$). The crown wall was modelled as non-porous linear elastic unit placed on top of the RM. The sheet piles were modelled as steel plates of steel plates of thickness 3 mm. The gabions were 2 m × 1 m × 1 m meshed boxes filled with crushed stones and a layer of impermeable membrane one face.

Results and Discussions

The seepage flow occurring through the top layer of seabed soil is shown in Fig. 3.2 for the unreinforced breakwater model. The RM breakwater was subjected to a tsunami impact of 10 m height. The sudden increase of height in seaside of breakwater caused the seepage of water from the seaside to harbour side. The higher permeability of the RM has resulted in denser flow paths through the structure. However, a corresponding seepage was observed thorough the top layer of seabed. The denser soil ($D_r = 90\%$)

Table 3.1 Material properties used for modelling the breakwater and seabed

Material	Armour	Core	Crown wall	Top sand layer	Bottom sand layer
Model	Hardening soil	Hardening soil	Linear elastic	Mohr–Coulomb	Mohr–Coulomb
Drainage type	Drained	Drained	Non-porous	Drained	Drained
Unsaturated density, γ_u	18	16	50	17	17
Dry density, γ_d	20	17	50	18	18
Initial void ratio, e_{int}	0.5	0.5	0.5	0.5	0.5
Reference secant modulus, ${E_{50}}^{\text{ref}}$	130×10^3	110×10^3	30×10^6	30×10^3	55×10^3
Reference constrain modulus, ${E_{\text{oed}}}^{\text{ref}}$	120.7×10^3	110×10^3			
Reference unloading reloading modulus, ${E_{\text{ur}}}^{\text{ref}}$	270×10^3	230×10^3			
Poisson's ratio, υ'	0.2	0.2	0.15	0.3	0.35
Power of stress level dependency (m)	0.5	0.5			
Cohesion, c'	0	0		1	1
Angle of internal friction, ϕ'	45	41		35	42
Dilatancy angle, ψ	5	10		0	0
Coefficient of earth pressure at rest, ${K_0}^{\text{nc}}$	0.316	0.35			
Flow parameters					
Dataset	Standard	Standard		Standard	Standard
Soil type	Coarse sand	Coarse sand		Coarse sand	Coarse

layer present beneath the top layer has limited the seepage through the top loose sand layer.

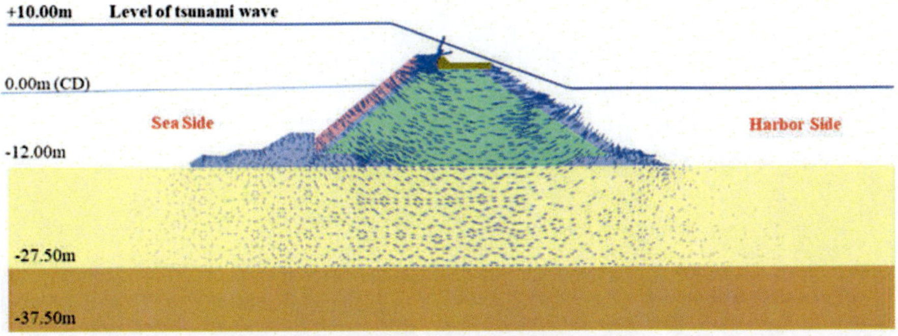

Fig. 3.2 Seepage in rubble mound breakwater under a tsunami impact

Effect of Adding Sheet Piles

When sheet piles were introduced into the seabed, notable changes can be seen in the seepage flow paths as shown in Fig. 3.3. Seepage through seabed across the sheet piles has stopped or at least reduced up to great extent. The sheet piles act as a cut-off wall to stop the seepage flow. Hence, the seepage path has deviated around the sheet pile. Thus, the provision of sheet piles beneath the rubble mound structure was found to be effective in reducing the seepage of water through the seabed beneath breakwater. However, seepage was observed through the zone of soil between the sheet pile walls.

The number of sheet piles was increased to four numbers as shown in Fig. 3.4. It was observed that the newly added sheet piles were effective in reducing the seepage through the soil profile. However, the seepage of water through the breakwater mound was not completely removed. This in turn resulted in the seepage of water within the zone of seabed between the inner sheet piles. The addition of inner piles had arrested the seepage of water through the seaside seabed profile.

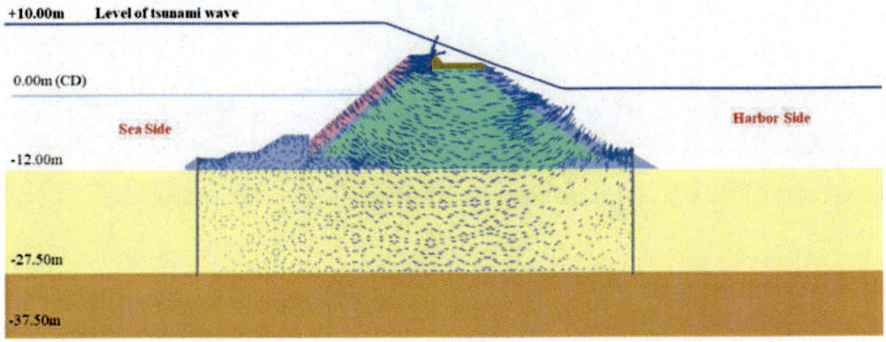

Fig. 3.3 Seepage through mound and seabed with two rows of sheet piles beneath the mound

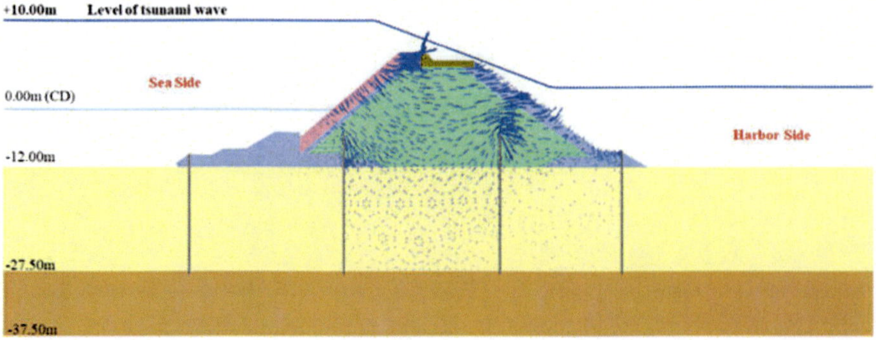

Fig. 3.4 Seepage through mound and seabed with four rows of sheet piles beneath the mound

Effect of Adding Special Gabions

The provision of gabions with impermeable membrane at the facing side was found to be effective in preventing the seepage through the RM. The special type of gabions was stone filled cuboid wire meshes with the seaward facing side covered with an impermeable membrane like geomembrane or steel sheets. The reduction in seepage through mound had also reduced the flow paths through seabed considerably as shown in Fig. 3.5. Even though the seepage through sea side has been completely arrested, a small amount of seepage was observed through the harbour side of the mound.

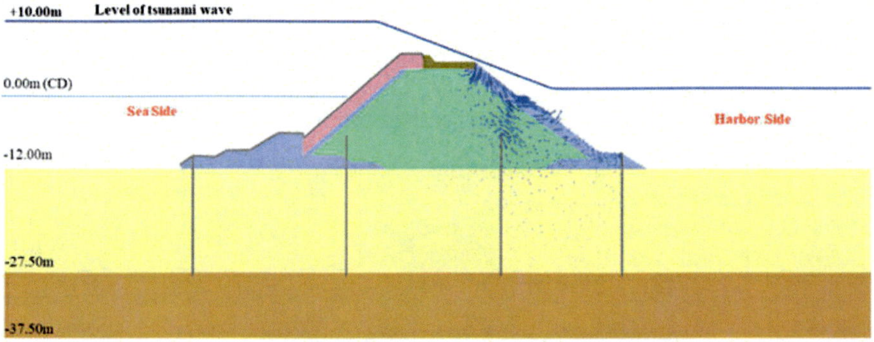

Fig. 3.5 Seepage through mound and seabed with sheet piles and special gabions

Conclusions

From the numerical analysis conducted, it could be concluded that the provision of sheet pile walls in the foundation soil of a RM breakwater can reduce the seepage flow through the seabed soils, that occurs due to large upraise of seawater at the sea side during a tsunami impact. The effect of sheet piles in reducing the seepage through seabed was prominent as the number of sheet piles was increased. However, the provision of sheet piles alone was not successful in arresting the flow paths through seabed completely. The higher permeability of the RM created denser flow paths through the mound through which seepage continued. The provision of gabions with impermeable membrane was effective in reducing the flow through the mound. Thus, combined effect of the impermeable gabions on the sea side slope and sheet piles provided beneath the RM breakwater was found to be effective in reducing the seepage and thereby improving the stability of the RM breakwater under tsunami impacts.

References

1. Lovholt F, Griffin J, Salgado-Gálvez MA (2015) Tsunami hazard and risk assessment on the global scale. In: Meyers RA (ed) Encyclopedia of complexity and systems science. Springer, Berlin Heidelberg, pp 1–34
2. Khattar MD (2001) Longest breakwater in India. In: PORTS conference 2001. American Society of Civil Engineers, pp 1–17
3. Gupta GVM, Murthy MVR, Subramanian BR (2005) Preliminary assessment of impact of Tsunami in selected coastal areas of India, Chennai
4. Kozai K, Kobayashi E, Kubo M (2005) Tsunami disaster assessment of ports in India by using QuickBird images. In: Asian association on remote sensing, pp 1–5
5. Chaudhary B, Hazarika H, Nishimura K (2017) Effects of reinforcement on the performance of breakwater foundation subjected to earthquake loadings. Int J Geotech Eng 11(2):186–197
6. Chaudhary B, Hazarika H, Murakami A, Fujisawa K (2018) Geosynthetic-sheet pile reinforced foundation for mitigation of earthquake and tsunami induced damage of breakwater. Geotext Geomembr 46(5):597–610
7. Chaudhary B, Hazarika H, Manafi Khajeh Pasha S (2018) Countermeasures for breakwater foundation subjected to foreshocks and main shock of earthquake loading. Marine Georesour Geotechnol 36(3):308–322
8. Chaudhary B, Hazarika H, Murakami A, Fujisawa K (2018) Countermeasures for enhancing the stability of composite breakwater under earthquake and subsequent tsunami. Acta Geotech 13(4):997–1017
9. Chaudhary B, Hazarika H, Murakami A, Fujisawa K (2018) Geotextiles and geomembranes geosynthetic-sheet pile reinforced foundation for mitigation of earthquake and tsunami induced damage of breakwater. Geotext Geomembr 46(5):597–610
10. Chaudhary B, Hazarika H (2018) Centrifuge modelling for stability evaluation of a breakwater foundation subjected to an earthquake and a tsunami. Ocean Eng 148:169–181
11. Takahashi H, Sassa S, Morikawa Y, Takano D, Maruyama K (2014) Stability of caisson-type breakwater foundation under tsunami-induced seepage. Soils Found 54(4):789–805
12. Cihan K, Yuksel Y (2011) Deformation of rubble-mound breakwaters under cyclic loads. Coast Eng 58(6):528–539

13. Cihan K, Yuksel Y (2013) Deformation of breakwater armoured artificial units under cyclic loading. Appl Ocean Res 42:79–86
14. Bijen H (2000) Ennore breakwater construction. Terra et Aqua 79:9–15
15. Recio J, Oumeraci H (2008) Hydraulic permeability of structures made of geotextile sand containers: laboratory tests and conceptual model. Geotext Geomembr 26(6):473–487

Chapter 4
Analytical and Numerical Modelling of Combined Pile-Raft Foundation for Tall Wind Turbine in Various Soils

N. Gaihre, P. K. Kolay, V. K. Puri, and S. Kumar

Introduction

The wind energy industry market is booming rapidly in recent years globally. The need of green and sustainable energy source is supporting the wind farm construction. Increase in wind power generation creates challenge in foundation construction in different types of soil. The foundation of wind turbine significantly affects the total cost of the turbines [1]. Generally, the most popular types of foundations used for wind turbine are raft foundations or pile foundations. These systems when implemented individually will fulfil the design requirements; however, in most cases they become conservative and not economical. On the other hand, these systems can be combined, when the conditions are suitable, and it could be more efficient, economical, and safe design, known as combined pile-raft foundation [2]. The feature that distinguished this foundation from others is load sharing between raft and piles. The piled raft foundation consists of three bearing elements such as piles, raft, and subsoil. Lang et al. [3] described about the wide use of piled raft foundation due to its comparatively small size foundation and much economical design and construction. Poulos and Davis [4] stated the process for estimating the load-settlement behaviour with the extension given by Randolph [5]. Poulos [6] studied extensively to get characteristics

N. Gaihre · P. K. Kolay (✉) · V. K. Puri · S. Kumar
School of Civil, Environmental and Infrastructure Engineering, 1230 Lincoln Drive, MC 6603, Carbondale, IL 62901, USA
e-mail: pkolay@siu.edu

N. Gaihre
e-mail: nirajan.gaihre@siu.edu

V. K. Puri
e-mail: puri@engr.siu.edu

S. Kumar
e-mail: kumars@ce.siu.edu

K. Muthukkumaran et al. (eds.), *Soil Dynamics, Earthquake and Computational Geotechnical Engineering*, Lecture Notes in Civil Engineering 300,
https://doi.org/10.1007/978-981-19-6998-0_4

of piled raft foundation behaviour with respect to number of piles, the nature of loading, raft thickness, and applied load. The increase in number of piles was found to improve foundation performance up to a certain point, beyond which makes no big difference. The effective piled raft foundation was designed by using the considerable part of the available capacity of the piles under the raft. The piles were described as the settlement reducer in the study [6]. Sönmez [7] studied the concept of design and load sharing mechanism of piled raft foundations to identify the contribution of raft to the total capacity of piled raft foundation. Kumar et al. [8] stated in research that design of piled raft foundation can reduce the number, length, and diameter of piles than conventional pile group foundation. The axial vertical load on pile-raft foundation was shared by raft varied from 23 to 31% of total axial load [8]. Bhartiya et al. [9] studied piled raft foundation to estimate the settlement for initial design. The systematic linear elastic finite element analyses on a series of unpiled raft, pile groups, and piled raft with different geometry and pile configuration were performed to determine the stiffness of raft, pile group, and piled raft. The pile-soil interaction was found to depend on the pile length, diameter, and elastic properties of the subsurface and foundation system. The piles were found to carry 70–90% of the total vertical load applied to the piled raft foundation system. Due to the complex soil-structure interaction between raft-soil and pile-soil elements, numerical modelling using finite element software is getting popular. Among different finite element software, the significant amount of research was carried out in PLAXIS 3D software due to its quick, reliable, and precise solution to complex soil-structure interaction. PLAXIS 3D is a three-dimensional finite element software to study deformation, stability, and groundwater flow analysis which is commonly used in geotechnical applications [28]. Phung [10] performed settlement analysis for piled raft foundation using PLAXIS 3D. The complex pile-cap-soil interaction can be modelled in PLAXIS 3D to predict the settlement of foundation due to large load from high-rise building [10]. Shrestha [11] modelled piled raft foundation in PLAXIS 3D software for wind turbine load in the different layer of soil. The effectiveness of pile-raft foundation in reducing settlement can be justified by using optimum number of piles [11]. Ravichandran et al. [12] researched about the procedure to design and optimize piled raft foundation to support wind turbine load in both clayey and sandy soil. The effect of soil strength parameters and wind speed in the design of piled raft foundation was examined during the study. The final design of foundation is controlled by differential settlement and rotation rather than by bearing capacity or total settlement [12]. Pham et al. [13] utilized $FLAC^{3D}$ software and concluded that piled raft foundation was efficient in terms of minimum surface settlement and raft rotation from currently available wind turbine foundation solutions.

Material Properties and Wind Tower Specification

A steel wind turbine tower of 100 m hub height of capacity 1.5 MW from Illinois, US was taken as the research problem to study the piled raft foundation. The wind

Table 4.1 Soil properties used in the present study

Parameter	Soft clayey soil	Medium-stiff clayey soil	Stiff clayey soil	Sandy soil	Unit
Dry unit weight (γ_{unsat}) [16]	11.9	15.0	16.7	15.6	kN/m^3
Saturated weight (γ_{sat}) [16]	17.3	18.0	20.3	19.5	kN/m^3
Modulus of elasticity (E)[a]	7460	14,918	29,643	53,626	kN/m^2
Poisson's ratio (ν) [17]	0.40	0.35	0.30	0.30	–
Cohesion (c) [18]	18.19	36.38	72.30	0	kN/m^2
Friction angle (ϕ) [17]	0	0	0	37	°

[a]Modulus of elasticity for clayey soil was taken from Das [19] and sandy soil was taken from USACE [20]

tower size and dimensions were taken from National Renewable Energy Laboratory (NREL) manual report under U.S. Department of Energy [14]. The dead load of the tower including all components was found to be 3581.07 kN (365,043 kg) [2]. The ASCE 7-10 [15] standard for 'Occupancy Category III and IV Buildings and Other Structures' was followed to get the design wind speed around the research region. The design wind speed was taken as 190 kmph (120 mph) for Illinois region. The shear force (lateral load) and bending moment acting at the base of the tower were found to be 645 kN and 30,705.13 kN m, respectively [2].

The soil properties for the research were taken from the literature as tabulated in Table 4.1. The groundwater was assumed to be at the ground surface. The concrete unit weight, modulus of elasticity, and Poisson's ratio were taken as 25 kN/m^3, 2.5 $\times$ 10^7 kN/m^2, and 0.15, respectively.

Calculation and Numerical Modelling

Design Literature

Piled raft foundation consists of three components: namely piles, raft, and subsoil. The applied loads are transmitted to the subsoil through both raft and piles. The load sharing between raft and piles distinguishes piled raft foundation from any other types. The design procedures outlined in Poulos [6] along with Hemsley [21], Poulos and Davis [4], and Randolph [5] are followed in the research study. The general bearing capacity equation proposed by Meyerhof [22] was used to calculate the ultimate load bearing capacity of the raft only ($Q_{u(r)}$). The circular raft was considered for the study so that piled raft foundation has equal capacity in all directions during wind turbine rotor rotation [12]. The effective area due to eccentric loading due to wind was calculated following the procedure outlined in the Guidelines for Design of Wind Turbines by DNV/Risϕ [23]. The load capacity of pile group ($Q_{u(p)}$) was calculated following the procedure compiled in Gaihre [2]. The ultimate axial capacity of the piled raft foundation ($Q_{u(\text{pr})}$) was considered lesser of: (i) sum of ultimate capacities of raft plus pile group, i.e. $Q_{u(\text{pr})} = Q_{u(p)} + Q_{u(r)}$, and (ii) ultimate capacity of the block. The factor of safety for the axial load capacity was calculated as $\text{FS} = Q_{u(\text{pr})}/Q$, where Q is the design axial load. The procedure to calculate the moment capacity of the piled raft foundation was given by Hemsley [21] compiled in Gaihre [2]. The ultimate moment capacity of the raft, pile group, and piled raft system was calculated based on the equations provided by Hemsley [21].

The lateral load capacity of piled raft foundation $(V_u)_{\text{pr}}$ was checked using the procedure outlined in Gudmundsdottir [24] for individual pile in cohesive and cohesionless soil. It is assumed that all the piles in the group will have similar lateral behaviour. The factor of safety for the lateral capacity is calculated as $\text{FS} = (V_u)_{\text{pr}}/V$, where V is the design shear force (lateral load).

The method proposed by Poulos [6] with an extension of estimating the load sharing between the raft and piles by Randolph [5] and Bhartiya et al. [9] was used to evaluate the vertical load-settlement behaviour of the piled raft. The load sharing among raft and piles in a piled raft foundation was assessed based on stiffness of the raft, piles, and piled raft as a block. The stiffness of piled raft as a block, K_{pr}, was determined by the equation as proposed by Randolph [5]. The equations involved in determining the stiffness are well explained in Gaihre [2].

The differential settlement equation for piled raft foundation was calculated based on the equations given by Randolph [5].

Bhartiya et al. [9] carried out the load-settlement analysis of a piled raft foundation. The equations involved in the calculation of settlement and load sharing proportion between piles and raft can be assessed through the respective paper.

Numerical Modelling

The PLAXIS 3D 2017, a finite element software, was used to perform the numerical modelling of the piled raft foundation. The model domain was fixed considering 10 times higher than the radius of the raft. The design model was normally fixed (horizontal movement restricted) on the sides and fully fixed at the bottom of the model. The material properties of different soils are shown in Table 4.1, whereas material properties of structural components such as raft and piles are shown in Tables 4.2 and 4.3, respectively. The materials were modelled as linear elastic Mohr–Coulomb's model based on data availability in the software.

Table 4.2 Raft properties used in the present study

Parameter	Symbol	Raft	Unit
Material model	–	Linear elastic	–
Drainage type		Non-porous	
Unit weight, *Typ*	γ	25	kN/m^3
Modulus of elasticity, *Typ*	E	25×10^6	kN/m^2
Poisson's ratio, *Typ*	N	0.15	

Table 4.3 Pile properties used in the present study

Parameter	Embedded beam	Unit
Material model	Elastic	–
Modulus of elasticity (E)	25×10^6	kN/m^2
Unit weight (γ)	0	kN/m^3
Beam type	Predefined massive circular beam	
Diameter (D)	0.6	m

Piled Raft Foundation. The plate element of thickness 0.5 m was created by assigning properties to the surface element created at the zero level (ground). The piles were modelled as embedded beam elements with rigid connection to the raft. The skin and toe resistance value were assigned to each embedded beam calculated using equations separately. The axial load, lateral load, and bending moment were assigned to the model at the centre, edge, and centre of the raft, respectively. The model was discretized by creating a finite element mesh with 10-noded elements with coarseness factor of 1.5. The finite element model was analysed in last stage for the assigned loads and boundary conditions after successfully defining each calculation phases.

Design Results

The results are obtained from the analytical solution as well as numerical modelling using PLAXIS 3D on each type of soil following the above-mentioned procedure. The piled raft foundation was modelled with a raft thickness of 0.5 and 15 m diameter rigidly connected to the pile group with 23, 32, and 46 piles. The parametric study was performed by varying the wind speeds, type of soil, and number of piles to cover possible uncertainty in the design. The modelled piled raft foundation along with deformed shape is shown in Fig. 4.1.

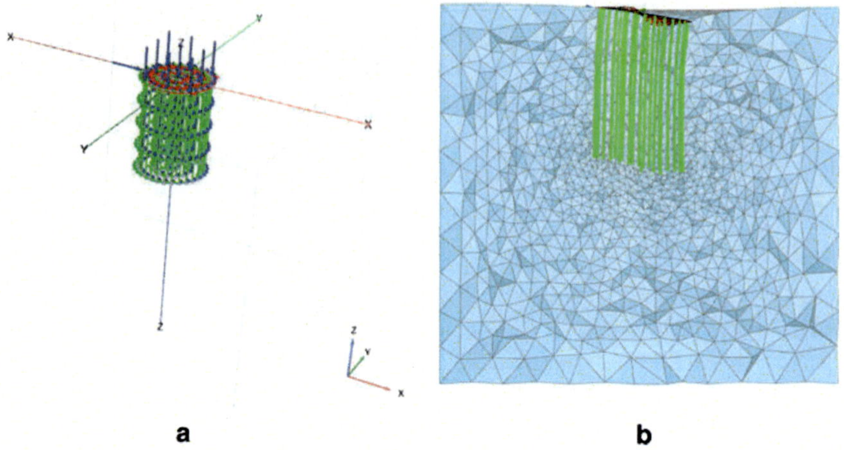

Fig. 4.1 **a** Modelled pile-raft foundation and **b** deformed piled raft model after applying all loads

Variation in Number of Piles

The variation in number of piles which can also be related to area improvement ratio (i.e. percentage of pile cross-section area over total foundation area) was studied to calculate the total settlement of piled raft foundation analytically using two methods: Randolph [5] and Bhartiya et al. [9]. From Fig. 4.2, the variation in settlement with increase in number of piles is found to be significantly less. The settlement is found to be high in soft clayey soil and gradually decreases with the increase in number of piles. The settlement in medium-stiff clayey soil follows the same trend with increase in number of piles like in soft clayey soil. But the settlement remains almost constant for stiff clayey soil and sandy soil, despite increasing the number of piles. The settlement in soft clayey soil is found to be 3.5 times higher than the settlement in stiff clayey soil.

The differential settlement and rotation of the piled raft foundation are calculated based on Randolph [5] method as tabulated in Table 4.4.

The differential settlement and rotation of piled raft foundation are found to be more in soft clayey soil but less in sandy soil but found to be increasing with number of piles in each soil. The allowable rotation for piled raft stated by Matsumoto [25] under static loading ranges from 1×10^{-3} to 2×10^{-3} rad. The design for the current study is safe for design axial load. The permissible differential settlement of piled raft foundation with 15 m diameter raft is 45 mm based on acceptable maximum inclination of tower of 3 mm/m [26].

The load sharing by pile and raft in piled raft foundation is calculated by using Bhartiya et al. [9] equation. The pile group is found to carry about 60–94% of the total vertical load while raft carry about 6–40% of the total vertical load. Raft is found to carry more load (40.2%) in soft clayey soil than other soils studied. Piles are found to share more proportion of load (93.57%) in stiff clayey soil than other

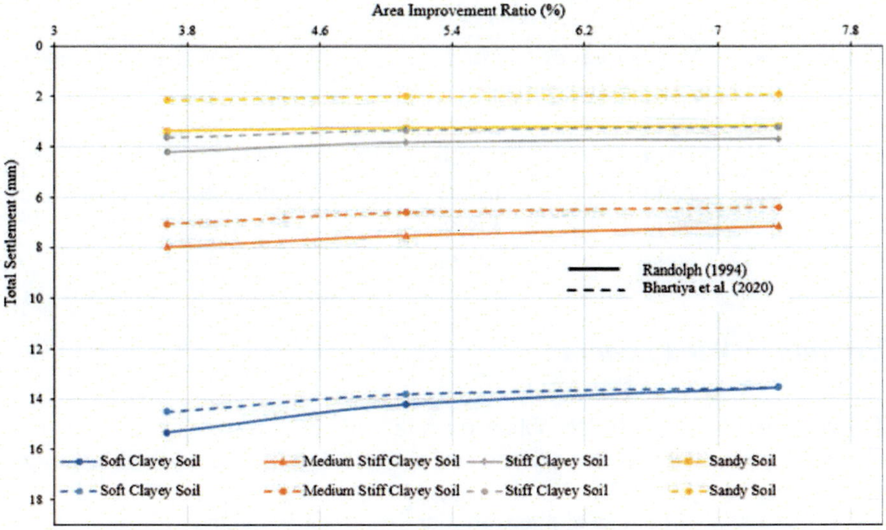

Fig. 4.2 Variation of settlement with area improvement ratio (%)

Table 4.4 Calculation of differential settlement and rotation of piled raft foundation

Settlement	Soft clayey	Medium-stiff clayey	Stiff clayey	Sandy soil
23 piles				
Δ (mm)	2.05	1.07	0.56	0.43
θ (radians)	10.3×10^{-5}	5.34×10^{-5}	2.82×10^{-5}	2.18×10^{-5}
32 piles				
Δ (mm)	2.25	1.19	0.61	0.42
θ (radians)	11.2×10^{-5}	5.96×10^{-5}	3.03×10^{-5}	2.11×10^{-5}
46 piles				
Δ (mm)	2.56	1.35	0.57	0.49
Δ (radians)	12.8×10^{-5}	6.78×10^{-5}	2.86×10^{-5}	2.46×10^{-5}

soils studied. These values are found to be similar with the study result published by Bhartiya et al. [9] and Kumar et al. [8].

The settlement obtained from PLAXIS 3D was compared to the settlement calculated analytically. The discrepancy ratio of S_{cal} [5]/S_{PLAXIS} and S_{cal} [9]/S_{PLAXIS} was evaluated as nearly 1.0 except for sandy soil using Bhartiya et al. [9] method. The more discrepancy in sandy soil can be assumed because of incompetence of the formulated equation for calculating the settlement in cohesionless soil. The effectiveness of modelling in PLAXIS 3D for piled raft foundation can be justified.

Table 4.5 Design load for various wind speed

Wind speed (mph)	Vertical load (kN)	Shear force or lateral load (kN)	Bending moment (kN m)
80	3581.02	286.91	13,646.72
100	3581.02	448.30	21,323.00
120	3581.02	645.55	30,705.13
140	3581.02	878.66	41,793.09
160	3581.02	1147.64	54,586.89

Variation in Wind Loading

The wind speed was varied to obtain the value of shear force (lateral load) and bending moment acting at the edge and centre of piled raft, respectively. The design load is tabulated in Table 4.5.

The differential settlement of piled raft foundation was evaluated with the variation in wind speed with different types of soil. The differential settlement was found to be increasing with increase in wind speed but decreased with increase in number of piles in studied soils. The stiff clayey soil and sandy soil behave approximately same in terms of differential settlement variation with wind speed. The piled raft foundation in soft clayey soil results in high differential settlement. When wind speed is 160 mph, the differential settlement is found to be more than the permissible different settlement of 45 mm for 15 m diameter raft. The differential settlement data in soft clayey soil for 46 number of piles when wind speed was 80 mph could not be obtained due to the computation error in software. The variation of differential settlement with wind speed for 23 piles is shown in Fig. 4.3. The similar type of trend was seen with 32 and 46 number of piles.

Variation of Lateral Load with Wind Speed

The analytical calculation of lateral deflection for piled raft was found to be similar to pile group foundation. So, the calculation is not discussed here. However, based on the separate study done by Gaihre [2], the analytically calculated lateral deflection was found to be more than PLAXIS 3D modelled deflection. The variation of lateral deflection with lateral load due to various wind speed for 32 number of piles is shown in Fig. 4.4. The similar type of trend was seen with 23 and 46 number of piles. The lateral deflections of foundation in soft clayey soil for wind speed from 100 to 160 mph were found to be more than the permissible value of 6.35 mm given by AISC [27]. With 23 number of piles in medium-stiff clayey soil when wind speed was 160 mph, the foundation was deflecting more than the permissible value. The lateral deflection was found to be increased with increasing wind speed but decrease with an increase in number of piles in studied soil. The lateral deflection in soft clayey

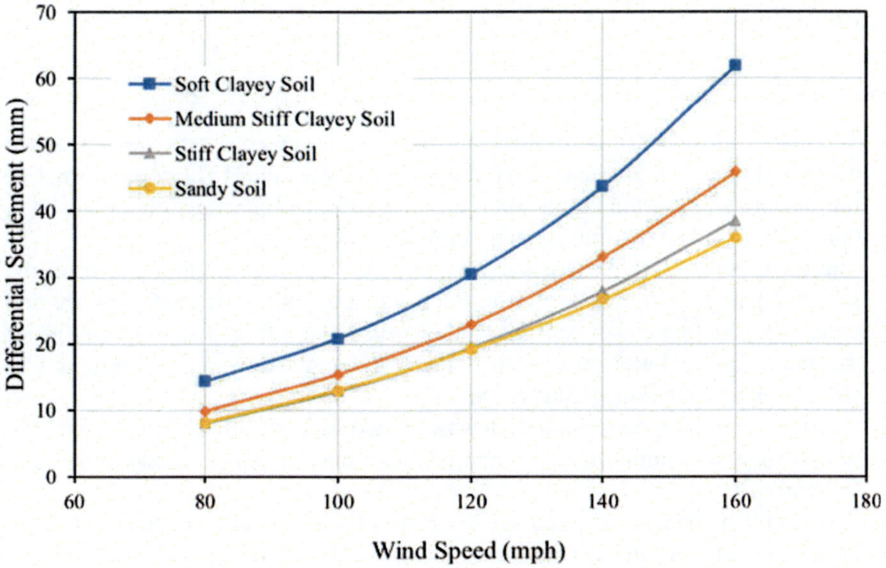

Fig. 4.3 Variation of differential settlement with wind speed for 23 number of piles

soil was found to be 3 times higher than that in sandy soil. In stiff clayey and sandy soil, the lateral deflection was found to be approximately same. The lateral deflection data for soft clayey soil with 46 piles when wind speed was 80 mph could be not obtained due to computation error in software.

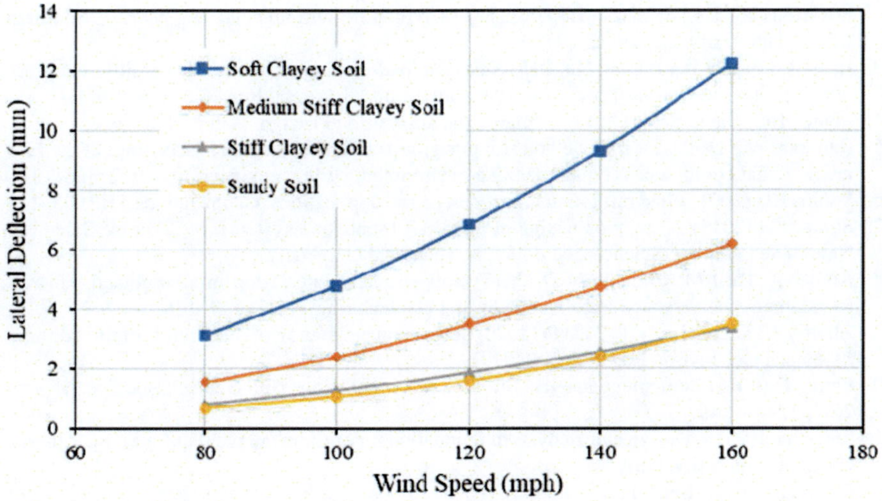

Fig. 4.4 Variation of lateral deflection with wind speed for 32 number of piles

Conclusion

Based on the limited study, the following conclusions may be made.

- The load bearing capacity of piled raft foundation increased with the increase in number of piles. The studied piled raft foundation was found to be good alternative for onshore wind turbine foundation for all wind speeds except 160 mph used in this study based on bearing capacity criteria.
- The importance of raft in load sharing of axial load was highlighted from the study of load sharing proportion in piled raft foundation. The raft was sharing more loads in soft clayey soil than other soils. The raft was sharing 6–40% of the total axial load with least in stiff clayey soil and sandy soil and more in soft clayey soil and medium-stiff clayey soil.
- The design of the piled raft fountain was found to be controlled by the differential settlement and rotation of the foundation rather than bearing capacity or the total settlement.
- The lateral deflection of piled raft foundation was found to decrease with an increase in number of piles but increased with the increase in wind speed.

References

1. Byrne BW, Houlsby GT (2006) Assessing novel foundation options for offshore wind turbines. In: Proceedings of the world maritime technology conference. The Institute of Marine Engineering, Science and Technology (IMarEST) (2006), London, pp 1–10
2. Gaihre N (2020) Analytical and numerical modeling of foundations for tall wind turbine in various soils. M.S. thesis, Department of Civil and Environmental Engineering, Southern Illinois University, Carbondale, USA
3. Lang R, Liu R, Lian J, Ding H (2015) Study on load-bearing characteristics of different types of pile group foundations for an offshore wind turbine. J Coastal Res 73:533–541
4. Poulos HG, Davis EH (1980) Pile foundation analysis and design. Wiley, New York
5. Randolph MF (1994) Design methods for pile groups and piled raft. In: Proceedings of the 13th international conference on man-machine-environment system engineering, vol 5, pp 61–82
6. Poulos HG (2001) Piled raft foundations: design and applications. Géotechnique 51(2):95–113
7. Sönmez N (2013) A study on design of piled raft foundation systems. M.S. thesis, School of Natural and Applied Sciences, Middle East Technical University
8. Kumar A, Patil M, Choudhury D (2017) Soil–structure interaction in a combined pile–raft foundation—a case study. Proc Inst Civil Eng Geotech Eng 170(2):117–128
9. Bhartiya P, Chakraborty T, Basu D (2020) Settlement estimation of piled raft for initial design. J Geotech Geoenviron Eng 146(2):04019127
10. Phung D (2016) Settlement analysis for piled raft foundations. Jpn Geotech Soc Special Publ 2(34):1244–1249
11. Shrestha S (2015) Design and analysis of foundation for onshore tall wind turbines. M.S. thesis, Clemson University, USA
12. Ravichandran N, Shrestha S, Piratla K (2018) Robust design and optimization procedure for piled raft foundation to support tall wind turbine in clay and sand. Soils Found 58(3):744–755
13. Pham H, Dias D, Miranda T, Cristelo N, Araújo N (2018) 3D numerical modeling of foundation solutions for wind turbines. Int J Geomech 18(12):04018164

14. LaNier M (2004) LWST phase I project conceptual design study: evaluation of design and construction approaches for economical hybrid steel/concrete wind turbine towers, 28 June 2002–31 July 2004
15. American Society of Civil Engineers: Minimum Design Loads for Buildings and Other Structures (ASCE/SEI 7-10). American Society of Civil Engineers (2013)
16. Peck RB, Hanson WE, Thornburn TH (1974) Foundation engineering, vol 10. Wiley, New York
17. MnDOT: Pavement Design Manual. Chapter 3. Section 3.2. Subgrade Soils (2007)
18. Terzaghi K, Peck RB, Mesri G (1967) Soil mechanics in engineering practice. Wiley, New York
19. Das BM (2016) Principles of foundation engineering, 8th edn. Cengage Learning, New Delhi
20. U.S. Army Corps of Engineers: Engineering and Design: Settlement Analysis. EM 1110-1-1904 (1990)
21. Hemsley JA (ed) (2000) Design applications of raft foundations. Thomas Telford
22. Meyerhof GG (1963) Some recent research on the bearing capacity of foundations. Can Geotech J 1(1):16–26
23. DNV/Risø (2002) Guidelines for design of wind turbines, 2nd ed. Jydsk Centraltrykkeri, Denmark. ISBN 87-550-2870-5
24. Gudmundsdottir B (1981) Laterally loaded piles. M.S. thesis, University of Alberta, Edmonton, Alberta, Canada
25. Matsumoto T (2014) Implication for design of piled raft foundations subjected to lateral loading. In: Proceedings of international symposium on advances in foundation engineering (ISAFE). Singapore, pp 113–136
26. Grunberg J, Gohlmann J (2013) Concrete structures for wind turbines. Wilhelm Ernst & Sohn, Berlin, Germany. ISBN 987-3-433-03041-7
27. American Institute of Steel Construction (AISC) (1973) Highway structures design handbook. AISC Marketing, Inc., Pittsburgh, PA
28. PLAXIS 3D: Reference Manual, 2017

Chapter 5
Prediction of Strength Parameters of Fibre Reinforced Soil Using Machine Learning Algorithms

G. Kannan and Evangelin Ramani Sujatha

Introduction

A variety of soil stabilization methods are in practice in the modern construction industry. This includes conventional chemical additives like cement, lime and fly ash stabilization to modern polymer-based treatment like resin, biopolymer, and synthetic polymers. These type of additive-based stabilization techniques usually improves the soil strength to many folds, but will impart a brittle nature to the treated soil [1]. Hence, a suitable alternative is to introduce fibre reinforcement to the soil [1]. Fibre reinforcement enhances the strength of the treated soil and offers a better post-failure response in contrast to the brittle behaviour of chemically treated soil. Many researchers proposed works on a natural fibres like jute fibre, sisal fibres, bamboo fibre, human hair, etc. and on synthetic fibres like polypropylene fibre, glass fibre and other waste material-based fibre.

Effect of fibre reinforcement on strength depends on the nature of the soil and the characteristics of the fibre material. A detailed study on fibre and soil is always necessary before assessing the characteristics of treated soil. But an elaborate experimental programme is not a practical option in the construction scenario. A convenient and handy method is therefore necessary to predict the approximate strength of treated soil before selecting the most suitable fibre for a particular construction project.

With such high uncertainties, developing an empirical relationship among the variables would be a complicated process. Machine Learning (ML) methods are the

G. Kannan
School of Civil Engineering, SASTRA Deemed to be University, Thirumalaisamudram, Thanjavur 613401, India
e-mail: rajkannan93@gmail.com

E. R. Sujatha (✉)
Centre for Advanced Research on Environment, School of Civil Engineering, SASTRA Deemed to be University, Thirumalaisamudram, Thanjavur 613401, India
e-mail: sujatha@civil.sastra.edu

K. Muthukkumaran et al. (eds.), *Soil Dynamics, Earthquake and Computational Geotechnical Engineering*, Lecture Notes in Civil Engineering 300,
https://doi.org/10.1007/978-981-19-6998-0_5

recent advancements that would be useful in such predictions. ML methods would be very useful if there is a complicacy involved in establishing a function-based relationship between independent and dependent variables and also when the bounds of the parameters are vast [2]. Initially, ML methods train the system to observe the pattern in the variable set and try to establish a relation among them through relevant algorithms, validate them and finally the performance of a ML model is tested with the unseen data on training and validation dataset to ensure its generalization capability.

In general, supervised ML techniques can be categorized as regression methods and classification methods. The outcome of classification algorithm will be a class label whereas the outcome of regression algorithm will be a numerical value. The regression models taken up for analysis in the present study are the artificial neural network (ANN) and the support vector regression (SVR).

Artificial Neural Network (ANN) is a computational model that mimics processing of information in human brain. The functional unit of ANN is called neuron. The type of ANN taken up for study is multilayer perceptron (MLP). A MLP has three layers: (a) input layer, (b) hidden layer, and (c) output layer. The input layer will be fed with features as inputs and the output layer will produce a numerical outcome. The hidden layer is present between the input and output layers. The hidden layer extracts the implicit information present in features. The hidden layer captures more complex relationships in the input by discovering the relationships between the features in the input. ANN learns by a mechanism called backpropagation. The base idea of backpropagation is that–adjust the weights associated with each node in each layer such that the mean squared error between the actual output and the predicted output is minimized.

SVR is a supervised learning algorithm that is used to predict discrete values. The basic idea behind SVR is to find the best fit line. In SVR, the best fit line is the hyperplane that has the maximum number of data points. Kernel functions are used to project the datapoints in the lower dimensional space to a higher dimensional data space. It applies for problems where the best fit line could not be found in lower dimension space, but it is possible to find it in higher dimension space.

The present work utilizes three variants of ANN namely, Levenberg–Marquardt (LM), Bayesian Regularization (BR) and Scaled Conjugate Gradient (SCG) algorithm and two variants of SVR namely, SVR with linear kernel (LK) and SVR with Gaussian kernel (GK) using MATLAB R2020b software to predict the UCS and CBR of fibre treated soil.

Methodology

Data

The accuracy of machine learning methods depends to a great extent on the quantum of available data. Hence a large numbers of data has been collected from various literary sources using natural fibres such as bamboo fibre [3], jute fibre [4], coir fibre [5], and palm fibre [6], and artificial fibres like glass fibre [7–10], polyester fibre [11], polypropylene fibre [12–15], basalt fibre [12, 16], waste plastic and waste tyre fibre [17].

Parameters

Two sets of analysis have been carried out separately for prediction of UCS and CBR of fibre treated soil. The input to the regression models taken up for study are the independent variables (features) and the output will either be UCS (for set 1) or CBR (for set 2). The input consists of the factors influencing the strength of the fibre reinforced soil and they are fibre length (mm), fibre diameter (mm), fibre dosage (%), plasticity index of the soil (%), maximum dry density (MDD) of soil (kN/m^3), and optimum moisture content (OMC) of the soil (%).

Procedure

The flow diagram for the present study is shown in Fig. 5.1. MATLAB curve fitting tool in deep learning toolbox is used for the ANN regression analysis. The dataset taken up for study consists of 100 samples each for UCS and CBR analysis. With this dataset of samples, the training-validation-testing split up is 70%–15%–15%, respectively. The analysis was done for both set 1 (UCS prediction) and set 2 (CBR prediction) separately with three different algorithms for prediction. A separate coding has been created for the two kernel variants of SVR. The performance of the developed regression models was evaluated using coefficient of prediction (R^2 value), root mean squared error (RMSE), and mean absolute error (MAE).

Algorithms

Levenberg–Marquardt algorithm of ANN works faster but consumes more memory space. Training in the Levenberg–Marquardt algorithm ceases when the MSE of validation starts increasing. Bayesian regularization is a type of ANN algorithm which works effectively for small and complicated datasets. It reduces the need for longer cross-verifications and avoids the necessity of the validation process. Using the scaled conjugate gradient algorithm in ANN makes it faster by replacing the conventional line search method for updation of weights with conjugate direction search. SVR with linear kernel is a traditional SVR technique where the goal is to find the best fit line for the datapoints. In scenarios where it is not possible to find the best fit line for the datapoints in lower dimensional space, kernel trick (with polynomial

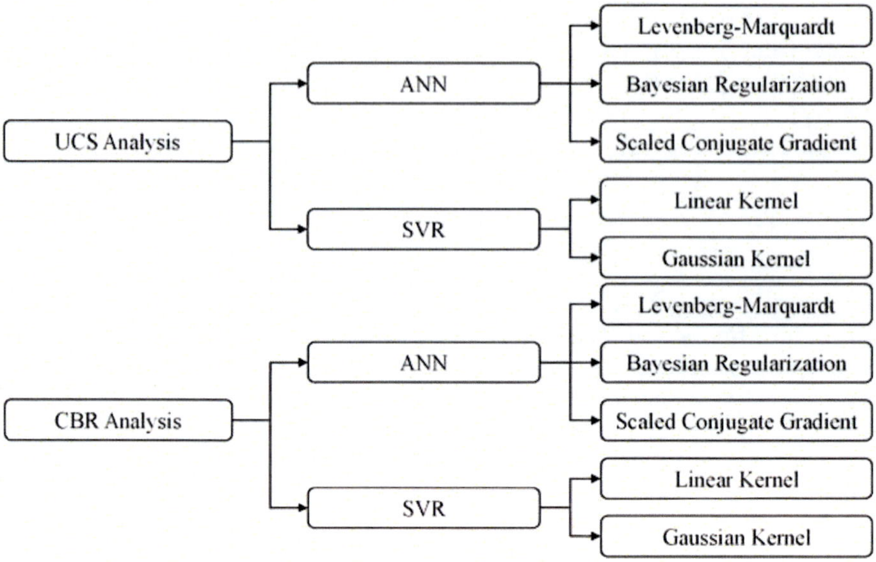

Fig. 5.1 Work plan for the study

kernel or Gaussian kernel or sigmoidal kernel) is used. SVR with Gaussian kernel is used in this study.

Results and Discussion

The performance of the regression models for this study in estimating the UCS and CBR when unseen samples during training and validation phase (test dataset) were presented to them was evaluated using coefficient of prediction (R^2 value), root mean squared error (RMSE), and mean absolute error (MAE). Lower values of MAE and RMSE indicate that the predicted outcome by the developed regression models is closer to the actual outcome. On the other hand, R^2 is a statistical measure of fit that indicates the percentage of the variance in the dependent variable that the independent variables in the developed regression model explain collectively. R^2 measures the strength of the relationship between the developed regression model and the dependent variable on a 0–1 scale. Higher R^2 value indicates smaller difference between the observed data and the predicted values. The formula for calculating R^2, MAE, and RMSE is given by (5.1), (5.2), and (5.3), respectively.

$$R^2 = 1 - \left[(\text{Actual value} - \text{Predicted value})^2 / (\text{Actual value} - \text{Mean actual value})^2\right] \quad (5.1)$$

$$\text{MAE} = \sum \text{Abs[Predicted value} - \text{Actual value]}/n \tag{5.2}$$

$$\text{RMSE} = \left[\sum \text{[Predicted value} - \text{Observed value]}^2/n\right]^{1/2} \tag{5.3}$$

Unconfined Compression Strength

The fit obtained by using five ML models taken up for the study on the test dataset is presented in Fig. 5.2. The red dots in Fig. 5.2 indicates the test dataset points and the line indicates the best fit line for these data points. The three performance evaluation metrics for the five regression models in prediction of UCS is presented in Table 5.1.

It is evident from Table 5.1 that all the algorithms of ANN captured the complex functional relationship between the dependent and the independent variables as observed through their R^2 values. However, the scatter plots of test dataset in Fig. 5.2 and the RMSE and MAE values in Table 5.1 reveals that only Bayesian regularization method maintained a proximity to the actual value during the analysis. An absolute error of 10.99 indicates that the average deviation of UCS was nearly 10.99 kPa for the test dataset. However, this value could be improved with further with increase in training dataset.

California Bearing Ratio

The performance evaluation metrics for the ML models for the predicting the CBR of fibre reinforced soil are tabulated in Table 5.2. The fit obtained for the selected ML models on the test dataset is shown in Fig. 5.3. Levenberg–Marquardt and Bayesian regularization of ANN and Gaussian kernel of SVR captured the most appropriate relationship between CBR and the independent variables and this is evident in their R^2 value. SVR with Gaussian kernel and ANN with Bayesian regularization exhibited lesser MAE and RMSE values. From the Table 5.2, it is evident that ANN with Bayesian regularization showed better performance in predicting the CBR of fibre reinforced soil since it has the least MAE (0.81) and RMSE (0.81) and high R^2 (0.91).

Correlation Heatmaps

A correlation coefficient (Spearman rank correlation coefficient) has been established between the input and the output variables (UCS and CBR) in order to exhibit the dependency among them. Unlike Pearson coefficient, Spearman coefficient works on

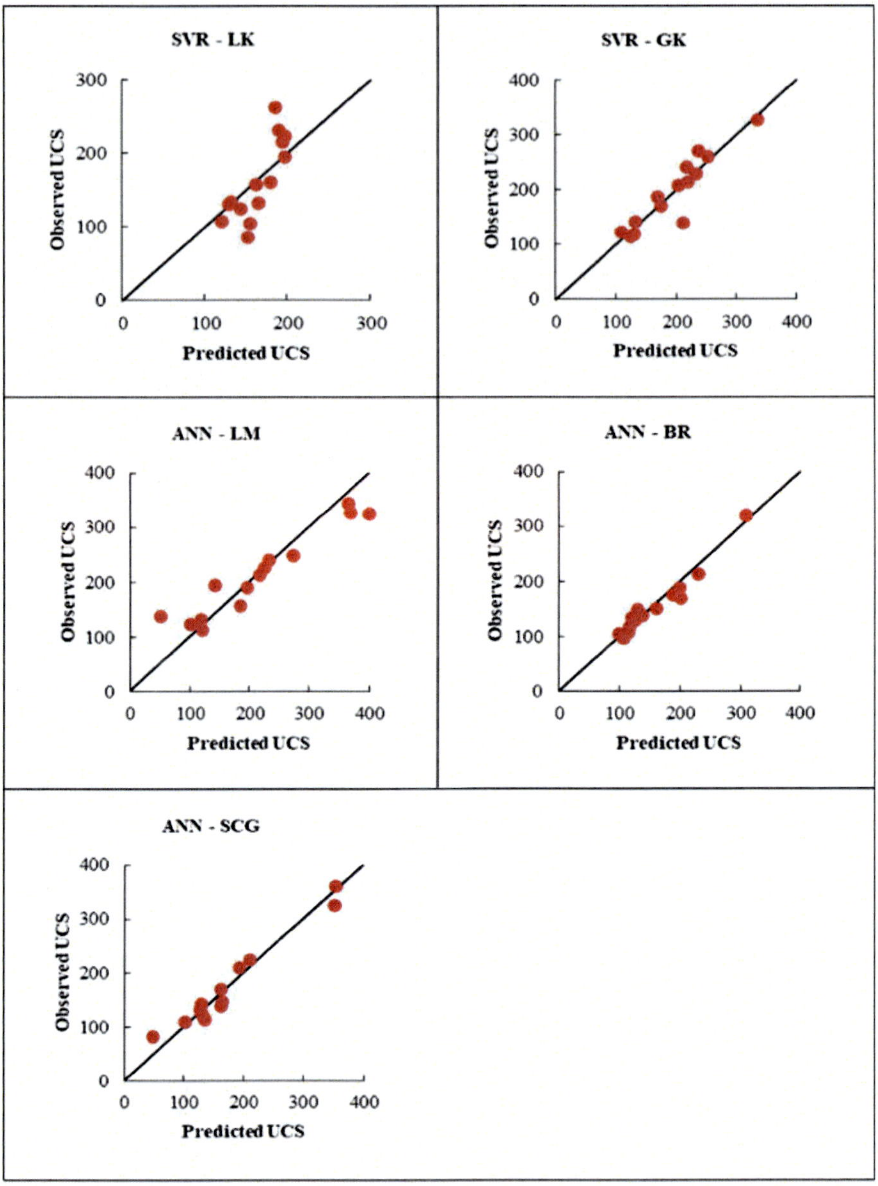

Fig. 5.2 Observed versus predicted value plots for UCS prediction through various algorithms

Table 5.1 Results of tested UCS prediction data through various ML algorithms

Method	Algorithm	R^2	RMSE	MAE
ANN	LM	0.92	37.83	27.45
ANN	BR	0.95	13.66	10.99
ANN	SCG	0.96	17.17	14.87
SVM	LK	0.64	35.48	27.06
SVM	GK	0.85	24.12	16.75

Table 5.2 Results of tested CBR prediction data through various ML algorithms

Method	Algorithm	R^2	RMSE	MAE
ANN	LM	0.90	1.89	1.28
ANN	BR	0.91	1.15	0.81
ANN	SCG	0.76	2.10	1.70
SVM	LK	0.68	2.85	2.34
SVM	GK	0.85	1.27	0.84

the basis of ranking of variables, thus identifying the monotony in the functional relationship rather than restricting it to linear function with Pearson coefficient. Figure 5.4 represents the correlation heatmap generated through MATLAB. Heatmaps helps in identifying the dominant and least influencing parameters under study using easy visual interpretation based on colour intensity. The correlation coefficient can hold any value between -1 and $+1$. Positive value indicates a direct relationship among variables and negative value an inverse relationship among variables. Correlation coefficient of zero indicates that there is no relationship among the variable under consideration. However, it should be noted that this correlation coefficient works purely on the basis of given input and output data. Further, dataset with precise input and output values are necessary to ensure the effectiveness of correlation prediction. In general, this can serve as a guide for identifying the most crucial factors under study, provided sufficient and precise data is available.

Based on the heatmap shown (Fig. 5.4), both CBR and UCS values has direct relationship with fibre dosage, length and maximum dry density of soil, whereas they maintain an inverse relation with fibre diameter, plasticity index and optimum moisture content of the soil. Although fibre length showed an almost direct relationship, it is close to zero. This could be due to the fact that in the supplied dataset, the length of the fibre is within a closer range and it did not impact the behaviour to a greater extent. Similarly, increasing the plasticity index and OMC showed a reduced UCS and CBR in the soil. Higher plasticity index shows more fines content in the soil and OMC indicates the water content at which soil shows maximum strength. Higher plasticity index and OMC indicates the nature of finer and softer soil thus reflecting a reduced strength.

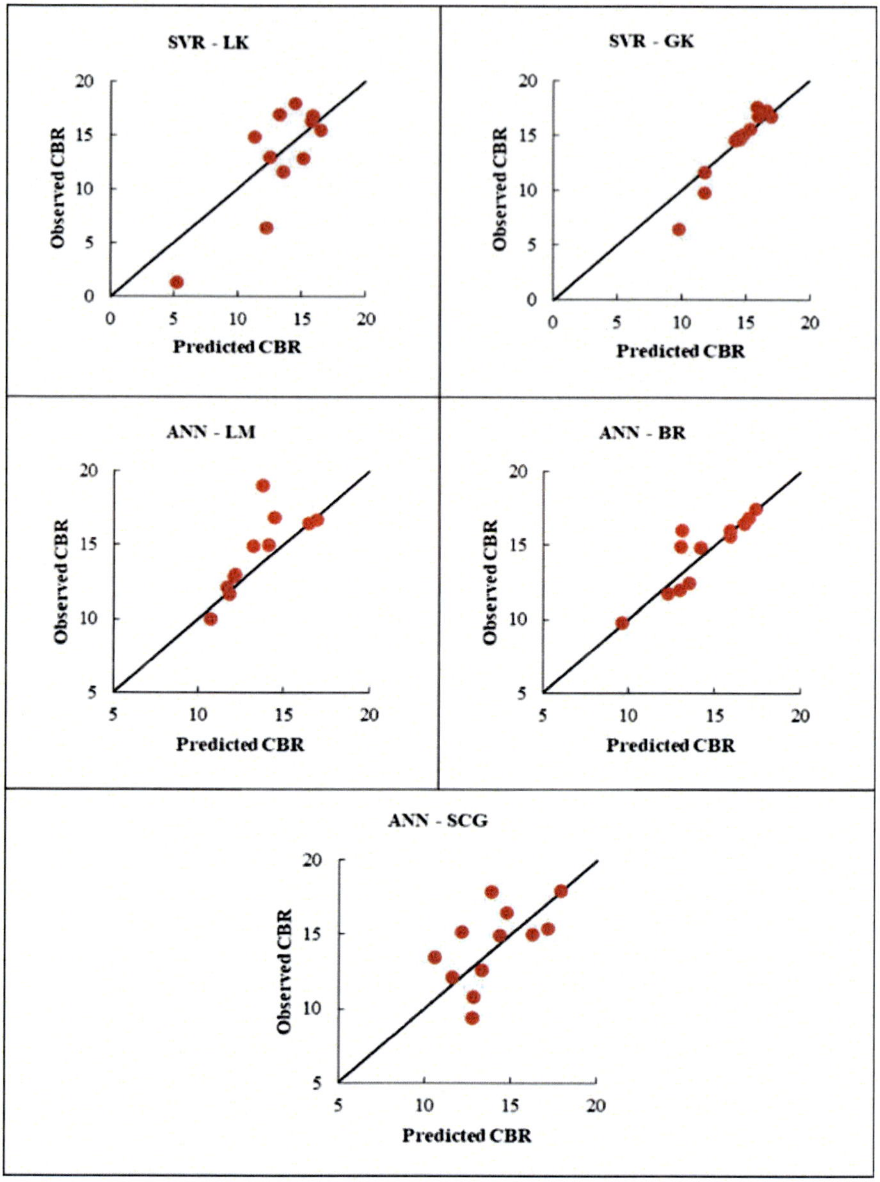

Fig. 5.3 Observed versus predicted value plots for CBR prediction through various algorithms

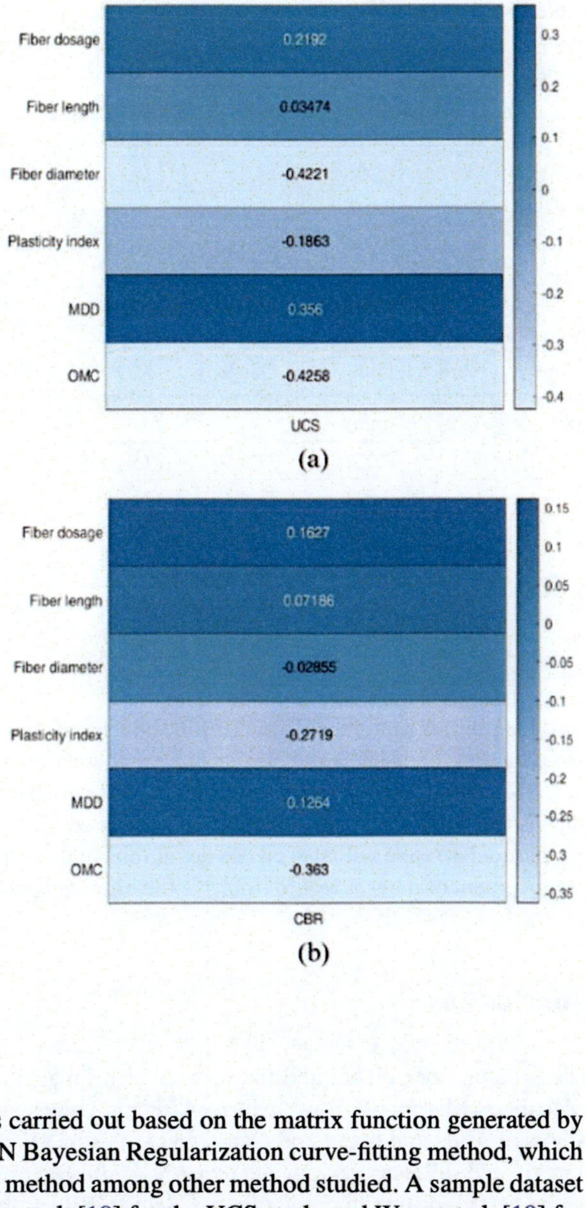

Fig. 5.4 Correlation coefficient heatmaps for **a** UCS dataset and **b** CBR dataset

Verification

A verification procedure was carried out based on the matrix function generated by MATLAB for testing the ANN Bayesian Regularization curve-fitting method, which evolved as a better prediction method among other method studied. A sample dataset taken from the works of Jiang et al. [18] for the UCS study and Wang et al. [19] for the CBR study was used for the verification procedure. The data was selected in such a way the values fairly fit into the study data range. The actual versus predicted outputs and their corresponding inputs for UCS and CBR verifications are shown in Tables 5.3 and 5.4, respectively. The comparison on actual versus predicted UCS in Table 5.3 shows that the deviation at lower UCS values is less and it increased with higher values of actual UCS. The CBR prediction showed higher deviations at a lower

Table 5.3 Verification of UCS prediction efficiency

Fibre dosage (%)	Fibre length (mm)	Fibre diameter (mm)	Plasticity index (%)	MDD (kN/m^3)	OMC (%)	Actual UCS (kPa)	Predicted UCS (kPa)
0.1	10	0.036	17.3	16.68	15.4	269	255.25
0.1	15	0.036	17.3	16.68	15.4	280	244.47
0.1	20	0.036	17.3	16.68	15.4	237	231.65
0.1	25	0.036	17.3	16.68	15.4	218	214.68

Table 5.4 Verification of UCS prediction efficiency

Fibre dosage (%)	Fibre length (mm)	Fibre diameter (mm)	Plasticity index (%)	MDD (kN/m^3)	OMC (%)	Actual CBR	Predicted CBR
0.3	6	0.022	37.3	15.29	22.1	6.4	5.02
0.6	6	0.022	37.3	15.22	21.8	8.1	6.86
0.9	6	0.022	37.3	15.17	21.3	9.2	8.72
0.3	12	0.022	37.3	15.4	21.9	5.8	5.53
0.6	12	0.022	37.3	15.28	21.7	7.8	7.54
0.9	12	0.022	37.3	15.19	21.6	8.51	9.1

fibre length of 6 mm than 12 mm. This shows that the training set would have very few data on higher UCS and similarly very few data on soil with lesser fibre length on CBR prediction. These observations shows that the quantity of data for training plays a crucial role in the behaviour of such prediction algorithms. Missing dataset in a particular range will also create deviations in the approximations. Hence, further enhancement of dataset would improve the efficiency and accuracy in the prediction.

Conclusions

The present work elaborated the idea of using machine learning methods such as Levenberg–Marquardt, Bayesian regularization and scaled conjugate gradient algorithms of ANN and linear and Gaussian kernels of SVR to predict UCS and CBR of fibre treated soil. Results indicate that ANN with Bayesian regularization is found to be the best algorithm to predict the UCS and CBR of fibre reinforced soil. Based on the input data given, heatmap reveals that there is a fairly good correlation among the variables. The study was carried out as a preliminary analysis to test the use ML algorithms in estimation of strength of fibre reinforced soil. The performance of the ML models for predicting geotechnical properties, particularly UCS and CBR could be improved by increasing the volume of the training dataset.

Acknowledgements The authors thank the Vice Chancellor of SASTRA Deemed to be University, Thanjavur, Tamil Nadu, India for the support and the facilities provided to carry out the work successfully.

References

1. Soleimani Kutanaei S, Janalizadeh Choobbasti A (2017) Effects of nanosilica particles and randomly distributed fibers on the ultrasonic pulse velocity and mechanical properties of cemented sand. J Mater Civ Eng 29:04016230. https://doi.org/10.1061/(asce)mt.1943-5533.0001761
2. Al Bataineh A, Kaur D (2018) A comparative study of different curve fitting algorithms in artificial neural network using housing dataset. In: Proceedings of the 2011 IEEE national aerospace and electronics conference (NAECON), July 2018, pp 174–178. https://doi.org/10.1109/NAECON.2018.8556738
3. Brahmachary TK, Rokonuzzaman M (2018) Investigation of random inclusion of bamboo fiber on ordinary soil and its effect CBR value. Int J Geo-Eng 9.https://doi.org/10.1186/s40703-018-0079-x
4. Singh KL, Jamatia M (2020) Study on pavement soil subgrade properties with reinforced fibres. Indian Geotech J 50:300–306. https://doi.org/10.1007/s40098-020-00425-3
5. Sujatha ER, Geetha AR, Karunya SR (2018) Strength and mechanical behaviour of coir reinforced lime stabilized soil. Geomech Eng 16:627–634. https://doi.org/10.12989/gae.2018.16.6.627
6. Qu J, Zhu H (2021) Function of palm fiber in stabilization of alluvial clayey soil in Yangtze River Estuary. J Renew Mater 9:767–787. https://doi.org/10.32604/jrm.2021.013816
7. Sujatha ER, Atchaya P, Darshan S, Subhashini S (2020) Mechanical properties of glass fibre reinforced soil and its application as subgrade reinforcement. Road Mater Pavement Des 0:1–12.https://doi.org/10.1080/14680629.2020.1746387
8. Patel SK, Singh B (2017) Strength and deformation behavior of fiber-reinforced cohesive soil under varying moisture and compaction states. Geotech Geol Eng 35:1767–1781. https://doi.org/10.1007/s10706-017-0207-y
9. Patel SK, Singh B (2019) Investigation of glass fiber reinforcement effect on the CBR strength of cohesive soil. Lect Notes Civ Eng 14:67–75. https://doi.org/10.1007/978-981-13-0559-7_8
10. Walker PJ (1995) Strength, durability and shrinkage characteristics of cement stabilised soil blocks. Cem Concr Compos 17:301–310. https://doi.org/10.1016/0958-9465(95)00019-9
11. Gupta G, Gupta DAK (2014) Effect of fibre length on polyester fibre reinforced clay. Int J Eng Technol Manag Appl Sci 3:4430–4434
12. Ayothiraman R, Singh A (2017) Improvement of soil properties by basalt. In: Joint conference on piled foundations and ground improvement technology for the modern building and infrastructure sector, pp 403–412
13. Tang C, Shi B, Gao W et al (2007) Strength and mechanical behavior of short polypropylene fiber reinforced and cement stabilized clayey soil. Geotext Geomembr 25:194–202. https://doi.org/10.1016/j.geotexmem.2006.11.002
14. Agarwal BK, Hathiwala SA, Solanki CH (2021) Strength and compaction behavior of randomly distributed polypropylene fiber-reinforced expansive clay. Lect Notes Civ Eng (LNCE) 136:63–74. https://doi.org/10.1007/978-981-33-6444-8_6
15. Taha MMM, Feng CP, Ahmed SHS (2020) Influence of polypropylene fibre (PF) reinforcement on mechanical properties of clay soil. Adv Polym Technol 2020:1–15. https://doi.org/10.1155/2020/9512839
16. Gao L, Hu G, Xu N et al (2015) Experimental study on unconfined compressive strength of basalt fiber reinforced clay soil. Adv Mater Sci Eng 2015.https://doi.org/10.1155/2015/561293

17. Gardete D, Luzia R, Sousa M et al (2019) Soil stabilization with waste plastic and waste tyre fibres. Stabilisation des sols avec déchets de plastique et fibres de pneus usages, pp 1–6. https://doi.org/10.32075/17ECSMGE-2019-0894
18. Jiang H, Cai Y, Liu J (2010) Engineering properties of soils reinforced by short discrete polypropylene fiber. J Mater Civ Eng 22:1315–1322. https://doi.org/10.1061/(asce)mt.1943-5533.0000129
19. Wang JY, Hughes PN, Augarde CE (2019) CBR strength of London Clay reinforced with polypropylene fibre. In: 17th European conference on soil mechanics and geotechnical engineering (ECSMGE)—Proceedings, September 2019. https://doi.org/10.32075/17ECSMGE-2019-0296

Chapter 6
Hydraulic Conductivity of Fly Ash–Bentonite Mixture Exposed to Salt Solutions: ANN Model and Sensitivity Analysis

K. Anjali Jayadas, G. Suneel Kumar, Suvendu Kumar Sasmal, and Rabi Narayan Behera

Introduction

Delineation of hydraulic conductivity of fly ash and related environment influencing products has been established by a number of notable researchers in an attempt to study a key geoenvironmental problem. Akgun [1] studied the geotechnical characterization of bentonite–sand mixture. Alawaji [2] studied the swelling and compressibility characteristics of sand–bentonite mixtures. Alonso et al. [3] studied the effect of wetting and drying on expansive bentonite–sand mixtures. Cuevas et al. [4] observed thermohydraulic gradients on bentonite. Ebina et al. [5] reported hydraulic conductivity of compacted clay sand specimens. Komine [6] observed the hydraulic conductivities of sand–bentonite mixtures. Mishra et al. [7] studied swelling characteristics of soil–bentonite mixtures. A key observation recorded in the process of studying available literatures related to hydraulic conductivity suggests lack of application of soft computing. Hence, this paper is an approach to apply the well-established Levenberg–Marquardt technique in a specific geo-environmental problem of noteworthy importance. Mere observation of experimental results will not be helpful, to establish a relationship among influencing input parameters and the output. The Levenberg–Marquardt model is selected as it has been successfully implemented in the field of geotechnical engineering by researchers [8–12].

In this work, first a neural network model is established to estimate hydraulic conductivity of fly ash–bentonite mixture. The neural network model is finalized based on number of neurons present in the hidden layer. An empirical model equation is developed based on the weights and biases obtained from the statistical analysis. Sensitivity analysis is carried out to observe the most important parameter. The

K. Anjali Jayadas · G. Suneel Kumar · S. K. Sasmal · R. N. Behera (✉)
National Institute of Technology Rourkela, Rourkela, India
e-mail: rnbehera82@gmail.com

K. Muthukkumaran et al. (eds.), *Soil Dynamics, Earthquake and Computational Geotechnical Engineering*, Lecture Notes in Civil Engineering 300,
https://doi.org/10.1007/978-981-19-6998-0_6

outcomes will be helpful in speedy calculation of hydraulic conductivity where the date set lies within specific range. A future scope is also suggested.

Details of Experiments

The basic geotechnical characteristics of both fly ash and bentonite are found out separately by conducting different tests in laboratory [19] which include; specific gravity test by density bottle method (IS: 2720 Part-3/SEC-I [13]), particle size analysis (both sieve and hydrometer analysis) (IS: 2720 Part 4 [15]), liquid and plastic limit (IS: 2720 Part-5 [17]) (only for bentonite), differential free swell index (IS: 2720 Part 40 [16]).

Different trial mixtures of fly ash and bentonite are taken in which bentonite content is varied from 5 to 15% at an increment of 5%. The water content-dry density relation of these mixtures is obtained by light compaction (IS: 2720 Part 7 [14]). For a trial mixture to qualify as a liner material, it needs to have hydraulic conductivity in terms of 1×10^{-7} cm/s. So, permeability tests are carried out on these trial mixes with normal water as permeant. Later, the permeant is replaced with different chemical solutions of varying concentrations to check its effect on the trial mixtures permeability characteristics. The different chemicals used as permeants include HCl, NaCl, NaOH, H_2SO_4, $CaCl_2$, and $FeCl_3$.

Dataset Preprocessing

The database for the ANN analysis was generated from the experimental investigation of chemical compatibility of a fly ash–bentonite-based hydraulic barrier by Kumar [18]. Total ninety numbers of experimental data were available. Out of which 72 numbers of data are used for training and the remaining data are used testing. NaCl, NaOH, HCl, H_2SO_4, $CaCl_2$, and $FeCl_3$ at different concentrations, i.e. 0, 0.01, 0.1, 0.5, and 1 M are the different permeant solutions used for the analysis. Both the nature and valency of the permeant solution influence the hydraulic conductivity. It was observed that with an increase in valency, hydraulic conductivity also increases. As well as the acidic permeant will have higher hydraulic conductivity. So, a weightage is assigned to the permeant solutions according to its valency and nature as shown in Table 6.1.

From the data available, weightage (W), concentration of permeant solution (C_p), and concentration of bentonite (C_b) are the input parameters chosen and hydraulic conductivity (k) is taken as output. The statistical values of the parameters of the dataset are presented in Table 6.2. In this case, both the input parameters and output parameters are normalized between $[-1, 1]$. A feed-forward back propagation network with hyperbolic tangent sigmoid function and linear function as transfer functions is used for this analysis.

Table 6.1 Weightage assigned to the permeant solutions

		Weightage		
Valency	Mono (1)	HCl (3)	NaCl (2)	NaOH (1)
	Di (2)	H_2SO_4 (6)	$CaCl_2$ (4)	
	Tri (3)	$FeCl_3$ (9)		
	Nature	Acidic (3)	Neutral (2)	Basic (1)

Table 6.2 Statistical values of the parameters

Parameter	Maximum value	Minimum value	Average value	Standard deviation
W	9	1	4.17	2.67
C_p (M)	1	0	0.32	0.39
C_b (M)	15	5	10.00	4.08
k (cm/s)	6.97×10^{-05}	3.6×10^{-08}	5.12×10^{-06}	1.04×10^{-05}

Development of Neural Network Model

To train the network, 75% of the whole data was used and the remaining 25% data was used to test the neural network developed. The normalized inputs and outputs are trained using the Levenberg–Marquardt (LM) algorithm. Total of 13 trial models were analysed using various training and testing data. The best of the models obtained is discussed here.

By varying the number of neurons in the hidden layer, the maximum R-value and minimum MSE was obtained. For 6 numbers of neurons in the hidden layer, the minimum MSE was found to be 0.0075 and maximum R is 0.9594. The variation of MSE with the number of hidden layer neurons is shown in Fig. 6.1. The neural network architecture is fixed as 3-6-1.

Comparison between predicted hydraulic conductivity (k) with experimentally observed k for training and testing data is shown in Fig. 6.2. The R^2 value between the observed and predicted data for training data is 0.9204 and that for testing data is 0.7958 which indicates that the model is fit for analysis. Figure 6.3 shows the histogram of residuals for the entire dataset. From the histogram, it can be seen that the residual data follows a normal distribution with a peak of curve adjacent to zero. It indicates the predicted values are well accorded with actual values. After training the neural network, the hidden layer-output weights, hidden layer-output weights, input bias, and output bias are obtained. The corresponding connection weights and biases are shown in Table 6.3.

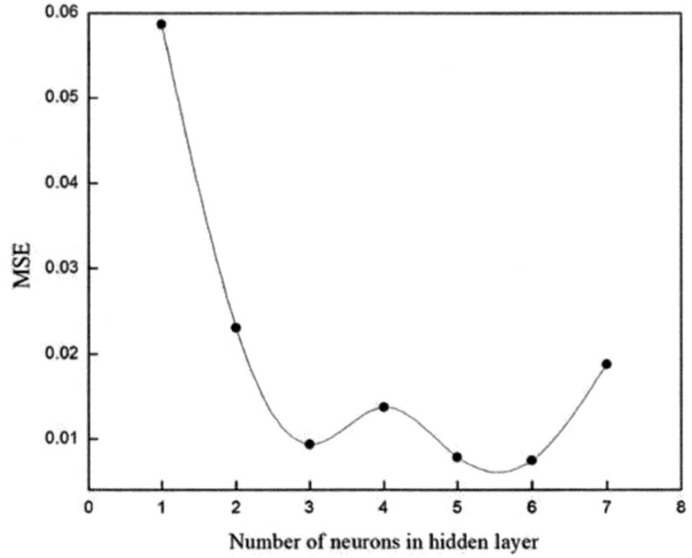

Fig. 6.1 Variation of MSE with number of hidden layer neurons

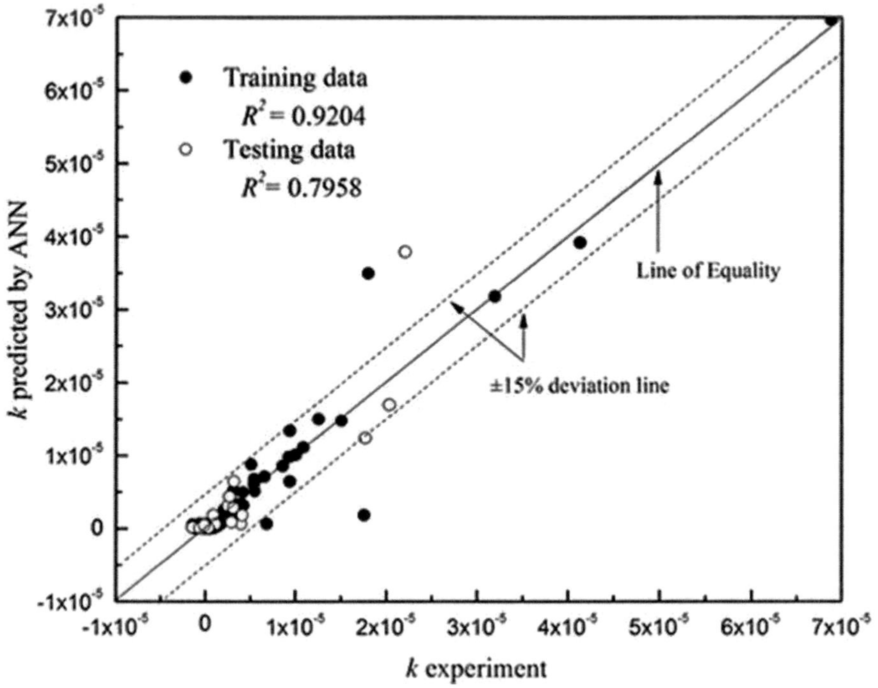

Fig. 6.2 Comparison between predicted hydraulic conductivity (k) with experimental k for training and testing data

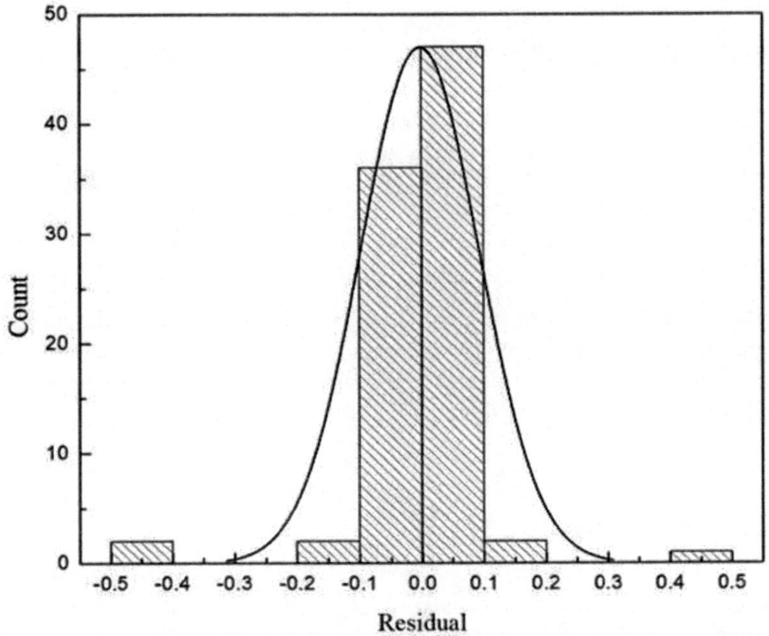

Fig. 6.3 Histogram showing the residual count along with the normal distribution curve for the dataset

Table 6.3 Values of connection weights and biases

Neuron	Weight				Bias	
	w_{ik}			w_k		
	W	C_p	C_b	k	b_{hk}	b_0
Hidden Neuron 1	−0.57	48.94	−48.84	0.08	48.55	3.12
Hidden Neuron 2	−1.75	−0.09	−2.20	1.05	1.93	
Hidden Neuron 3	1.33	0.19	−0.82	3.33	−1.79	
Hidden Neuron 4	2.56	0.03	0.32	10.34	−0.56	
Hidden Neuron 5	−2.08	0.01	−0.30	19.95	0.74	
Hidden Neuron 6	−1.45	0.05	−0.44	−11.56	1.02	

Sensitivity Analysis

Pearson's correlation, Spearman's rank correlation, variable perturbation method, Garson's algorithm, connection weight approach, and weight magnitude analysis are the methods used for sensitivity analysis in this study. They are based on the trained weights and biases from the neural network model. The cross-correlation of inputs with the output 'k' using Pearson's correlation and Spearman's correlation is

Table 6.4 Cross-correlation of inputs and output

	Pearson's correlation				Spearman's correlation			
Variables	W	C_p	C_b	k	W	C_p	C_b	k
W	1	0	0	0.249	1	0.03	0.03	0.30
C_p		1	0	0.250		1	0.05	0.32
C_b			1	−0.387			1	−0.76
K				1				1

shown in Table 6.4. The correlation indices of both methods give similar results. The hydraulic conductivity of the fly ash–bentonite mixed clay liner is highly correlated with the concentration of bentonite (C_b), followed by concentration of permeant (C_p) and weightage given according to valency and nature of permeant (W).

Variable Perturbation

The most influencing input parameter is found out by perturbing the input variables in the range of $\pm 20\%$ @10%. The sensitivity index (S_i) is calculated using Eq. (6.1). From Fig. 6.4, it can be seen that W is deviating the most from the base value followed by C_b and Cp.

$$S_i = \frac{1}{N}\sum_{i=1}^{N}\left(\frac{\%\text{ change in output}}{\%\text{ change in input}}\right) \times 100 \tag{6.1}$$

where N = number of data = 90.

Weight Methods

The sensitivity index and relative importance using the weight methods, Garson algorithm, connection weight approach and weight magnitude analysis are presented in Table 6.5. And the ranking of the variables analysed using all methods used for sensitivity analysis is showed in Table 6.6. According to variable perturbation and Garson's algorithm, W has the highest rank followed by C_b and C_p. By connection weight and weight magnitude analysis, C_b has the highest rank followed by C_p and W.

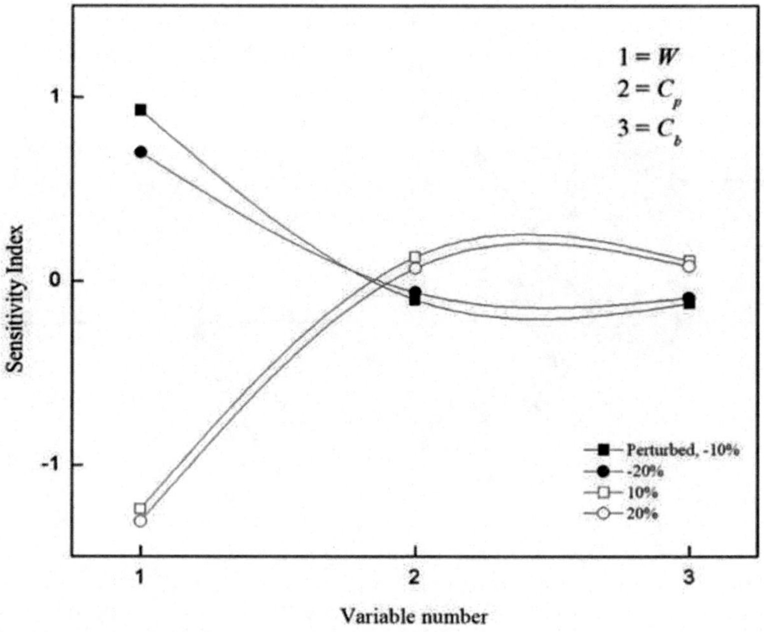

Fig. 6.4 Variable perturbation

Table 6.5 Sensitivity index and relative importance of input parameters

	Sensitivity index				Relative importance		
Parameters	−20%	−10%	10%	20%	Garson algorithm	Connection weight	Weight magnitude analysis
W	0.70	0.93	−1.24	−1.31	58.40	4.21	0.2
C_p	−0.06	−0.10	0.13	0.07	10.68	4.26	1.01
C_b	−0.09	−0.12	0.11	0.08	30.92	−6.27	1.08

Table 6.6 Ranking of input variables

Parameters	Method adopted			
	Variable perturbation	Garson's algorithm	Connection weight	Weight magnitude analysis
W	1	1	3	3
C_p	3	3	2	2
C_b	2	2	1	1

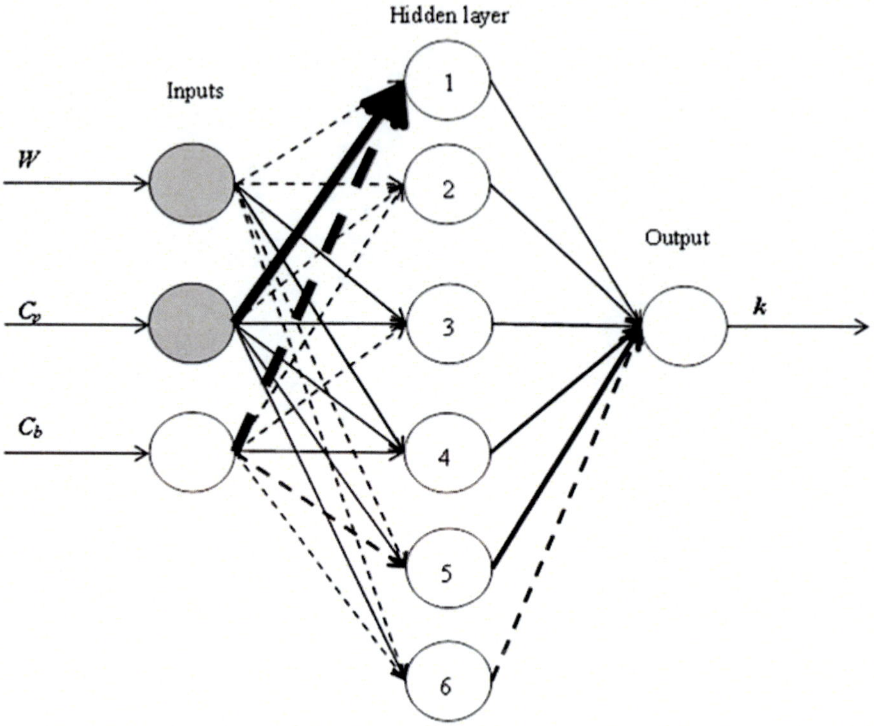

Fig. 6.5 Neural interpretation diagram

Neural Interpretation Diagram

With the connection weights shown in Table 6.3, the NID for the neural network for predicting the hydraulic conductivity of fly ash–bentonite mixture is shown in Fig. 6.5. From the NID, the conclusion can be drawn that W and C_p are directly related and C_b is inversely related to k.

Development of ANN Model Equation

A mathematical equation relating the input parameters W, C_b, and C_p with the output parameter 'k' is developed. The model equation for predicting the hydraulic conductivity of fly ash–bentonite mixture is formulated using the values of the weights and biases as shown in Table 6.3. The following expressions are obtained:

$$A_1 = -0.57\,W + 48.94\,C_p - 48.84\,C_b + 48.55 \tag{6.2}$$

$$A_2 = -1.75\,W - 0.09\,C_p - 2.20\,C_b + 1.93 \tag{6.3}$$

$$A_3 = 1.33\,W + 0.19\,C_p - 0.82\,C_b - 1.79 \tag{6.4}$$

$$A_4 = 2.56\,W + 0.03\,C_p + 0.32\,C_b - 0.56 \tag{6.5}$$

$$A_5 = -2.08\,W + 0.01\,C_p - 0.30\,C_b + 0.74 \tag{6.6}$$

$$A_6 = -1.45\,W + 0.05\,C_p - 0.44\,C_b + 1.02 \tag{6.7}$$

$$B_1 = 0.08\left(\frac{e^{A_1} - e^{A_1}}{e^{A_1} + e^{A_1}}\right) \tag{6.8}$$

$$B_2 = 1.05\left(\frac{e^{A_2} - e^{A_2}}{e^{A_2} + e^{A_2}}\right) \tag{6.9}$$

$$B_3 = 3.33\left(\frac{e^{A_3} - e^{A_3}}{e^{A_3} + e^{A_3}}\right) \tag{6.10}$$

$$B_4 = 10.34\left(\frac{e^{A_4} - e^{A_4}}{e^{A_4} + e^{A_4}}\right) \tag{6.11}$$

$$B_5 = 19.95\left(\frac{e^{A_5} - e^{A_5}}{e^{A_5} + e^{A_5}}\right) \tag{6.12}$$

$$B_6 = -11.56\left(\frac{e^{A_6} - e^{A_6}}{e^{A} + e^{A_6}}\right) \tag{6.13}$$

$$k_n = B_1 + B_{2_1} + B_3 + B_4 + B_5 + B_6 + 3.12 \tag{6.14}$$

Equation (6.14) produces the normalized value of k. It can be denormalized using the maximum and minimum values of k with the help of the following expression:

$$k = 0.5(k_n + 1)(k_{\max} - k_{\min}) + k_{\min} \tag{6.15}$$

where $k_{\max} = 6.97 \times 10^{-05}$ cm/s and $k_{\min} = 3.64 \times 10^{-08}$ cm/s.

These empirical equations can be used to find out the hydraulic conductivity of fly ash–bentonite mixture used as liner material with reasonable accuracy. It has to be noted that this equation is valid for only if the value of the parameters lies in this range of the dataset.

Conclusions and Future Recommendations

The database for the ANN analysis was generated from the experimental investigation of chemical compatibility of a fly ash–bentonite-based hydraulic barrier. Using ninety number of experimental dataset, empirical equation is developed to predict the hydraulic conductivity of fly ash–bentonite mixture. It is expressed as a function of its associated factors like the weightage (W) given according to the valency and nature of the permeant solution, the concentration of permeant solution (C_p), and concentration of bentonite (C_b). The statistical computing based on Artificial Neural Network suggests the following major inferences.

- In the residual analysis using the histogram, the normal distribution curve drawn over the histogram shows that a maximum number of residuals are nearer to zero. So, it can be said that the training of network is done well and can be used for prediction with considerable accuracy.
- From Pearson's correlation coefficient and Spearman's correlation, it is found that, the hydraulic conductivity of the fly ash–bentonite mixed clay liner is highly correlated with the concentration of bentonite (C_b), followed by concentration of permeant (C_p) and weightage given according to valency and nature of permeant (W).
- An empirical equation in the form of Eq. (6.15) is derived to predict the hydraulic conductivity for this dataset, with reasonable accuracy.
- From the NID, W and C_p are directly related and C_b is inversely related to k.

Using the LM algorithm, some of the hydraulic conductivity predicted are negative which is not realistic. So, a better algorithm has to be implemented for this dataset in the future.

References

1. Akgün H (2010) Geotechnical characterization and performance assessment of bentonite/sand mixtures for underground waste repository sealing. Appl Clay Sci 49(4):394–399
2. Alawaji HA (1999) Swell and compressibility characteristics of sand–bentonite mixtures inundated with liquids. Appl Clay Sci 15(3–4):411–430
3. Alonso EE, Romero E, Hoffmann C, García-Escudero E (2005) Expansive bentonite–sand mixtures in cyclic controlled-suction drying and wetting. Eng Geol 81(3):213–226
4. Cuevas J, Villar MV, Martın M, Cobena JC, Leguey S (2002) Thermo-hydraulic gradients on bentonite: distribution of soluble salts, microstructure and modification of the hydraulic and mechanical behaviour. Appl Clay Sci 22(1–2):25–38
5. Ebina T, Minja RJ, Nagase T, Onodera Y, Chatterjee A (2004) Correlation of hydraulic conductivity of clay–sand compacted specimens with clay properties. Appl Clay Sci 26(1–4):3–12
6. Komine H (2004) Simplified evaluation on hydraulic conductivities of sand–bentonite mixture backfill. Appl Clay Sci 26(1–4):13–19
7. Mishra AK, Ohtsubo M, Li L, Higashi T (2011) Controlling factors of the swelling of various bentonites and their correlations with the hydraulic conductivity of soil-bentonite mixtures. Appl Clay Sci 52(1–2):78–84

8. Das SK, Samui P, Sabat AK, Sitharam TG (2010) Prediction of swelling pressure of soil using artificial intelligence techniques. Environ Earth Sci 61(2):393–403
9. Das SK, Samui P, Sabat AK (2011) Application of artificial intelligence to maximum dry density and unconfined compressive strength of cement stabilized soil. Geotech Geol Eng 29(3):329–42
10. Behera RN, Patra CR, Sivakugan N, Das BM (2013) Prediction of ultimate bearing capacity of eccentrically inclined loaded strip footing by ANN, part I. Int J Geotech Eng 7(1):36–44
11. Behera RN, Patra CR, Sivakugan N, Das BM (2013) Prediction of ultimate bearing capacity of eccentrically inclined loaded strip footing by ANN: part II. Int J Geotech Eng 7(2):165–72
12. Sasmal SK, Behera RN (2018) Prediction of combined static and cyclic load-induced settlement of shallow strip footing on granular soil using artificial neural network. Int J Geotech Eng. https://doi.org/10.1080/19386362.2018.1557384
13. IS 2720: (1980 Part 3/sec 1) Method of test for soil, Determination of specific gravity
14. IS 2720: (1980 Part 7) Determination of water content-dry density relation using light compaction
15. IS 2720: (1985 Part 4) Method of test for soil, Grain size analysis
16. IS 2720: (1985 Part 40) Method of test for soil, Determination of free swell index of soils
17. IS 2720: (1985 Part 5) Methods of test for soils: determination of liquid and plastic limit
18. Kumar GS (2018) Chemical compatibility of a fly ash-bentonite based hydraulic barrier chemical compatibility of a fly ash-bentonite based hydraulic barrier. M. Tech thesis, National Institute of Technology Rourkela, India
19. Kumar GS, Balaiah KV, Behera RN (2020, March) Chemical compatibility of fly ash–bentonite based hydraulic barrier. In: Advances in computer methods and geomechanics: IACMAG symposium 2019 Volume 2, vol 56. Springer Nature, p 429

Chapter 7
Influence of Inclined Loads on the Behaviour of Piles—A Numerical Study

Sai Manasa Chirlam Charla, Teja Munaga, and Kalyan Kumar Gonavaram

Introduction

General

Pile foundations in off-shore structures and bridges are subjected to inclined loads and moments due to simultaneous application of vertical loads and lateral loads. Due to the complexities involved in predicting the pile group response, the piles subjected to vertical loads are often considered separately for analysis from piles subjected to lateral loads. The load capacity of piles obtained by considering the effect of combined loading is generally on the conservative side [1, 2]. Hence, for more rational design of pile foundations, the presence of both vertical and lateral loads, i.e. inclined loading need to be considered.

Extensive theoretical and experimental investigations were conducted on piles subjected to inclined loading by researchers. Meyerhoff et al. [3] studied the behaviour of rigid piles embedded in uniform and stratified soils subjected to inclined loads. The reduction in load capacity of rigid piles is observed as 30 and 50% for inclination of 30° and 60°, respectively [4, 5].

Most of the studies in the literature [6, 7] discussed about the single pile and group of piles subjected to the vertical loads or lateral loads separately without paying much attention to combined vertical and horizontal loads resulting in inclined loading on

S. M. Chirlam Charla · T. Munaga · K. K. Gonavaram (✉)
Department of Civil Engineering, National Institute of Technology, Warangal 506004, India
e-mail: kalyan@nitw.ac.in

T. Munaga
e-mail: mteja@student.nitw.ac.in

T. Munaga
Research Division, Vatsalya Consulting Services, Aurangabad, India

K. Muthukkumaran et al. (eds.), *Soil Dynamics, Earthquake and Computational Geotechnical Engineering*, Lecture Notes in Civil Engineering 300,
https://doi.org/10.1007/978-981-19-6998-0_7

the piles. The current study is carried out to analyse the piles under inclined loading using numerical method. A comprehensive approach is presented in this study, in which numerical solution is obtained by three-dimensional analysis using PLAXIS 3D software. In this approach, behaviour of model piles embedded in stratified soil under different inclinations of load is studied. The effect of inclination angle on the pile group response due to the possibility of changes in the lateral load direction during the lifetime on the pile structure is studied.

Methodology and Material Properties

PLAXIS 3D is a finite element (FE) software used to simulate the soil behaviour for geotechnical applications. PLAXIS 3D is enabled with a convenient graphical user interface (GUI) to generate a geometry model and create FE mesh for analysis. Staged construction mode is used in the analysis for realistic simulation of excavation, load application, and change in water levels. The various stages involved in FE analysis using PLAXIS 3D are shown in Fig. 7.1.

Overview of the Project

The dimensions of the soil volume are chosen as 1.52 × 0.76 × 1.2 m. These dimensions were selected such that there are no boundary effects on piles; since the distance between the edge of the pile group and boundary is more than 10*D* (*D* is pile diameter) in the direction of application of load [8, 9]. The thickness of clay

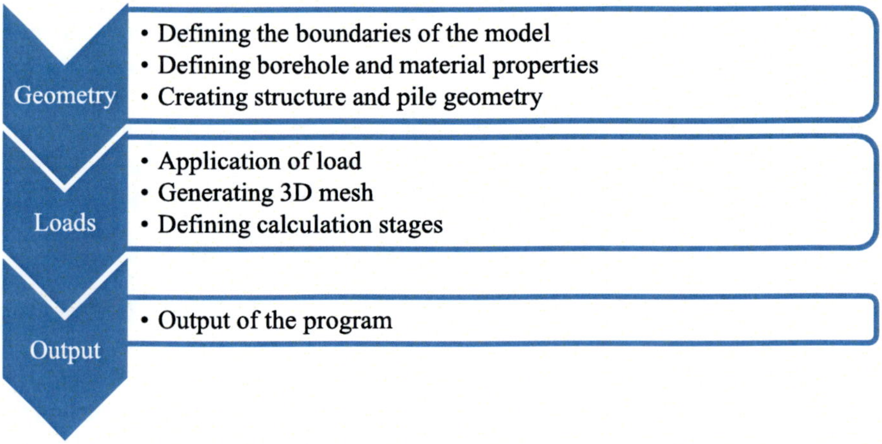

Fig. 7.1 Stages involved in FE analysis using PLAXIS 3D

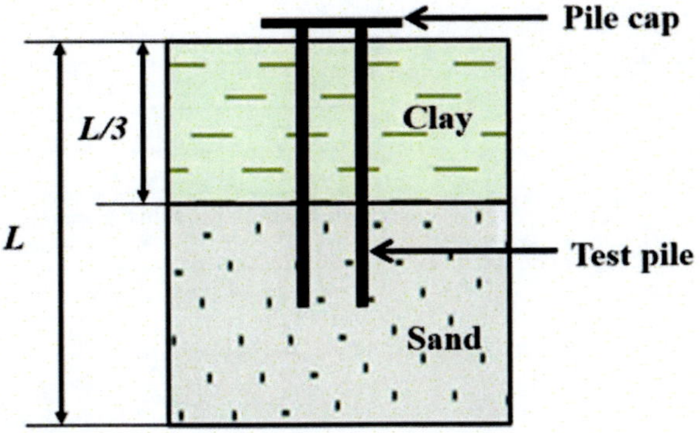

Fig. 7.2 Schematic of pile group in stratified soil

Table 7.1 Properties of sand and clay

Property	Sand	Clay
Saturated unit weight (kN/m^3)	15	17
Undrained cohesion, C_u (kPa)	–	70
Poisson's ratio, μ_s	0.35	0.45
Modulus of elasticity, E (MPa)	36	11
Angle of internal friction, ϕ	35°	–
Interface factor, R_{inter}	0.75	0.35
Model	Mohr–Coulomb	Mohr–Coulomb

and sand layers in stratified soil was chosen according to the length of pile. The clay layer is of thickness $L/3$ (L is length of model pile) followed by sand in the bottom portion. An interface factor (R_f) of 0.8 is considered in the analysis to account for soil-pile interaction. Typical representation of clay and sand is shown in Fig. 7.2. The properties of sand and clay used for FE analysis in PLAXIS 3D are listed in Table 7.1.

Properties of Pile and Pile Cap

The dimensions of model piles used in the current study were adopted from experimental study [10] on piles embedded in stratified soil. Accordingly, model aluminium piles of 16 mm diameter were chosen. For piles embedded in sand and normally loaded clays,

Table 7.2 Properties of pile and pile cap

Pile	Pile cap
Material—Aluminium	Material—Aluminium
Diameter = 16 mm	Thickness = 16 mm
Unit weight, γ = 25.8 kN/m^3	Unit weight, γ = 25.8 kN/m^3
Elasticity modulus, E = 70 Gpa	Elasticity modulus, E = 70 Gpa
Model—Linear elastic	$\mu_s = 0.323$
	Model—Linear elastic

$$\text{Stiffness factor, } T = \sqrt[5]{\frac{E_p I_p}{\eta_h}} \tag{7.1}$$

where

E_p Young's modulus of pile material, in MN/m^2
I_p moment of inertia of the pile cross section, in m^4
η_h modulus of subgrade reaction, in MN/m^3.

Upon substitution and calculation, T is obtained as 0.135 m; $L \geq 4\ T$ represents long flexible pile. Hence, a pile length of 640 mm corresponding to aspect ratio of 40 is chosen to represent a long flexible pile according to IS 2911 (Part 1/s 1): 2010 [11]. The spacing between the piles in the pile group used for the analysis was 3*D*, 5*D*, and 7*D*. The thickness of pile cap was 16 mm and the dimensions of the pile cap were considered according to the spacing between piles. The edge distance equivalent to pile diameter, 16 mm is considered between the edges of corner pile and pile cap. Properties of pile and pile cap used for the analysis are listed in Table 7.2.

Results and Discussion

Validation of PLAXIS 3D

The FE software PLAXIS 3D used in the current study is validated with experimental study [8] on piles embedded in stratified soil. It can be inferred from Fig. 7.3 that the deviation of experimental results with FE results is less than 5%. The pile head deflection obtained from FE analysis is relatively more due to the consideration of modified stress strain curve for safety factors.

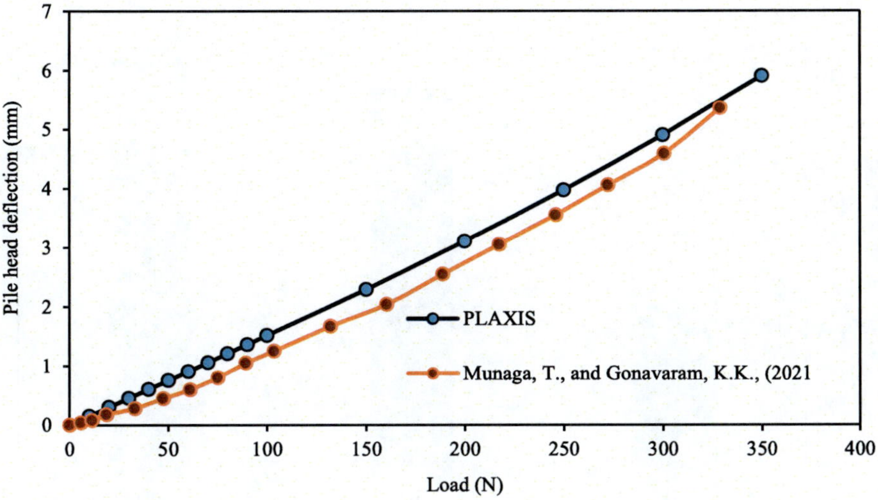

Fig. 7.3 Validation of PLAXIS 3D

Pile Group Response Under Pure Lateral Load

The response of pile groups under the application of pure lateral load is discussed in this section. Lateral capacity of pile group is considered as load corresponding to pile cap deflection of 0.2*D* [12]. The variation of lateral capacities of pile groups with change in pile spacing is shown in Fig. 7.4. It can be inferred from Fig. 7.4 that there is a proportionate increase in the lateral capacity as the spacing increases. The pile spacing of 5*D* is considered as optimum spacing for the current study, and hence, it is adopted for the further study in inclined loading condition.

Response of Piles to Inclined Load

The response of pile groups under the application of inclined lateral load is evaluated in terms of lateral deflection and bending moment (BM) in the pile. The magnitude of inclined load applied in each case is equivalent to the lateral capacity of the corresponding pile group. The inclined load is modelled in PLAXIS 3D by dividing the load into x-component and z-component (i.e. in lateral direction and vertical direction). The lateral deflection of 2×2 pile group with 5*D* spacing is monitored for load with varying inclinations (0°, 15°, 30°, 60°, 90°). The variation of lateral deflection of pile with increase in the load inclination is shown in Fig. 7.5. It is interested to note that the lateral deflection of the pile decreased with increase in inclination, even when the magnitude of load is constant. This decrease in lateral deflection is accompanied by an increase in vertical deflection of pile groups. The

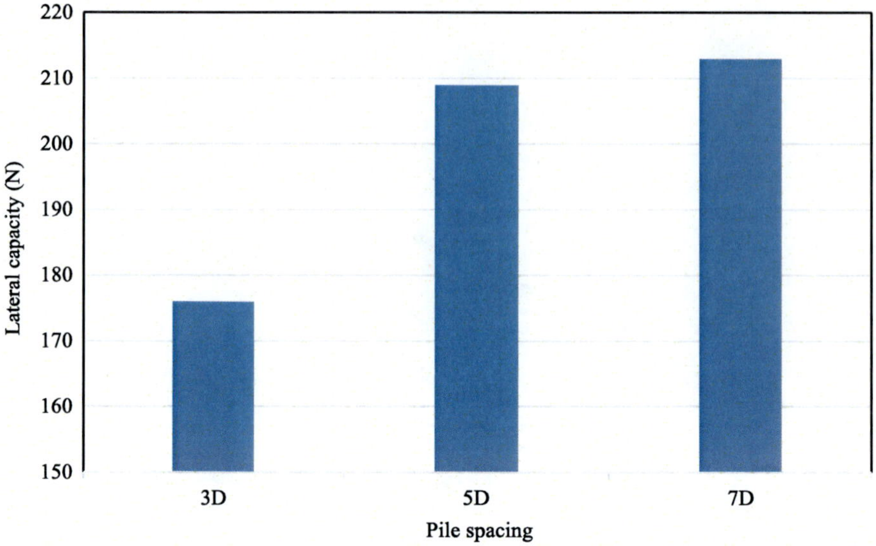

Fig. 7.4 Variation of lateral capacity of pile groups for spacing of piles long piles

decrease in lateral deflection can be attributed to partial transfer of load from lateral direction to axial direction. Similar response is reported by researchers [13] in case of pile groups subjected to inclined loads.

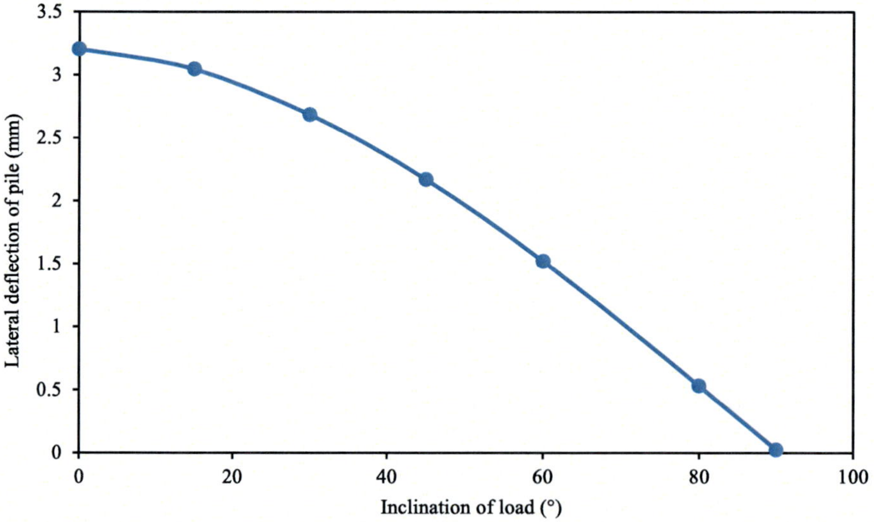

Fig. 7.5 Variation of lateral deflection of 2 × 2 pile groups with 5*D* spacing for varying inclined load

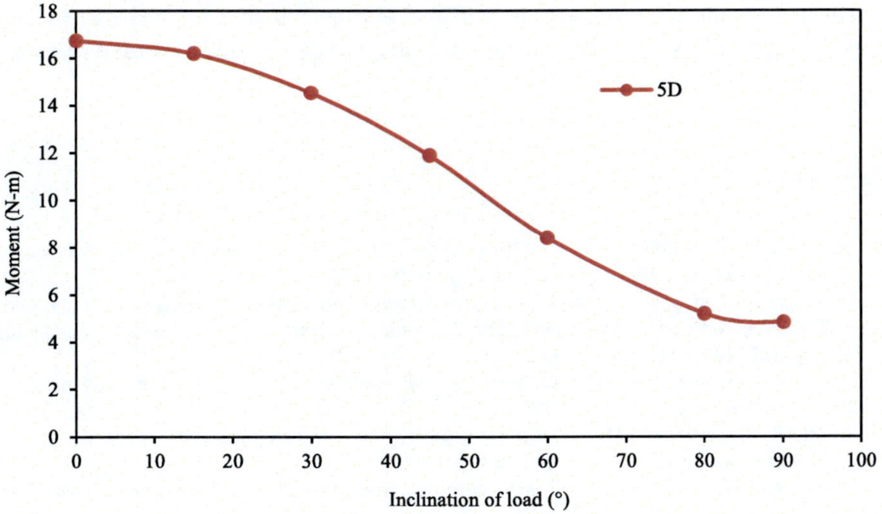

Fig. 7.6 Variation of moment in the 2 × 2 pile group with 5D spacing for varying load inclination

Variation of BM with Inclination

The BM in the pile is significantly influenced by the type of the load acting on the pile. The variation of maximum BM in the pile with load inclination is shown in Fig. 7.6. It can be inferred that the maximum BM experienced by the pile is considerably reduced as the load inclination increases. This is due to the reduction in the lateral component of the applied inclined load leading to reduced BM in the pile.

Conclusions

The behaviour of laterally loaded piles embedded in stratified soil consisting of sand overlain by clay under static lateral load and inclined load is analysed in the present study using FE software, PLAXIS 3D. The effect of spacing and load inclination on the response of pile groups is presented. The following conclusions can be drawn from the current study:

- The lateral capacity of pile group increases with an increase in pile spacing. However, there is no significant increase in lateral capacity beyond spacing of *5D*; hence, *5D* is considered as optimum spacing for inclined loading.
- The lateral deflection in the piles decreased with an increase in the inclination of the load; however, this decrease in lateral deflection is accompanied by increase in vertical deflection.

- The magnitude of maximum BM in the pile decreased with increase in the inclination of load due to the reduction in magnitude of load in lateral direction.

References

1. Subramanian RM, Boominathan A (2016) Dynamic experimental studies on lateral behaviour of batter piles in soft clay. Int J Geotech Eng 10(4):317–327
2. Madhusudan Reddy, K., and R. Ayothiraman. Experimental studies on behavior of single pile under combined uplift and lateral loading. Journal of Geotechnical and Geoenvironmental Engineering 141.7: 04015030 (2015)
3. Meyerhof GG, Sastry VVRN (1985) Bearing capacity of rigid piles under eccentric and inclined loads. Can Geotech J 22(3):267–276
4. Chari TR, Meyerhof GG (1983) Ultimate capacity of rigid single piles under inclined loads in sand. Canadian Geotech J 20(4):849–854
5. Sastry VVRN, Meyerhof GG (1990) Behaviour of flexible piles under inclined loads. Can Geotech J 27(1):19–28
6. Munaga T, Khan MM, Gonavaram KK (2020) Axial and lateral loading behaviour of pervious concrete pile. Indian Geotech J 50(3):505–513
7. Huchegowda BK, Teja M, Kalyan Kumar G (2020) Evaluation of lateral capacity of pile foundation using finite Element method in layered soil. In: Advances in geotechnical and transportation engineering, Springer, pp 79–84
8. Kim Y, Jeong S (2011) Analysis of soil resistance on laterally loaded piles based on 3D soil–pile interaction. Comput Geotech 38(2):248–257
9. Kim BT, Yoon GL (2011) Laboratory modeling of laterally loaded pile groups in sand. KSCE J Civ Eng 15(1):65–75
10. Munaga T, Gonavaram KK (2021) Influence of stratified soil system on behavior of laterally loaded pile groups: an experimental study. Int J Geosynthet Ground Eng 7(2):1–14
11. IS: 2911 (Part 1/Sec 1) (2010) IS code of practice for design and construction of pile foundations. Bureau of Indian Standards, New Delhi
12. Chandrasekaran SS, Boominathan A, Dodagoudar GR (2010) Group interaction effects on laterally loaded piles in clay. J Geotech Geoenviron Eng 136(4):573–582
13. Conte E, Troncone A, Vena M (2015) Behaviour of flexible piles subjected to inclined loads. Comput Geotech 69:199–209

Chapter 8
3D Numerical Analysis of Screw Pile Subjected to Axial Compressive and Lateral Load

P. V. Pavan Kumar, Shantanu Patra, Sumanta Haldar, Michael John Brown, Jonathan Adam Knappett, and Yaseen Umar Sharif

Introduction

Screw piles usually consists of a central shaft attached to one or more helical plates and are installed by applying torque and axial thrust or crowd force. Until now, screw piles have typically used to support the structures like transmission towers, pipelines, buildings, etc. [1]. Recent studies show that screw piles can be a potential alternative to the existing foundations supporting the wind turbines in offshore environment [2, 3]. As they provide several advantages compared to the concrete foundations including less noise during installation, fast installation, spoil production and can be used immediately after installation [4, 5]. Although the screw piles application for the offshore wind turbine foundations has been appraised [3, 6], there is a limited quantity of studies on its applicability as a foundation for onshore wind turbines. The foundations typically used for onshore wind turbines include octagonal gravity base foundations, precast concrete foundations, and pile group foundations [7]. These foundation options involve high material cost produce large amounts of spoil during excavation and involve vibration in case of driven piles which may cause disturbance to the adjacent structures [4]. Considering these advantages, the screw pile foundation can be a potential alternative to concrete foundations.

The foundations of the wind turbines are typically subjected to axial compressive load due to the self-weight of the superstructure and high lateral loads due to wind forces [2]. Past studies involve the lateral [6, 8, 9], axial compressive [10, 11] and tensile [11, 12] behaviour of screw piles. The above studies reported that screw pile capacity is significantly higher under the compressive and tensile load in comparison

P. V. Pavan Kumar (✉) · S. Patra · S. Haldar
Indian Institute of Technology, Bhubaneswar, India
e-mail: pvk11@iitbbs.ac.in

M. J. Brown · J. A. Knappett · Y. U. Sharif
University of Dundee, Dundee, Scotland, UK

K. Muthukkumaran et al. (eds.), *Soil Dynamics, Earthquake and Computational Geotechnical Engineering*, Lecture Notes in Civil Engineering 300,
https://doi.org/10.1007/978-981-19-6998-0_8

to straight shaft piles. To obtain the maximum benefit of the upper helix in terms of the lateral behaviour, it has been suggested that the helix should be placed near to the ground surface [6, 9]. However, the combined lateral and vertical loading behaviour of the screw piles is not yet clearly defined and is limited to [13]. They reported that the lateral capacity of screw pile is enhanced with the increase in vertical compressive load. However, the location of the helix is not optimized for the enhancement in the lateral load of the screw pile. The screw piles behaviour under combined loading is still not clearly understood when they are used as a foundation for the wind turbines.

The present study aims at defining the behaviour of screw piles in sand under the axial compressive and lateral loading. A 3D numerical analysis is carried out using the commercially available software ABAQUS 2018 on the single-helix screw piles (SHSP) and double-helix screw piles (DHSP), and solid straight shaft concrete piles (CP). Yield envelopes were developed for the concrete straight shafted piles and the screw piles which aid in determining the failure load for the piles considered in the study. The optimum position of top helix for screw pile is determined by placing the helix at different locations below the ground surface and choosing the location that provides the maximum lateral load. The combined load on the pile is applied in terms of the axial and the lateral displacement to get the compressive load (V) and lateral load (H) along with the gravity load generating initial stresses.

Soil Properties and Screw Pile Geometry

The present study considers a loose sand (relative density = 25%) with the properties shown in Table 8.1. The diameter of the shaft (D_s), thickness of the shaft (t_s), diameter of helix (D_h), thickness of the helix (t_h) are kept constant and the spacing of the helix (S) is varied. The screw piles considered are single and a double-helix design and have the geometric (Fig. 8.1) and material properties are shown in Table 8.2. The soil is modelled assuming an elastoplastic behaviour with Mohr–Coulomb failure criteria and linear elastic behaviour is assumed for the screw pile. In order to minimize numerical instability, a small amount of apparent cohesion (1 kPa) is considered for the soil. The screw piles are assumed to be made of mild steel. The solid straight shafted concrete pile is also considered in this study, and its properties are shown in Table 8.2.

Table 8.1 Properties of soil [14]

Property	Value
Unit weight, γ (kN/m^3)	15
Modulus of elasticity, E (kN/m^2)	20,000
Internal frictional angle, φ (°)	30
Angle of dilation, ψ (°)	0
Cohesion, c (kN/m^2)	1
Poisson's ratio, μ	0.3

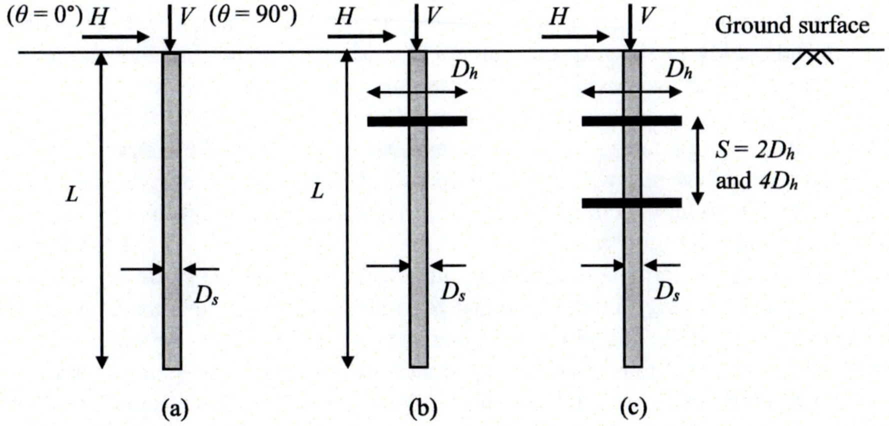

Fig. 8.1 Geometry of pile: **a** straight shaft concrete pile **b** single-helix pile and **c** double-helix pile

Table 8.2 Properties of pile

Property	Steel (Screw pile)	Concrete
Modulus of elasticity, E (kN/m^2)	200×10^6	25×10^6
Poisson's ratio, μ	0.27	0.15
Density, ρ (kg/m^3)	7850	2400
Length of the pile, L (m)	6	6
Diameter of shaft, D_s (m)	0.3	–
Thickness of shaft, t_s (m)	0.012	–
Thickness of helix, t_h (m)	0.025	–
Diameter of helix, D_h (m)	$4D_s$	–
Spacing of helix, S (m)	$2D_h$, $4D_h$	–

The screw pile configurations considered for the study are shown in Fig. 8.1. For the single-helix and the double-helix screw piles the top helix position is optimized in the present study to get the maximum benefit of the helix in terms of its lateral behaviour. The location of the helix is chosen based on the maximum lateral load obtained and is discussed in the further section (i.e. 1.3 m from ground surface). For the double-helix screw piles, two helix spacing are considered in the study ($S = 2D_h$ and $4D_h$) by keeping the top helix at 1.3 m below the ground level.

Finite Element Modelling

In this study, the behaviour of screw pile and the straight shafted concrete pile under combined load is determined using a 3D finite element analysis software ABAQUS CAE (2018). The soil, concrete pile and screw pile are modelled with eight-noded, linear brick with reduced integration elements (C3D8R). Certain assumptions are made in this study to simplify the numerical modelling. These include, the piles are considered to be wished-in-place (installation process is not considered) and the helical plates are modelled as flat plates. Considering the helix as a flat plate however has little influence on the pile response [13]. The finite element model of single-helix screw pile with meshing and soil-screw pile assembly is shown in Fig. 8.2. Soil boundary is chosen such that it has no influence on results according to [15]. The top soil is considered to be free in all directions, the vertical boundaries are restrained in the horizontal directions and are allowed to move vertically and the bottom boundary is fully fixed. The distance to the bottom and the side boundaries are shown in Fig. 8.2. The pile-soil interaction is provided with the contact pair algorithm defining the master and slave surfaces in Abaqus. The friction factor for the steel and the soil is considered to be 0.7 (ALA 2001) resulting in the interface friction angle of 21°.

In the present numerical analysis, two loading stages are involved. Wherein the first stage involves the application of gravity load to generate the in-situ stresses due to the weight of the soil and pile. In the second stage, the combined compressive and lateral displacement is applied on pile top to determine the compressive and lateral capacities. The load inclination angle is varied from $\theta = 0°$ to $\theta = 90°$, where 0° corresponds to a purely horizontal load and 90° corresponds to a purely vertical load as shown in Fig. 8.2. The inclination angle is varied as 0°, 20°, 40°, 60° and 90° in the present study. The lateral deflection criteria involve the loads corresponding to 50 mm [16] pile head deflection. In case of compression, the failure criteria is corresponding to the 10% of pile diameter, which is 30 mm.

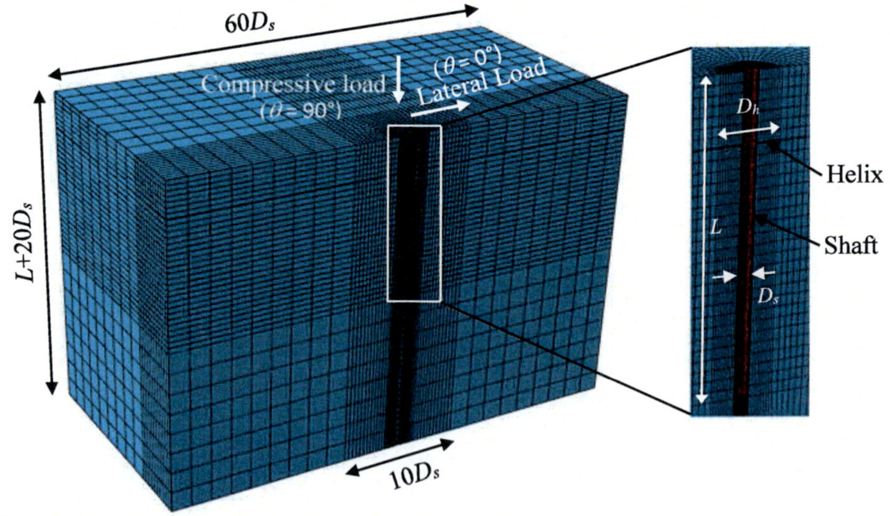

Fig. 8.2 Finite element mesh of soil-screw pile assembly

Numerical Model Validation

The finite element model developed was validated against the lateral load test of [8] and the compressive load tests of Elsherbiny and El Naggar [10]. Sakr [8] carried out field tests upon single-helix and double-helix screw piles subjected to lateral load. The screw pile has a shaft diameter and wall thickness of 324 mm and 9.5 mm, respectively. The screw pile was embedded to a depth of 5.6 m and is installed in dense sand. The helix is located at 400 mm from the pile tip with a diameter of 610 mm. The site soil conditions are adopted according to [8]. Figures 8.3a, b show the load–deflection profile of single-helix and double-helix piles from field tests and present study. From Fig. 8.3 it can be found that the numerical results are in good agreement with the field test data. Similarly [10] carried out the field on the screw piles subjected to compressive load installed in cohesionless soil. From the field test data, the soil is found to be medium to dense sand and the soil properties for the numerical model are adopted according to the field soil [10]. Screw piles with single-helix has a shaft diameter of 273 mm and helix diameter of 610 mm. The installation depth of the screw pile is 5.5 m. Figure 8.3c shows the compressive load–deflection graphs from the present study and the field tests. The deviation in case of the compressive load is observed because, during installation the soil below the pile is densified resulting in higher capacity. In the numerical model, the pile is wished-in-place and cannot capture this installation behaviour. That said, this validation provides the confidence that the 3D numerical model can be used to obtain the response of screw piles.

Determining the Optimum Location of Helix

The optimum location of the upper helix was determined by varying the position of helix. This was done by observing the lateral load developed while varying the position of helix. Figure 8.4 shows the lateral load variation with the change in helix depth (z) corresponding to 50 mm lateral displacement. The maximum lateral capacity was found at a helix position of 1.3 m below the ground. Therefore, in present study the top helix is positioned at a distance of 1.3 m below the ground level for both the single-helix and double-helix piles to determine the behaviour of screw pile under the combined load.

Results and Discussion

Combined loading on the screw piles and straight shafted concrete piles are carried out with different inclinations of 0°, 20°, 40°, 60° and 90°. Horizontal and vertical loads corresponding to all these inclinations were measured whether it is with respect

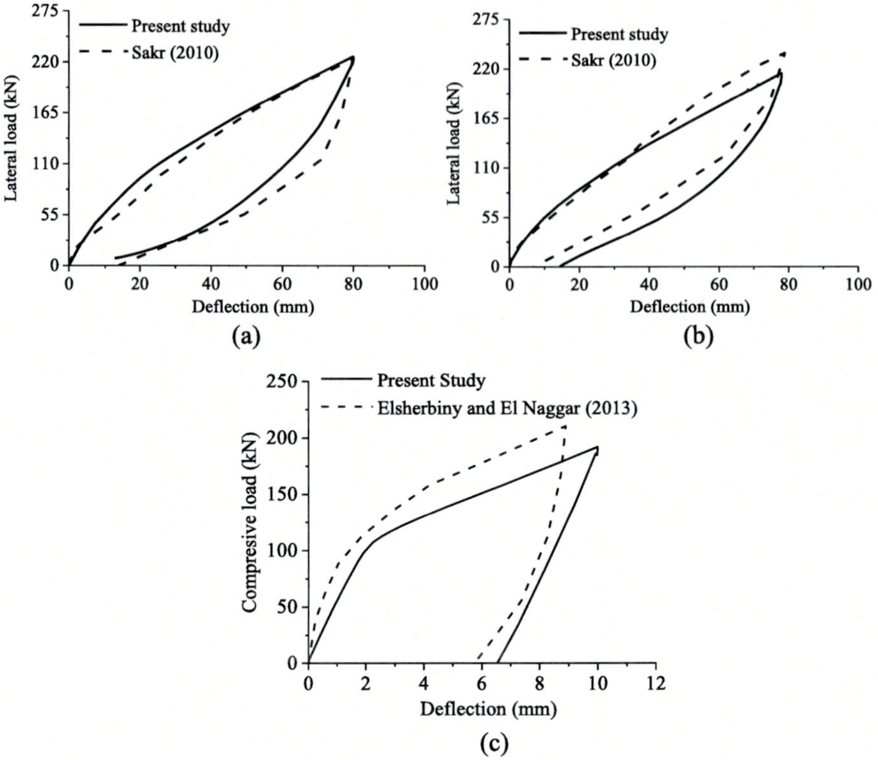

Fig. 8.3 Measured and predicted load–deflection profiles **a** single-helix pile [8] **b** double-helix pile [8] **c** single-helix pile [10]

Fig. 8.4 Variation of lateral load with the change in the position of helix

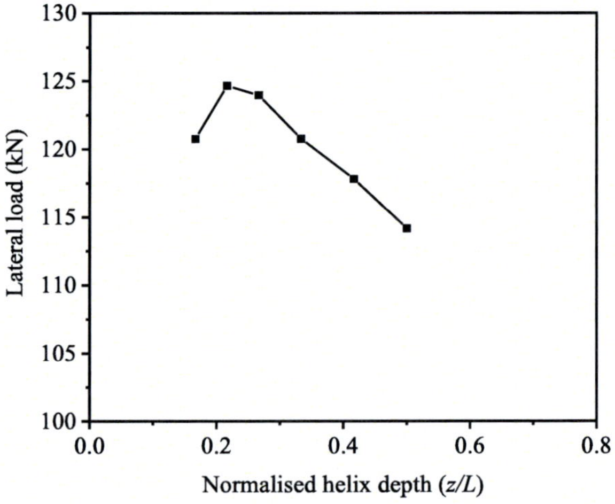

Table 8.3 Comparison of single-helix and double-helix screw piles capacities

θ (°)	SHSP		DHSP ($S = 2D_h$)	
	Horizontal load, H (kN)	Vertical load, V (kN)	Horizontal load, H (kN)	Vertical load, V (kN)
0	120.9	0	125.0	0
20	214.9	78.2	316.0	115.0
40	230.1	193.0	363.3	304.8
60	148.3	256.9	236.0	408.8
90	0	295.4	0	475.8

to the horizontal displacement criteria or the vertical displacement criteria whichever is minimum.

Effect of Helix

Screw piles with single-helix and double-helix are analysed under the combined load. It is observed that with the increase in the inclination angle both for single-helix and double-helix screw piles, the horizontal capacity under the compressive load is increased up to an inclination of 40°. For 0° and 20° the horizontal displacement criteria is predominant (50 mm) and the load corresponding to this displacement is considered. When the inclination is greater than 40°, the vertical displacement criteria is predominant and the loads corresponding to this displacement criteria is chosen. The percentage increase in the horizontal capacity of the DHSP with spacing of $2D_h$ is increased with the inclination angle (i.e. under higher compressive load) and is found to be 59% more than single-helix screw pile. In case of DHSP, the bottom helix contributes to the increase in vertical capacity there by increases the horizontal load compared to the SHSP. Loads obtained corresponding to different inclination angles for the single-helix and double-helix screw piles are shown in Table 8.3. Based on these loads obtained, yield envelopes are developed for the DHSP with spacing of $2D_h$ is shown in Fig. 8.5. The vertical and horizontal loads are normalized with the loads corresponding to $\theta = 0°$ (H_0) and $\theta = 90°$ (V_0). The range of the yield envelope is more for the DHSP showing that under combined loading condition the DHSP performs better rather than SHSP.

Effect of Spacing of Helix

Double-helix screw piles are analysed for two different spacing under the combined load. Figure 8.6 shows the yield envelopes developed for the DHSP with different spacing. It is observed from Fig. 8.5 that the increase in the helix spacing improves

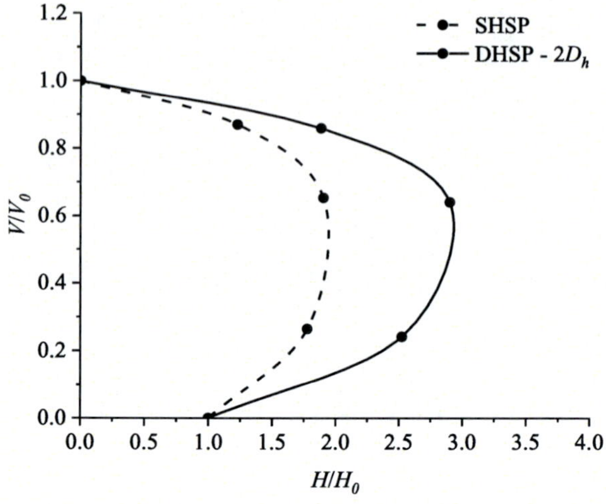

Fig. 8.5 Yield envelope for the single-helix and double-helix screw piles

the screw pile performance in terms of its capacity. As the helix spacing is increased from 2 to $4D_h$, the influence of compressive load is more for the higher spacing helix. This improves the screw pile lateral capacity for higher helix spacing, provided the top helix position is constant. Table 8.4 shows the load values obtained for the DHSP with two different spacing. Lateral load of the screw piles having $4D_h$spacing is around 28% higher than the screw pile with $2D_h$ spacing at an inclination of 60°. This shows that for the efficient use of screw piles subjected to combined load, the spacing between the helixes should be more. However, the present study is limited to only two spacing configurations and other helixes spacing need to be further studied.

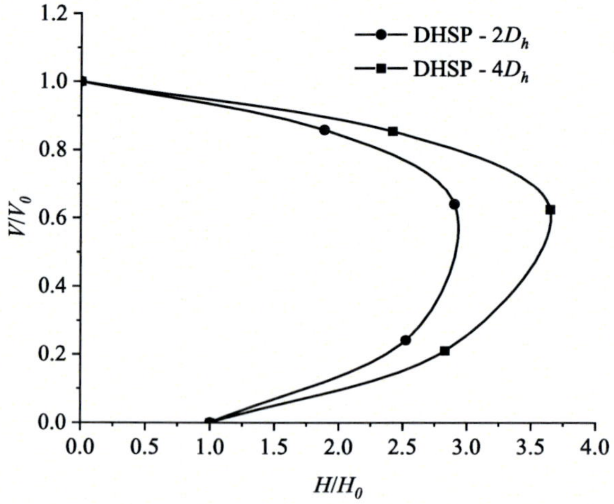

Fig. 8.6 Yield envelope of double-helix screw piles with different spacing

Table 8.4 Comparison of load values of double-helix screw pile with varied spacing

θ (°)	DHSP ($S = 2D_h$)		DHSP ($S = 4D_h$)	
	Horizontal load, H (kN)	Vertical load, V (kN)	Horizontal load, H (kN)	Vertical load, V (kN)
0	125.0	0	125.2	0
20	316.0	115.0	354.1	128.9
40	363.3	304.8	456.5	383.1
60	236.0	408.8	302.5	523.9
90	0	475.8	0	612.8

Table 8.5 Comparison of load values for single-helix screw pile with concrete pile

θ (°)	SHSP		CP	
	Horizontal load, H (kN)	Vertical load, V (kN)	Horizontal load, H (kN)	Vertical load, V (kN)
0	120.9	0	85	0
20	214.9	78.2	177.7	64.7
40	230.1	193.0	134.6	112.9
60	148.3	256.9	77.0	133.4
90	0	295.4	0	125.8

Behaviour of Screw Pile with Concrete Pile

The performance of the solid straight shaft concrete pile under the combined load is analysed and is compared with that of the screw pile having same shaft diameter. Table 8.5 shows the load values obtained for the SHSP and the CP. It is observed that load values obtained for the screw pile is significantly more compared to the concrete pile. The maximum improvement in the lateral load of the screw pile is found to be at 60° and is around 92% higher than the concrete pile. At higher inclinations, the effect of compressive load is more thereby resulting in the higher percentage increase in lateral capacity. Considering the weight of piles for the given geometry (7.2 kN for the screw pile and 10.4 kN for the concrete pile), the capacities of the screw piles per unit weight is also observed to be more than that of the concrete piles. It shows that the screw pile efficiency in terms of the load carrying capacity is better than concrete pile and could be used as a substitute to the conventional concrete piles.

Conclusions

The behaviour of the screw piles and the solid straight shafted concrete piles under combined compressive and lateral load are studied using 3D finite element method in the present study. Single-helix and double-helix screw piles with the variation of spacing of the helix are also studied. The developed numerical model is validated against the field tests. From the study, it is observed that to increase the lateral load of the screw piles, the placement of the helix should be optimized. Optimum location of the helix is found to be 1.3 m below the ground level for the screw pile considered. Maximum improvement in the lateral load for the SHSP and DHSP is observed at 40° and is reduced with the increase in inclination angle. Yield envelopes are developed for all the piles considered in the study which can be useful in determining the failure load. It is observed that the DHSP performs better (59% higher lateral capacity under 60° inclination for $S = 2D_h$) under combined load conditions compared to SHSP. In case of DHSP, screw piles with helix spacing of $4D_h$ shows 28% higher lateral capacity for inclination of 60°. It is also observed that the screw pile capacity per unit weight is more than the straight shaft concrete piles. Present study, shows that the screw pile behaves well under the combined load. This gives the assurance that the screw piles can be used as a potential alternative to the conventional concrete piles in case of the wind turbines.

References

1. El Sharnouby MM, El Naggar MH (2018) Field investigation of lateral monotonic and cyclic performance of reinforced helical pulldown micropiles. Canad Geotech J 55(10)
2. Byrne BW, Houlsby GT (2015) Helical piles: an innovative foundation design option for offshore wind turbines. Philos Trans Royal Soc a: math Phys Eng Sci 373(2035):20140081
3. Davidson C, Brown MJ, Cerfontaine B, Al-Baghdadi T, Knappett J, Brennan A, Augarde C, Coombs W, Wang L, Blake A, Richards D, Ball JD (2020) Physical modelling to demonstrate the feasibility of screw piles for offshore jacket supported wind energy structures. Géotechnique 1–50 (2020)
4. Buhler R, Cerato AB (2010) Design of dynamically wind-loaded helical piers for small wind turbines. J Perform Constr Facil 24(4):417–426
5. Perko HA (2009) Helical piles: a practical guide to design and installation. Wiley
6. Al-Baghdadi TA, Brown MJ, Knappett JA, Ishikura R (2015) Modelling of laterally loaded screw piles with large helical plates in sand. In: Frontiers in offshore geotechnics III: proceedings of the 3rd international symposium on frontiers in offshore geotechnics (ISFOG 2015). Taylor & Francis Books Ltd, pp 503–508
7. Shrestha S (2015) Design and analysis of foundation for onshore tall wind turbines. Clemson University, USA, All Theses
8. Sakr M (2010) Lateral resistance of high capacity helical piles–case study. In: Proceedings of the 63rd Canadian geotechnical and 6th Canadian permafrost conference. Calgary, Alberta, pp 12–16
9. Patra S, Kumar PVP, Haldar S (2019) Forensic investigation of laterally loaded screw pile using finite element analysis. In: 16th Asian regional conference on soil mechanics and geotechnical engineering, Taipei, Taiwan
10. Elsherbiny ZH, El Naggar MH (2013) Axial compressive capacity of helical piles from field tests and numerical study. Can Geotech J 50(12):1191–1203
11. Sakr M (2009) Performance of helical piles in oil sand. Can Geotech J 46(9):1046–1061
12. Hao D, Wang D, O'Loughlin CD, Gaudin C (2019) Tensile monotonic capacity of helical anchors in sand: interaction between helices. Can Geotech J 56(10):1534–1543

13. Al-Baghdadi TA, Brown MJ, Knappett JA, Al-Defae AH (2017) Effects of vertical loading on lateral screw pile performance. Proc Inst Civ Eng-Geotech Eng 170(3):259–272
14. Bowles JE (1995) Foundation analysis and design. McGraw-hill
15. Karthigeyan S, Ramakrishna VVGST, Rajagopal K (2007) Numerical investigation of the effect of vertical load on the lateral response of piles. J Geotech Geoenviron Eng 133(5):512–521
16. Salgado R (2008) The engineering of foundations. McGraw-Hill, Europe

Chapter 9
Finite Element Modelling of Laboratory One-Dimensional Consolidation of Soft Clays

A. S. Kasyap Vasudevan and Sridhar Gangaputhiran

Introduction

For any civil engineering structure, settlement of soil below the foundation threatens its serviceability in its design life. Hence, consolidation of soil is an importance phenomenon that must be studied upon. When it comes to soft clays, the behaviour under different loads and site conditions is very complex. In order to understand the behaviour of soils using numerical analysis, various kinds of material models were developed by geotechnical researchers [1]. Validation of such numerical tools and material models is very important for accurate predictions. The existing constitutive models for soil are not completely successful in predicting the pore pressure dissipation and development of effective stress during the process of consolidation [2, 3]. The settlements of a one-dimensional consolidation test conducted in laboratory are simulated using ABAQUS/CAE [4] and the settlements of both the approaches are compared in this study.

One-Dimensional Consolidation

Terzaghi in 1943 made assumptions on the process of consolidation and designed the first consolidation apparatus for consolidation test. Many of the geotechnical engineers use the Terzaghi's theory to solve various types of problem related to soils. The clay layer was assumed to be homogeneous and fully saturated. The soil grains and water within was considered incompressible. Validity of Darcy's law was

A. S. Kasyap Vasudevan (✉) · S. Gangaputhiran
Department of Civil Engineering, National Institute of Technology Karnataka, Surathkal 575025, India
e-mail: kasyap.vasudevan@gmail.com

K. Muthukkumaran et al. (eds.), *Soil Dynamics, Earthquake and Computational Geotechnical Engineering*, Lecture Notes in Civil Engineering 300,
https://doi.org/10.1007/978-981-19-6998-0_9

assumed, and the coefficient of permeability was considered constant throughout the layer. He came up with the differential equation relating the rate of change of pore water pressure with respect to time and depth [5] as shown in Eq. (9.1).

$$\frac{\partial u}{\partial t} = \frac{k}{\gamma_w m_v} \frac{\partial^2 u}{\partial z^2} = C_v \frac{\partial^2 u}{\partial z^2} \tag{9.1}$$

This gave birth to the coefficient of consolidation C_v, which is an important factor which signifies the rate consolidation of a soil. C_v depends on the coefficient of permeability k, coefficient of volume compressibility, m_v, and weight density of pore fluid γ_w. The theory turned out to be pivotal in geotechnical engineering.

Modified Cam Clay Model and ABAQUS/CAE

The first critical state models for describing the behaviour of soft soils such as clay, the cam clay model was first developed by Kenneth H Roscoe of University of Cambridge [6]. His team conducted a number of studies, which resulted in a model suitable for clays behaviours. The model was a plastic strain-hardening model based on critical state theory [7]. The model was further modified by changing the shape of the yield surface to an ellipse. This was named modified cam clay model. The model is formulated within the plane of effective mean stress and deviatoric stress; it assumes associated plasticity behaviour between the plastic yield function and the plastic potential function.

The equation for yield surface for modified cam clay model is given as:

$$f(p, q, p_c) = \left[\frac{q}{M}\right]^2 + p(p - p_c) \leq 0 \tag{9.2}$$

where q is the equivalent stress, p is the pressure, p_c is the pre-consolidation pressure, and M is the slope of the critical state line in p–q space. This study uses modified cam clay model for modelling kaolin clay.

Numerical modelling tool used in this study is ABAQUS/CAE 2019 Student version by Dassault Systems (2013) [4]. ABAQUS/CAE is an engineering simulation programme-based on finite element method. It is widely known for modelling and analysis of different problems in mechanical and civil engineering. The analysis usually involves three stages. These three stages are preprocessing (or modelling), processing (evaluation or simulation), and postprocessing (results and visualisation).

Numerical Modelling

The numerical modelling was done in such that the laboratory conditions were simulated.

The model is made of two parts. The dimensions of the specimen are 40 mm height and 100 mm diameter as shown in Fig. 9.1. The loading plate is also defined above the specimen with a thickness of 10 mm. Hence, for an axisymmetric model, the dimensions in metres become 0.04 m height and 0.05 m radius. The material behaviour is defined for the given clay, out of which elastic behaviour and plastic behaviour define the modified cam clay model parameters. The loading plate, which is made of steel is defined as linearly elastic.

Properties of commercially available kaolin clay used in the numerical analysis are given in Table 9.1. To find the consolidation properties of the soil, conventional consolidation test with two-way drainage was conducted by Sridhar [8]. Since coefficient permeability varies as consolidation progress, it must be defined with respect to variation of void ratio. The coefficient of permeability is obtained from Fig. 9.2, which was obtained through laboratory experiments conducted by Sridhar [8]. The Young's modulus and Poisson's ratio of the steel are assumed as 210 GPa and 0.3, respectively.

Initially, a seating load of 12.5 kPa is applied in the first step. Steps are defined to simulate the loading cycles for the loads of 50, 100, and 200 kPa. The step for simulating consolidation is defined with a longer duration so that the consolidation is complete. The boundary conditions are assigned so as to simulate the laboratory conditions. The displacements and rotations are restrained appropriately along the boundaries. The base is restrained for horizontal and vertical displacement. The sides of the soil specimen and loading plate are restrained to move horizontally. To simulate the drainage boundary at top and bottom, the pore pressure at top and bottom is set to zero. The element type specified is 8-node axisymmetric quadrilateral, biquadratic displacement, and bilinear pore pressure with reduced integration. (CAX8RP) as

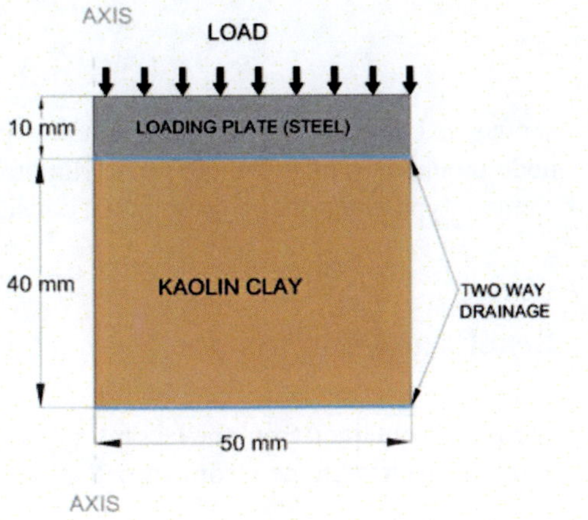

Fig. 9.1 One-dimensional consolidation model

Table 9.1 Properties of the kaolin clay for one-dimensional consolidation [8]

Properties	Units	Values
(1) Density, ρ	kg/m^3	1300
(2) Elastic behaviour: Porous elastic		
(i) Log bulk modulus, κ	–	0.025
(ii) Poisson's ratio, μ	–	0.3
(iii) Tensile limit		0
(3) Plastic behaviour: clay plasticity		
(i) Log plastic bulk modulus, λ	–	0.108
(ii) Stress ratio, *M*	–	1.113
(iii) Initial yield surface size, *a*	N/m^2	6250
(iv) Wet yield surface size, *β*		1
(v) Flow stress ratio	–	1
(4) Initial void ratio, *e*	–	1.2611
(5) Specific weight of wetting fluid, γ_w	N/m^3	9810

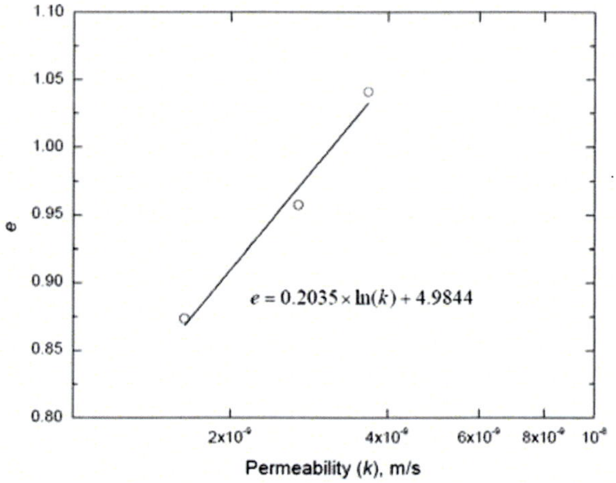

Fig. 9.2 Relationship between void ratio and permeability [6]

shown in Fig. 9.3. For the steel loading plate, the element type used was an 8-node biquadratic axisymmetric quadrilateral, reduced integration (CAX8R). In this element type, the pore pressure is not considered.

Results and Discussions

The results obtained from the laboratory experiment done by Sridhar [8] is compared with the results obtained from numerical analysis for the loads 50, 100, and 200 kPa.

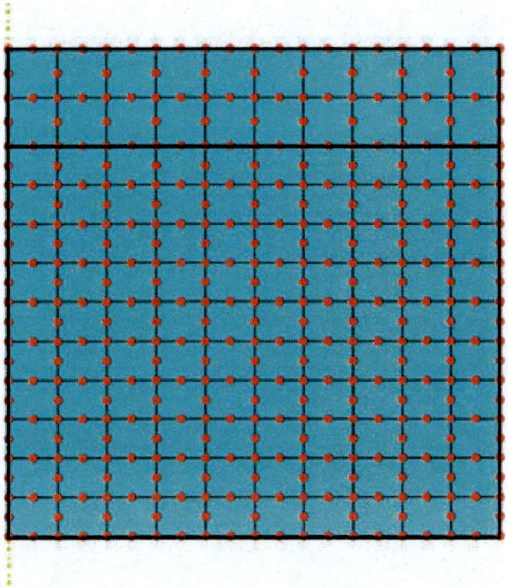

Fig. 9.3 Meshing of the consolidation test model in ABAQUS/CAE

The settlement time data from ABAQUS is imported to a spreadsheet, and graphs are plotted for comparisons. The settlement data of the node which is at the top right corner of the model is considering the position of dial gauge which was used to measure the settlement in the laboratory. Figures 9.4, 9.5 and 9.6 show the time settlement curve for different loads. The difference in the trends of laboratory results and ABAQUS results could be attributed to the approximation in the numerical analysis done by the software. The variations of settlement when compared to the laboratory results were found to be 0.86%, 1.33%, and 4.9% for pressure load of 50 kPa, 100 kPa, and 200 kPa, respectively. Figure 9.7 shows the pore pressure distribution along the depth of soil in between consolidation. The top and bottom face of the soil has zero pore pressure due to the two-way drainage provided during consolidation. There is a very small residual undissipated pore pressure of around 0.15 kPa at the mid-height of the sample. Initially, the pore water takes most of the load and as consolidation progress, it reduces to zero gradually. This is visible in Fig. 9.8. Since the pore pressure at top and bottom boundary is set to zero, throughout the test, it remains zero since it act as the drainage boundary. Pore pressure dissipation for the load increment 100–200 kPa with respect to time at different heights from the base is shown in Fig. 9.9. Since base and top acts as a drainage boundary, at any time, pore pressure is highest at the mid-height (i.e. at 20 mm above base). As we move towards a drainage boundary (top or bottom), pore pressure dissipates faster and reaches zero. The development of effective stress at different heights from base for the load increment 100–200 kPa is show in Fig. 9.10. The rate of pore pressure dissipation is reflected here. The effective stress develops faster as we move towards the drainage boundary. It is evident that the clay at the mid height (20 mm above base) consolidated the slowest along the depth.

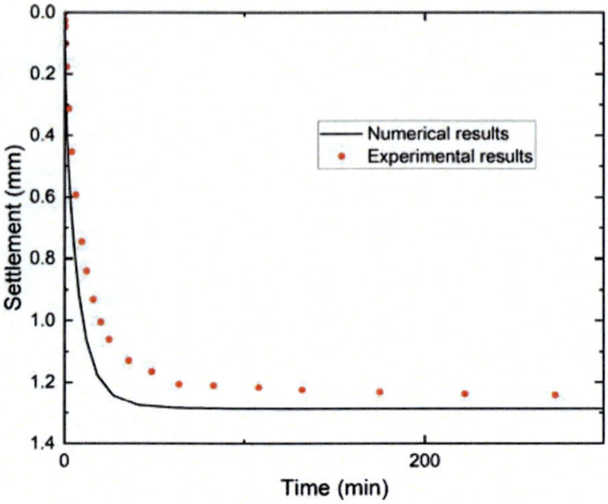

Fig. 9.4 Time settlement curve for a load of 50 kPa

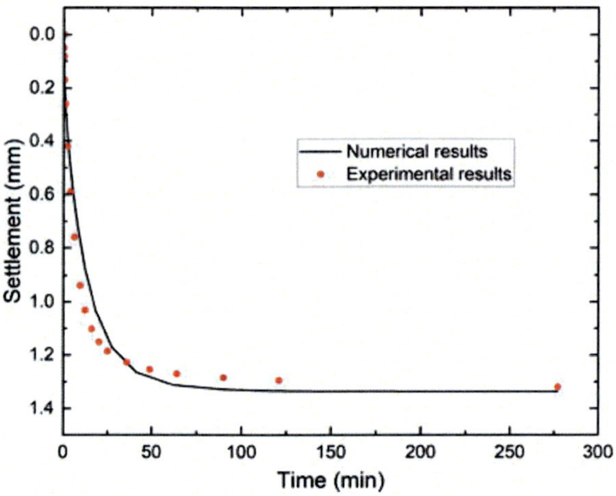

Fig. 9.5 Time settlement curve for a load of 100 kPa

Conclusions

Finite element analysis tool called ABAQUS/CAE was used to model one-dimensional consolidation test conducted in laboratory. Modified cam clay model and linear elastic model were used to represent the behaviour of kaolin clay and steel, respectively. The results from the numerical analysis were compared with the experimental results from [8]. The settlements predicted by the numerical analysis was very much in agreement with the experimental result. The deviations were just 0.86%, 1.33%, and 4.9% for consolidation pressures of 50 kPa, 100 kPa, and 200 kPa, respectively. The modified cam clay model predicted the settlement quite agreeably. The

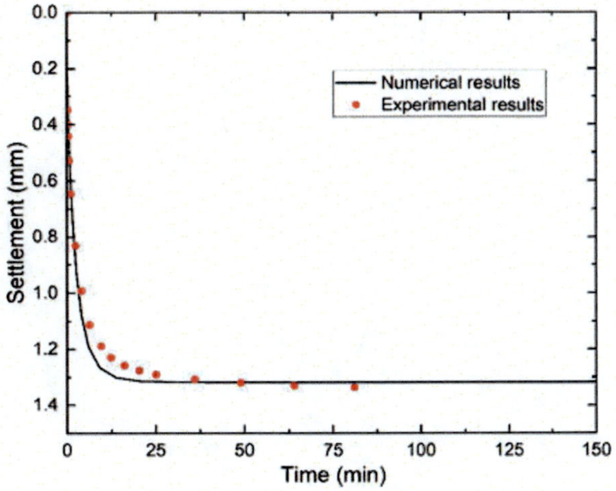

Fig. 9.6 Time settlement curve for the load of 200 kPa

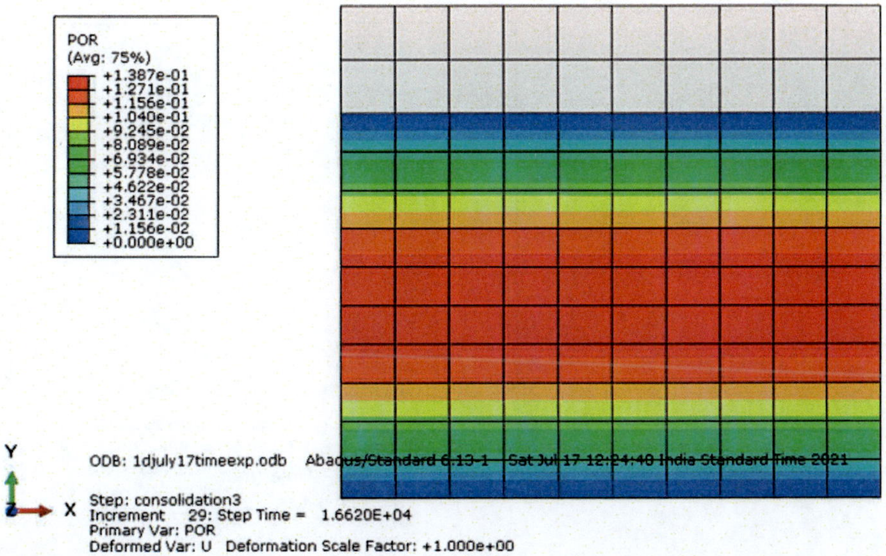

Fig. 9.7 Contours of pore pressure at the end of consolidation (pore pressure in Pa) for the load increment 50–100 kPa

model also gave expected trends for pore pressure dissipation, pore pressure distribution, and effective stress development. This study can be further extended by studying parameters like pore pressure, effective stress, etc., by conducting controlled consolidation tests with complex loading and unloading patterns to explore and compare the elastic and plastic behaviour of the clay. The present study can act as a precursor to modelling preloading problems with vertical drains.

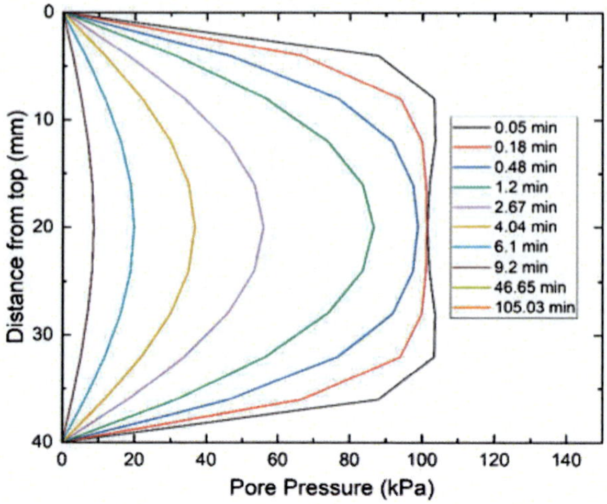

Fig. 9.8 Dissipation of pore pressure for the load increment 100–200 kPa

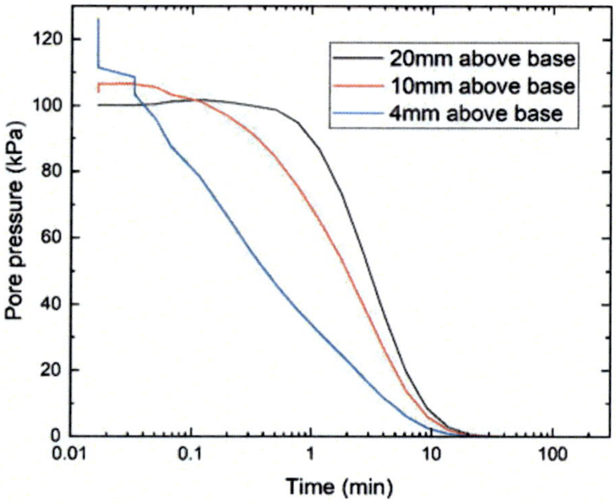

Fig. 9.9 Dissipation pore pressure at different depths for the load increment 100–200 kPa

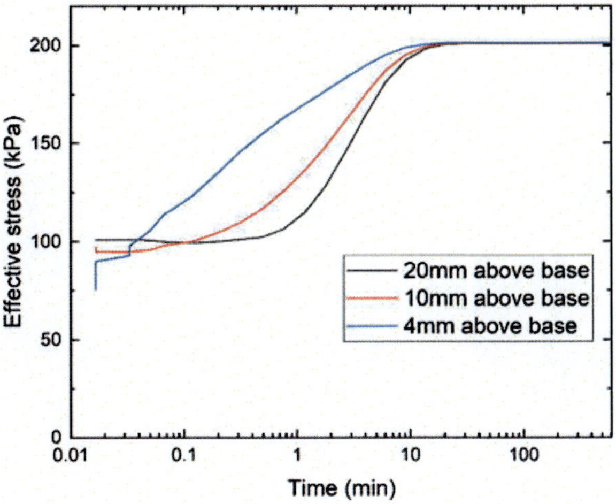

Fig. 9.10 Development of effective stress at different depths for the load increment 100–200 kPa

References

1. Ti KS, Huat BB, Noorzaei S, Jaafar S, Sew GS (2009) A review of basic soil constitutive models for geotechnical application. Electron J Geotech Eng 97(2):375–391
2. Hird CC, Pyrah IC, Russell D, Cinicioglu F (1995) Modelling the effect of vertical drains in two-dimensional finite element analyses of embankments on soft ground. Canad Geotech J 32(5):795–807
3. Indraratna B, Redana IW (1997) Plane-strain modeling of smear effects associated with vertical drains. J Geotech Geoenviron Eng ASCE 123(5):474–478
4. Dassault Systems, ABAQUS Theory Guide 6.13 Documentation. http://130.149.89.49:2080/v6.13/books/stm/default.htm. Last accessed 02 Apr 2021
5. Terzaghi K (1943) Chapter XIII Theory of consolidation, Theoretical soil mechanics. Wiley, p 265.
6. Roscoe KH, Schofield AN, Wroth CP (1958) On the yielding of soils. Geotechnique 8(1):22–53
7. Wikipedia, https://en.wikipedia.org/wiki/Critical_state_soil_mechanics. Last accessed 02 Sept 2020
8. Sridhar G (2015) Studies on vacuum preloading of soft clay deposit. PhD Thesis, Indian Institute of Technology Madras- National University of Singapore

Chapter 10
Numerical Study on Uplift Capacity of Helical Pile Embedded in Homogeneous and Layered Soil

S. Vijayakumar and M. Muttharam

Introduction

To resist compression, uplift force, overturning moment, and lateral forces, helical pile foundation systems are commonly used. Helical piles are widely used by engineers and designers and are considered one of the most significant advances in foundation engineering of the second half of nineteenth century [1]. In many countries, it is a popular foundation choice. When the uplift force acting on the foundation is more than the vertical compressive load, an alternative economic foundation should be selected for the ideal performance of the foundation. In that situation, helical pile would be one of the choices. Generally used for structures like solar panels, light houses, transmission towers, offshore windfarms, upward seepage pressure on foundation, structures subjected to seismic force experiences the uplift force and other light structures which are predominantly subjected to uplift loads.

Several researchers have been performed by various numerical and experimental investigations to predict the helical pile's uplift capability. Few research studies [2–7], Saurabh et al. [8] have been conducted to determine the ultimate uplift capacity of helical piles or helical anchors theoretically and experimentally. The ultimate uplift capacity of a helical pile is determined by the number of helical plates, the depth of the upper and lowermost helical plates, the ratio of spacing between the helical plates to the diameter of the plate, and the embedment depth, as per the reported investigations.

S. Vijayakumar (✉)
PG Student, Division of Soil Mechanics and Foundation Engineering, Department of Civil Engineering, College of Engineering Guindy, Anna University, Chennai 600025, India
e-mail: vijayakumarce2498@gmail.com

M. Muttharam
Professor, Division of Soil Mechanics and Foundation Engineering, Department of Civil Engineering, College of Engineering Guindy, Anna University, Chennai 600025, India
e-mail: muttharam@gmail.com

K. Muthukkumaran et al. (eds.), *Soil Dynamics, Earthquake and Computational Geotechnical Engineering*, Lecture Notes in Civil Engineering 300,
https://doi.org/10.1007/978-981-19-6998-0_10

These researches concentrated on calculating the vertical uplift capacity of helical piles buried in homogeneous ground.

Using three-dimensional finite element software, PLAXIS 3D, a numerical analysis on the uplift capacity of helical pile embedded in layered soil system, was carried out in this study.

Materials and Geometry

Material

A single-layer homogenous soil system and layered soil system are considered to determine the helical pile's uplift capacity. To develop the finite element model, the properties of the soil deposit reported by Salem andHussein [6] are considered and are presented in Table 10.1. The properties of pile and helix anchor reported by Srinivasan et al. [9] are used in the present analysis and are reported in Table 10.2.

Table 10.1 Soil properties [6]

Soil	Used for validation	Used for analysis	
	Very stiff silty clay	Dense aand	Stiff silty clay
Unit weight, γ (kN/m^3)	18	18	18
Modulus of elasticity, E (MPa)	10.5	140	9.5
Poisson's ratio, ν	0.3	0.32	0.3
Cohesion, C (kN/m^2)	105	0	95
Friction angle, ϕ (°)	0	39	0

Table 10.2 Pile and helix properties [9]

Use for both analysis and validation	Properties	Unit weight, γ (kN/m^3)	Modulus of elasticity, E (MPa)	Poisson's ratio, ν
	Pile and helix	78.5	210×10^3	0.25

Helical Pile Configurations

The validated helical pile consists of two helical steel bearing plates of 200.0 mm diameter each, attached to a central steel shaft of 45.0 mm diameter. The spacing between the plates is approximately two and half of the times the helix diameter, i.e. S/D = 2.50, in which S is the helix spacing and D is the helix diameter. The helical pile length is 2.0 m.

The helical pile adopted in parametric analysis consists of two helical steel bearing plates of 0.5, 0.75, 1 m diameter each, attached to a central steel shaft of 0.25 m diameter. The spacing between the plates is varied from 0.5 to 3.5 m and consequently increases with 0.5 m. S/D = 0.50, 0.75, 1.00, 1.25, 1.50, 1.75, 2.00, 2.25, 2.50, 2.75, and 3.00 are used in the analysis, in which S is the helix spacing and D is the helix diameter. The helical pile length is 8.0 m.

Problem Definitions

It is cumbersome to create the helical curve in PLAXIS 3D software. So, the pile with the helical surface is created in AUTOCAD software, and the model is imported in PLAXIS software. The model was created using AUTOCAD and imported to the structure mode. Problem definition for the model is shown in Fig. 10.1.

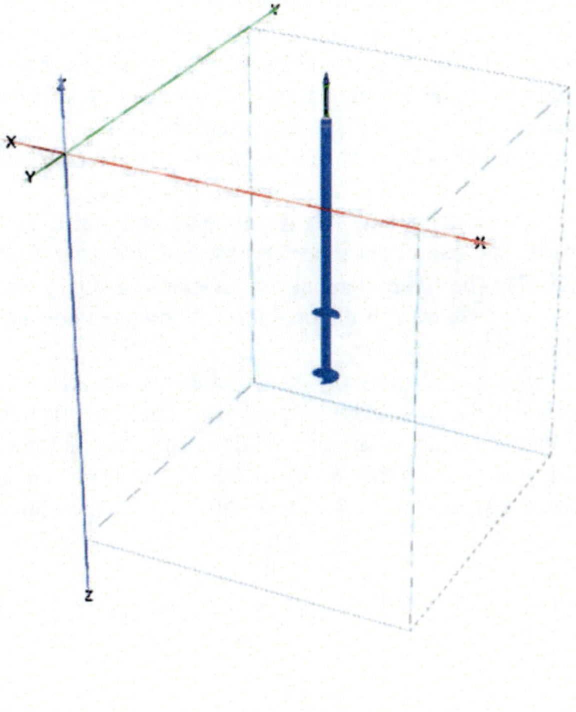

Fig. 10.1 Problem definition

Whenever a structure is imported to the PLAXIS 3D from AUTOCAD, proper limit should be set in AUTOCAD which is same as limit used for Soil contour in PLAXIS 3D. Most of the cases, there is a failure in mesh generation due to improper scaling of the structural element. Proper elemental properties should be assigned to prevent the failure of mesh generation.

Analysis

Validation

The results obtained from the full-scale helical pile load tests as reported by Salem and Hussein [6] were selected as the benchmark problem for the present study. The testing program consisted of tensile/pullout load tests on steel piles. The test helical pile consists of two helical steel bearing plates of 200.0 mm diameter each, welded to a central steel solid shaft. The test was carried out on site with cohesive deposit. The test results are considered for the validation purpose in the present study.

All the surface loose materials are removed, and then the helical pile was placed vertically in place. A wooden beam was placed horizontally at a special groove at the pile head. Two men began rotating the wooden rod and the helical pile screw soil until it reached its final position to be ready for testing. For the test program, adequate capacity was delivered by a hydraulic jack of 16 ton capacity. The load applied by the jack was controlled by supplying fluid pressure through a manual hydraulic pump. The measurement of the axial tensile load applied to the test pile was based on the pressure gauge reading on the hydraulic pump as a backup. The vertical pile movements during the tension test are measured using displacement dial gauges. Average displacement readings from the two dial gauges are plotted against the applied jack loads to yield load–displacement curve for tested pile. The curve was used to establish the ultimate pile capacity.

In that study, field test results obtained during helical pile testing were used to verify the numerical model performed using the finite element software ADINA, (2017). The finite element model main objective was modeling helical pile load-transfer behavior for the field tests. Upon the model calibration and verification with the experimental data.

Figure 10.2 plots the load–deflection curves obtained from field pull-out test conducted on helical piles founded in cohesive soil along with the same obtained from numerical analysis through ADINA software and numerical analysis carried out in the present study through PLAXIS 3D software. The maximum difference between the measured and computed uplift loads at the maximum allowable displacement is about 1.11%. Results obtained from the numerical analysis by using medium coarseness mesh matches with the experimental results and the variation was less than 10%. Hence, the medium coarseness mesh was adopted for the further parametric analysis.

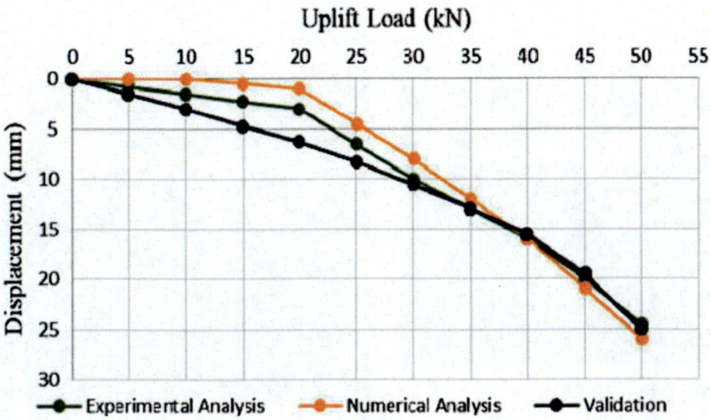

Fig. 10.2 Comparison of load–displacement curve

Parametric Study

PLAXIS 3D software is used for the present numerical study. The soil material is modelled as a ten-noded tetrahedral element using the Mohr–Coulomb model. The Pile shaft and helical blade were modelled using plate element, and the parametric study was carried out.

Limits of soil counter is defined as 10 × 10 × 12 m. Units are kN and m. The pile with the helical surface model was created using AUTOCAD and imported to the PLAXIS in structure mode. Medium coarseness mesh was created, and the mesh connectivity plot is shown in Fig. 10.3.

Calculation consists of three phases, including Initial phase. The initial phase is gravity loading; Phase 1 is the installation of the pile where the installation effects are neglected; in this phase, the piles are assumed to be embedded in the soil, and Phase 2 is the application of uplift load.

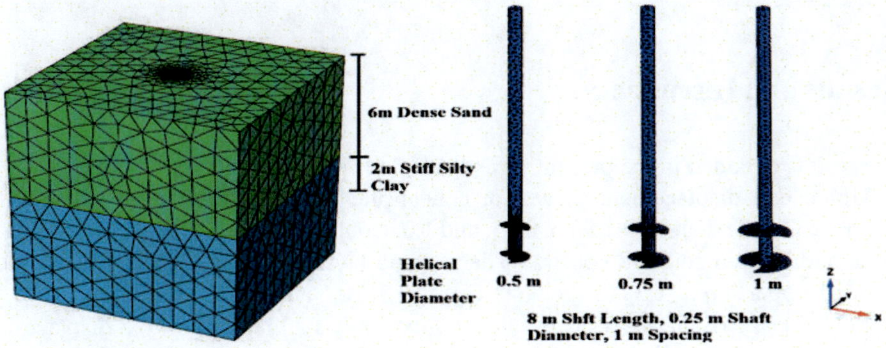

Fig. 10.3 Mesh connectivity plot used for parametric study

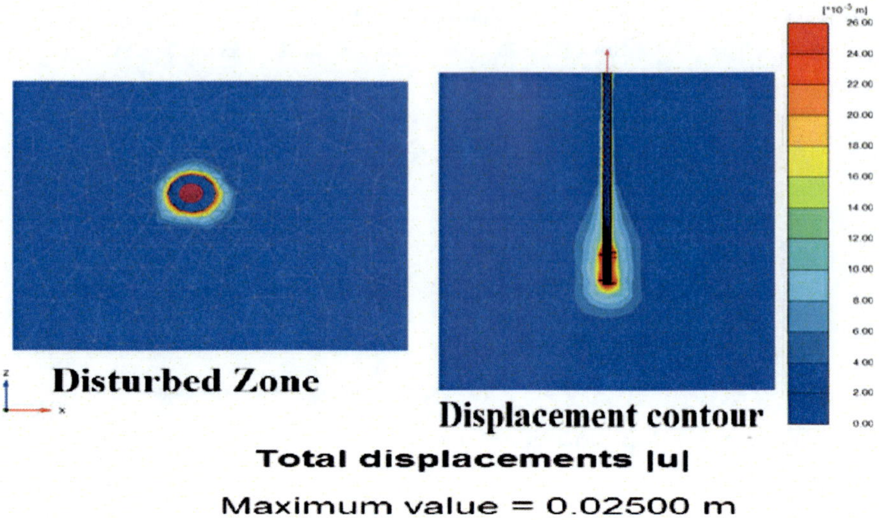

Fig. 10.4 Displacement contour and disturbed zone

In this present study, the results are generated up to the 20 mm displacement because the IS 2911-IV-1985 recommends [10],

The safe load shall be at least one of the following,

1. Two-thirds of the load with a total displacement is 12 mm or load that corresponds to a specified permissible uplift and
2. The half-load point at which the load–displacement curve shows the definite break (downward trend).

Initially, parameters such as the spacing between two helical plates, diameter of helical plate, soil type, and soil layer type are varied. A typical displacement contour and disturbed zone for helical pile subjected to uplift load are shown in Fig. 10.4. It shows that soil gets disturbed more near the helical plate.

Results and Discussion

The results obtained in the present investigation are discussed in this section.

Uplift load–displacement curves for helical pile with 0.25 m as shaft diameter and helical blade diameters of 0.5, 0.75, and 1.0 m embedded in stiff silty clay, dense sand, and layered soil bed consist of dense sand layer followed by stiff silty clay layer in proportion of 1:3 are shown in Figs. 10.5, 10.6 and 10.7, respectively.

In this analysis, the load–displacement curve shows a break prior to a displacement of 2 cm. By considering the displacement criteria, the uplift load at 12 mm displacement is compared with different parameters.

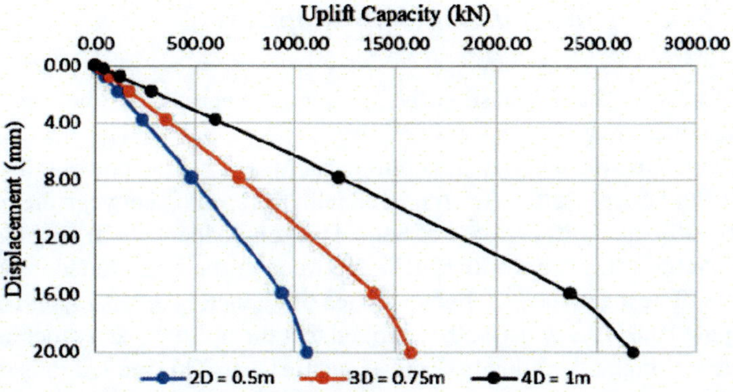

Fig. 10.5 Load–displacement curve (stiff silty clay)

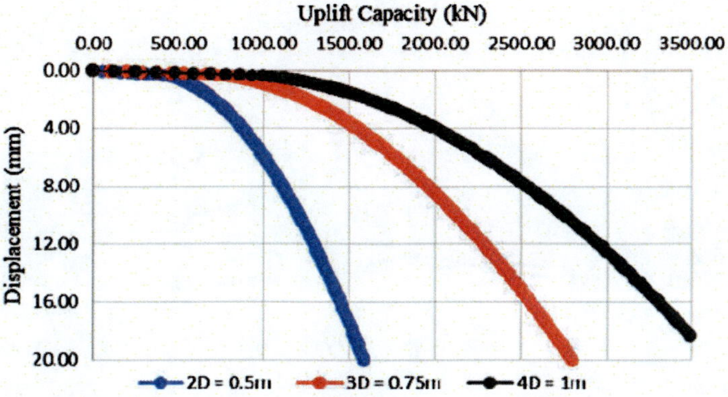

Fig. 10.6 Load–displacement curve (dense sand)

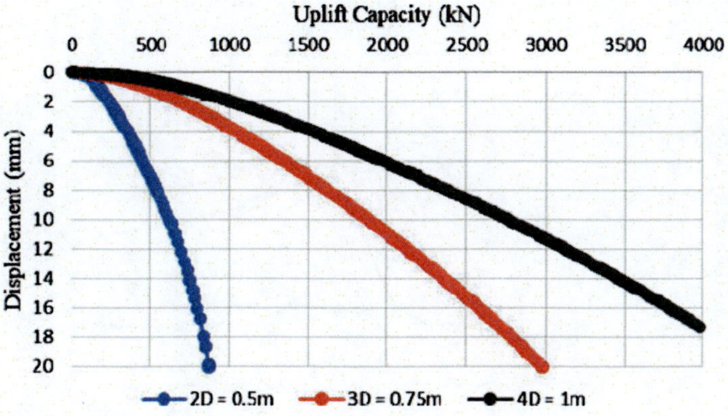

Fig. 10.7 Load–displacement curve (6 m dense sand layer followed by 2 m stiff silty clay layer)

Effect of Spacing of Helix on Uplift Capacity

The analysis shows that increase in the spacing between the two helices the uplift capacity also increases.

After the parametric analysis, the uplift capacities of helical pile of 0.5, 0.75, and 1.0 m diameter helical plate were compared with the spacing between the helix. The variation of ultimate uplift capacity of helical pile embedded in stiff silty clay, dense sand, and layered soil system with spacing between the helix is plotted in Figs. 10.8, 10.9 and 10.10, respectively, for helical plates of diameters 0.5, 0.75, and 1.0 m.

From Figs. 10.8, 10.9 and 10.10, it derived that the uplift capacity increases up to a particular spacing and gradually decreasing after that. This reduction is due to the transition of cylindrical shear mode to individual bearing mode. In cylindrical shear mode, the helixes act together, whereas in individual bearing mode, the helical plates

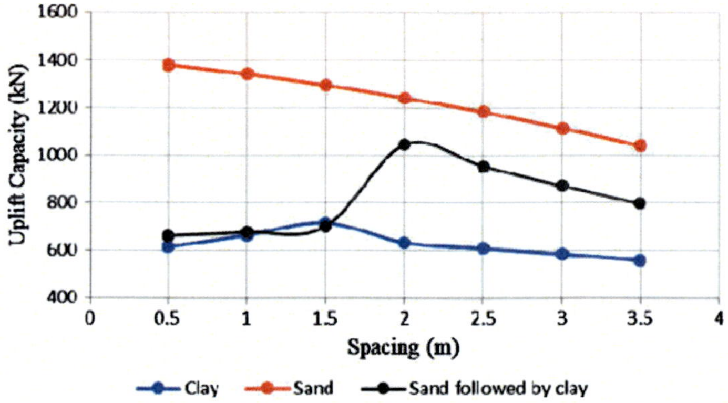

Fig. 10.8 Variation of uplift capacity due to spacing (0.5 m dia helix)

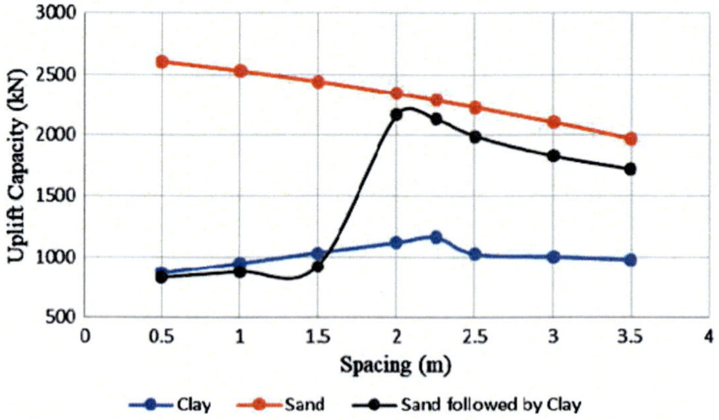

Fig. 10.9 Variation of uplift capacity due to spacing (0.75 m dia helix)

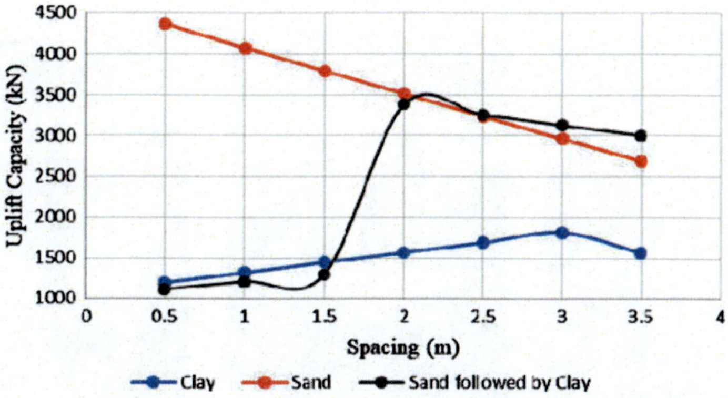

Fig. 10.10 Variation of uplift capacity due to spacing (1 m dia helix)

are acting individually. It will happen depending upon the spacing and diameter of the helices. In this study, uplift capacity decreases when the spacing to a diameter of helix ratio is 2 to 4.

The displacement contours of 0.5 and 1 m diameter helical plate due to uplift load for various spacing are shown in Fig. 11a, b. It shows that soil gets disturbed more near the helical plate. It clearly shows the helical plate acts individually.

Effect of Uplift Capacity Due to Diameter of Helical Plates

Diameter of the helix plays a vital role in the uplift capacity of the helical pile. Variation of uplift capacity for the 0.5, 0.75, and 1.0 m diameter helix is shown in Figs. 10.12, 10.13 and 10.14 for different soil and layered soil system.

From Figs. 10.12, 10.13 and 10.14, it is inferred that increases in the diameter of the helix increase the uplift capacity. Diameter of the helix also decides the critical spacing of the helix, which is also clearly indicates in these figures. In this study, for layered system with top layer 6 m dense sand and bottom layer 2 m stiff silty clay, the critical S/D ratio was varied 2 to 4, and for stiff silty clay and dense sand the critical S/D ratio was 3.

Conclusion

1. Increasing the diameter of the helix from 0.5 to 0.75 m and 0.75 to 1 m increases the uplift capacity by about 20–30% for piles embedded in uniform clay layer, but 50–100% for piles embedded in uniform sand layer.
2. Spacing of helixes shows more significant variation in the uplift capacity; it decides the mode of resistance in the soil.

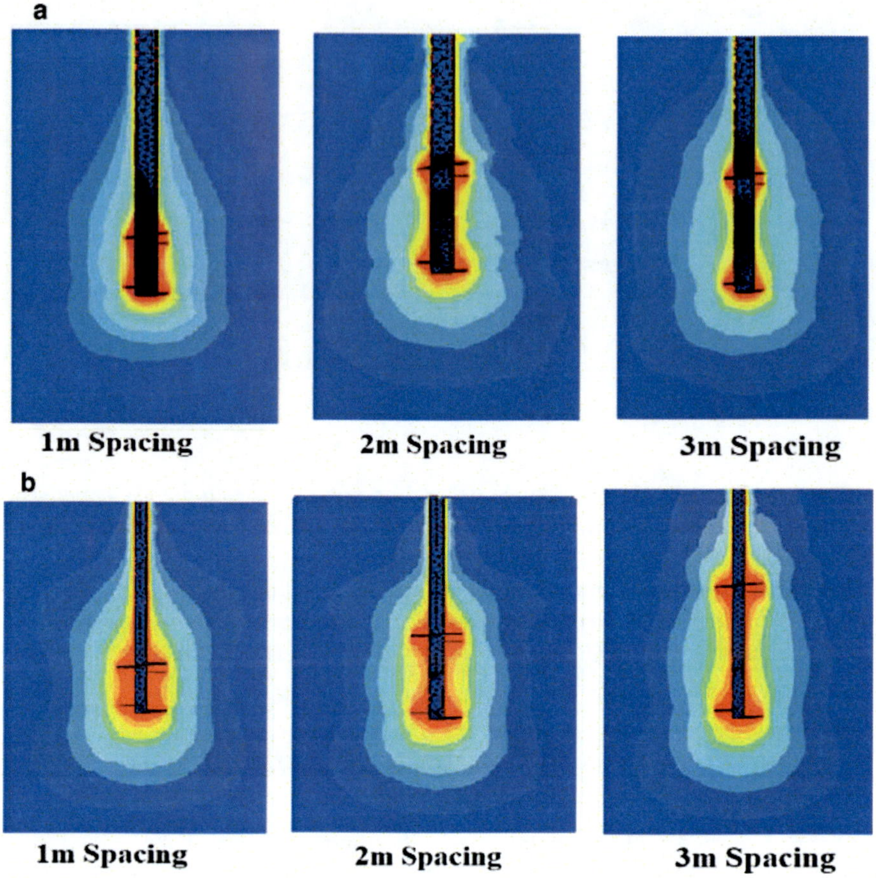

Fig. 10.11 **a** Displacement contour (0.5 m diameter helix). **b** Displacement contour (1 m diameter helix)

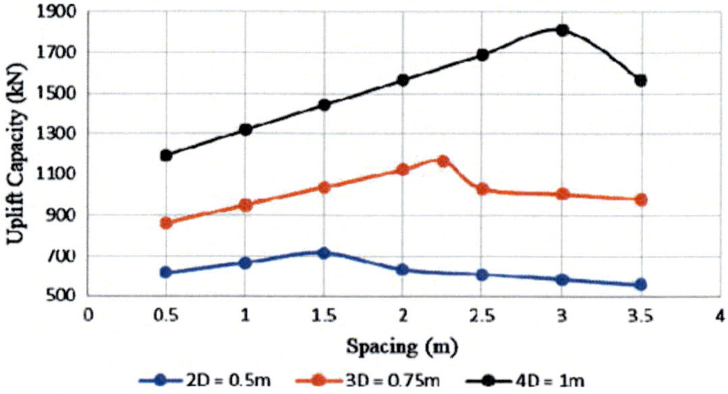

Fig. 10.12 Uplift capacity for pile embedded in stiff silty clay

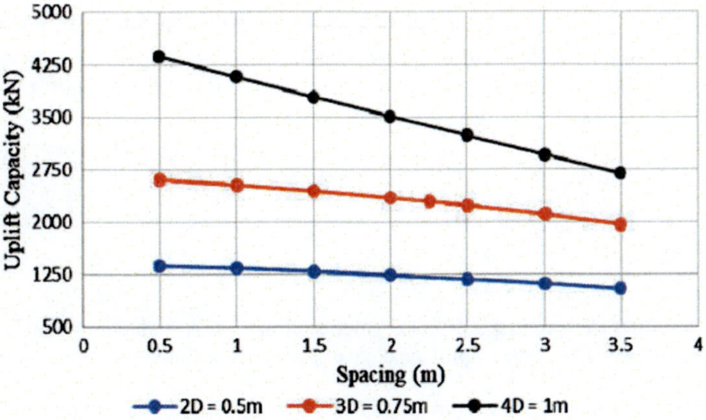

Fig. 10.13 Uplift capacity for pile embedded in dense sand

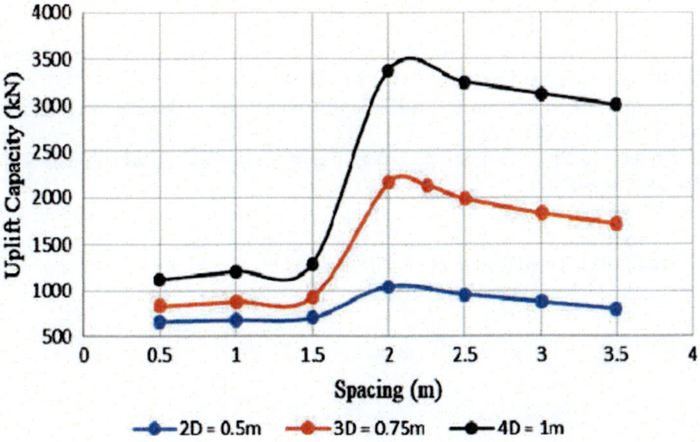

Fig. 10.14 Uplift capacity for pile embedded in layered system with top layer 6 m dense sand and bottom layer 2 m stiff silty clay

3. Varying the spacing of helix from 0.5 to 3.5 m in a pile shows an increment in uplift capacity up to spacing of 1.5 m for 0.5 m diameter helical pile, 2.25 m for 0.75 m diameter helical Pile after that a small decrement in the uplift capacity, and 3 m for 1 m diameter helical Pile after that a small decrement in the uplift capacity.
4. Uplift capacity decreases when the spacing to a diameter of helix ratio is 2–4. This decrement is due to the transition from cylindrical shear resistance to the individual bearing.
5. When the spacing between the helixes is large, the pile will resist the uplift capacity by an individual bearing, so the uplift capacity decreases accordingly.

6. The spacing at which the transition occurs is called critical spacing. The diameter of helix greatly influences this critical spacing.
7. Spacing of helixes shows significant variation in the uplift capacity; it decides the mode of resistance in the soil.

References

1. Perko HA (2009) Helical piles a practical guide to design and installation. Wiley
2. Bouazza A, Finlay TW (1990) Uplift capacity of plate anchors buried in a two-layered sand. J Geotech 40(2):293–297
3. Ilamparuthi K, Dickin EA, Muthukrisnaiah K (2002) Experimental investigation of the uplift behaviour of circular plate anchors embedded in sand. Can Geotech J 39(3):648–664
4. Karthick Kumar V (2020) Experimental and numerical study on uplift capacity of helical pile in clay. M.E. thesis, Submitted to Anna University, Chennai
5. Merifield RS, Sloan SW (2006) The ultimate pullout capacity of anchors in frictional soils. Canadian Geotech J 43:852–868
6. Salem TN, Hussein MM (2017) Axial tensile capacity of helical piles from field tests and numerical study. J Port Said Eng Res 21(2):111–119
7. Wang D, Merifield RS, Gaudin C (2013) Uplift behaviour of helical anchors in clay. Can Geotech J 50(6):575–584
8. Rawat S, Gupta AK (2017) Numerical modelling of pullout of helical soil nail. J Rock Mech Geotech Eng 9(4):648–658
9. Srinivasan V, Ghosh P, Santhoshkumar G (2019) Embedded in homogeneous and layered cohesionless soil. Int J Civ Eng
10. IS 2911-Part 4 (1985) Load test on piles—design and construction of pile foundations, Bureau of Indian Standards, New Delhi

Chapter 11
Simplified Plane Strain Consolidation Modelling of Stone Column

Puspendu Ray, Ambarish Ghosh, and Dipankana Bhattacherjee

Introduction

The use of stone column to stabilized soft ground is a very common practice, especially in India. This method is used to improve bearing capacity and to reduce differential settlement of superstructure constructed on weak ground. Stone column improves ground condition by two different ways. First, due to high permeability of stone column, it dissipates excess pore water pressure rapidly and increases bearing capacity. Second, due to high stiffness of column material, it reduces vertical stresses and settlement of soft ground. The dissipation of pore pressure or consolidation mechanism is very similar to vertical drain, and this part has been well researched [2, 5, 6] over the years. In case of sand drain, stiffness of surrounding soil and sand drain is considered equal, but in case of stone column, column stiffness makes an important role in stress distribution in soft soil. Therefore, ignorance of column stiffness may not give appropriate result in consolidation of stone column reinforced ground. Generally, stiffness value of stone column material is about 10–20 times more than soft soil [3].

Although, axisymmetric unit cell model is very popular to determine long-term consolidation effect of reinforced ground, but it is restricted to the centerline of the embankment. In case of multi-drain analysis considering full width of the embankment plane, strain analysis is very important, and additionally, it reduces analysis time and cost as compared to three-dimensional analysis.

The objective of this study is to develop a simplified plane strain model of stone column reinforced ground to calculate the consolidation effect of the reinforced ground with reasonable accuracy as compared to axisymmetric model. Stone column

P. Ray (✉) · A. Ghosh · D. Bhattacherjee
Department of Civil Engineering, Indian Institute of Engineering Science and Technology, Howrah, West Bengal, India
e-mail: pray.rs2019@civil.iiests.ac.in

K. Muthukkumaran et al. (eds.), *Soil Dynamics, Earthquake and Computational Geotechnical Engineering*, Lecture Notes in Civil Engineering 300,
https://doi.org/10.1007/978-981-19-6998-0_11

is considered as a free drainage path and to avoid complexities effect of smear zone and well resistance has not been taken into this study.

Brief of Previous Work

The radial flow of water into the vertical drain and consolidation effect of soft ground were described by Barron [2]. The basic assumption was: (a) the saturated pore water is incompressible in nature, and vertical loads are carried by excess pore water pressure, (b) vertical drain has circular influence zone, (c) soil deforms in vertical direction only, and applied load distribution is uniform over the soil region. Considering decrease of soil volume is equal to discharge of pore water from the soft soil, equation of change of pore water pressure due to radial axisymmetric flow was developed.

$$\frac{\partial \overline{u}}{\partial t} = c_r\left(\frac{1}{r}\frac{\partial u_r}{\partial r} + \frac{\partial^2 u_r}{\partial r^2}\right) \tag{11.1}$$

where c_r = coefficient of consolidation in radial flow, $\overline{u}$ = average excess pore water pressure, u_r = excess pore water pressure due to radial flow, r = arbitrary distance from centre of drain.

According to Barron [2], average rate of consolidation due to radial flow

$$U_r = 1 - \mathrm{e}^{\left(-8T_r/F(n)\right)} \tag{11.2}$$

where $T_r = c_r t/d_e^2$ = time factor of radial flow, $F(N) = [N^2/(N^2-1)]\ln(N) - (3N^2-1)/4N^2$ (ignoring smear zone and well resistance), $N = d_e/d_c$, d_c and d_e diameters of a drain well and its influence zone, respectively.

Hird [5] and Indraratna [6] both incorporated plane strain modelling in vertical sand drain consolidation. Hird [5] considered vertical drain with well resistance whereas Indraratna [6] considered both smear zone and well resistance effect in calculation. According to Indraratna [6], the average rate of consolidation in plane strain condition is

$$\overline{U_{hp}} = 1 - \mathrm{e}^{\left(-8T_{hp}/\mu_p\right)} \tag{11.3}$$

where T_{hp} = time factor in plane strain condition, μ_p = factor depends on dimensions of drain and soft soil.

Based on Barron's drain well consolidation theory, Han [3] developed axisymmetric unit cell modelling of stone column. Stone column is assumed as drained layer at the centre of the cylindrical soft soil. For simplification of analytical model, Han [3] ignored well resistance and smear zone effect. Equal vertical strain theory is also incorporated in the model. Considering the dissipation of pore water pressure from soft soil equal to reduction of soil volume in cylindrical unit cell, Han [3] proposed equation for vertical and radial consolidation.

$$\frac{\partial \overline{u}}{\partial t} = c'_r\left(\frac{1}{r}\frac{\partial u_r}{\partial r} + \frac{\partial^2 u_r}{\partial r^2}\right) + c'_v\frac{\partial^2 u}{\partial z^2} \tag{11.4}$$

where $c'_r = (k_r/\gamma_w)\left[m_{v,c}(1-a_s) + m_{v,s}a_s/m_{v,c}m_{v,s}(1-a_s)\right]$, modified coefficient of consolidation of radial flow.

$c'_v = (k_v/\gamma_w)\left[m_{v,c}(1-a_s) + m_{v,s}a_s/m_{v,c}m_{v,s}(1-a_s)\right]$, modified coefficient of consolidation of vertical flow. a_s = represents area replacement ratio = A_c/A, A and A_c = total area of unit cell and area of column, respectively.

Due to very high permeability of column, radial flow of soft soil is more dominating as compared to vertical flow. Therefore, consolidation of unit cell stone column is more reasonable due to radial flow. So, ignoring vertical consolidation, Han's equation can be rewrite as per Tan and Oo [6] as: (Fig. 11.1)

$$\overline{U_r} = 1 - \mathrm{e}^{\left(-8T'_r/F(N)\right)} \tag{11.5}$$

where $F(N) = [N^2/(N^2-1)]\ln(N) - \left(3N^2-1\right)/4N^2$ and $T'_r = c'_r t/d_e^2$

Plane Strain Modelling

Hansbo [4] axisymmetric analysis of vertical drain can be modified to an equivalent plane strain modelling as follows. Consider all around impervious boundaries except for the outlets at the centre, stone column. Soil and stone column are loaded vertically. The soil is fully saturated.

According to Darcy's law, the velocity of pore water flow (v_x) in horizontal direction can be calculated by the following equation

$$v_x = \left(\frac{k_{xp}}{\gamma_w}\right) \times \left(\frac{\partial u}{\partial x}\right) \tag{11.6}$$

where k_{xp} = coefficient of permeability in horizontal direction, subscripts x and p denote direction of flow and plane strain condition, respectively; γ_w = unit weight of water; u = pore water pressure.

Now, considering total inflow of pore water pressure into the column through an imaginary small slice of width dz, at a distance x from the centre of the column is equal to change of volume of soil block of width *(B_s–x)*

$$v_x = \left(\frac{\partial \varepsilon}{\partial t}\right) \times (B_s - x) \tag{11.7}$$

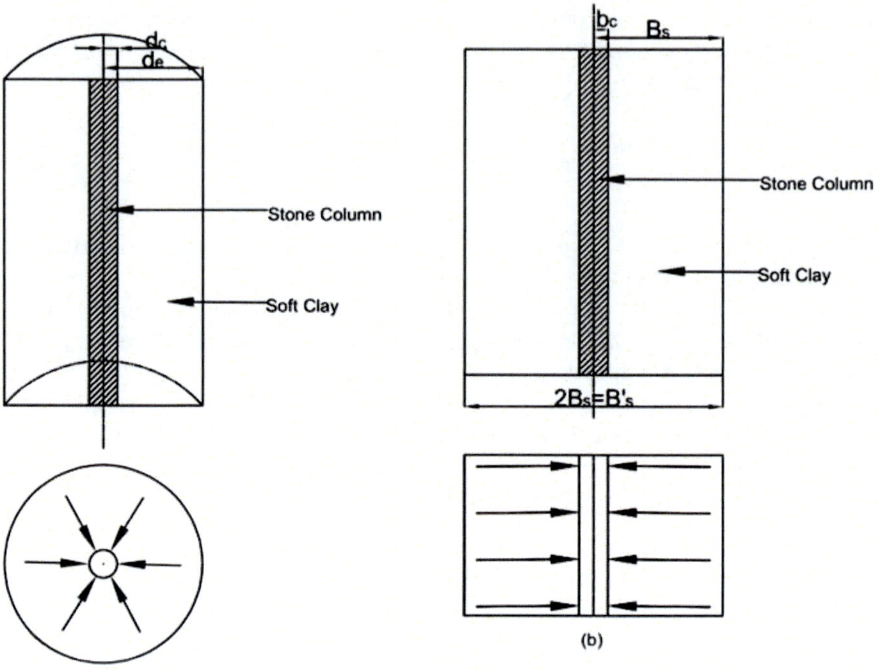

Fig. 11.1 **a** Axisymmetric model **b** plane strain model

where ε = strain in vertical direction. Now, substituting Eq. (11.7) in Eq. (11.6), we can get,

$$\frac{\partial u}{\partial x} = \left(\frac{\gamma_w}{k_{xp}}\right)\left(\frac{\partial \varepsilon}{\partial t}\right)(B_s - x)B_s \geq x \geq b_c \tag{11.8}$$

Integrating Eq. (11.8) with the boundary condition $u = 0$, $x = b_c$

$$u = \left(\frac{\gamma_w}{k_{xp}}\right)\left(\frac{\partial \varepsilon}{\partial t}\right)\left\{B_s(x - b_c) - \frac{1}{2}\left(x^2 - b_c^2\right)\right\} \tag{11.9}$$

Let $\overline{u}$ be the average excess pore water pressure. Therefore,

$$\overline{u} = \frac{\int_{b_c}^{B_s} u\mathrm{d}x}{B_s} \tag{11.10}$$

Putting Eq. (11.9) in Eq. (11.10), we get

$$\overline{u} = \left(\frac{\gamma_w}{k_{xp}}\right)\left(\frac{\partial \varepsilon}{\partial t}\right) \times \frac{B_s^2}{2} \times \mu_p \tag{11.11}$$

$$\mu_p = \frac{2}{3}\left(1 - {}^{b_c}/_{B_s}\right)^3 \tag{11.12}$$

Now,

$$\frac{\partial \varepsilon}{\partial t} = m_v \times \frac{\partial \overline{\sigma'}}{\partial t} = -m_v \times \frac{\partial \overline{u}}{\partial t} \tag{11.13}$$

where $\overline{\sigma'}$ = average effective overburden stress, m_v = coefficient of volume compressibility.

Substituting Eq. (11.13) in Eq. (11.11) and applying boundary condition, $\overline{u} = \overline{u_0}$ at $t = 0$, we get,

$$\overline{u} = \overline{u_0} \mathrm{e}^{\left(-8T_{xp}/\mu_p\right)} \tag{11.14}$$

where time factor $T_{xp} = c_{xp}t/B_s'^2$ and coefficient of consolidation $c_{xp} = k_{xp}/m_v\gamma_w$

So, average degree of consolidation

$$\overline{U_{xp}} = 1 - \frac{\overline{u}}{\overline{u_0}} = 1 - \mathrm{e}^{\left(-8T_{xp}/\mu_p\right)} \tag{11.15}$$

Equation (11.15) is suitable to determine plane strain consolidation effect in sand drain where compressibility of soft soil and drainage material is considered same. Therefore, the constant μ_p in plane strain can be compared with axisymmetric constant *F(N)* of Barron's [2] solution.

In case of stone column, compressibility of soft soil and stone column is significantly different. So, at any time applied load on stone column supported soft ground can be written as

$$\overline{\sigma_s} A_s + \overline{\sigma_c} A_c = fA \tag{11.16}$$

where $\overline{\sigma_s}$ = average total stress on soft soil, $\overline{\sigma_c}$ = average total stress on stone column, A_s and A_c = cross-sectional area of soft soil and stone column, respectively, f = average applied pressure on soil block, $A = A_s + A_c$

According to Han et al. [3], in case of stone column, Eq. (11.13) can be modified as

$$\frac{\partial \varepsilon}{\partial t} = -\frac{m_{v,c} m_{v,s} A_s}{m_{v,c} A_s + m_{v,s} A_c} \frac{\partial \overline{u}}{\partial t} \tag{11.17}$$

$$\frac{\partial \varepsilon}{\partial t} = -\frac{m_{v,c} m_{v,s} (1-n)}{m_{v,c}(1-n) + m_{v,s} n} \frac{\partial \overline{u}}{\partial t} \tag{11.18}$$

where $m_{v,c}$ and $m_{v,s}$ are coefficient of volume compressibility of stone column and soft soil, respectively; n = area replacement ratio = A_c/A.

The coefficient of compressibility can also be calculated from Poisson's ratio (ν) and modulus of elasticity (E)

$$m_v = \frac{(1+\nu)(1-2\nu)}{E(1-\nu)} \tag{11.19}$$

Due to difference of stiffness between stone column and soft soil stress concentration phenomenon will occur. The stress concentration ratio (s) can be calculated as

$$s = \frac{\overline{\sigma_c}}{\overline{\sigma_s}} = \frac{m_{v,s}}{m_{v,c}} \tag{11.20}$$

Now, using Eqs. (11.18) and (11.15) can be rewritten as

$$\overline{U_{xp}} = 1 - \frac{\overline{u}}{\overline{u_0}} = 1 - \mathrm{e}^{\left(-8T'_{xp}/\mu_p\right)} \tag{11.21}$$

where T'_{xp} = modified time factor = $c'_{xp}t/B'^2_s$, c'_{xp} = modified coefficient of consolidation = $(k_{xp}/\gamma_w)\left[m_{v,c}(1-n) + m_{v,s}n/m_{v,c}m_{v,s}(1-n)\right]$

So, Eq. (11.21) defines the degree of consolidation of plane strain stone column model.

At a particular time frame, the rate of consolidation of stone column in axisymmetric model and plane strain model should be same. Therefore, degree of consolidation in plane strain model ($\overline{U_{xp}}$) can be compared to degree of consolidation of Han's axisymmetric model (Eq. 11.5) ($\overline{U_r}$).

$$\overline{U_{xp}} = \overline{U_r} \tag{11.22}$$

Now, rearranging Eq. (11.22)

$$\frac{T_{xp}}{T_r} = \frac{\mu_{xp}}{F(N)} = \frac{k_{xp}}{k_r}\frac{d_e^2}{B'^2_s}\frac{m_{v,c}(1-n) + m_{v,s}n/m_{v,c}m_{v,s}(1-n)}{m_{v,c}(1-a_s) + m_{v,s}a_s/m_{v,c}m_{v,s}(1-a_s)} \tag{11.23}$$

Now, rearranging Eq. (11.23) and eliminating the common terms

$$\frac{k_{xp}}{k_r} = \frac{\mu_{xp}}{F(N)}\frac{B'^2_e}{d_e^2} \tag{11.24}$$

$$\frac{k_{xp}}{k_r} = \frac{B'^2_s}{d_e^2} \times \frac{\frac{2}{3}\left(1 - {}^{b_c}/_{B_s}\right)^3}{[N^2/(N^2-1)]\ln(N) - \left(3N^2-1\right)/4N^2} \tag{11.25}$$

Equation (11.25) denotes the correlation of permeability between plane strain and axisymmetric model.

Numerical Modelling and Verification

To verify the developed analytical correlation, finite element analysis of unit cell model has been carried out for axisymmetric and equivalent plane strain geometry (Fig. 11.2). Two different numerical models are used to compare the simulation results. For axisymmetric model, data has been collected from Tan's [7] paper. Due to symmetricity, half of the model has been used which also reduces the simulation time. The stone column has a length of 10 m with 0.85 m column diameter and 2.55 m outer soil diameter. Other properties are as follows: unit weight of soft soil and column is 15 kN/m^2, Poisson's ratio is 0.3, elasticity modulus of soil and column are 3000 kPa and 30,000 kPa, respectively, horizontal and vertical permeability of soil and column is 3.47×10^{-9} and 1.16×10^{-9} m/s, respectively, effective cohesion of soil and column are 0.1 and 1 kPa, respectively, whereas effective angle of internal friction of soil and column are 22° and 40° respectively.

In case of plane strain model, axisymmetric unit cell geometry has been transformed to an equivalent plane strain geometry based on conversion of vertical drain system into plane strain wall studied by Indraratna [6]. The method has been developed based on equal cross section of column and surrounding soil in both cases. Therefore, applying the conversion technique the width of the column was found 0.56 m and width of the soil was found 5.1 m. Applying this geometry in Eq. (11.25),

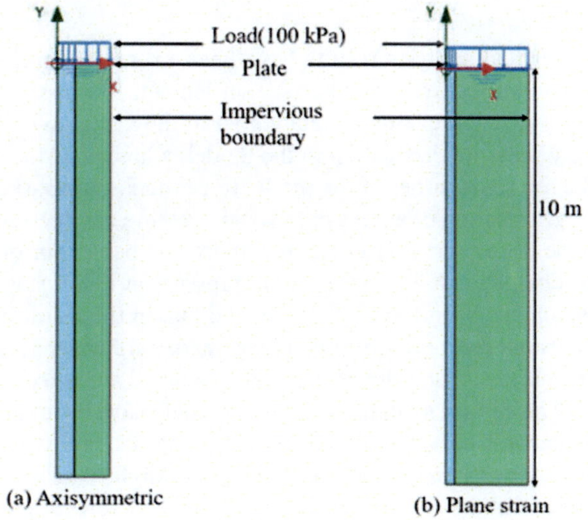

Fig. 11.2 Finite element model

modified permeability of the soil was found 1.54×10^{-8} m/s. Other parameters are remained unchanged in this study.

Both the models have been developed using finite element software PLAXIS 2D (version 20.0). The 15-node triangular element has been used in this study. The size of the mesh has been finalized based on mesh convergence study. In each model, water table has been considered at the ground level. Only ground level has been considered to be permeable, whereas other sides of the cell has been chosen impermeable. The models are horizontally fixed, fully restrained at the bottom and fully free at the ground level.

To idealize the consolidation behaviour of the soft soil and column, liner elasto-plastic Mohr coulomb failure criteria has been adopted. The column has been modelled as drained material, whereas soft clay has been modelled as undrained material. To simulate the horizontal flow properly, horizontal permeability of the soil was taken slightly more than vertical permeability as well as permeability of the stone column was kept very high compared to soil permeability. The stress concentration behaviour and equal strain condition have been modelled by applying loading on a rigid plate having axial stiffness of 1×10^7 kN/m and bending stiffness of 1×10^4 kN m^2/m. Total consolidation has been simulated in three phases, i.e. Phase 1: generation of initial effective stress and pore water pressure due to self-weight, Phase 2: activation of plate and loading. This phase is instantaneous, so there is no consolidation or dissipation of excess pore pressure, Phase 3 is consolidation phase. Pore water pressure dissipation was simulated in this stage until the excess pore pressure reaches 0.01 kPa.

Results and Discussion

Numerical analysis results have been expressed in terms of dissipation of excess pore water pressure and surface settlement. The pore water pressure dissipation was measured at the bottom right corner of the model to understand the effect of slowest pore pressure dissipation in the model, whereas surface settlement was measured at the top right corner of the model to get maximum surface settlement with time.

Dissipation of excess pore water pressure has been shown in Fig. 11.3. The initial node of the curve describes the undrained behaviour of the soft clay under 100 kPa surcharge load. The excess pore pressure at initial stage is expected to be 100 kPa, which has been found very close to that in the result in both the cases. It has been observed that both the model reach zero pore water pressure at the same time which represents great efficiency of the developed analytical model.

The surface settlements of two models have been shown in Fig. 11.4. The surface settlement curve of the plane strain model represents good agreement with the axisymmetric model at the starting but slightly disagreement has been observed at the end of the curve. Axisymmetric model shows slightly more (5–10%) settlement as compared to plane strain model.

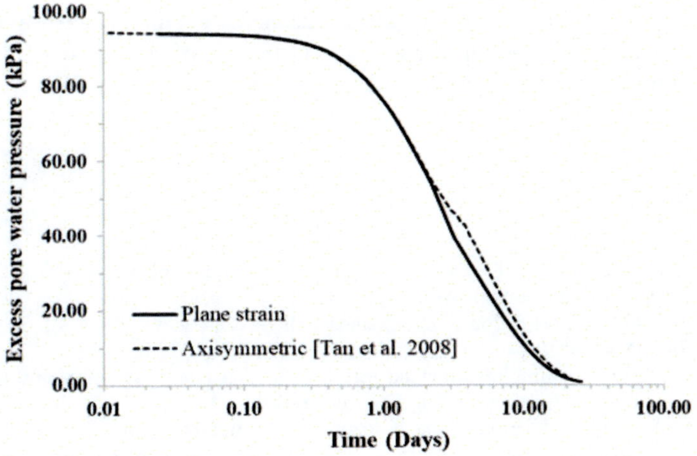

Fig. 11.3 Simulated dissipation of pore water pressure with time

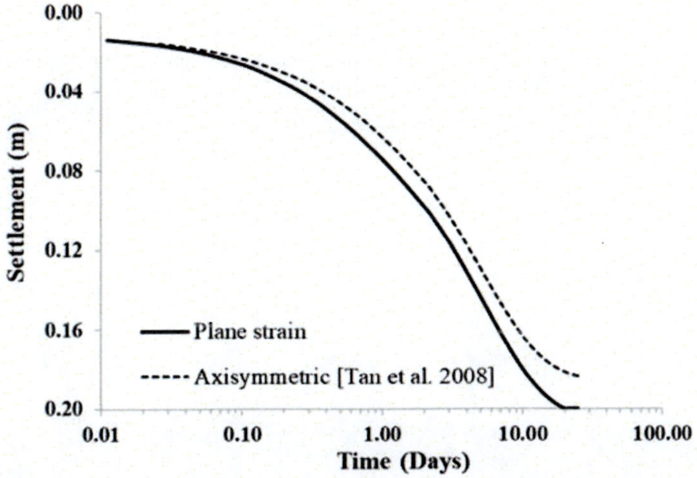

Fig. 11.4 Simulated surface settlement with time

Conclusion

This study proposed a simple conversion technique to obtain equivalent plane strain model from axisymmetric model. This method is based on conversion of equivalent geometry and soil permeability. Equivalent soil permeability can be determined based on Eq. (11.25), whereas equivalent geometry can be determined from Indraratna's study [6].

Validity of this method has been done based on axisymmetric unit cell stone column data available in Tan's [7] paper. Therefore, based on finite element analysis, it can be concluded that this method shows reasonably good agreement in terms of pore water pressure dissipation calculation, but in case of surface settlement, this method underpredicts 5% to 10% as compared to axisymmetric model.

References

1. Barksdale RD, Bachus RC (1983) Design and construction of stone columns. FHWA/RD-83/026, Federal Highway Administration, Washington, D.C.
2. Barron RA (1948) Consolidation of fine-grained soils by drain wells. Trans Am Soc Civ Eng 113:718–742. Paper No. 2346
3. Han J, Ye S-L (2001) Simplified method for consolidation rate of stone column reinforced foundations. J Geotech Geoenviron Eng 1277:597–603
4. Hansbo S (1981) Consolidation of fine-grained soils by prefabricated drains. In: Proceeding of 10th international conference soil mechanics and foundations engineering, vol 3, Balkema, Rotterdam, The Netherlands, pp 677–682
5. Hird CC, Pyrah IC, Russell D (1992) Finite element modelling of vertical drains beneath embankments on soft ground. Geotechnique 423:499–511
6. Indraratna B, Redana IW (1997) Plane-strain modeling of smear effects associated with vertical drains. J Geotech Geoenviron Eng 1235:474–478
7. Tan SA, Tjahyono S, Oo KK (2008) Simplified plane-strain modeling of stone-column reinforced ground. J Geotech Geoenviron Eng 134(2):185–194

Chapter 12
Lateral Displacements of Soft Ground Treated with PVDs Under Embankment Loading

C. Abhinay Kumar, J. Y. V. Shiva Bhushan, and Madhav Madhira

Introduction

India is a country with large tracts of coastal lands consisting of highly soft soils. Soft soils are known for high water content, low undrained shear strength, low permeability, high compressibility resulting in excessive settlements that pose many geotechnical challenges. Due to their time-dependent characteristics, more time is required to complete consolidation. Improvement of soft ground is the only solution to overcome the problem of excessive deformations.

Several ground improvement techniques are available to improve the ground. Vertical drains are most commonly suited for fine-grained, inorganic high water content, low strength soils. Sand drains were used in the past but PVDs are more effective, faster to install, and economical [1]. PVD with preloading is used to get faster results by creating an artificial drainage path to the water present in the soils [2]. The disturbance caused while installing affects the performance of PVD-treated ground [3]. The rate of construction, diameter, spacing, and discharge capacity of vertical drains play important role in achieving ultimate settlements [4, 5].

Soft soil under embankment loading settles due to both consolidation and lateral deformation due to plastic flow of subsoil [3]. Treatment of soft ground with PVD along with preloading increases both the rates of vertical and horizontal consolidation. Loading rate and embankment load are the main factors that cause lateral displacements in the soil. During the construction of embankment, if the shear deformation exceeds the rate of vertical consolidation, failure of structure is likely. The present study considers the embankment loading on the PVD-treated soft ground.

C. Abhinay Kumar · J. Y. V. Shiva Bhushan (✉) · M. Madhira
Department of Civil Engineering, VNR Vignana Jyothi Institute of Engineering and Technology, Hyderabad 500090, India
e-mail: jyvshivabhushan@gmail.com

K. Muthukkumaran et al. (eds.), *Soil Dynamics, Earthquake and Computational Geotechnical Engineering*, Lecture Notes in Civil Engineering 300,
https://doi.org/10.1007/978-981-19-6998-0_12

The objective of this paper is to quantify the effect on the reduction of lateral deformations at the toe of embankment alongside settlements beneath the embankment for different spacing of PVDs.

Problem Statement

An ultra-mega power project is constructed at Krishnapatnam, Nellore, Andhra Pradesh [6]. The subsoil is mostly clayey, while top soils are medium dense to very dense sand. Figure 12.1 shows the soil profile at the location which consists of very dense and medium dense sand up to 7.5 m below ground level, and a soft clayey soil is underlying which is extended up to 20 m from the surface. Table 12.1 presents the subsoil and embankment soil properties at Krishnapatnam ultra-mega power project (KUMPP) from [7]. PVDs of length 20.0 m were installed in a triangular pattern.

Construction of Embankment

Embankment of height 4 m is constructed with a base width of 33 m and with side slopes of 1 V:2H. A sand mat of 0.3 m is placed above the ground surface as a drainage layer. The rate of construction (Fig. 12.2) is small at the initial stage and rapid after 15 days of construction. Construction of embankment started from 24th day and finished by the end of 54th day after the installation of PVDs. The total period of construction is 30 days for 4 m high embankment for a design load of 68 kPa.

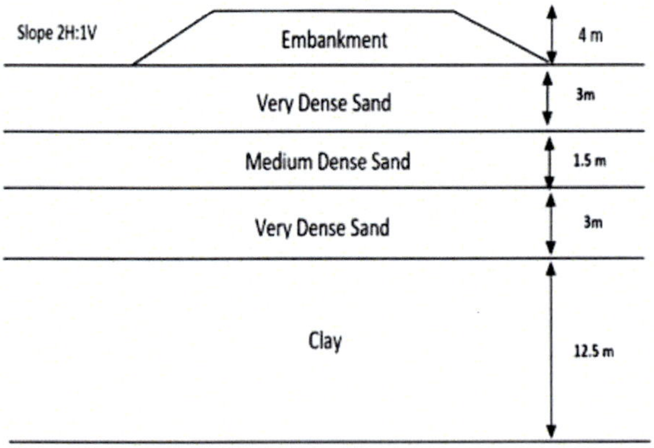

Fig. 12.1 Soil profile at KUMPP

Table. 12.1 Properties of soils

Parameter	Embankment	Very dense sand-1	Medium dense sand	Very dense sand-2	Clay
γ_{unsat} (kN/m^3)	17	17	17	17	15
γ_{sat} (kN/m^3)	19	20	20	20	17
e (void ratio)	0.5	0.5	0.5	0.5	0.5
k_x (m/day)	0.05	$1.73 * 10^{-2}$	$1.7 * 10^{-2}$	$1.7 * 10^{-2}$	$1.6 * 10^{-6}$
k_y (m/day)	0.05	$4.3 * 10^{-3}$	$4 * 10^{-3}$	$4 * 10^{-3}$	$4 * 10^{-6}$
Poisson's ratio	0.3	0.3	0.25	0.25	0.35
Deformation modulus (MPa)	30	40.6	42.1	40.6	1.059
Cohesion (kPa)	1	3	3	3	25
Angle of shearing resistance	30	41.5	36	41.5	0

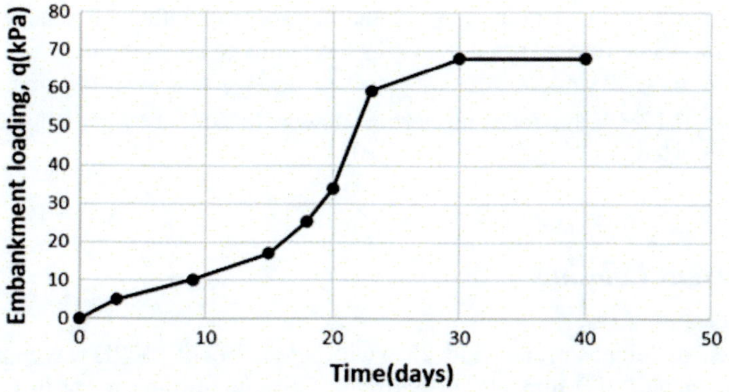

Fig. 12.2 Embankment loading with time

Finite Element Analysis

Finite element software—PLAXIS 2D, 2020—was used to analyse the problem. As the loading is symmetrical, only half the embankment and the ground were modelled for plane strain condition. Lateral boundaries are at 1.5 times the base width of an embankment. Sand layer extends from the surface to 7.5 m depth and clay layer from 7.5 to 20 m below the ground surface. Numerical analysis of the embankment with soil is conducted in plane strain model in PLAXIS 2D with 15-noded triangular elements.

Model is validated first for Terzaghi's 1D consolidation theory [8] for untreated ground. Mohr–Coulomb model is used to define the embankment and sandy soils [9]. A conventional equation given by Hird et al. [10] is used to convert axisymmetric

Table 12.2 Soil properties used in the model

Soil property	Model properties
Poisson's ratio (v)	0.3
Tangent of critical state line (M)	0.8
Coefficient of consolidation C_v (m^2/year)	0.7–1.2
Modified swelling index (λ)	0.1130
Modified creep index (κ)	0.1617

model to plane strain one as

$$\frac{k_{\text{pl}}}{k_{\text{ax}}} = \frac{2}{3[\ln \ln[n] + \left[\frac{k_{ax}}{r_s}\right] \ln \ln[s] - \left[\frac{3}{4}\right]} \tag{12.1}$$

where k_{pl} is the horizontal permeability of undisturbed zone in plane strain unit cell, k_{ax} is the horizontal permeability of undisturbed zone in axisymmetric unit cell, k_s is the horizontal permeability of smear zone in axisymmetric unit cell, n is the influence ratio r_e/r_w, s is the smear ratio r_e/r_w., r_e is the radius of influence zone, r_w is the equivalent radius of vertical drain, and r_s is the radius of smear zone.

Where $M = 1.73 \sin\varphi$, $\lambda = \frac{C_c}{2.3(1+e)}$, $\kappa = \frac{2C_s}{2.3(1+e)}$ and swelling or reloading index, $C_s = 0.12e^{1.13e_0}$. Model properties shown in Table 12.2 are adopted from Radhakrishnan [6].

Embankment Loading

Embankment of height 4 m and width 25 m with side slopes of 1 V:2H was constructed and a sand mat of 300 mm placed above the ground surface to drain the water (Fig. 12.3). PVDs of length 20 m were installed using a triangular pattern. The bottom boundary is fixed in both vertical and horizontal displacements, and the side boundaries are fixed in the horizontal direction while the top boundary is free.

Finite element mesh of the soft ground, embankment, and the PVDs is shown in Fig. 12.4. Finer meshing has been considered so that more precise results could be obtained. To increase the accuracy of the results, 15-noded triangular patterns are considered which provides a fourth-order interpolation.

Validation

Numerical results for Terzaghi 1D consolidation theory [8] are compared with the results from the present FE model in Fig. 12.5. The results are comparable but with some difference.

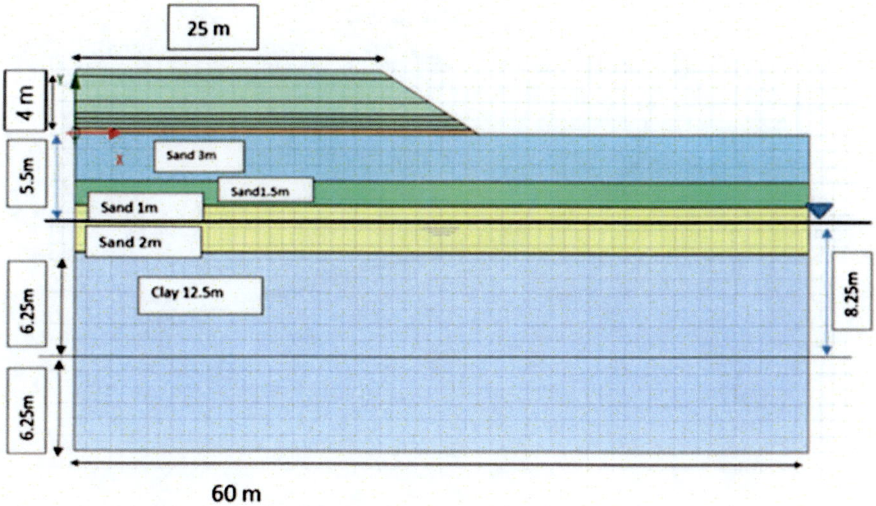

Fig. 12.3 Embankment over stratified soils

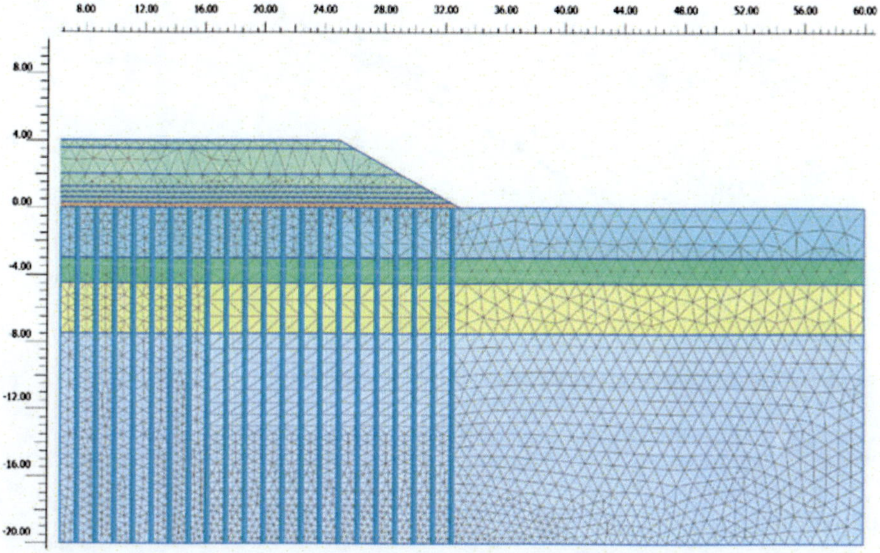

Fig. 12.4 Finite element mesh of soft ground under embankment loading

Final Settlement

Asaoka method [11] is used for interpreting and extrapolating the settlements observed in the field. The intersection points of 45° line and the plot of settlements

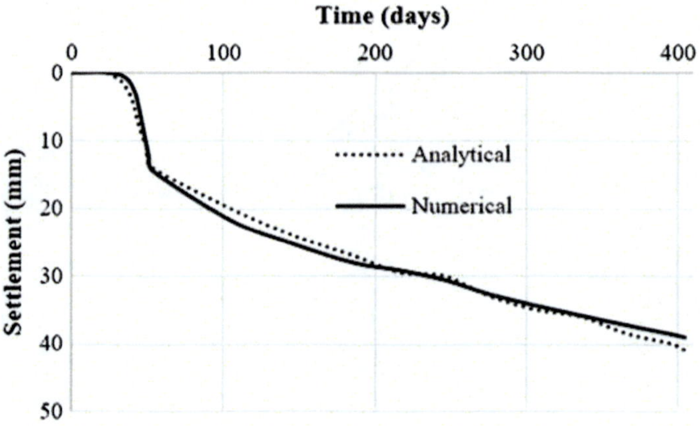

Fig. 12.5 Comparison of settlement curves with analytical solution

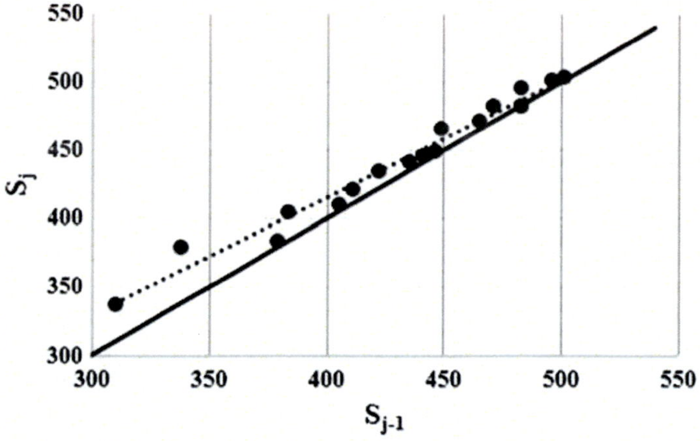

Fig. 12.6 Final settlement from Asaoka plot for 1.25 m c/c PVDs

give the final settlement. Observed results at the location for 1.25 m PVD @c/c and 2.5 m PVD spacing are analysed (Figs. 12.6 and 12.7) and final settlement (S_f) of 507 and 506 mm is obtained, respectively.

Results

Finite element models of stratified sand overlying soft clay with and without PVD are analysed and are compared to illustrate the effectiveness of PVDs on the rate and

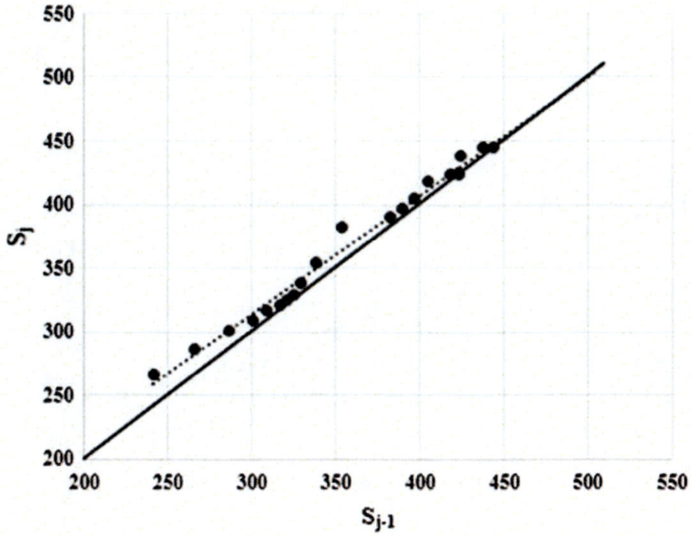

Fig. 12.7 Final settlement from Asaoka plot for 2.5 m c/c PVDs

magnitude of vertical and lateral deformations. Construction is completed at end of 54 days.

Settlements

Numerical results for the variations of surface settlements beneath the centre of embankment with time for PVD-treated soft ground for different spacing are obtained. Settlement of 507 mm is attained at 405 days for 1.25 m PVD @ c/c spacing. Deformation analysis is performed till 424 days for PVDs @ 2.5 m c/c spacing. Comparison of FEM results with measured surface settlements for PVD-treated ground @ 1.25 m c/c and 2.5 m c/c is shown in Fig. 12.8. The results predicted are comparable with the field observations.

At the end of 424 days, settlements beneath the centre of embankment for 2.5 m PVD @ c/c spacing and untreated ground are, respectively, 444 and 44 mm. Rate of consolidation is more and rapid in the case of PVD-treated ground for 1.25 m c/c spacing, in which case 99% degree of consolidation (U) is achieved in 405 days.

Based on the properties used for 1.25 and 2.5 m c/c spacing, settlements for 1.5 m PVD spacing are predicted by FEM and is shown in Fig. 12.9. A settlement of 488 mm is achieved in 405 days. With the provision of PVD treatment, 99, 96, and 87% degrees of consolidation are achieved for PVD-treated ground @1.25 m, 1.5 m, and 2.5 m c/c spacing at the end of 405 days.

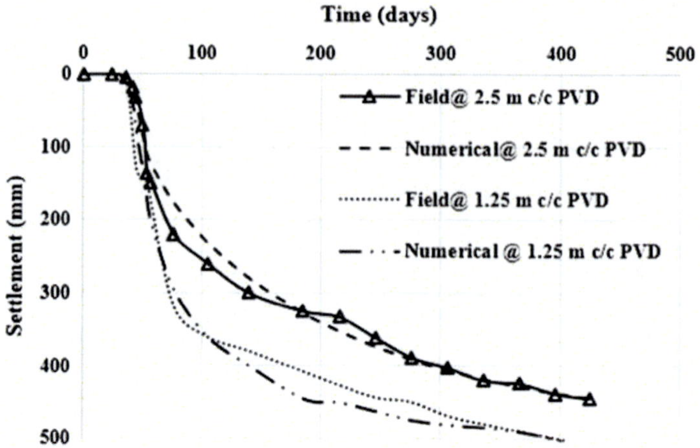

Fig. 12.8 Comparison of settlement of PVD-treated ground for different spacing

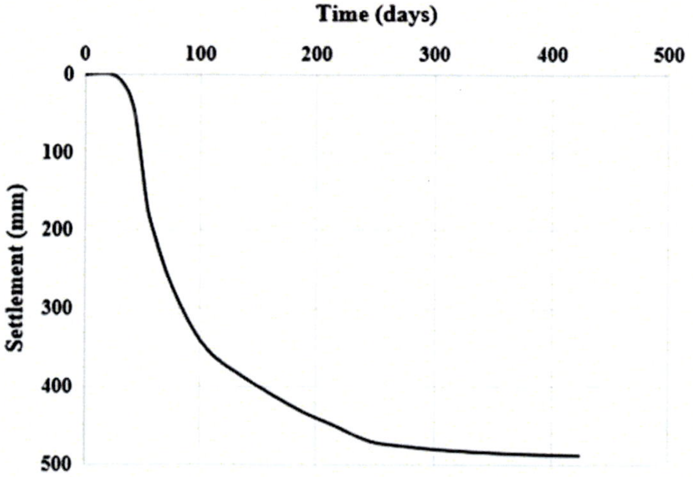

Fig. 12.9 Surface settlement with respect to time for PVD @ 1.5 m c/c

Figure 12.10 shows the surface settlement profiles beneath the embankment at the end of 405 days, obtained from FEM results. Maximum settlement for untreated, PVD-treated ground with @ 2.5 m c/c, 1.5 m c/c, and 1.25 m c/c spacing are, respectively, 41 mm, 440 mm, 488 mm, and 503 mm, while settlements attained at the toe are, respectively, 3, 121, 133, and 135 mm.

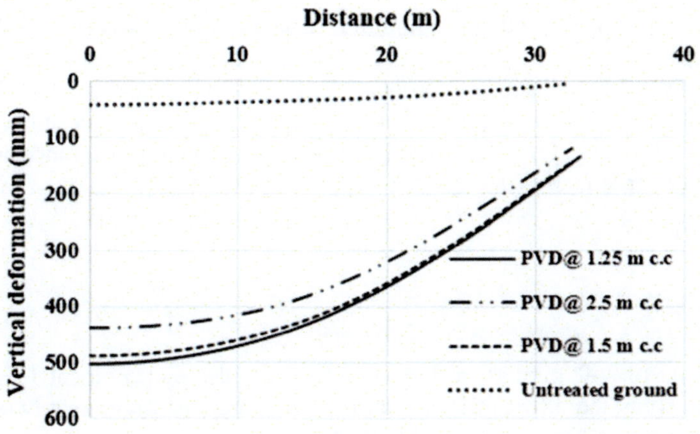

Fig. 12.10 FEM results of surface settlement profiles at the end of 405 days

Lateral Deformations

Lateral movement of soft ground at the toe of an embankment is studied by using FEM, at 405 days for PVD at different spacing. Figure 12.11 shows the variations of lateral displacements with depth. For soft ground treated with PVD, the top few metres of sandy strata deform inwards towards the embankment centerline and gradually deforms outwards and increases till the sand strata ends.

Lateral deformations at the end of 405 days are 6.92–8.18 times with respect to untreated ground for PVD with considered spacing. At depth of 7.5 m from ground

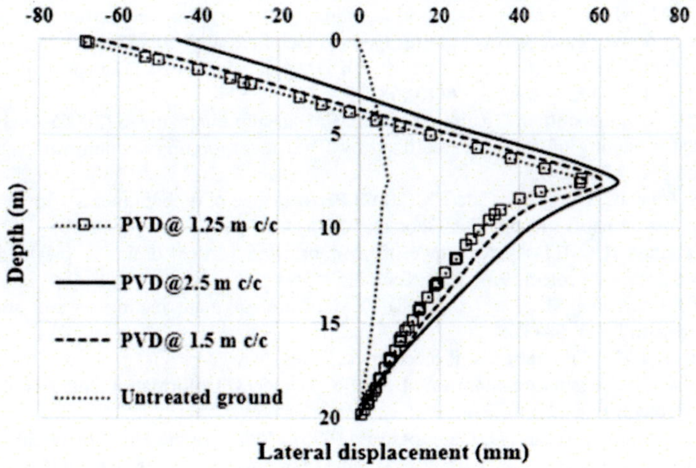

Fig. 12.11 Variation of lateral displacement at the toe of embankment with depth

in soft clay, lateral displacement is 55 mm in case of PVD-treated ground at 1.25 m c/c spacing.

Larger inward movement is seen near the top of the ground as since both the vertical and horizontal consolidations are significant. In the top layer of clay, lateral displacement variation for PVD-treated ground is large, and the difference decreases minimally as depth increases.

Conclusions

Deformations of soft ground treated with PVDs under embankment loading are analysed to study the effect of PVDs on rate of vertical and horizontal consolidation. Asaoka method is used to predict final settlements from measured settlement data for PVD-treated ground, and the results are compared with FEM findings. The results of lateral deformations of PVD-treated soft ground at various spacing are presented. It is observed that rate of settlements is higher in case of less spacing, and a gradual decrease in lateral displacements is observed with less spacing of PVD. Under embankment loading, PVD treatment of soft ground has a significant effect on the rate of lateral deformations.

References

1. Mehdizadeh A, Fakharian K (2015) Field instrumentation of a preloading project with prefabricated vertical drains. Ninth international symposium on field measurements in geomechanics FMGM Sydney, pp 453–464
2. Bergado DT, Manivannan R, Balasubramaniam AS (1996) Proposed criteria for discharge capacity of prefabricated vertical drains. Geotext Geomembr 14(9):481–505
3. Mahmoud SH, Ayman Lf, Yasser ME-M (2019) Application of PVD in soft soil for ground improvement. Int J Eng Appl Sci 6(10):68–77
4. Indraratna B, Sathananthan I, Rujikiatkamjorn C, Balasubramaniam AS (2005) Analytical and numerical modeling of soft soil stabilized by PVD incorporating vacuum preloading. Int J Geomech 5(2):114–124
5. Shui LS, Chai JC, Hong ZS, Caia FX (2002) Performance of embankments with and without reinforcement on soft subsoil. Can Geotech J 39(4):838–848
6. Radhakrishnan R (2011) Performance of pre-fabricated vertical drains & pre-loading for a major construction project. Proc IGC Kochi 357–360
7. Rajesh BG, Chukka SK, Dey A (2018) FEM of embankment resting on soft soil ground stabilized with PVD. SAGES
8. Terzaghi K (1943) Theoretical soil mechanics. Wiley, New York
9. Loganathan N, Balasubramaniam AS, Bergado DT (1993) Deformation analysis of embankment. J Geotech Eng 119(8)
10. Hird CC, Pyrah IC, Russell D, Cinicioglu F (1995) Modelling the effect of vertical drains in two-dimensional finite element analyses of embankments on the soft ground. Can Geotech J 32(5):795–807
11. Asaoka A (1978) Observational procedure of settlement prediction. Soils Found 18(4):87–101

Chapter 13
Sensitivity Study of the Pressure-Dependent Soil Model Based on the Abutment-Backfill Pushover Behaviour

Aritra Bagchi and Prishati Raychowdhury

Introduction

Bridge abutments are the essential part of the bridge system whose behaviour strictly relies on the characteristics of the backfill soil. When bridge is subjected to any monotonous or cyclic loading, the generated force gets transmitted to the abutments through the bridge deck. The movement of the wall thus creates a sufficient amount of lateral force into the backfill which gets opposed by the deformation in the backfill soil. After a certain amount of allowable lateral displacement, it will lead to the failure of the end support system of the bridge and reduces the overall stability of the bridge failure get reduced. Seismic anomalies can create a serious menace to the abutment-backfill system in the form of unseating of the deck from the abutment. Excessive earth pressure due to thermal expansion can also get transmitted to the deck and foundation through the bridge abutment [2, 9]. Proper constitutive modelling of soil can capture the deformation characteristics of the abutment. The relative influence of each parameter of the constitutive model can also be estimated via intensive numerical simulations. Generally, stiffness and maxi um allowable displacement are the two controlling parameters which define the hyperbolic nature of the curve. Hence, there is a need to investigate the force–displacement behaviour of the abutment-backfill system through numerical simulation which helps us to gain deeper knowledge about the failure mechanism of the system.

Experimental investigations along the numerical simulation of the tests were carried out by several researchers in the past to capture the abutment failure mechanism when subjected to static and dynamic loadings. Dicleli [3] developed a model of a typical integral bridge considering the nonlinear behaviour of abutment-backfill system in order to carry out static pushover analyses of the bridge under positive

A. Bagchi (✉) · P. Raychowdhury
Department of Civil Engineering, Indian Institute of Technology, Kanpur 208016, India
e-mail: aritra@iitk.ac.in

K. Muthukkumaran et al. (eds.), *Soil Dynamics, Earthquake and Computational Geotechnical Engineering*, Lecture Notes in Civil Engineering 300,
https://doi.org/10.1007/978-981-19-6998-0_13

thermal variations. Stewart et al. [7] tested a full-scale wall-backfill system for cyclic loading in order to observe the force–displacement behaviour of the abutment. Lemnitzer et al. [4] developed a full-scaled bridge abutment setup with granular backfill material to check the lateral performance and deformation pattern of the wall. Shamsabadi et al. [10] made an attempt to validate the UCLA and UCD abutment experimental results with log-spiral hyperbolic model (LSH) along with the detailed FEM models. Wilson and Elgamal [12] documented the passive earth pressure force–displacement relationship in two large-scale experiments and in finite element simulations. Tehrani et al. [11] presented closed-form relationships for lateral response of abutment backwalls with uniform backfills obtained by performing extensive parametric studies with a previously validated limit equilibrium model coupled with hyperbolic soil stress–strain relations.

Based on the UCLA abutment experiment, we develop a two-dimensional finite element model of the abutment-backfill setup. The pressure-dependent multi-yield material available in OpenSees platform is used to capture the constitutive behaviour of the dense silty sand placed behind the abutment wall in UCLA experiment. After calibration of the model parameters, the pushover curve for the corresponding abutment-backfill system is obtained via static pushover analysis which is compared with the result available in the literature [8, 10]. Further sensitivity analyses such as First-Order-Second-Moment (FOSM) analysis along with the preliminary sensitivity study (Tornado diagram) are carried out to examine the dependence of the maximum lateral force on the soil constitutive model parameters.

Numerical Modelling of UCLA Experiment

Brief Description of UCLA Experiment

University of California, Los Angeles (UCLA) researchers [8] performed a test on a seat-type abutment wall using a silty sand as a backfill material under monotonic and cyclic loading. The backfill was extended about sufficient times the wall height in the horizontal direction to eradicate the boundary effects. The incremental thrust was then applied horizontally between two wingwalls taking care of almost no vertical movement. Gapping at the wall-backfill interface was not allowed. Smooth plywood wrapped with plastic sheets was used in constructing wingwalls to minimize frictional force in the interior faces. This is done to achieve a plane-strain condition experimentally.

Modelling Aspects

Dimensions of the Domain: For a representative highway bridge, due to the greater extension of the backwall in the lateral direction compared to its height, the abutment-backfill-structure (ABS) interaction turns out to be a plane-strain problem. The extent of the backfill soil domain behind the abutment wall was chosen based on the fact where the differences observed in computed lateral load–displacement curve became close to negligible [10]. Although some increased dimensions were chosen in the study for conservative purpose. The right and left vertical boundaries were placed at sufficient distances from the wall to eradicate the boundary effects on the pushover curve. Figure 13.1 represents the two-dimensional finite element mesh configuration of the abutment wall-backfill system where H refers to the height of the abutment wall.

Modelling of the Abutment: The abutment back wall was modelled as an elastic beam-column element with three degrees of freedom (3DOF) per node. The high stiffness was taken for the wall element to avoid the influence of wall distortion on the resulted force–displacement curves. The bottom of the backwall was kept fixed in vertical direction in order to simulate the foundation resisting the vertical movement of the wall.

Modelling of the Soil Domain with Boundary Conditions: The soil domain was discretized into a finite element mesh which consists of four-nodded quadrilateral elements having two degrees of freedom (2DOF) per node. At the base boundary of the soil domain, both vertical and horizontal movements were restricted while at two lateral boundaries on either side of the model, the horizontal degree of freedom was fixed only allowing free to move vertically. The mesh size was selected on the basis of comparison with the theoretical vertical stress at low and higher depths

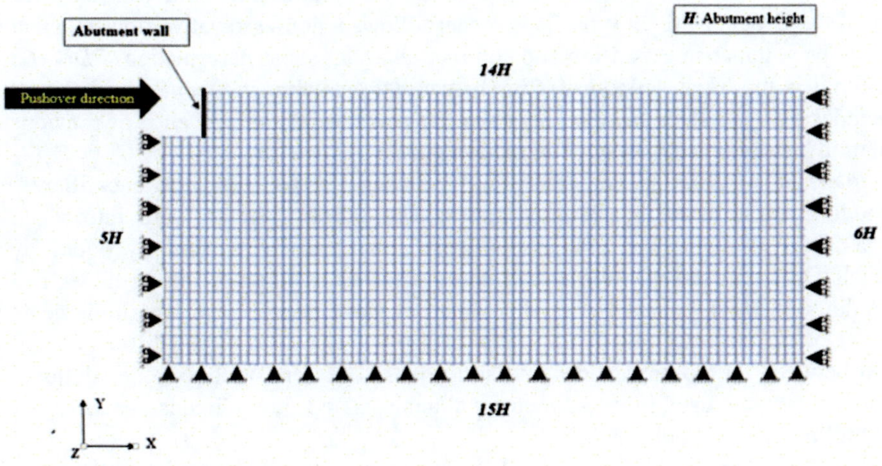

Fig. 13.1 2D FEM abutment-backfill system

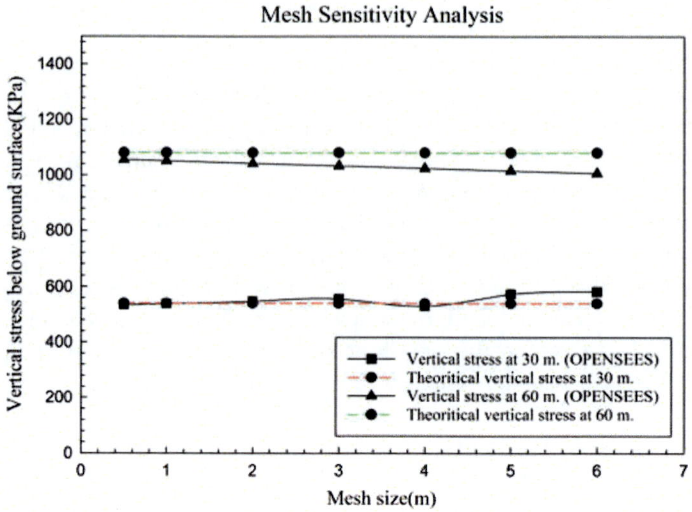

Fig. 13.2 FEM mesh sensitivity check

from the soil surface. Finally, the mesh size was chosen as 0.3 m by 0.3 m based on the observations where the fluctuation level reached to the minimal. The variation in computed vertical stress below the soil surface with mesh size is presented in Fig. 13.2.

Constitutive Modelling of Soil: The pressure-dependent multi-yield (PDMY02) soil constitutive material which was developed by Yang et al. [13, 14] and before that by Parra [5] is used in the context to model the behaviour of the silty sand backfill. It is an elastic–plastic material for simulating the essential response characteristics of pressure sensitive soil materials under general loading conditions. The yield surfaces are of the Drucker-Prager type. This model follows a non-associative flow rule that develops volumetric expansion and contraction under shear deformation. *PDMY02* material is modified version of *PDMY* material available in *OpenSees* library. It includes the K_σ effect and also parameter to account for the influence of previous dilation history on subsequent contraction phase.

Interface Modelling: The interaction between the wall and the backfill was modelled by introducing dummy nodes. The dummy nodes were introduced in a 2DOF domain. The wall nodes were connected to the dummy nodes by equalDOF command where the wall nodes acted as retained or master nodes and the dummy nodes behaved like constrained or slave nodes. Zero length elements (*zeroLengthContact2D* available in *OpenSees*) were used to establish the connection between the dummy nodes and soil nodes. They are node-to-node frictional contact elements used in two-dimensional analysis and three-dimensional analysis that follow Mohr–Coulomb law.

Stage-wise Analysis: The whole 2D FEM model of the abutment-backfill system was first analysed to establish the geostatic stress. During the application of geostatic

Table 13.1 Parameters used in FE simulation of the UCLA abutment test [10]

Properties	Value
Model used	Hardening soil model (available in PLAXIS)
Unit weight (ϒ) (KN/m^3)	20
Cohesion (c) (kPa)	14
Friction angle (ϕ) (degrees)	40
Poisson's ratio (υ)	0.3

stress, the soil was treated as a linear elastic material initially and subsequently considered as a plastic material. In the second step, the wall element was added to the soil domain using specific interface arrangements stated earlier. After applying the wall's self-weight, gravity analysis was done further to reach the equilibrium condition. In the third step, a active thrust was applied to the wall in a form of known rate of lateral displacement (static pushover) uniformly distributed along the height of the wall until peak lateral reaction occurred at certain lateral displacement. The resultant force–displacement curve was generated after post-processing the output data in Matlab 2017b.

Calibration Phase of the Soil Model: Shamsabadi et al. [8, 10] made an attempt to model the UCLA abutment test using *PLAXIS* [1] software package. The shear strength parameters were obtained by conducting the triaxial tests after obtaining the soil samples from the field. The cyclic triaxial tests were performed at different confining pressures to obtain the deviatoric stress versus axial strain (loading–unloading) curves. The properties of silty sand used as a backfill material along with the constitutive soil model variables are itemized down in Table 13.1.

In this study, the pressure-dependent soil model (*PDMY02*) parameters are calibrated to match the monotonic pushover curve available in the literature. The two parameters, i.e. cohesion (c) and friction angle (ϕ), are kept similar to that obtained from the triaxial tests. Reference mean effective confining pressure (P_{ref}) is selected as 100 kPa (default). A 3D finite element model simulating the cyclic consolidated-undrained (CU) test is developed in *OpenSees* complying proper boundary conditions. After applying the displacement history (loading–unloading) on the top surface of the model, the displacement-controlled analysis is performed and finally the deviatoric stress (kPa) versus the axial strain (%) is obtained after post-processing the output data. The reference elastic modulus (E_{ref}) is selected based on how close the output, i.e. deviatoric stress versus axial strain curve, resembles that of the triaxial ones. The calibrated parameters are also listed down in Table 13.2.

Finally, the numerically obtained force–displacement curve along with the monotonic pushover curve available in the literature is presented in Fig. 13.3 that shows a quite good resemblance ($R^2 = 0.987$).

Table 13.2 Calibrated PDMY02 constitutive model parameters (OpenSees)

Parameters	Value
rho (t/m^3)	2.0
Reference elastic modulus (kPa) (at $p_r = 100$ kPa)	2.38e5
Reference shear modulus (kPa) (at $p_r = 100$ kPa)	9.153e4
Reference bulk modulus (kPa) (at $p_r = 100$ kPa)	1.983e5
Poisson's ratio (υ)	0.3
Friction angle (ϕ) (degrees)	40
Phase transformation angle (degrees)	29
Reference pressure (p_r) (kPa) (default)	100
Pressure-dependent coefficient (d)	0.5
Cohesion (c) (kPa)	14

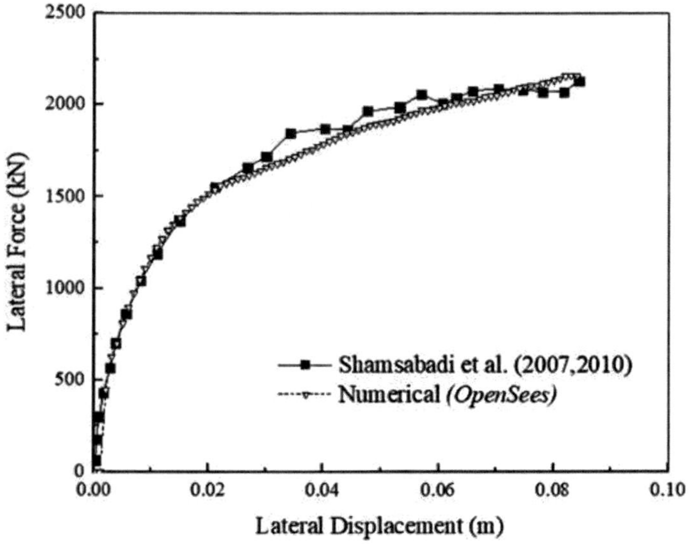

Fig. 13.3 Validation of numerical pushover curve ($R^2 = 0.987$)

Sensitivity Analysis of PDMY02 Model

Two different types of sensitivity studies, i.e. preliminary sensitivity study (expressed via Tornado diagram) and First-Order-Second-Moment (FOSM) method, were carried out to examine the influence of the considered constitutive model variables on the maximum lateral force demand. Six model parameters were considered for the sensitivity study such as the reference elastic modulus (E_{ref}), cohesion (c), density of soil (ρ), reference mean effective confining pressure (P_{ref}), friction angle (ϕ), and phase transformation angle (ϕ_{PT}). Reference elastic modulus was taken hereby instead of reference shear modulus and bulk modulus which are connected by Poisson's ratio (υ) through mathematical relationship.

Preliminary Sensitivity Study

In preliminary sensitivity study, the relative sensitivity of the parameters is obtained in terms of 'sensiscores' (Sandstrom 2016) and represented in a Tornado diagram in a descending order. For each parameter, a reference value, x_i, a maximum value, $x_{i,\max}$, and a minimum value, $x_{i,\min}$, are defined. A group of r criteria C_j, $C_{j+1} \ldots C_{r-1}$, C_r are also chosen. From the analysis, output values f_{Ci} ($x_{i,\max}$) and f_{Ci} ($x_{i,\min}$) are obtained with respect the criterion C_j. In the context, the criterion C_j was chosen as the maximum lateral force corresponding to the specific lateral displacement. The difference in the results from the maximum and minimum input values $\Delta x_{i,Cj}$ can then be expressed as:

$$\Delta x_{i,Cj} = \left| f_{Ci}\left(x_{i,\max}\right) - f_{Ci}\left(x_{i,\min}\right) \right| \tag{13.1}$$

The sum of $\Delta x_{i,Cj}$ for all input parameters with respect to C_j can be stated as:

$$\text{Sum}_j = \sum_{i=1}^{n} \Delta x_i, C_j \tag{13.2}$$

The relative sensitivity or 'sensiscore', α (x_i), describes the effect of varying the input variable x_i on the overall outcome for all criteria relative to the effect of the other input variables.

$$\alpha(x_i) = 100 \times \left(\frac{\Delta x_i, c_j}{\text{Sum}} + \frac{\Delta x_i, c_{j+1}}{\text{Sum}_{j+1}} + \cdots + \frac{\Delta x_i, c_{r-1}}{\text{Sum}_{r-1}} + \frac{\Delta x_i, c_r}{\text{Sum}_r} \right) \times \left(\frac{1}{n} \right) \tag{13.3}$$

A threshold value was chosen to be 2.5%. Finally, the relative importance of each input model parameter is represented via Tornado diagram in a descending order from left to right as shown in Fig. 13.4. It can be observed from the Tornado diagram that sensiscore for the three parameters (E_{ref}, c, and ρ) is found to exceed the considered threshold value, hence expected to have a significant influence on the output response compared to others.

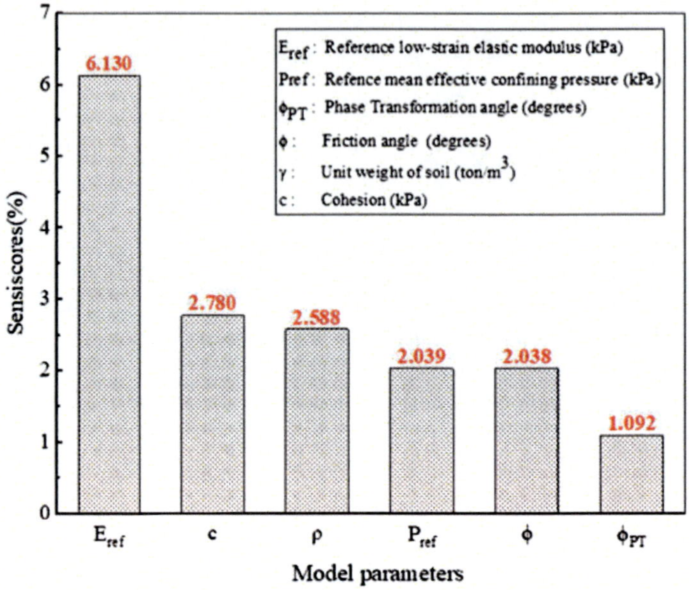

Fig. 13.4 Tornado diagram

First-Order Second Moment (FOSM) Analysis

In FOSM, all the model parameters are assumed to follow Gaussian probability distribution ('truncated') with all positive values possible and having a specific coefficient of variation (COV). Due to consideration of first order expansion, the correlation between any two variables is assumed to be negligible. Furthermore, the relationship between the response variables and the random variables is approximated to be linear or slightly nonlinear [6]. Let Y be a random variable representing the output demand parameter expressed in the following form:

$$Y = f(X_1, X_2 \ldots, X_n) \tag{13.4}$$

The stochastic variable X_i can be defined by its statistical properties, i.e. mean (μ_x) and variance ($\sigma^2{}_x$). Now, Y can be expanded using Taylor series with higher terms being neglected, and it can be written as:

$$Y \approx f(\mu_{X_1}, \mu_{X_2}, \ldots, \mu_{X_n}) + \sum_{i=1}^{n} (X_i - \mu_{X_i}) \frac{\delta f}{\delta X_i} \tag{13.5}$$

The mean μ_Y and variance ($\sigma^2{}_Y$) can be approximated as:

$$\mu_Y \approx f(\mu_{X_1}, \mu_{X_2}, \ldots, \mu_{X_n}) \tag{13.6}$$

$$\sigma_Y^2 \approx \sum_{i=1}^{n} \sigma_{X_i}^2 \left(\frac{\delta f(X_1, X_2, \ldots, X_n)}{\delta X_i} \right)^2 \tag{13.7}$$

In Eq. (13.7), the cross-correlation terms are neglected. The partial derivatives can be determined numerically using central difference method as follows:

$$\frac{\delta f(X_1, X_2, \ldots, X_n)}{\delta X_i} = \frac{f(X_1, \ldots, \mu_i + \Delta_{X_i}, X_n) - f(X_1, \ldots, \mu_i + \Delta_{X_i}, X_n)}{2\Delta X_i} \tag{13.8}$$

Simulations are performed changing each input parameter individually and keeping the other variables to its mean value for the approximate estimation of the partial derivatives as shown in Eq. (13.8). The FOSM sensitivity result is shown in Fig. 13.5 via bar diagram representing the importance of parameters. Each parameter following truncated Gaussian distribution was sampled randomly to establish the linear to slightly nonlinear relationship between input parameters and output, i.e. maximum lateral force. The statistical properties of the *PDMY02* constitutive model parameters considered for FOSM sensitivity analysis are given in Table 13.3.

The relative variance for each parameter can be calculated by the following expression.

$$\text{Relative variance (RV)} = \sigma_{Xi}^2/\sigma_Y^2 \tag{13.9}$$

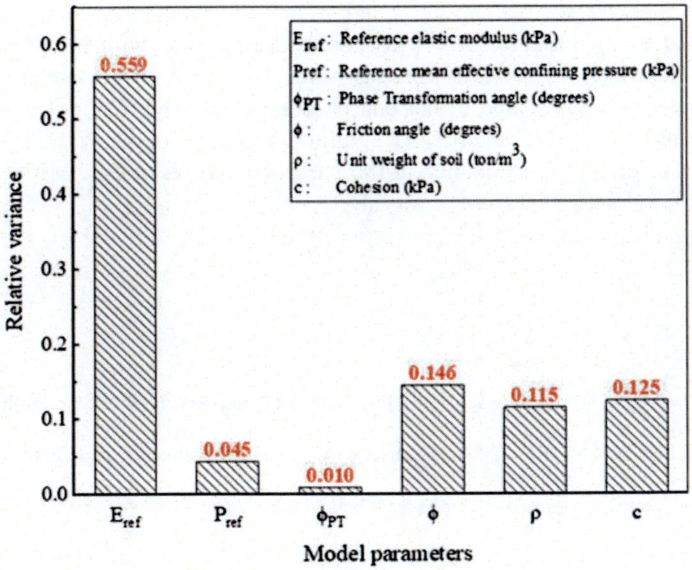

Fig. 13.5 Relative variance for maximum lateral force demand

Table 13.3 Statistical properties of considered *PDMY02* model parameters for FOSM sensitivity analysis

Parameters	Unit	Mean	Coefficient of variation (COV) (%)
Reference elastic modulus	kPa	2.38e5	30
Friction angle (ϕ)	degrees	40	30
Phase transformation angle	degrees	29	30
Reference pressure (p_r)	kPa	70	30
Cohesion (c)	kPa	14	30
Density (ρ)	ton/m^3	2	30

Summary and Conclusions

It can be observed that the effects of the parameters reference elastic modulus (E_{ref}) and phase transformation angle (ϕ_{PT}) are found to be maximal and minimal on the lateral force demand, respectively, where both the methods stated earlier agree well with each other. Also, the sensitivity of the lateral force demand to cohesion (c) and density (ρ) are quite high that can be observed by both analyses. The effect of friction angle (ϕ) is shown to be high in case of FOSM analysis compared to preliminary sensitivity analysis (Tornado diagram). Therefore, few conclusions can be drawn from the study that are stated as follows:

1. The dissimilarity that occurs is probably due to the approximate estimation of the partial derivatives used in calculating output variance and without considering the mutual dependence of the parameters in FOSM analysis.
2. Due to increase in E_{ref}, ϕ, and c individually, the maximum lateral force is observed to increase linearly with positive slope, but with increase in P_{ref}, decrease in maximum lateral force is noticed. A strong cross-correlation thus exists among E_{ref}, ϕ, and c that can be concluded. The relationship between elastic modulus and mean effective confining pressure for the *PDMY02* material follows inverse proportionality in the case of fixed reference elastic modulus while the reference pressure is varying.

References

1. Brinkgreve RBJ, Vermeer PA (1998) Plaxis manual version 7. Balkema Publisher, The Netherlands
2. Duncan JM, Mokwa RL (2001) Passive earth pressures: theories and tests. J Geotech Geoenviron Eng 127(3):248–257
3. Dicleli M (2005) Integral abutment-backfill behavior on sand soil-pushover analysis approach. J Bridg Eng 10(3):354–364
4. Lemnitzer A, Ahlberg ER, Nigbor RL, Shamsabadi A, Wallace JW, Stewart JP (2009) Lateral performance of full-scale bridge abutment wall with granular backfill. J Geotech Geoenviron Eng 135(4):506–514

5. Parra E (1996) Numerical modelling of liquefaction and lateral ground deformation including cyclic mobility and dilation response in soil systems. Ph.D. Dissertation, Department of Civil Engineering, RPI, Troy, NY
6. Raychowdhury P (2009) Effect of soil parameter uncertainty on seismic demand of low-rise steel buildings on dense silty sand. Soil Dyn Earthq Eng 29:1367–1378
7. Stewart JP, Taciroglu E, Wallace JW (2007) Full scale cyclic testing of foundation support systems for highway bridges. Part II: abutment backwalls. Department of Civil and Environmental Engineering University of California, Los Angeles
8. Shamsabadi A (2007) Three-dimensional nonlinear seismic soil abutment-foundation-structure interaction analysis of skewed bridges. PhD Dissertation, Department of Civil Engineering, University of Southern California
9. Shah BR (2007) 3D finite element analysis of integral abutment bridges subjected to thermal loading. MSc. Dissertation, Civil Engineering College of Engineering of Kansas State University, Manhattan
10. Shamsabadi A, Tehrani PK, Stewart JP, Taciroglu E (2010) Validated simulation models for lateral response of bridge abutments with typical backfills. J Bridge Eng 15(3)
11. Tehrani PK, Shamsabadi A, Stewart JP, Taciroglu E (2016) Backbone curves with physical parameters for passive lateral response of homogeneous abutment backfills. Bull Earthq Eng 14:3003–3023
12. Wilson P, Elgamal A (2010) Large-scale passive earth pressure load-displacement tests and numerical simulation. J Geotech Geoenviron Eng (ASCE) 136(12):1634–1643
13. Yang Z, Elgamal A, Parra E (2003) Computational model for cyclic mobility and associated shear deformation. J Geotech Geoenviron Eng 129(11):1119–1127
14. Yang Z, Lu J, Elgamal A (2008) OpenSees soil models and solid fluid fully coupled elements: user's manual. Department of Structural Engineering, University of California, San Diego

Chapter 14
Behaviour of Skirted Foundation Under Different Loading Conditions Using FEM Approach

Piyush Katariya and Mayur G. Vanza

Introduction

In present era, India is working on natural source of Renewable Energy (viz. wind energy, solar energy, etc.), so in that case there is a need to accommodate geotechnical perspective in line of design of foundation for the stability of these structures. One of the major problem for these structures like wind turbine is angle of rotation and differential settlement of foundation due to moment loading. When the combined load gets transferred to the soil below the footing, foundation rotates from its position due to moment loading.

Despite the fact that number of approaches for ground improvement are available and well-developed, still in certain site conditions where the loading requirements of structures are different, they are hard to apply to existing footing. For these type of situations, foundation with vertical skirts becomes a great unconventional approach for reducing the differential settlement. Foundation with skirts have been widely used for offshore structures like wind turbine because of simple installation as compare to pile foundation. Skirted foundation works on the principle that the soil between the skirts gets confined and thus the applied load transferred to the hard strata. Capacity to encounter tilt due to moment generated by wind load also gets enhanced by providing the circumferential skirt. Also, skirts helps to minimize the embedment depth and dimensions of the foundation. Shallow foundation with vertical skirts have been now used in structures for oil and gas industry.

To investigate the behaviour of foundation under combined loading, i.e., vertical and moment loading when the soil profile consists of layered soil, the conventional approach based on the effective footing area may underestimate/overestimate load carrying capacity. By considering this, a three-dimensional finite element modelling

P. Katariya (✉) · M. G. Vanza
Applied Mechanics Department, L. D. College of Engineering, Ahmedabad, India
e-mail: piyushkataria49@gmail.com

K. Muthukkumaran et al. (eds.), *Soil Dynamics, Earthquake and Computational Geotechnical Engineering*, Lecture Notes in Civil Engineering 300,
https://doi.org/10.1007/978-981-19-6998-0_14

approach is selected. Additionally, the findings are compared with non-skirted foundation.

Literature Study

Saleh and Alsaeid [1] performed laboratory investigation and numerical modelling for strip footing which shows that increasing the length of the skirt improve the load-settlement behaviour of foundation. The rate of enhancement in load carrying capacity increases with the increase of angle of inclination of load and reached its peak value while skirt length equal to 0.5B. Eid [2] clearly presented that with increase in length of skirts and with decrease in relative density of soil the improvement in bearing capacity of shallow foundation increases. Bhatkar and Puduskar [3] analysed the effect of raft foundation with one-sided and two- sided skirts using FEA which indicate that for raft with two sided skirts improvement in bearing capacity increases with skirts depth and settlement decreases. For raft with one-sided skirt, both bearing capacity and settlement increases with increase in depth and width of raft. Sharma and Chetia [4] studied the behaviour of raft foundation provided at surface with and without skirts for two different soil model, i.e., hardening soil model and Mohr-coulomb model using a finite element software. It was found that with the increase in depth of skirts, the bearing capacity increases, and settlement decreases for both the model. Kumar and Dutta [5] carried out plate load tests on three different soil which describes that providing foundation with double skirt proved more beneficial in comparison with the singly skirted footing to reduce the settlement and increasing the bearing capacity. Santosh Kumar and Ghosh [6] presented a plasticity model based on slip line theory to evaluate the bearing capacity enhancement by providing skirts to the foundation for cohesion-less soil under the effect of seismic waves. The results showed the lesser effect in bearing capacity during earthquake because of skirts provision.

The literature shows that by provision of circumferential skirts below square, strip and circular footing, the soil beneath the footing and between the skirts gets confined and thus improves the bearing capacity and reduces the settlement of foundation under vertical as well as inclined loading. In order to give an optimum solution for octagonal gravity base raft foundation with skirt and to provide cost-effective foundation, there is need to study of skirted raft foundation with different parametrical aspect and for different loading combination. Keeping in view this aspect, this work lead to study parametric effect on behaviour of skirted raft foundation with structural skirt using PLAXIS 3D software.

Numerical Modelling and Material

Definition of the Problem

An extensive three-dimensional FEM analysis is carried out using PLAXIS 3D software to study the effects of skirts to counter moment loading in addition to axial loading. To be able to calculate the bearing capacity of the soil some parameters have to be set. These are the soil properties of the soil considered for analysis, material properties and geometry of foundation, material properties of skirts and also the forces (i.e., vertical and moment load with respect to self-weight and wind conditions) acting on the foundation.

Soil Condition

The geometry of the finite element soil model taken for the analysis is 100 × 100 × 40 m. The three soil layers shown in Fig. 14.1 were added to the borehole and their respective material properties listed in Table 14.1 were assigned. The soils were modelled as hardening soil model. The key input parameters for the hardening soil model are unsaturated unit weight, saturated unit weight, Young's modulus, Cohesion, Angle of Internal Friction.

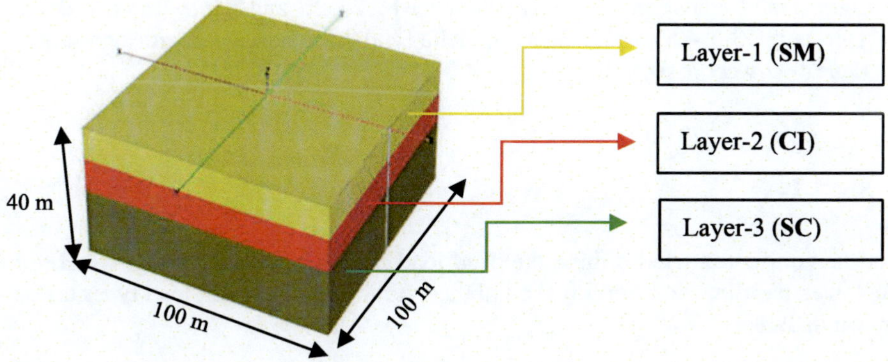

Fig. 14.1 Soil profile

Table 14.1 Soil model properties

Properties	SM	CI	SC	Unit
h	0.0–10.0	10.0–15.0	15.0–40.0	M
Material model	Hardening soil model			–
Material behaviour	Drained			–
Dry density $\gamma_{(d)}$	15.80	16.70	16.80	KN/m^3
Saturated density $\gamma_{(s)}$	17.60	20.20	19.70	KN/m^3
Secant stiffness (E_{50})	12,500/15,000/17,500	16,800	22,000	KN/m^2
Poisson's Ratio (υ)	0.2	0.3	0.2	–
Cohesion (C')	0.01	2.32	0.14	Kg/cm^2
Friction angle (φ')	27/29/31	5	29	°

Table 14.2 Foundation material properties

Properties	Foundation	Skirts	Unit
Material	Elastic		–
Unit weight γ	25.00	78.50	KN/m^3
Young's modulus (E)	30×10^6	210×10^6	KN/m^2
Poisson's ratio (ν)	0.15	0.30	–

Footing Geometry and Material Properties

Octagonal raft foundation of different diameter 16, 24, 26, and 30 m with skirts depth varying as 1.5, 3.0, 4.5, and 6.0 m are modelled and the material properties presented in Table 14.2 were assigned.

Load Cases

For the considered wind turbine the dead load is 51.71 MN consists of weight of the tower, nacelle, and rotor. Final combined vertical and moment load types are as shown in Table 14.3.

Table 14.3 Combined loads applied in FEM analysis

Loading conditions	I	II	III	Unit
Vertical load (V)	51.71			MN
Moment load (M)	$M_1 = 155.93$	$M_2 = 180.84$	$M_3 = 207.60$	MN m

Finite Element Modelling and Meshing

Figure 14.2 clearly shows the typical finite element model for the skirted footings on the layered soil. The modelling in PLAXIS is to be done under four stages. In stage-1, the soil boundary and soil properties are to be specified. Then, in next stage, the geometry and properties of foundations are to be implied. The foundation surface and skirts are to be modelled as plate element. In third stage, the meshing is to be done. There are five meshing patterns (i.e. very coarse, coarse, medium, fine and very fine mesh), thus allowing the programmer to refine an area, line or point further. Figure 14.3 displays the standard mesh attained for the finite element model. It is observed that very fine meshing requires too much computation time to identify the optimal configuration of mesh for any model. So, medium type of mesh is used in this work as shown in Fig. 14.3. In the last stage of modelling the loading conditions are to be specified as per staged construction because when the plasticity of soil gets involved in finite element analysis, the equations which are going to be solved becomes nonlinear to accommodate that problem needs to solve in a series of calculation steps.

Result Analysis and Discussion

By using above modelling and soil properties total forty eight numbers of models are analysed in PLAXIS 3D for varying loading conditions, length of skirts, diameter of foundation and for different strength and stiffness parameters of soil for both skirted and non-skirted octagonal raft foundation.

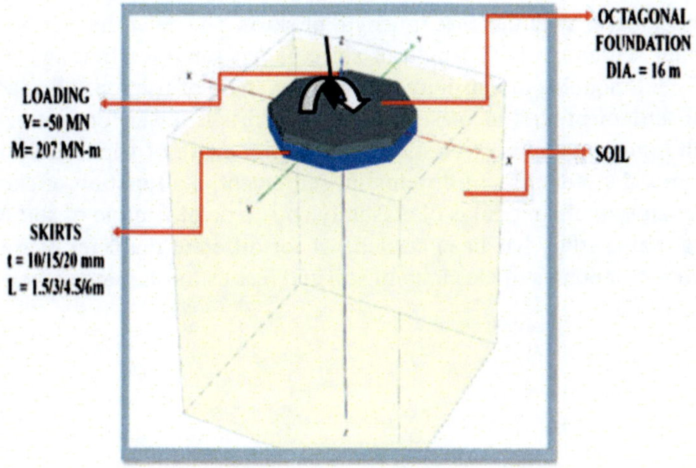

Fig. 14.2 Foundation model with skirts

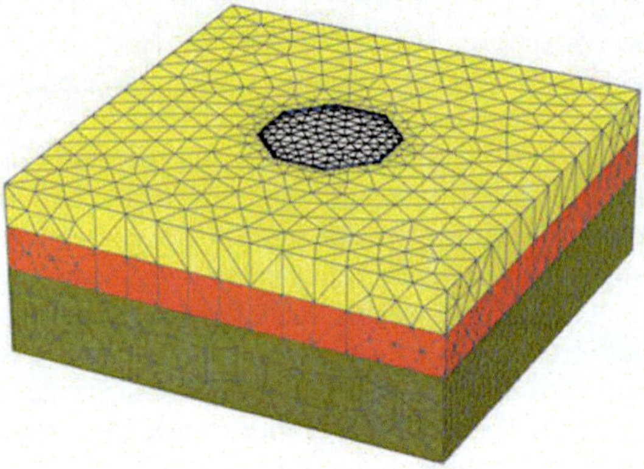

Fig. 14.3 3D view of the meshed model for soil with medium mesh

A quantitative analysis of observed effect of length of skirts in magnitude of settlement and load carrying capacity of foundation with variables is presented in the following subsections:-

Effect of Length of Skirts on Differential Settlement Due to Combined Loading

The failure pattern., deformation pattern and the variation of settlement is plotted against distance along the diameter of foundation for a typical 16 m dia. footing and for different eccentric loading and for varying length of skirts 1.5, 3.0, 4.5, and 6.0 m in Figs 14.4, 14.5, 14.6, 14.7, 14.8 and 14.9, respectively. The differential settlement decreases with increase in length of skirts. However the rate of decrease in differential settlement with length of skirts is not same for footings resting on soil of different modulus of elasticity and varying angle of friction. The decrease in differential settlement with increase in length of skirts is lower for footings resting in soil with higher modulus of elasticity and higher angle of friction. So it can be concluded that the reduction of differential settlement is a function of skirt length and also depends on the modulus of elasticity and angle of friction of soil on which it is resting. Same study has been carried out for different diameter of foundation and for different modulus of elasticity of soil and results for the same given in Table 14.4 (Fig. 14.10).

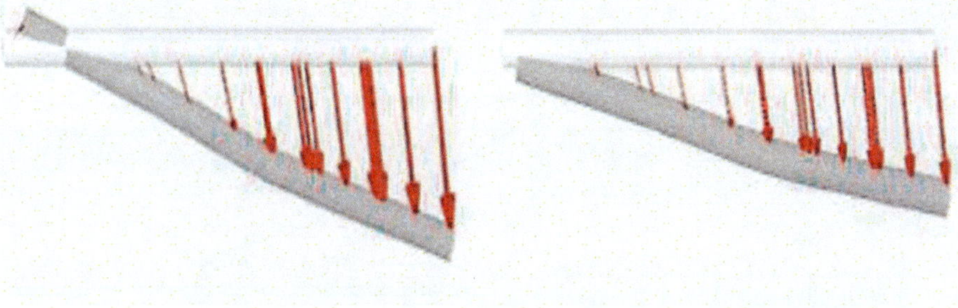

a.) Non-Skirted b.) Skirted

Fig. 14.4 Deformation pattern of foundation for combined loading condition

Fig. 14.5 Failure pattern of non-skirted foundation for configuration with soil layer $E = 12{,}500$ KN/m^2, Dia. of octagonal foundation = 16 m under combined loading of ($V = 51.71$ MN and $M3 = 207.60$ MN m) 16 m

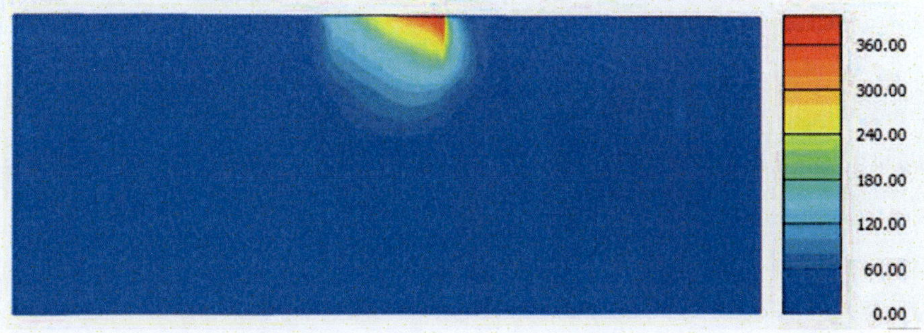

Fig. 14.6 Failure pattern of skirted foundation (length of skirts 6.0 m) for configuration with soil layer $E = 12{,}500$ KN/m^2, Dia. of octagonal foundation = 16 m under combined loading of ($V = 51.71$ MN and $M3 = 207.60$ MN m)

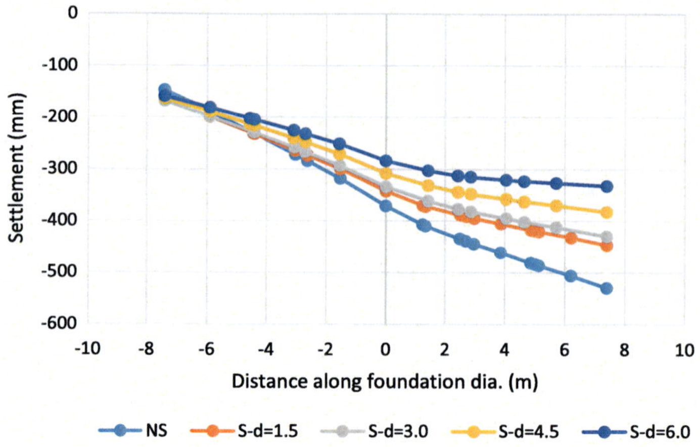

Fig. 14.7 Settlement of foundation plate along the distance of diameter of foundation for E = 12,500 KN/m^2, Dia. of octagonal foundation = 16 m under combined loading of (V = 51.71 MN and $M1$ = 155.93 MN m)

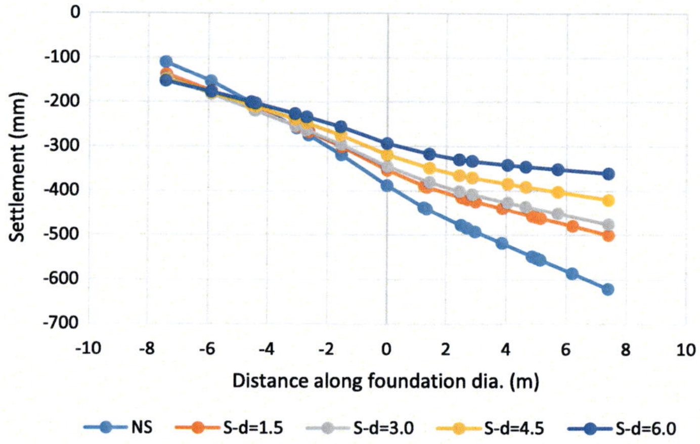

Fig. 14.8 Settlement of foundation plate along the distance of diameter of foundation for E = 12,500 KN/m^2, Dia. of octagonal foundation = 16 m under combined loading of (V = 51.71 MN and $M2$ = 180.84 MN m)

Effect of Length of Skirts on Settlement Due to Vertical Loading

Displacement pattern for non-skirted and skirted foundation under constant vertical load is shown in Fig. 14.10 where skirts provided with a length of 6.0 m. It is clear from pattern that by the inclusion of skirts 40 mm of settlement decreased and study describes this reduction is effective after 3.0 m length of skirt. Typical load-settlement

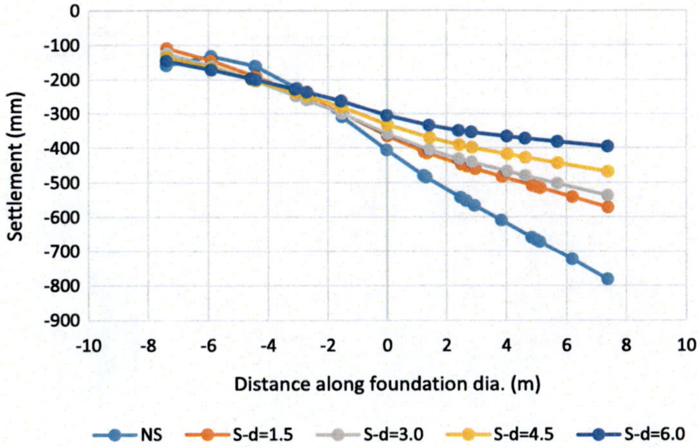

Fig. 14.9 Settlement of foundation plate along the distance of diameter of foundation for E = 12,500 KN/m^2, Dia. of octagonal foundation = 16 m under combined loading of (V = 51.71 MN and $M3$ = 207.60 MN m)

Table 14.4 Reduction in differential settlement of foundation for E = 12,500 KN/m^2 and Dia. of octagonal foundation = 16 m under different combined loading

Loading condition	Foundation	Length of skirts (m)	Diff. settlement (mm)	Reduction in settlement (mm)	Reduction (%)
$V + M_1$	Non-Skirted	0	382	–	–
	Skirted	1.5	280	103	27
		3	261	121	32
		4.5	219	163	43
		6	173	209	55
$V + M_2$	Non-Skirted	0	511	–	–
	Skirted	1.5	363	148	29
		3	329	182	36
		4.5	270	241	47
		6	208	303	59
$V + M_3$	Non-Skirted	0	621	–	–
	Skirted	1.5	462	159	26
		3	412	209	34
		4.5	332	289	47
		6	248	373	60

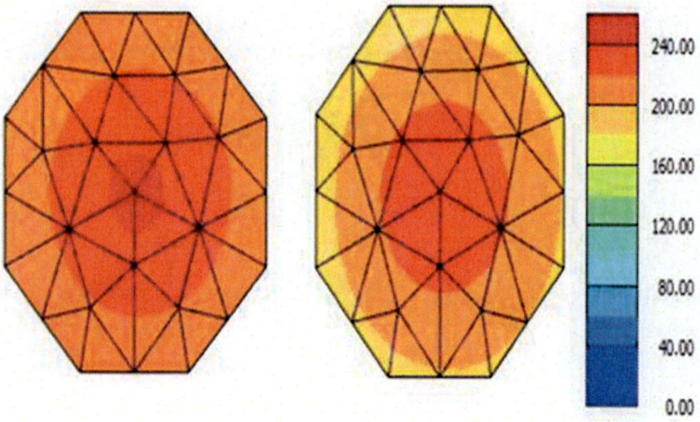

Fig. 14.10 Displacement pattern for non-skirted and skirted foundation of dia. 16 m and length of skirts 6.0 m

curves for non-skirted and skirted octagonal raft foundation with length of skirts in the range of 1.5 to 6.0 m are shown in Figs. 14.12, 14.13, and 14.14, respectively, with constant vertical load of 51.71MN. Results obtained from the analysis revealed that provision of skirts reduce the settlement due to vertical load as shown in Table 14.5. The reduction in settlement increases with increasing the length of skirts and decreases with increasing the value of modulus of elasticity and angle of friction of soil. Though comparing the skirted foundation with non-skirted, it is revealed that settlement reduces not in a major way by inclusion of skirts upto 3.0 m of length. To check this criteria study was carried out for various diameter of foundation.

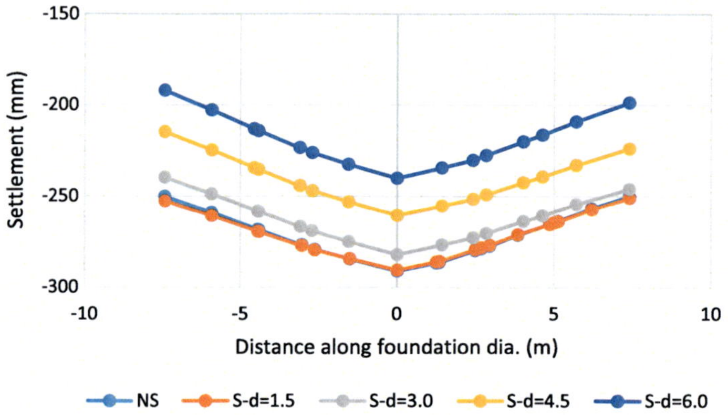

Fig. 14.11 Settlement of foundation plate along the distance of diameter of foundation for E = 12,500 KN/m^2, Dia. of octagonal foundation = 16 m under vertical loading of (V = 51.71MN)

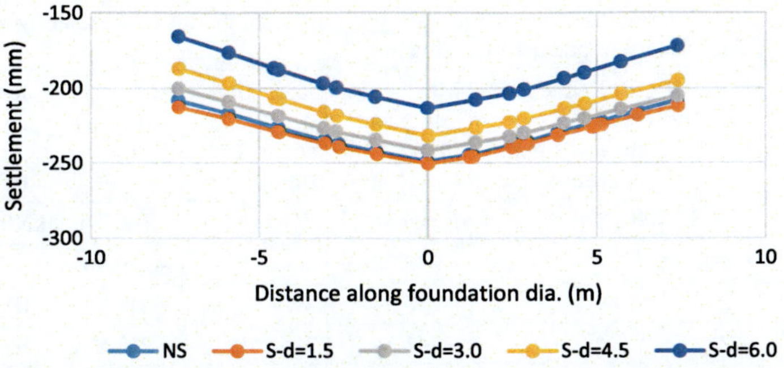

Fig. 14.12 Settlement of foundation plate along the distance of diameter of foundation for $E = 15{,}000$ KN/m^2, Dia. of octagonal foundation = 16 m under vertical loading of ($V = 51.71$MN)

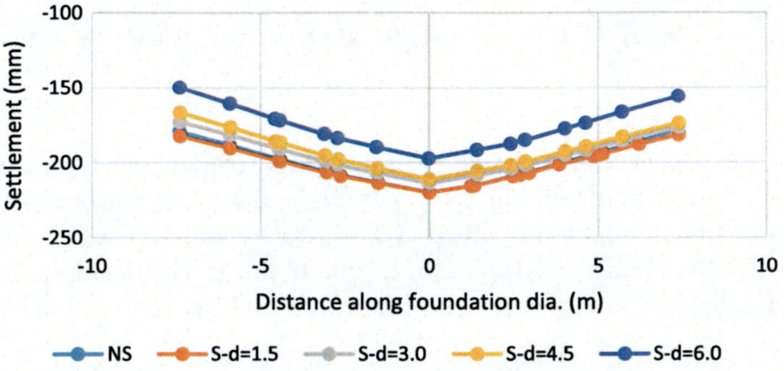

Fig. 14.13 Settlement of foundation plate along the distance of diameter of foundation for $E = 17{,}500$ KN/m^2, Dia. of octagonal foundation = 16 m under vertical loading of ($V = 51.71$MN)

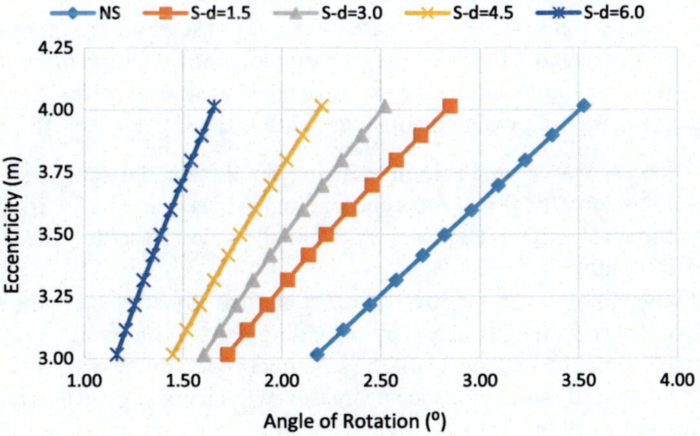

Fig. 14.14 Angle of rotation for varying eccentricity with increase in length of skirts for 16 m diameter of foundation

Table 14.5 Reduction in settlement of foundation for different modulus of elasticity and Dia. of octagonal foundation = 16 m under vertical loading

Length of skirts (m)	Max. settlement (mm)	Reduction in settlement (mm)	Max. settlement (mm)	Reduction in settlement (mm)	Max. settlement (mm)	Reduction in settlement (mm)
	$E = 12{,}500$ kN/m^2		$E = 15{,}000$ kN/m^2		$E = 17{,}500$ kN/m^2	
0.00	−291.02	–	−248.98	–	−219.90	–
1.50	−290.55	0	−250.55	2	−219.93	0
3.00	−282.05	9	−241.74	7	−213.67	6
4.50	−260.64	30	−231.97	17	−210.99	9
6.00	−240.09	51	−213.35	36	−197.17	23

Effect of Length of Skirts on Angle of Rotation with Respect to Eccentricity

The variation of eccentricity is plotted against angle of rotation for a typical octagonal raft foundation of 16 m dia. in Fig. 14.4. The angle of rotation has been determined for each moment loading. In this situation, also a related observance was seen same as in case of differential settlement with length of skirts. The decrease in angle of rotation is found to be more with increase in length of skirts as shown in Table 14.6.

Conclusion

The main objective of this study was to assess the response of octagonal raft foundation with and without skirts for different eccentric loading condition through a series of finite element modelling and their respective analysis. Based on the data obtained from the analysis, the following conclusions can be made.

- Length of skirts play a vital role in reducing the differential settlement induced due to combined vertical and moment loading. As the length of skirts increase, the differential settlement decreases. And, the reduction of settlement seen in the range of 15 to 60%.
- The settlement due to vertical load does not get majorly affected by inclusion of skirts upto 3.0 m in length. Also, by providing skirts with length varying after 3.0 m upto 6.0 m, the improvement in settlement reduction was seen.
- Angle of rotation of foundation also decreases with increasing the length of skirts. The reduction of 50 to 75% in angle of rotation has been observed. For 16 m diameter of foundation, reduction in angle of rotation of 47% was observed by providing 6.0 m length of skirts for maximum eccentricity of 4.01 m in this study

Table 14.6 Reduction in angle of rotation of foundation plate with increase in length of skirts for 16 m diameter of foundation with respect to varying eccentricity

Skirts length(m)	Eccentricity (m)	Difference settlement (mm)	tan (β)	Angle of rotation β (°)
0.0	3.02	−281.53	−0.04	2.18
	3.50	−364.69	−0.05	2.82
	4.01	−455.75	−0.06	3.53
1.5	3.02	−222.59	−0.03	1.72
	3.50	−287.02	−0.04	2.22
	4.01	−367.98	−0.05	2.85
3.0	3.02	−207.12	−0.03	1.61
	3.50	−260.14	−0.04	2.02
	4.01	−325.74	−0.04	2.52
4.5	3.02	−186.67	−0.03	1.45
	3.50	−230.64	−0.03	1.79
	4.01	−283.98	−0.04	2.20
6.0	3.02	−150.34	−0.02	1.17
	3.50	−178.94	−0.02	1.39
	4.01	−214.17	−0.03	1.66

- It is also observed that as the strength and stiffness parameters like modulus of elasticity and angle of internal friction of sub soil increases, the effectiveness of skirts decreases.

References

1. Saleh NM, Alsaeid AE (2008) Performance of skirted strip footing subjected to eccentric inclined load. Electronic J Geotech Eng 13
2. Eid HT (2013) Bearing capacity and settlement of skirted shallow foundation on sand. Int J Geomech 13
3. Bhatkar T, Puduskar S (2014) Behavior of raft foundation with vertical skirt using plaxis 2d. Int J Eng Res Dev
4. Sharma K, Chetia N (2016) Load-displacement behaviour of skirted raft foundations on sand using PLAXIS 2D. Int J Adv Eng Res Sci
5. Kumar A, Dutta RK (2020) A study on bearing capacity of skirted square footings on different sands. Indian Geotech J
6. Santosh Kumar G, Ghosh P (2020) Ultimate bearing capacity of skirted foundation on cohesionless soil using slip line theory. Comput Geotech
7. Aghbari A, Dutta RK (2008) Performance of square footing with structural skirt resting on sand. Geomech Geoeng Int J
8. Randolph MF, Bransby MF (1998) Combined loading of skirted foundations. Geotechnique 48:637–655

Chapter 15
Numerical Modelling for Prediction of Ground Subsidence Over Room and Pillar Mining in an Underground Coal Seam

Nisheeth Shekhar, Supriya Pal, Arpan Jaiswal, and Pratip Hazara

Introduction

Coal is the most essential and abundant fossil fuel in India. It accounts for 55% of the country's energy needs. Commercial primary energy consumption in India has grown by about 700% in the last four decades. India's current per capita commercial primary energy consumption is about 350 kg/year (kilograms of oil equivalent per year) (https://coal.nic.in/). Jharkhand, Odisha, Chhattisgarh, West Bengal, Madhya Pradesh, Telangana, and Maharashtra are the major coal-producing states in India (Geological Survey of India). India is the second-largest producer and consumer of coal worldwide, just after China (International Energy Agency, IEA). Among all the mining techniques, the underground mining method dominates the world's opencast method, as more than 60% of the total coal production in the world is extracted from underground mines. However, it acquires less than 10% of overall coal production (Coal Directory of India, 2016–17).

Room and pillar's (R&P) method of underground mining is the most common method used in India. In this extraction method, untouched coal 'pillars' of large dimensions are left to support the overlying strata leaving open areas, i.e. 'rooms and galleries' for the mobility of workers and various mining equipment. The key to successful room and pillar mining is selecting the optimum pillar size. If the pillars are too small, the mine would collapse. If the pillars are too large, then significant quantities of valuable material would be left behind, reducing the profitability of the

N. Shekhar (✉) · S. Pal · A. Jaiswal · P. Hazara
Department of Civil Engineering, National Institute of Technology, Durgapur, West Bengal 713209, India
e-mail: pratik.96.ns@gmail.com

S. Pal
e-mail: supriya.pal@ce.nitdgp.ac.in

K. Muthukkumaran et al. (eds.), *Soil Dynamics, Earthquake and Computational Geotechnical Engineering*, Lecture Notes in Civil Engineering 300,
https://doi.org/10.1007/978-981-19-6998-0_15

mine. The percentage of material mined varies depending on many factors, including the material mined, the height of the pillar, and roof conditions.

The total coal production from U.G. mines in India is around 80% by the R&P method only. Different types of support systems are used to make a safe working condition for the workers. But due to some geomining constraints, only 30–35% recovery of coal is made possible by using the R&P method (developed state), and also a massive amount of energy resources (coal or any other minerals) are left behind in the form of pillars of immense dimensions [1].

Depillaring is coal extraction from pillars and is used, as vast quantities of minerals are left in those intact pillars. It can be done either with (i) caving or (ii) stowing. In depillaring with caving, the roof of the panel is allowed to collapse into the voids or caved area known as goaf. As roof strata are allowed to collapse, the ground surface develops cracks and subsides to a greater extent. In depillaring with stowing, the goaf is completely packed with some incombustible materials like river sand and pond ash. After depillaring for mineral extraction from an underground mine, it is the safest method as the probable subsidence is reduced to the greatest extent.

Despite all precautionary measures, surface subsidence due to underground mining activity is an unavoidable problem as soon as a cavity is formed to extract ores. 70 to 90% of the subsidence occurs immediately, and 10 to 30% occurs in a period of 300 to 500 days for an undisturbed rock mass, whereas for a disturbed rock mass, about 90 to 95% of subsidence occurs during excavation and 5 to 10% occurs in a period of 50 to 100 days [2]. Therefore, the forecasting of surface subsidence due to underground mining activities through numerical modelling techniques would be highly beneficial to the mining authorities in protecting surface properties and, most importantly, to increase the safety of miners [3].

Keeping this in view, the present study was undertaken to assess the quantum of surface subsidence over an underground coal mine at Raniganj coalfield area, Paschim Bardhaman District, West Bengal, India, through the application of numerical modelling techniques. The different cases such as pillar formation, depilating, and stowing were considered in numerical modelling through Plaxis 3D FEM software to assess the ground displacement and probable surface tensile crack formation.

Site Details

In this study, an underground mine in the Raniganj coalfield area, Paschim Bardhaman District, West Bengal, India, has been chosen, which is proposed to be developed by room and pillar mining method. The block lies between the latitudes 23°-38′-10″ to 23°-39′-45″ E and longitudes of 87°-12′-00″ to 87°-14′-10″.

The total area of the mine geological block is 5.85 sq km. The project area of about 2.54 sq km is confined between fault $F1$-$F1$, fault $F3$-$F3$, fault $F4$-$F4$, and the colliery leasehold boundary. The mine area is roughly rectangular, bounded by $F1$-$F1$, $F3$-$F3$, $F10$-$F10$, and $F4$-$F4$, as shown in Fig. 15.1.

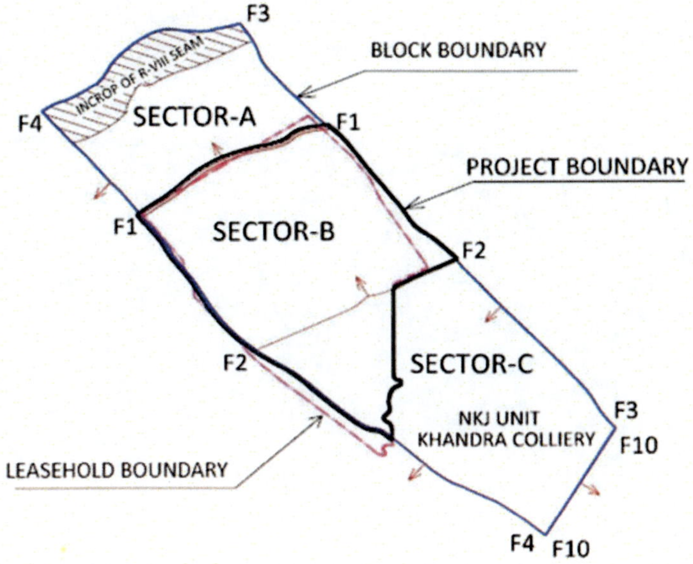

Fig. 15.1 Mine geological block

The roof of a coal seam, considered for subsidence prediction, is located at a depth of 121 m below the ground level. Panel 5 of the seam (as shown in Fig. 15.2) was considered for modelling in the present study. Square-shaped pillars of size 23 × 23 m were made during the development stage. Gallery of width 6 m was made for smooth mobility of mining equipment. The overburden consists of various layers of sandstones and shales with the top 4 m of silty sand. The considered coal seam has a thickness varying from 2.68 to 4.5 m. However, the height of the panel was considered 2.7 m thick for numerical modelling. Figure 15.3 shows the designed plan of the room and pillar developed panel used in numerical modelling.

Numerical Modelling

FEM (Finite Element Method) makes it possible to discretize the structure into single continuous elements. It takes into account the geometry of the structure, boundary conditions, and initial conditions. FEM or FDM-based software is a powerful computational tool for solving many engineering problems in the current scenario. In the present study, Plaxis 3D (FEM-based software) was used for numerical simulation of subsidence prediction over the panel as even the complex topography and geostrata of the site can easily be simulated using this software. The program uses a convenient graphical user interface that enables users to quickly generate an accurate three-dimensional finite element mesh based on a composition of horizontal cross-sections at different vertical levels.

Fig. 15.2 Plan of developed panel 5

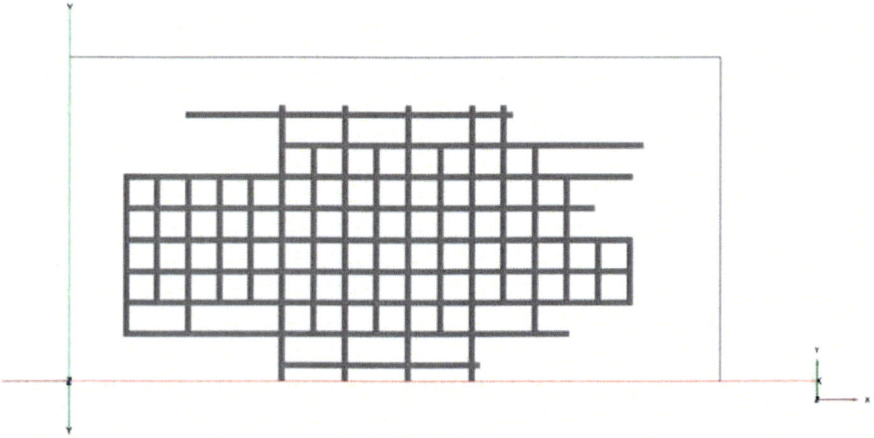

Fig. 15.3 Plan of R&P developed panel (white portion: coal; grey portion: voids)

All the overlying layers and coal seams were considered over the panel in numerical modelling in all the cases of pillar development and depillaring with simultaneous stowing. The model geometry is chosen so that it is closest to the actual geometry mine. A rectangular panel of area 600 × 300 m is considered for subsidence prediction. The linear elastic material model is selected for the overburden layers. Strength parameters of the soil layers used in the numerical modelling are given in Table

15.1. Strength parameters of the fill material (dense sand) used for stowing during depillaring are given in Table 15.2.

Medium quality of meshing was used to generate finite elements for the model. The elements generated by the software for the soil volume are 10-noded tetrahedral elements. Local mesh refinement is done at the surface, and the layer present above the coal seam is considered in the analysis. The mesh generated and connectivity plot of the model are shown in Fig. 15.4.

Table 15.1 Geotechnical parameters of soil layers used in the numerical modelling

Type of soil	*E* (MPa)	Poisson's ratio (ν)	γ_{unsat} (K N/m^3)	γ_{sat} (K N/m^3)
Laterite	710	0.35	20.6	21.2
Alluvium	890	0.31	20	20.5
Loose sand	16	0.3	13.5	16
Sandstone	31.2 e3	0.305	26.1	26.5
Shale	61 e3	0.31	25.4	25.85
Coal	91	0.4	16.85	17.2

Table 15.2 Geotechnical parameters of fill material used

Geotechnical parameters	Value
E (MPa)	20
Poisson's ratio (ν)	0.4
γ_{unsat} (K N/m^3)	19
γ_{sat} (K N/m^3)	20

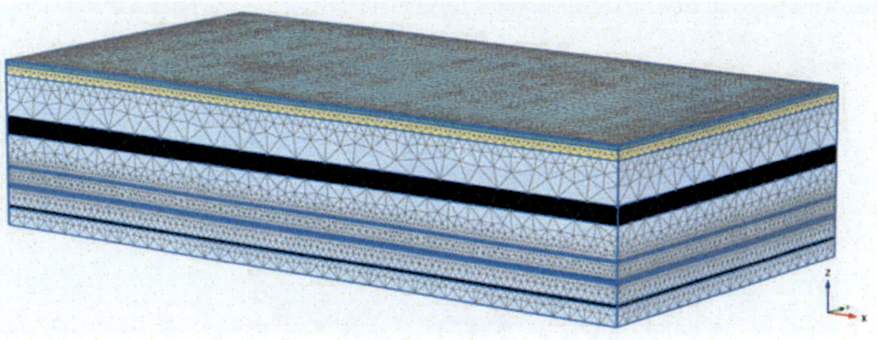

Fig. 15.4 Generated mesh (connectivity plot)

Subsidence Prediction

Subsidence is an inevitable consequence of underground mining because of the voids created due to coal extraction, so the movement of overlaying layers is simulated in different phases.

The numerical simulation for subsidence prediction is done in the following phases:

1. Initial phase: In-situ stresses of the model are considered.
2. Phase 1: Subsidence calculated for the developed panel.
3. Phase 2 (after phase 1): Complete depillaring with caving in the central portion of the panel.
4. Phase 3 (after phase 1): Complete depillaring with simultaneous stowing with a fill material.

Initially, in phase 1, the numerical analysis is carried out for a thoroughly developed panel where the coal pillars are left intact to support the overlying strata. Surface subsidence and coal seam roof movement are calculated.

Further, to increase coal recovery from a panel, complete depillaring is to be done by mining authorities which are analysed by both caving and simultaneous stowing with a filler material (dense sand). Thus, in phase 2, the complete removal of pillars from the central portion of the panel is analysed as an underground cavity. Figure 15.5 shows the cavity formed in the central portion of the panel after complete depillaring.

In phase 3, complete depillaring with simultaneous stowing with a fill material in the cavity is analysed. This is done to support the overlying strata and minimize the subsidence effect at the ground surface. Figure 15.6 shows the panel completely stowed with a fill material in the voids (the stowing material is red). This phase is

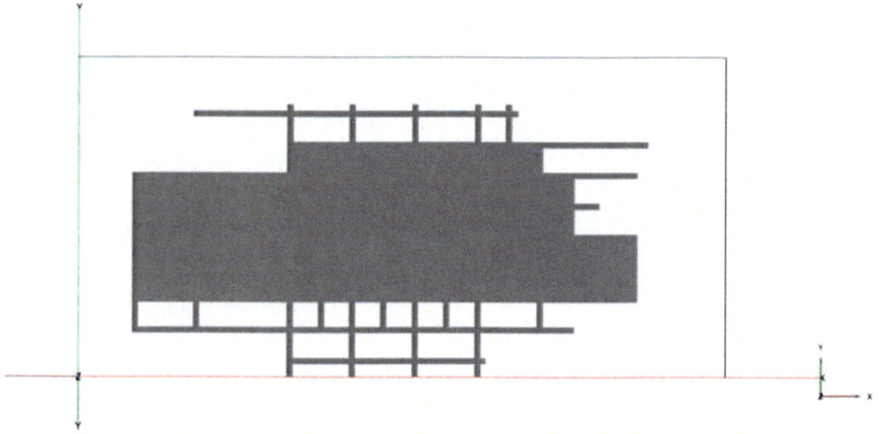

Fig. 15.5 Plan showing cavity formed after complete depillaring

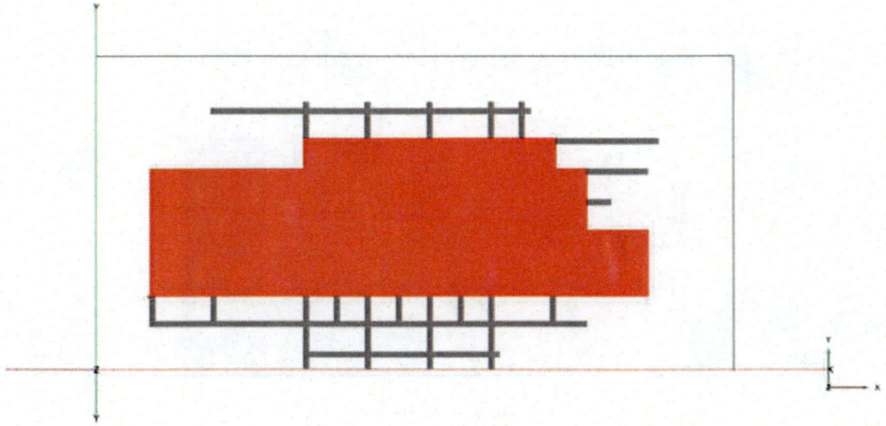

Fig. 15.6 Plan of a panel with stowed material (red portion: stowed material)

executed after phase 1 so that subsidence for both conditions can be compared easily (for phase 2 and phase 3).

Result and Discussion

Figures 15.7 and 15.8 show the vertical displacements shading the ground surface and seam roof, respectively, for phase 1.

Figures 15.9 and 15.10 show the vertical displacements shading the ground surface and seam roof, respectively, for phase 2.

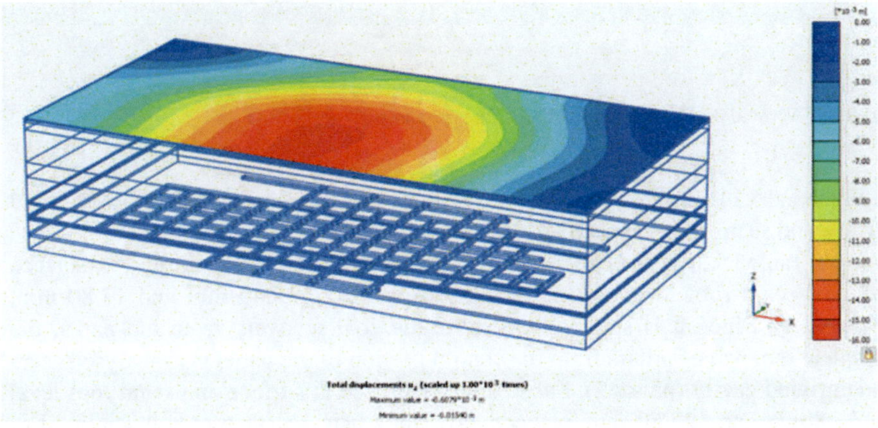

Fig. 15.7 Vertical displacements shading at the ground surface of the developed panel

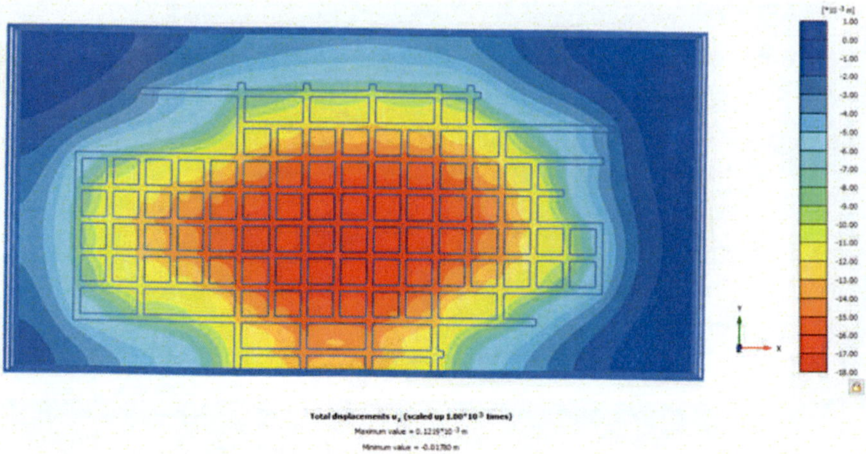

Fig. 15.8 Vertical displacements shading at seam roof level of the developed panel (top view)

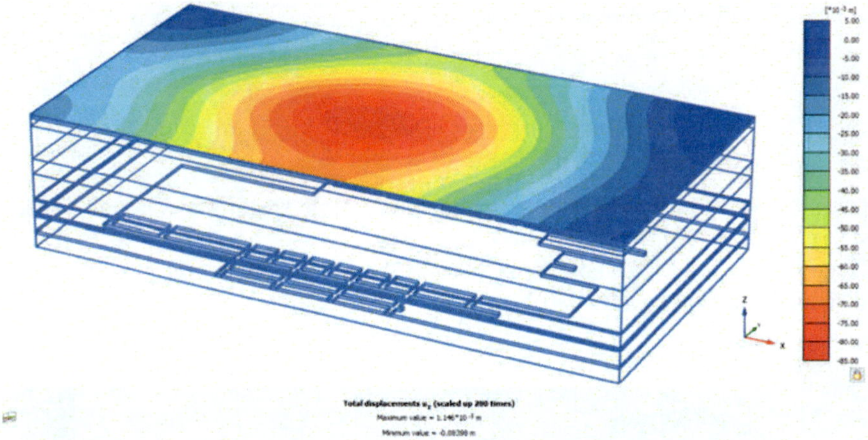

Fig. 15.9 Vertical displacements shading at the ground surface

Figures 15.11 and 15.12 show the vertical displacements shading the ground surface and seam roof, respectively, for phase 3.

From Table 15.3, it can be seen that the vertical displacements at the ground surface and coal seam roof level for the developed panel are 15.40 mm and 17.80 mm, respectively. Since it is stable, to increase the coal recovery from the panel, the complete removal of pillars from the central portion of the panel was analysed as an underground cavity (phase 2). Subsidence estimated at surface and seam roof level is 83.98 mm and 127.50 mm, respectively. But when the same case is analysed as depillaring with simultaneous stowing of a fill material (dense sand) in the voids

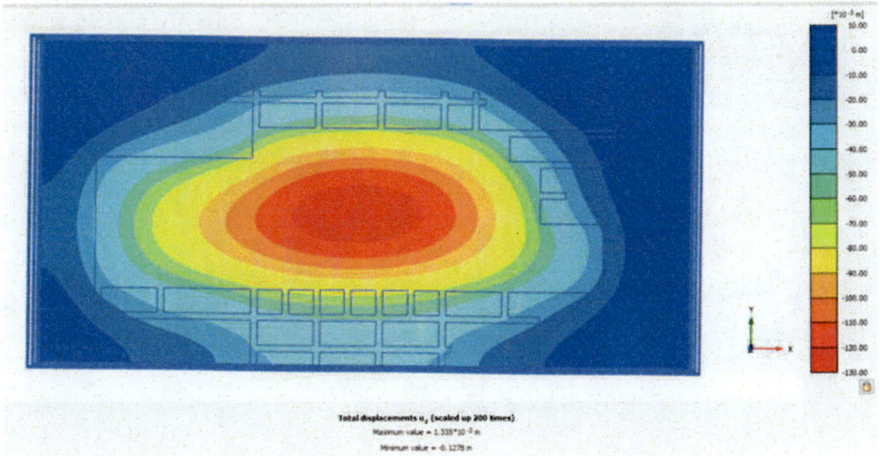

Fig. 15.10 Vertical displacements shading at seam roof level (top view)

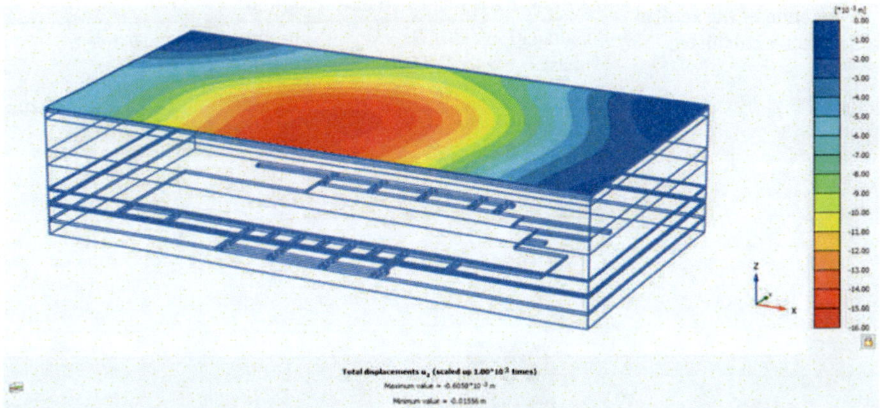

Fig. 15.11 Vertical displacements shading at the ground surface for the stowed panel

formed (phase 3), there was not much increase in the subsidence compared to the developed panel. At the ground surface, it is estimated as 15.56 mm, and at the coal seam roof level, 18.07 mm. Figure 15.13 shows the comparison of estimated surface subsidence for all phases, and it can be seen there is a considerable difference in subsidence for caving and simultaneous stowing.

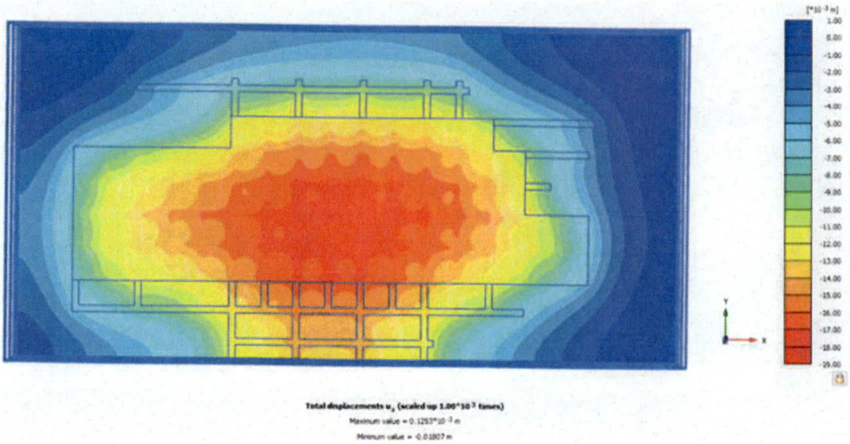

Fig. 15.12 Vertical displacements shading at seam roof level for stowed panel (top view)

Table 15.3 Tabular representation of the vertical displacements calculated

Total vertical displacements $\|U_z\|$ (mm)			
Calculation phases	Total displacements $\|U\|$ (mm)	$\|U_z\|$ @ $z = -$ 0.1 m	$\|U_z\|$ @ $z = -$ 121 m
Phase 1	25.23	15.40	17.80
Phase 2	128.20	83.98	127.50
Phase 3	25.33	15.56	18.07

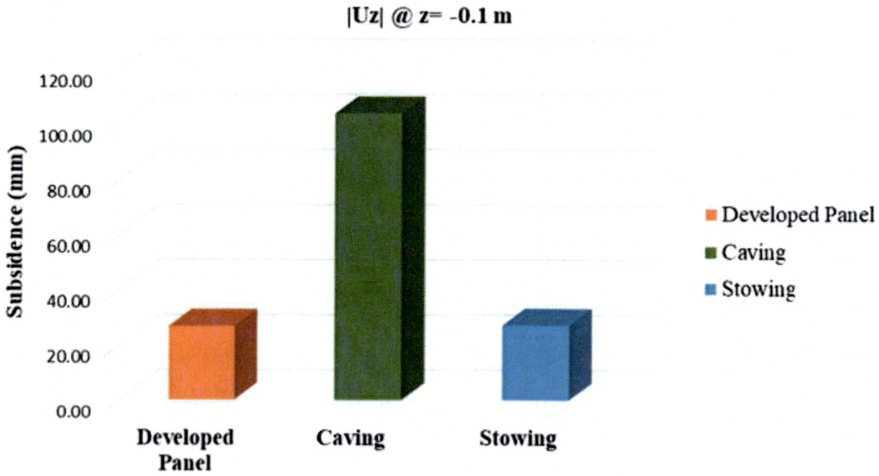

Fig. 15.13 Comparison of predicted ground surface subsidence

Conclusion

Results from the numerical model indicate that the developed panel is stable with just 15.40 mm subsidence at the surface and 17.80 mm at the roof of the coal seam. Ground surface subsidence for depillaring with simultaneous stowing was estimated at approximately 16 mm. When the same was done with the caving method, estimated subsidence was greater than 80 mm. Following the results obtained from numerical modelling, by adopting a depillaring operation with simultaneous stowing methodology of coal extraction from underground mines, surface subsidence can be arrested, especially for locations where surface topography or infrastructures are protected. By predicting the surface subsidence before the extraction using numerical simulations, we get an idea of the stability, which can be used afterwards with proper roof support systems to ease the coal extraction operations from the pillars of either abandoned or working mines. This study is not just limited to coal mines. It can also be used for other minerals extracted from underground mines with the same mining method. But it is necessary to include the complex geological conditions present in the mining area.

Acknowledgements The authors express their heartfelt gratitude to the Department of Civil Engineering, NIT Durgapur, West Bengal, India, for providing all necessary support and assistance to undertake the present research work.

References

1. Pandit K, Chourasia A, Bhattacharyya SK (2012) Depillaring of coal mine and roof supports. In: Conference paper—October 2012
2. Prakash A, Kumar N, Kumbhakar D et al (2018) A safe depillaring design for shallow depth of cover with the influence of surface ground movements: a study in Jharia Coalfield. Arab J Geosci 11:164. https://doi.org/10.1007/s12517-018-3508-4
3. Suresh D, Yarrakula K (2018) Subsidence monitoring techniques in coal mining: Indian scenario. Indian J Geo-Marine Sci 47(10):1918–1933
4. Nakache R, Boukelloul ML, Bouhedja A, Fredj M (2019) Stability analysis of the pillars of the underground mine Chaabet El-Hamra, Algeria by analytical and numerical methods. Naukovyi Visnyk Natsionalnoho Hirnychoho Universytetu 2:21–27
5. Zingano A, Weiss A (2019) Subsidence over Room and pillar retreat mining in a low coal seam. Int J Min Sci Technol 29:51–57
6. Mukherjee S, Pahari DP (2019) Underground and opencast coal mining methods in India: a comparative assessment. Space Culture, India 7(1):39–55
7. Hasim R, Jaiswal A, Shrivastava BK (2018) Support design during depillaring operation in bord and pillar panel using numerical simulation method. AMSE J-AMSE IIETA Publ 78:351–363
8. Plaxis 3D (2013)

Chapter 16
A Review on Numerical Simulation of Large Deformation Problems in Geotechnical Engineering

Kritesh Chouhan and Jitesh T. Chavda

Introduction

Many geotechnical engineering problems are considered as large deformation problems (LDPs) which include penetration of pile and under reamed pile, successive landslides, sinking of open caisson, tensile testing of geosynthetic, formation of soil plug in open-ended pipe, uplift anchor, installation effects in helical pile, installation of prefabricated vertical drain, penetration test, soil–soil and soil–structure interaction observed in case of offshore foundations like spudcan and anchor. The term large deformation is elusive in civil engineering practice [42]. In geotechnical engineering, the deformation of soil as well as the displacement of structure such as footing are considered as large deformation problem when the strains are more than 10% [43].

The assessment of LDPs is generally done using semi-empirical methods and engineering judgement skills obtained from field and physical model testing [39]. For the complex LDPs, physical model testing becomes expensive and time-consuming [61]. Numerical approach is another way to assess complex LDPs. In computational geomechanics, geometry changes either partially or fully, when encounter such extensive deformation [42]. In conventional finite element (FE) analysis, mesh distortion of the FE model occurs due to change in the geometry which prevents engineers to obtain reliable solutions of LDPs [42, 68, 73, 77]. Consequently, advance finite element formulations like coupled Eulerian–Lagrangian (CEL), arbitrary Lagrangian–Eulerian (ALE), remeshing and interpolation technique by small strains (RITSS), material point method (MPM), and particle finite element method (PFEM) are used to solve LDPs in geotechnical engineering [6, 18, 19, 25, 30, 31, 38, 39, 48, 51, 63, 68]. Due to advancement in computer technology, it becomes possible to attempt such LDPs in computational geomechanics [6, 40, 46, 70].

K. Chouhan (✉) · J. T. Chavda
Sardar Vallabhbhai National Institute of Technology, Surat 395007, India
e-mail: chouhankritesh@gmail.com

K. Muthukkumaran et al. (eds.), *Soil Dynamics, Earthquake and Computational Geotechnical Engineering*, Lecture Notes in Civil Engineering 300,
https://doi.org/10.1007/978-981-19-6998-0_16

In this paper, the various applications in geomechanics where large deformation encounters are presented. The various numerical approaches used to solve LDPs are reviewed. A cone penetration test (CPT) is simulated using conventional finite element approach (PLAXIS 2D) in order to observe the limitations in using conventional FE analysis to solve LDPs like CPT. Results are presented in the form of load versus displacement plots. Cone resistance is evaluated for various embedment depth. The limitations and future direction of conventional FE analysis are also discussed in the review.

Numerical Approaches

Several numerical approaches are used to solve large deformation geotechnical problems. In the continuum, Lagrangian and Eulerian descriptions are used to describe deformation. In Lagrangian description, mesh is deformed with material as loading is applied whereas in Eulerian description, position of mesh is fix, and material moves within the mesh [42, 56]. The conventional finite element approach is not able to quantify large deformation due to mesh distortion of FE mesh (see Fig. 16.1) at extensive deformation [42, 68]. The advance FE formulations like CEL, ALE, RITSS, MPM, and PFEM are developed to solve problems with large deformation. The CEL and ALE are widely used for assessing the large deformation problems in geotechnical engineering [6, 18, 19, 30, 31, 38, 39, 51, 68]. Some of the widely used large deformation finite element (LDFE) formulations are discussed in succeeding subsections.

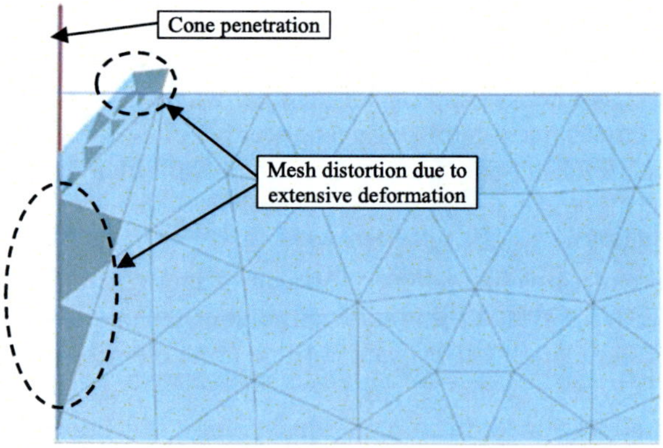

Fig. 16.1 Distorted mesh due to large deformation in conventional FE analysis

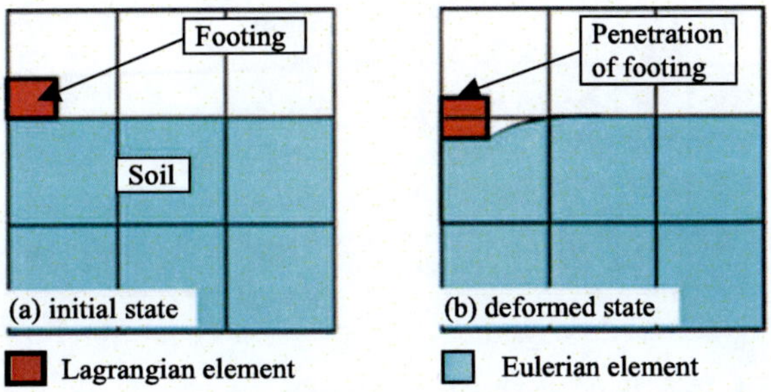

Fig. 16.2 Concept of CEL method

Coupled Eulerian–Lagrangian Method

The coupled Eulerian–Lagrangian (CEL) method is based on coupling of Eulerian and Lagrangian formulations. In the CEL method, both Eulerian and Lagrangian formulations are used to discretize the material depending on the stiffness of the material. The stiffer material is discretized by Lagrangian formulation, and less stiff material is discretized by Eulerian formulation. Figure 16.2 presents an example of penetration of footing in which the Lagrangian and Eulerian formulation are used for footing and soil, respectively. The soil material moves within the Eulerian mesh as footing penetrate. The interface between two materials is modelled by the contact algorithms [10, 42]. In the CEL method, the problem cannot be modelled as 2D (i.e. plane strain or axisymmetric) and is only modelled as 3D [68]. The CEL method is used and appreciated by many researchers in various geotechnical engineering problems like pile penetration [58], spudcan penetration [57], rectangular plate under uplift load [20], vane shear test [32], installation of giant deep-buried circular open [45], analysis of tunnel excavation [41], and blast analysis of tunnel [75, 76]. The disadvantage of the CEL method is that a fine Eulerian mesh is needed to obtain accurate results [42].

Arbitrary Lagrangian–Eulerian Method

The arbitrary Lagrangian–Eulerian (ALE) method is based on the operator split technique proposed by Benson [8]. The ALE method is a variant of r-adaptive finite element method which is designed to eliminate mesh distortion [68]. Figure 16.3 depicts the three basic steps of the ALE method, i.e. moving of material, remeshing, and transfer of the solution from old mesh to new mesh [23]. In the ALE method, the quality of the result depends on the mesh updating algorithm [42]. Many researchers have used the ALE method for the simulation of large deformation problems like consolidation problem [51], geotechnical construction process [59], dynamic analysis [50], pile installation [22], strip footing [7], and cone penetration test [26].

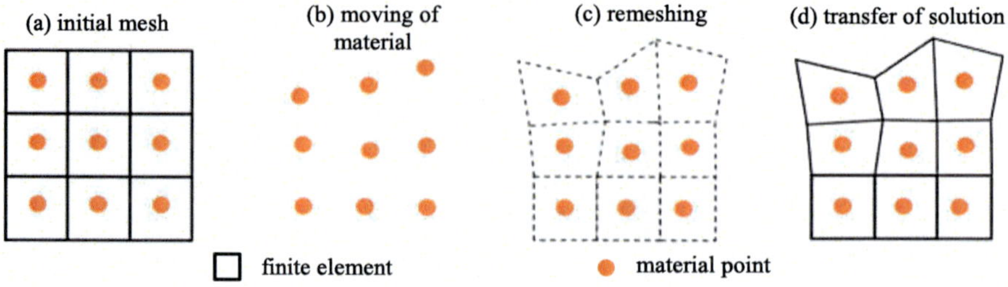

Fig. 16.3 Basic step in ALE method

Remeshing and Interpolation Technique by Small Strains Method

The remeshing and interpolation technique by small strains (RITSS) method was proposed by Hu and Randolph [36] to overcome mesh distortion in LDPs. The RITSS method comes under the ALE category. In the RITSS method, initial standard Lagrangian increment is applied with periodic remeshing of the model followed by the interpolation of all stress and strains. This approach makes RITSS a robust method [63]. The RITSS method is a versatile method for practical application like deep penetration of shallow foundation [36], strip and circular footing [67], and spudcan foundation [33], caisson installation [83], penetrometers [82], plate anchors [69], spudcan foundation [72], and rotation and pull out of plate anchor [1].

Material Point Method

The material point method (MPM) is an advance numerical method used to solve large deformation problem wherein the material is history dependent [48]. The MPM is modification of particle in-cell (PIC) method which was developed by Sulsky et al. [62]. It is a mesh-based particle method [42]. Figure 16.4 depicts the basic concept of the MPM in which the continuum is divided into a cloud of material points and is discretized as Lagrangian formulation. The cloud of material points is moved in a fixed Eulerian mesh to encounter the large deformations. The MPM is used to simulate numerous large deformation problems like cone penetration test [9, 14, 48], installation effect of driven piles [27, 28], and jacked piles [2, 54], etc.

Particle Finite Element Method

The particle finite element method (PFEM) is a meshless method. It was initially developed for applications of fluid dynamics [37, 52, 53]. In the PFEM, the domain is

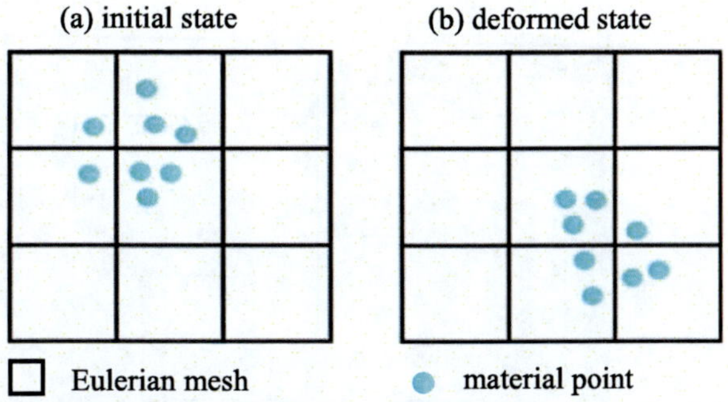

Fig. 16.4 Concept of MPM

discretized with the Lagrangian description. The nodes can move freely like a particle and form a cloud of points. The new position of nodes is identified and meshed. Then, the governing equation is solved using standard FEM. Many researchers have appreciated the use of PFEM to assess large deformation problems in geotechnical engineering like modelling of ground excavation [12], modelling of tunnel [13], study of granular column collapse [81], simulation of landslide [79], investigation of movement in granular matter [80], simulation of flows [21], simulation of undrained insertion problems [49], and smoothed particle finite element method [74, 78].

Simulation of Cone Penetration Using PLAXIS 2D

Numerical simulation of the cone penetration test (CPT) is an arduous task as it involves large deformation. Many researchers have used advance numerical methods to simulate the cone penetration like discrete element method [5], coupled Eulerian–Lagrangian method [24, 31], arbitrary Lagrangian–Eulerian method [26, 64, 66], particle finite element method [29, 49], material point method [9, 14, 48], and smoothed particle hydrodynamics method [44]. In this section, the penetration of cone is simulated using conventional finite element approach with PLAXIS 2D. PLAXIS 2D programme is a user-friendly programme which is used extensively in computational geomechanics.

Modelling of Soil Domain and Boundary Conditions

The simulation of cone penetration for dense sand is carried out using PLAXIS 2D programme. The dense sand obeys the non-associated flow rule. The axisymmetric model with 15-noded triangular element and 12-Gaussian quadrature points is used for the analysis. The size of the soil domain is considered as 1 m × 1 m on the basis of formation of plastic zone. The detail of FE mesh, element, and model dimension is depicted in Fig. 16.5. The roller supports are assigned at the vertical boundary

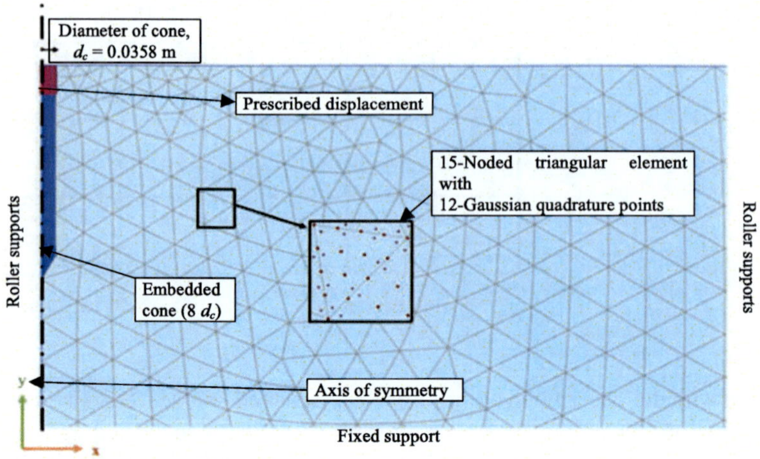

Fig. 16.5 FE model for CPT depicting the element type and boundary conditions

to allow movement in vertical direction, and fixed supports are assigned at lower boundary to restrict the movement in both vertical and horizontal directions. The ground water is not considered in the analysis.

Modelling of Soil and Cone

The Mohr–Coulomb and linear elastic models are used to simulate soil and rigid cone, respectively. The material and geometrical parameters of sand and cone are taken from the study of Gupta et al. [32] and are given in Table 16.1. The fine meshing is used for the FE analysis. A prescribed displacement is applied at the top of the cone to simulate the cone penetration. The ultimate failure load (Q_u) is determined for various embedment depths.

Simulation of CPT

The conventional FE analysis is not capable to solve large deformation due to mesh distortion. A large value of prescribed displacement cannot be applied to simulate cone penetration as there will be mesh distortion and analysis will stop. In the present study, a prescribed displacement is applied at the top of the cone for various embedment depths from 0.0 d_c (i.e. cone is at ground surface) to 20.0 d_c. The ultimate failure load is determined for various embedment depths of the cone. The total resistance (i.e. cone and sleeve resistance) is evaluated using rigid interface (i.e. $R_{\text{int}} =$ 1.0) for two cases i.e. cone embedded and 20 d_c embedment.

Table 16.1 Material and geometrical parameters [32]

Parameters	Symbol	Sand	Cone	Unit
Material model	–	Mohr–Coulomb	Linear elastic	–
Conditions	–	Drained	Non–porous	–
Unit weight	γ	20	20	kN/m^3
Modulus of elasticity	E	100	210×10^3	MPa
Poisson's ratio	ν	0.25	0.30	–
Cohesion	c	0.1	–	kN/m^2
Angle of friction	ϕ	37.5	–	Degree (°)
Dilation angle	ψ	7.5	–	Degree (°)
Diameter of cone	d_c	–	0.0358	m
Height of conical part	h_c	–	0.031	m
Apex angle	α	–	60	Degree (°)

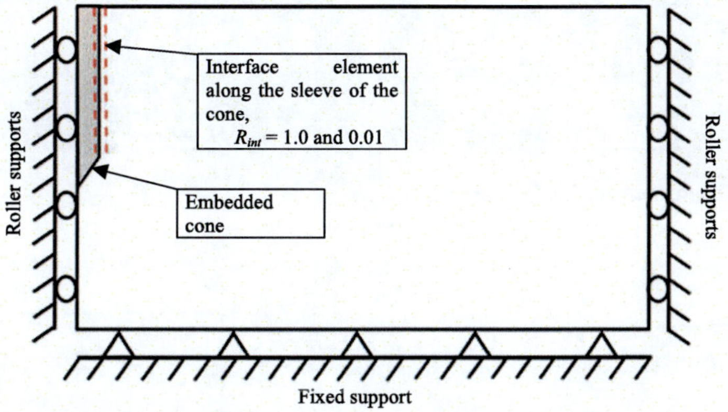

Fig. 16.6 Schematic view of model for cone penetration test

In PLAXIS programme, the interface elements are used to simulate interaction between the sleeve of cone and soil. For the study, the interface elements are assigned at the periphery of the sleeve of the cone (see Fig. 16.6). The cone resistance is evaluated by assigning an interface value, $R_{\text{int}} = 0.01$ for embedment of 20 d_c. Figure 16.6 depicts the boundary conditions and the interface element between cone sleeve and surrounding soil.

Results and Discussions

The numerical analysis is performed to simulate the penetration of a rigid cone into dense sand with an objective to evaluate the cone resistance corresponding to various embedment depth using PLAXIS 2D programme. Figure 16.7 depicts the total incremental displacement contours surrounding the cone tip and sleeve of the cone at an embedment depth of 0.6 and 8 d_c.

Figure 16.8 shows the variation of load at various embedded depths of the cone from 0.0 d_c (i.e. cone tip is at ground surface) to 10 d_c for rigid interface (i.e. R_{int} = 1.0). It is observed from the plot that the value of ultimate load is increased with increased in the embedment depth of the cone.

Figure 16.9a shows the various of load with displacement for three cases, i.e. cone embedded, 20 d_c embedment with $R_{int} = 1.0$ and 0.01. The prescribed displacement

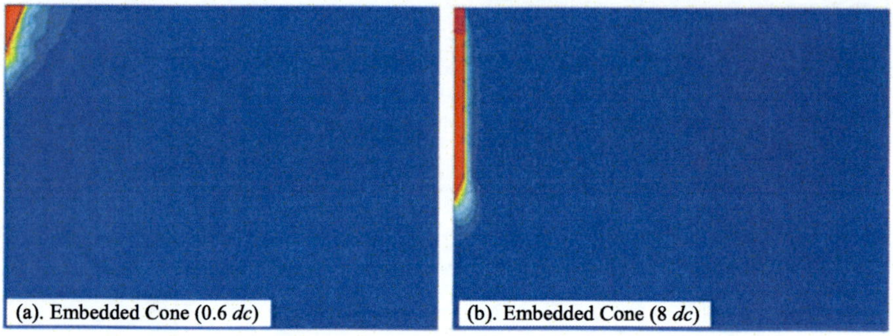

Fig. 16.7 Total incremental displacement contours

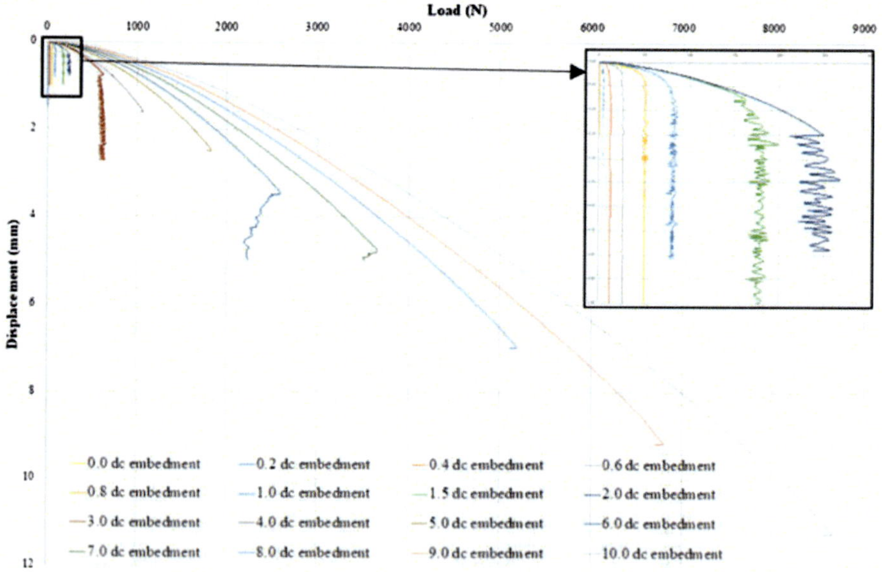

Fig. 16.8 Load versus displacements plot for various embedment depths of cone having $R_{int} = 1.0$ between soil and sleeve of the cone

is applied up to the failure to obtain the ultimate failure load. It is observed from the plot that ultimate load corresponding to $R_{int} = 0.01$ is less as compared to $R_{int} = 1.0$ due to elimination of sleeve resistance. Figure 16.9b depicts the comparison of conventional FE analysis (PLAXIS 2D) and the CEL results from Gupta et al. [32]. Numerically, the total resistance and cone resistance are evaluated separately. From the plot, it is seen that the cone resistance at lower embedment depth (i.e. only cone embedded) provides a good fit to the CEL method [32]. However, the cone resistance ($R_{int} = 0.01$) and total resistance ($R_{int} = 1.0$) are significantly higher than the CEL method with increase in embedment depth to 20 d_c.

Research Potential of Large Deformation Problems

Generally, the experimental and numerical simulations are used for quantification of large deformations in geomechanics. The experimental simulations of geotechnical problems with large deformation are attempted by many researchers, mainly for the offshore foundations [4, 11, 16, 17, 25, 33–35, 47, 60, 65, 71] (Chavda and Dodagoudar 2021). However, the experimental simulation becomes expensive and time-consuming depending on the complexity of the LDPs.

The numerical simulations of large deformation geotechnical problems are increased over decades. The large deformation finite element (LDFE) analysis does not require any prior assumptions of a failure mechanism and can reflect the natural development of the failure zone which provide reasonably good prediction of the bearing behaviour [39]. Numerous LDFE formulations are developed for the assessment of large deformations like CEL, ALE, MPM, RITSS, and PFEM which are appreciated by many researchers in computational geomechanics [1, 9, 51, 57, 81]. Generally, the ALE and CEL are widely used methods in the field of geotechnical engineering. It is observed that the ALE method is more flexible than the CEL method because the problem can be simulated 2D as well as 3D depending on the natural of problem (i.e. loading, geometry, and boundary condition). Whereas, the problem is simulated only as 3D in the CEL method [3].

However, the LDFE formulations have limitations like large number of remeshing in case of the ALE method, difficulties in free surface tracking in the CEL method, and the PFEM and MPM are computational expensive [42]. The thorough understanding of the different aspects of large deformation geotechnical problems, such as load-penetration response, soil flow mechanism, bearing capacity of foundation subject to loading, and slope stability simulation, help in the development of analytical, empirical, and semi-empirical solutions and mitigating the challenges faced in executing the project. However, the research for the assessment of LDPs in geotechnical engineering is carried out extensively only in the past decade, moreover for limited geotechnical problems. This gives a further scope in investigating other large deformation geotechnical problems using the LDFE formulations.

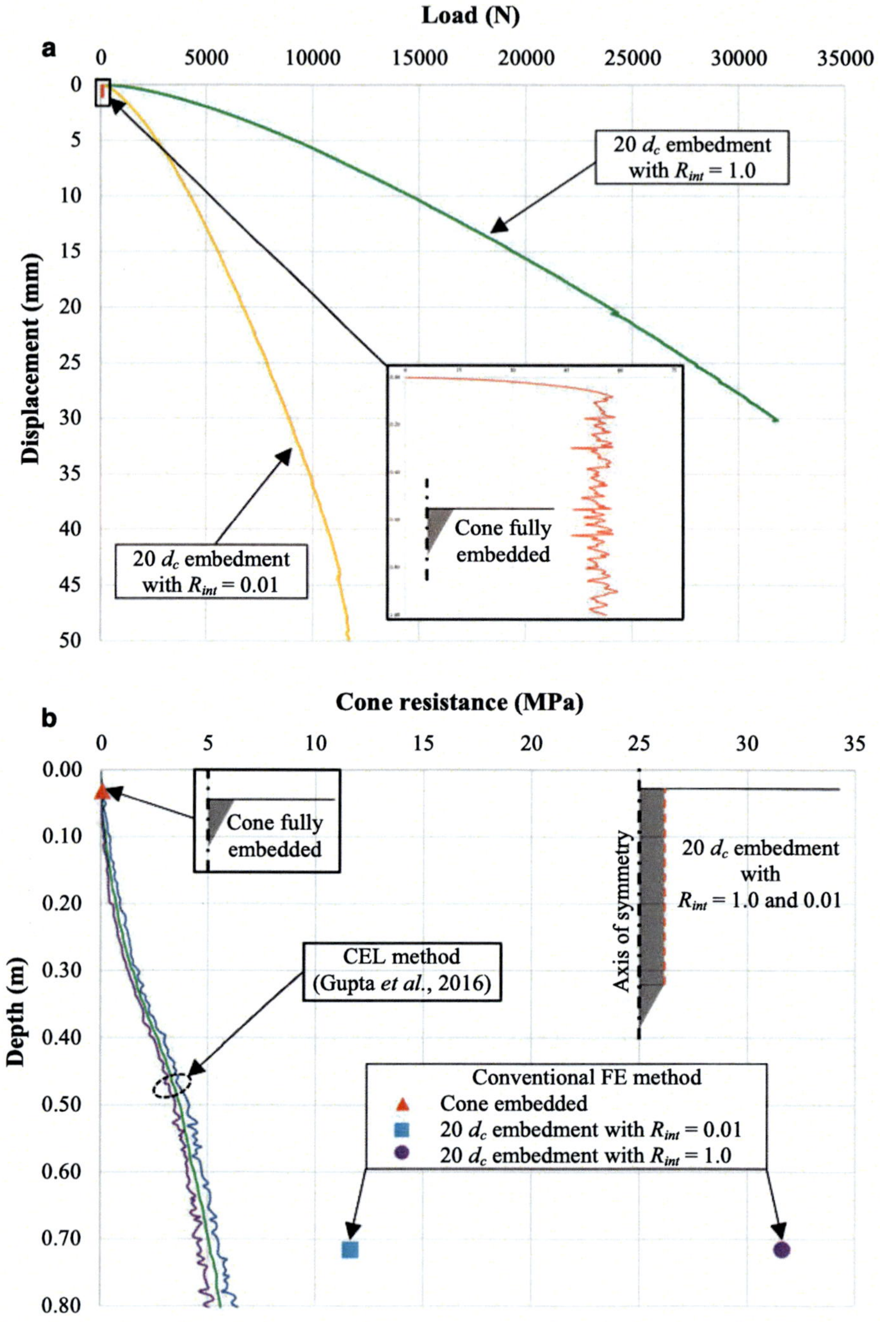

Fig. 16.9 **a** Load versus displacement plot. **b** Comparison between conventional FE method and the CEL method

Concluding Remarks

The present study reviews the various applications in geotechnical engineering in which large deformation occurs. The numerous advance finite element formulations used to solve large deformation problems in geomechanics are critically discussed. The simulation of cone penetration using conventional finite element analysis is also presented. This review provides a brief understanding of numerous large deformation problems and a basic idea about various computational formulations used in geotechnical engineering. Based on this review, the following conclusions are drawn:

- Large deformations are encountered in many geotechnical engineering problems which are quantified using experimental and numerical simulations. Due to limitations with experimental simulation, the numerical simulations are widely used to solve large deformation problems.
- Conventional finite element analysis is not suitable to quantify large deformation in geotechnical engineering due to mesh distortion. So, advance finite element formulations are used like CEL, ALE, RITSS, MPM, and PFEM.
- In the CEL method, the contact algorithm plays an important role. The CEL method is used to analyse the Lagrangian elements (like, pile, anchors, etc.). The advantage of geometry cannot be used in the CEL method as problem is simulated as 3D model.
- The arbitrary Lagrangian–Eulerian method uses strength of both Lagrangian and Eulerian elements. The accuracy of ALE method depends on the mesh updating algorithm. It is more flexible than the CEL method because the problem can be simulated 2D as well as 3D, and the advantage of geometry can be taken.
- From the simulation of cone penetration using PLAXIS 2D programme, it is opined that the conventional FE analysis is not capable to solve large deformation problems. Further study is needed to frame some guidelines regarding simulation of large deformation problem using conventional finite element analysis in geotechnical engineering. The work in this direction is underway.

References

1. Al Hakeem N, Aubeny C (2021) Numerical modeling of keying of vertically installed plate anchor in sand. Ocean Eng 223:108674. https://doi.org/10.1016/j.oceaneng.2021.108674
2. Al Kafaj KI (2013) Formulation of a dynamic material point method (MPM) for geomechanical problems. PhD thesis
3. Andresen L, Khoa HDV (2013) LDFE analysis of installation effects for offshore anchors and foundations. In: Installation effects in geotechnical engineering. Taylor & Francis Group London, UK
4. Ansari Y, Kouretzis G, Pineda J (2018) Sand-pipe interaction at fault crossings: experimental and numerical investigation. In: Proceedings of the 16th European conference on earthquake engineering, Thessaloniki, Greece, June
5. Arroyo M, Butlanska J, Gens A, Calvetti F, Jamiolkowski M (2011) Cone penetration tests in a virtual chamber. Geotechnique 61(6):525–531. https://doi.org/10.1680/geot.9.p.067

6. Aubram D (2015) Development and experimental validation of an arbitrary Lagrangian–Eulerian (ALE) method for soil mechanics. Geotechnik 38(3):193–204
7. Bakroon M, Daryaei R, Aubram D, Rackwitz F (2017) Arbitrary Lagrangian–Eulerian finite element formulations applied to geotechnical problems. In: Numerical methods in geotechnics, pp 33–44
8. Benson DJ (1989) An efficient, accurate and simple ALE method for nonlinear finite element programs. Comput Methods Appl Mech Eng 72(3):305–350
9. Beuth L, Vermeer PA (2013) Large deformation analysis of cone penetration testing in undrained clay. In: International conference on installation effects in geotechnical engineering (ICIEGE), Rotterdam 1–7. Taylor and Francis Group, London. ISBN 978-1-138-00041-4
10. Brown HK, Burns SP, Christon MA (2002) Coupled Eulerian–Lagrangian methods for earth penetrating weapon applications. Technical Report
11. Bruton DA, Bolton M, Carr M, White D (2008) Pipe-soil interaction with flowlines during lateral buckling and pipeline walking—the SAFEBUCK JIP. In: Offshore technology conference, January 2008. https://doi.org/10.4043/19589-ms
12. Carbonell JM, O Ate E, Su Rez B (2010) Modeling of ground excavation with the particle finite element method. J Eng Mech 136(4):455–463
13. Carbonell JM, O Ate E, Su Rez B (2013) Modelling of tunnelling processes and rock cutting tool wear with the particle finite element method. Comput Mech 52(3):607–629
14. Ceccato F, Beuth L, Vermeer PA, Simonini P (2016) Two-phase material point method applied to the study of cone penetration. Comput Geotech 80:440–452. https://doi.org/10.1016/j.compgeo.2016.03.003
15. Chavda JT, Dodagoudar GR (2021) Experimental evaluation of failure zone in sand beneath the ring footing and cutting edge of open caisson using image analysis. In: Proceedings of the Indian geotechnical conference 2019: IGC-2019, Volume I. Springer Nature, p 273
16. Chavda JT, Dodagoudar GR (2020) Experimental studies on circular open caisson: load-penetration response and soil flow mechanism. Int J Phys Model Geotech 1–53. https://doi.org/10.1680/jphmg.20.00050
17. Chavda JT, Mishra S, Dodagoudar GR (2020) Experimental evaluation of ultimate bearing capacity of the cutting edge of an open caisson. Int J Phys Model Geotech 20(5):281–294. https://doi.org/10.1680/jphmg.18.00052
18. Chen Y, Zhao W, Han J, Jia P (2019a) A CEL study of bearing capacity and failure mechanism of strip footing resting on c-φ soils. Comput Geotech 111:126–136
19. Chen X, Zhang L, Chen L, Li X, Liu D (2019b) Slope stability analysis based on the coupled Eulerian–Lagrangian finite element method. Bull Eng Geol Env 78(6):4451–4463
20. Chen Z, Tho KK, Leung CF, Chow YK (2013) Influence of overburden pressure and soil rigidity on uplift behavior of square plate anchor in uniform clay. Comput Geotech 52:71–81
21. DáValos C, Cante J, Hern Ndez JA, Oliver J (2015) On the numerical modeling of granular material flows via the particle finite element method (PFEM). Int J Solids Struct 71:99–125
22. Dijkstra J, Broere W, Heeres OM (2011) Numerical simulation of pile installation. Comput Geotech 38(5):612–622
23. Donea J, Huerta A, Pouthot JPh, Rodriguez-Ferran A (2004) Arbitrary Lagrangian–Eulerian methods. Wiley.https://doi.org/10.1002/9781119176817.ecm2009
24. Fallah S, Gavin K, Jalilvand S (2016) Numerical modelling of cone penetration test in clay using coupled Eulerian Lagrangian method. In: Proceedings of civil engineering research in Ireland 2016, Galway, Ireland, 29–30 August
25. Fan S, Bienen B, Randolph MF (2021) Effects of monopile installation on subsequent lateral response in sand. I: Pile installation. J Geotech Geoenviron Eng 147(5):04021021
26. Fan S, Bienen B, Randolph MF (2018) Stability and efficiency studies in the numerical simulation of cone penetration in sand. Geotech Lett 8(1):13–18. https://doi.org/10.1680/jgele.17.00105
27. Galavi V, Beuth L, Zuada Coelho B, Tehrani FS, Holscher P, Van Tol F (2017) Numerical simulation of pile installation in saturated sand using material point method. Procedia Eng 175:72–79. https://doi.org/10.1016/j.proeng.2017.01.027

28. Galavi V, Martinelli M, Elkadi A, Ghasemi P, Thijssen R (2019) Numerical simulation of impact driven offshore monopiles using the material point method. In: Proceedings of the XVII ECSMGE. https://doi.org/10.32075/17ECSMGE-2019-0758
29. Gens A (2019) Hydraulic fills with special focus on liquefaction. In: Proceedings of the XVII ECSMGE 2019, Reykjavik, Island. http://hdl.handle.net/2117/179180
30. Grabe J, Wu L (2016) Coupled Eulerian–Lagrangian simulation of the penetration and braking behaviour of ship anchors in clay. Geotechnik 39(3):168–174
31. Gupta T, Chakraborty T, Abdel-Rahman K, Achmus M (2015) Numerical modelling of finite deformation in geotechnical engineering. In: Advances in structural engineering. Springer, New Delhi, pp 689–701
32. Gupta T, Chakraborty T, Abdel-Rahman K, Achmus M (2016) Large deformation finite element analysis of static cone penetration test. Indian Geotech J 46(2):115–123
33. Hossain MS, Hu Y, Randolph MF, White DJ (2005) Limiting cavity depth for spudcan foundations penetrating clay. Geotechnique 55(9):679–690. https://doi.org/10.1680/geot.2005.55.9.679
34. Hu P, Stanier SA, Wang D, Cassidy MJ (2016) Effect of footing shape on penetration in sand overlying clay. Int J Phys Model Geotech 16(3):119–133. https://doi.org/10.1680/jphmg.15.00013
35. Hu P, Wang D, Stanier SA, Cassidy MJ (2015) Assessing the punch-through hazard of a spudcan on sand overlying clay. Geotechnique 65(11):883–896
36. Hu Y, Randolph MF (1998) A practical numerical approach for large deformation problems in soil. Int J Numer Anal Meth Geomech 22(5):327–350
37. Idelsohn SR, Onate E, Del Pin F (2004) The particle finite element method: a powerful tool to solve incompressible flows with free-surfaces and breaking waves. Int J Numer Meth Eng 61:964–989
38. Kardani M, Nazem M, Carter JP, Abbo AJ (2015) Efficiency of high-order elements in large-deformation problems of geomechanics. Int J Geomech 15(6):04014101
39. Khoa HDV, Jostad HP (2016) Application of coupled Eulerian–Lagrangian method to large deformation analyses of offshore foundations and suction anchors. Int J Offshore Polar Eng 26(03):304–314
40. Kim YH, Hossain MS, Edwards D, Wong PC (2019) Penetration response of spudcans in layered sands. Appl Ocean Res 82:236–244
41. Kim D (2021) Large deformation finite element analyses in TBM tunnel excavation: CEL and auto-remeshing approach. Tunn Undergr Space Technol 116:104081
42. Konkol J (2014) Numerical solutions for large deformation problems in geotechnical engineering. PhD Interdisc J 2014:49–55
43. Krabbenhoft K, Zhang X (2013) Particle finite element method for extreme deformation problems. In: Tectonomechanics Colloquium, Paris
44. Kulak RF, Bojanowski C (2011) Modeling of cone penetration test using SPH and MM-ALE approaches. In: 8th European LS-DYNA users conference, Strasbourg, May 2011
45. Lai F, Liu S, Deng Y, Sun Y, Wu K, Liu H (2020) Numerical investigations of the installation process of giant deep-buried circular open caissons in undrained clay. Comput Geotech 118:103322
46. Liu H, Xu K, Zhao Y (2016) Numerical investigation on the penetration of gravity installed anchors by a coupled Eulerian–Lagrangian approach. Appl Ocean Res 60:94–108. https://doi.org/10.1016/j.apor.2016.09.002
47. Liu M, Zhu Z (2012) Sand deformation around an uplift plate anchor. J Geotech Geoenviron Eng 138(6):728–737. https://doi.org/10.1061/(asce)gt.1943-5606.0000633
48. Martinelli M, Galavi V (2021) Investigation of the material point method in the simulation of cone penetration tests in dry sand. Comput Geotech 130:103923
49. Monforte L, Arroyo M, Carbonell JM, Gens A (2017) Numerical simulation of undrained insertion problems in geotechnical engineering with the particle finite element method (PFEM). Comput Geotech 82:144–156. https://doi.org/10.1016/j.compgeo.2016.08.013

50. Nazem M, Carter JP, Airey DW (2009) Arbitrary Lagrangian–Eulerian method for dynamic analysis of geotechnical problems. Comput Geotech 36(4):549–557. https://doi.org/10.1016/j.compgeo.2008.11.001
51. Nazem M, Sheng D, Carter JP, Sloan SW (2008) Arbitrary Lagrangian–Eulerian method for large-strain consolidation problems. Int J Numer Anal Meth Geomech 32(9):1023–1050. https://doi.org/10.1002/nag.657
52. Onate E, Idelsohn SR, Celigueta MA, Rossi R (2008) Advances in the particle finite element method for the analysis of fluid-multibody interaction and bed erosion in free surface flows. Comput Methods Appl Mech Eng 197:1777–1800
53. Onate E, Idelsohn SR, Del Pin F, Aubry R (2004) The particle finite element method an overview. Int J Comput Methods 1:267–307
54. Phuong NTV, Van Tol AF, Elkadi A, Rohe A (2015) Numerical investigation of pile installation effects in sand using material point method. Comput Geotech 73:58–71. https://doi.org/10.1016/j.compgeo.2015.11.012
55. PLAXIS 2D tutorial manual (2020)
56. Pucker T, Bienen B, Henke S (2013) CPT based prediction of foundation penetration in siliceous sand. Appl Ocean Res 41:9–18
57. Qiu G, Grabe J (2012) Numerical investigation of bearing capacity due to spudcan penetration in sand overlying clay. Can Geotech J 49(12):1393–1407
58. Qiu G, Henke S, Grabe J (2011) Application of a coupled Eulerian–Lagrangian approach on geomechanical problems involving large deformations. Comput Geotech 38(1):30–39
59. Savidis SA, Aubram D, Rackwitz F (2008) Arbitrary Lagrangian–Eulerian finite element formulation for geotechnical construction processes. J Theor Appl Mech 38(1–2):165–194
60. Ragni R, Bienen B, Stanier S, O'Loughlin C, Cassidy M (2019) Observations during suction bucket installation in sand. Int J Phys Model Geotech 20(3):1–18. https://doi.org/10.1680/jphmg.18.00071
61. Souli MH, Benson DJ (2013) Arbitrary Lagrangian Eulerian and fluid-structure interaction: numerical simulation. Wiley
62. Sulsky D, Chen Z, Schreyer HL (1994) A particle method for history-dependent materials. Comput Methods Appl Mech Eng 118(1–2):179–196. https://doi.org/10.1016/0045-7825(94)90112-0
63. Tian Y, Cassidy MJ, Randolph MF, Wang D, Gaudin C (2014) A simple implementation of RITSS and its application in large deformation analysis. Comput Geotech 56:160–167
64. Tolooiyan A, Gavin K (2011) Modelling the cone penetration test in sand using cavity expansion and arbitrary Lagrangian Eulerian finite element methods. Comput Geotech 38(4):482–490. https://doi.org/10.1016/j.compgeo.2011.02.012
65. Ullah SN, Stanier S, Hu Y, White D (2017) Foundation punch-through in clay with sand: centrifuge modelling. Geotechnique 67(10):870–889
66. Vanden Berg P (1994) Analysis of soil penetration. PhD thesis, Delft University Press
67. Wang CX, Carter JP (2002) Deep penetration of strip and circular footings into layered clays. Int J Geomech 2(2):205–232
68. Wang D, Bienen B, Nazem M, Tian Y, Zheng J, Pucker T, Randolph MF (2015) Large deformation finite element analyses in geotechnical engineering. Comput Geotech 65:104–114
69. Wang D, Hu Y, Randolph MF (2010) Three-dimensional large deformation finite-element analysis of plate anchors in uniform clay. J Geotech Geoenviron Eng 136(2):355–365
70. Wang T, Zhang Y, Bao X, Wu X (2020) Mechanisms of soil plug formation of open-ended jacked pipe pile in clay. Comput Geotech 118:103334
71. Xiao Y, Cao H, Luo G (2019) Experimental investigation of the backward erosion mechanism near the pipe tip. Acta Geotech 14(3):767–781. https://doi.org/10.1007/s11440-019-00779-w
72. Yu L, Hu Y, Liu J, Randolph MF, Kong X (2012) Numerical study of spudcan penetration in loose sand overlying clay. Comput Geotech 46:1–12
73. Yuan B, Xu K, Wang Y, Chen R, Luo Q (2017) Investigation of deflection of a laterally loaded pile and soil deformation using the PIV technique. Int J Geomech 17(6):04016138. https://doi.org/10.1061/(ASCE)GM.1943-5622.0000842

74. Yuan WH, Wang B, Zhang W, Jiang Q, Feng XT (2019) Development of an explicit smoothed particle finite element method for geotechnical applications. Comput Geotech 106:42–51
75. Zaid M, Sadique MR (2021) A simple approximate simulation using coupled Eulerian–Lagrangian (CEL) simulation in investigating effects of internal blast in rock tunnel. Indian Geotech J 1–18. https://doi.org/10.1007/s40098-021-00511-0
76. Zaid M, Sadique MR, Alam MM (2021) Blast analysis of tunnels in Manhattan-Schist and Quartz-Schist using coupled-Eulerian–Lagrangian method. Innovative Infrastruct Solutions 6(2):1–10. https://doi.org/10.1007/s41062-020-00446-0
77. Zhang L, Cai Z, Wang L, Zhang R, Liu H (2018a) Coupled Eulerian–Lagrangian finite element method for simulating soil-tool interaction. Biosys Eng 175:96–105. https://doi.org/10.1016/j.biosystemseng.2018.09.003
78. Zhang W, Yuan W, Dai B (2018b) Smoothed particle finite-element method for large-deformation problems in geomechanics. Int J Geomech 18(4):04018010
79. Zhang X, Krabbenhoft K, Sheng D, Li W (2015a) Numerical simulation of a flow-like landslide using the particle finite element method. Comput Mech 55(1):167–177
80. Zhang X, Sheng D, Kouretzis GP, Krabbenhoft K, Sloan SW (2015b) Numerical investigation of the cylinder movement in granular matter. Phys Rev E 91(2):022204
81. Zhang X, Krabbenhoft K, Sheng D (2014) Particle finite element analysis of the granular column collapse problem. Granular Matter 16(4):609–619
82. Zhou H, Randolph MF (2009) Resistance of full-flow penetrometers in rate-dependent and strain-softening clay. Géotechnique 59(2):79–86
83. Zhou H, Randolph MF (2006) Large deformation analysis of suction caisson installation in clay. Can Geotech J 43(12):1344–1357. https://doi.org/10.1139/t06-087

Chapter 17
Potential Use of Fine Fraction of Municipal Solid Waste as Replacement of Soil in Embankment

Parul Rawat and Supriya Mohanty

Introduction

Waste generation and its management have become a challenging problem for the whole world. Municipal solid waste (MSW) is the most important by-product of urbanization that is increasing as world cities generate about 1.3 billion ton of solid waste per year (2012) which is expected to increase to 2.2 billion ton by 2025 [1]. The heterogeneity of the waste is itself complex as the composition of the waste changes from place to place and required continuous correlation with the average standard of living. 100% utilization of MSW is not possible, but a part of it can be recovered, recycled, and reused in some other different form. It has been noticed in previous studies that around 60% of the waste can be categorized under fine soil-like material [2–4]. These MSW fines can be used as a soil replacement as road embankment or as backfill material [2, 5].

Numerical studies were done with the help of finite element-based software like Plaxis2D, GeoStudio, Geo5, etc. Seepage and stability analysis of the landfill sites in China was done with the help of SEEP/W and SLOPE/W, respectively [6]. A comparative study of the landfill stability was done with finite element and limit equilibrium analysis, both the analysis predicts same failure surface [7]. Reliability-based assessment of municipal solid landfill slope was studied by Rajesh et al. in 2016 [8]. A multi-phase numerical model study was conducted to stimulate the vertical settlement of MSW due to liquid and gas flow, and the proposed model can estimate the porosity, gas/liquid pressure, and saturation and stress distribution in settling landfill sites [9]. Although these numerical studies give an overview of the material

P. Rawat (✉) · S. Mohanty
Department of Civil Engineering, Indian Institute of Technology (BHU) Varanasi, Varanasi, Uttar Pradesh 221005, India
e-mail: parulrawat.rs.civ18@itbhu.ac.in

K. Muthukkumaran et al. (eds.), *Soil Dynamics, Earthquake and Computational Geotechnical Engineering*, Lecture Notes in Civil Engineering 300,
https://doi.org/10.1007/978-981-19-6998-0_17

under different load conditions, the field application of waste is not possible until it goes through strict environmental parameters.

Objective of the Study

The objective of the study is to use the fine fraction of MSW, i.e. fraction finer than 4.75 mm (MSW fines) for geotechnical applications. To check the reusability of MSW fines as a fill material for the road embankment, numerical analysis has been performed for a typical road embankment fill material with the help of FEM-based software Plaxis2D and stability analysis using Geo5. This study focuses on the utilization of MSW fines as a replacement for natural soils in the field.

Materials and Methodology

Site Description and Sample Collection

The samples were collected from Ramana (near Ramana-Varanasi sewage treatment plant), Varanasi (U.P.), India. The waste collected is thoroughly mixed and left for air drying. Once the dried waste was sieved through different size sieves, i.e. 45, 26.5, 8, and 4.75 mm IS sieves. The considered waste is classified under poorly graded silty sand (SM). The detailed physical, chemical, and geotechnical properties of the considered waste fines were discussed by Rawat and Mohanty [10].

Dimensioning of the Problem

The road embankment was designed similar to the design of the embankment dam. The design of the embankment is proposed for a 4-lane road near the same landfill site (Ramana, Varanasi) from where the MSW sample was collected. The proposed model includes 5 m high embankment of MSW with a cover of 0.5 m thickness of local soil. The cover layer was provided to reduce the erosion as well as to enhance the stability of the embankment. The embankment was subdivided into five layers. Each layer of thickness 1 m was compacted at 99% of relative compaction. The embankment of a total 5.5 m height, top width of 16 m with a side slope of 2H:1V was subjected to the traffic load of 10 kN/m^2. A typical profile of the subsoil can be seen in Fig. 17.1, which consists of 1.5 m of sandy silt layer, followed by 10.5 and 5 m of the fine sand layer at different densities. The water table was noticed at 13 m from the ground level.

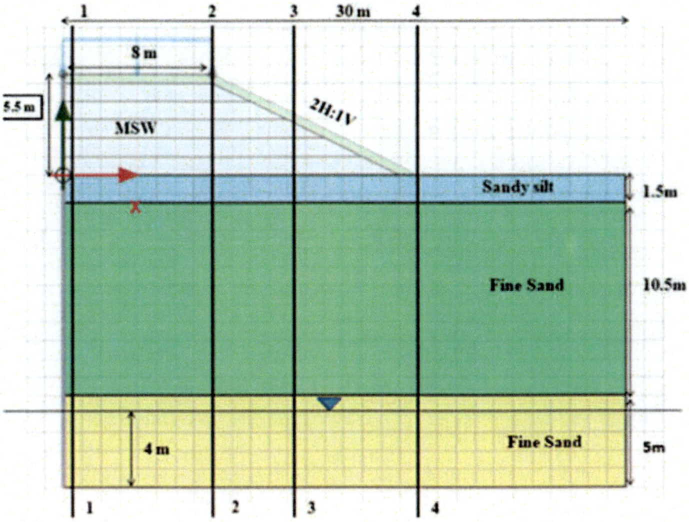

Fig. 17.1 Road embankment model showing 4 different sections considered for the analysis using Plaxis2D

The problem was solved as a two-dimensional plane-strain consolidation problem and because of the symmetry of the problem; the analysis was conducted for the half of the embankment section. The analyses for settlement, stress, and slope stability of the embankment were carried by using Plaxis2D and Geo5 software, respectively.

Municipal Solid Waste Embankment Modelling

Material Model and Parameters

In the present study, Mohr–Coulomb material model was used for simulating the MSW embankment. Mohr–Coulomb material model is a linear elastic perfectly plastic model and represents the first-order approximation of soil. The linear part is based on Hooke's law, and the plastic part is based on the Mohr–Coulomb failure criterion. This model can be used as the first analysis of the problem with limited soil parameters. The parameters required for the analysis are cohesion (c), friction angle (Φ), dilatancy angle (Ψ), Poisson's ratio (ν), and Young's modulus (E) of the material. The Poisson's ratio for most of the soil ranges from 0.3 to 0.4 and the dilatancy angle was assumed to be zero for all the soils. The Young's modulus for the subsoil was correlated with the SPT data of the field [11]. In this study, the horizontal permeability was considered approximately equal to that of the vertical permeability. The shear strength and density parameters of the MSW considered were determined in the laboratory as discussed above in the geotechnical characterization section.

Table 17.1 Parameters used for the embankment fill and subsoil

Parameters	Embankment fill	Subsoil 1	Subsoil 2	Subsoil 3
Soil type	MSW	Sandy silt	Fine sand	Fine sand
Material mode	Mohr–Coulomb	Mohr–Coulomb	Mohr–Coulomb	Mohr–Coulomb
$\gamma_{\text{unsaturated}}$ (kN/m^3)	14.678	15.4	15.59	18.148
$\gamma_{\text{saturated}}$ (kN/m^3)	17.1879	18.172	18.396	21.414
e_{initial}	0.55314	0.5	0.5	0.5
Cohesion (C) (kN/m^2)	42.1949	0	0	0
Angle of internal friction (Φ) (°)	30.746	27.5	28	32
Modulus of elasticity (E) (kN/m^2)	14,985.6	17,363.7	27,793.7	27,525.3
Permeability (m/day)	0.01011	0.08640	8.640	8.640

The parameters of the subsoil of the area into consideration were collected from the site investigation test (SPT) conducted by Vishwa Bhumi Technologies for Geotechnical survey for the preparatory survey of Ganga Rejuvenation Project [12]. The parameters of the local soil, as well as the embankment fill (MSW) are provided in Table 17.1.

Boundary Conditions

The ground below 17 m was considered as stiff and contributed negligible ground movement; therefore, the bottom boundary was considered completely rigid and impermeable. The embankment model was restrained in the horizontal direction as only vertical loads were activated during the analysis. The vertical boundaries were set free for drainage.

Mesh Generation

The 15-noded triangular element was used in Plaxis2D to discretize the MSW embankment model and subsoil. The 15-noded element provides fourth-order interpolation for displacements and numerical interpolation involves 12 stress points. The convergence study was conducted to determine the optimum mesh for the embankment model (Fig. 17.2). From the convergence result, the fine meshing was adopted as an optimum mesh with the 2012 number of elements. The construction stage of the embankment consists of 6 phases; the first 5 stages consist of 1 m thick MSW

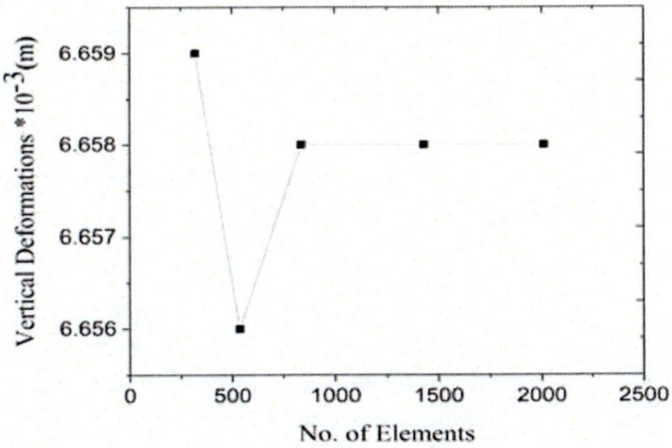

Fig. 17.2 Convergence plot of the road embankment with MSW

layer each and allowed to consolidate for 15 days after every stage of construction, whereas 6th phase consists of soil cover of 0.5 m.

Results and Discussions

Settlement Analysis

The settlement analysis was carried out using finite element-based software Plaxis2D. Plaxis2D is widely used to simulate the complex behaviour of soil. The present embankment model ranged from −6 to 17 m in vertical direction (where 0 m was considered as the ground level) and 0–30 m in the horizontal direction. The ground and the embankment were free to settle in the vertical direction with $y_{\min}$ fully fixed and horizontal boundaries normally fixed. After the end of the 6th phase, a phase was considered to apply a vertical uniform static load (10 kN/m^2) on the embankment cross-section representing the traffic loads. The variation of the settlement with depth was recorded at 4 different critical sections in the embankment model, i.e. sections 1–1, 2–2, 3–3, and 4–4 (Fig. 17.1).

Here, the overburden pressure and the traffic loads were applied in the vertical direction due to which very little displacement in the horizontal direction was noticed. Hence, only vertical displacements were considered for the further detailed settlement study. The ground profile under the embankment is of mainly sandy strata, the immediate settlement was computed analytically from Eq. 17.1 due to overburden pressure caused by embankment and applied vertical load. The immediate settlement observed to be 270.48 mm.

$$S_{\mathrm{i}} = qB\left(\frac{1-\mu^2}{E_{\mathrm{avg}}}\right)I_{\mathrm{f}} \tag{17.1}$$

S_i Immediate settlement
q Intensity of contact pressure
B Width of the embankment (32 m)
μ Poisson's ratio (considered 0.4)
E_{avg} Average of Young's modulus of soils (26,794.46 kN/m^2).
I_f Influence factor (considered 2.96).

In the contour plots of displacement, it can be seen (Fig. 17.3a) that the maximum vertical displacements are found to be within the embankment region. With the movement from the centre towards the toe of the embankment, i.e. from section 1–1 to 4–4, displacements reduced as overburden pressure of the above layers decreases (Fig. 17.3b). The maximum displacements in contour plots can be seen in the range of 56 mm. However, the total displacement calculated using analytical Eq. 17.2 shows displacement of 70.931 mm at the end of primary consolidation (Table 17.2), this variation in settlement indicates conservative results of the analytical approach as compared to the numerical approach. The primary settlement of the MSW embankment was also computed by Eq. 17.2 as given below.

$$S = \frac{H}{1 + e_o} \mathrm{Cc} \log \frac{\sigma_o + \Delta\sigma}{\sigma_o} \tag{17.2}$$

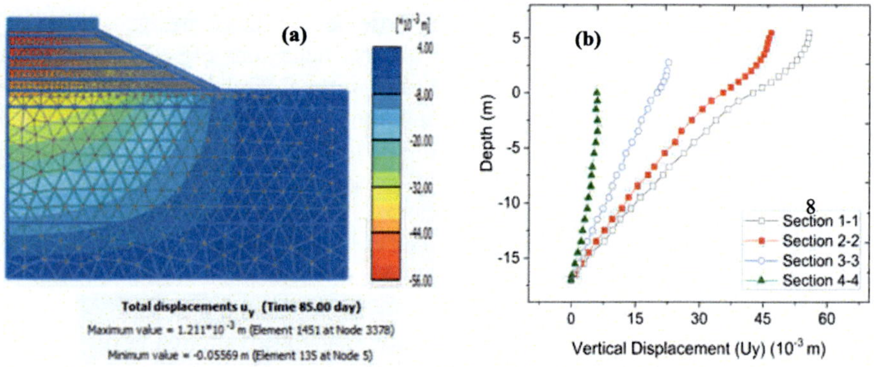

Fig. 17.3 Variation of vertical displacement: **a** contour fill and **b** variation with depth at different sections of the road embankment with MSW

Table 17.2 Results of primary settlement analysis

Height of the embankment (m)	5.5
Primary consolidation in the embankment (mm)	46.066
Primary consolidation in the subsoil (mm)	24.86
Total primary consolidation (mm)	70.931

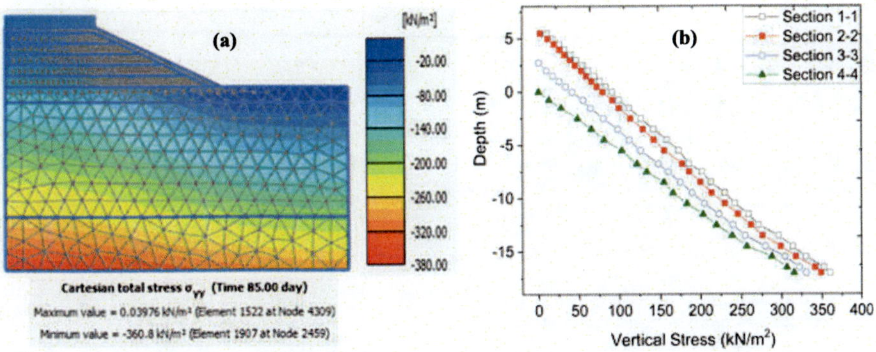

Fig. 17.4 Variation of vertical stress: **a** contour fill and **b** variation with depth at different sections of the road embankment with MSW

where

S	Primary compression in the layer under consideration, m
H	Height of the layer,
e_0	Initial void ratio, 0.55314 for MSW and 0.5 for subsoil,
Cc	Compression index, 0.12383 (MSW) and 0.1 (subsoil).

The settlement computed from the above equation for the MSW embankment and subsoil is given in Table 17.2. The total settlement computed theoretically and numerically is found to be below the allowable recommended settlement, according to IRC: 75 (1979) [13] for a design of low embankments (the allowable settlement range is 300–600 mm).

The variation in stresses along the depth of the embankment has also been recorded at different critical sections. The contour plots and the variation of stresses with depth are shown in Figs. 17.4, 17.5, and 17.6. It can be noted that the maximum stresses are induced in the bottom layers. The vertical stresses are more profound than that of the horizontal because of overburden pressure and assumed vertical traffic load. There is very little variation in horizontal stresses at different sections, but, the vertical stress decreases moving from section 1–1 to 4–4 (i.e. centre to toe) (Figs. 17.4b and 17.5b). The shear stresses can be seen maximum in section 3–3 which is in the middle of the embankment slope (Fig. 17.6b). The contour plot of the shear stress (Fig. 17.6a) shows that the maximum shear stress concentrated below the embankment towards the sloping side under the considered load.

Stability Analysis

The slope stability analysis was carried out by using Geo5 software which is a geotechnical software with analytical as well as finite element analysis solutions. It

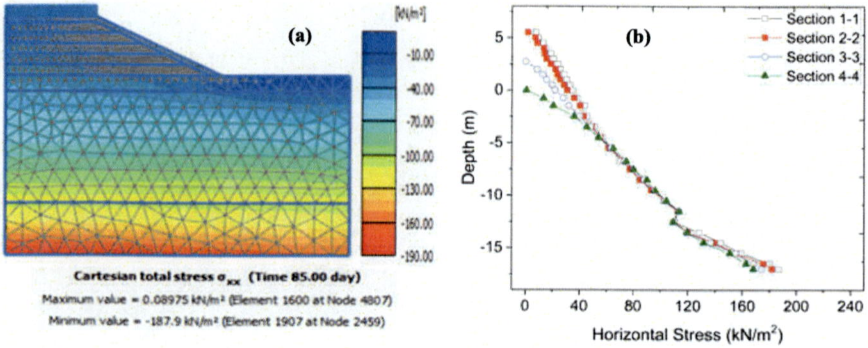

Fig. 17.5 Variation of horizontal stress: **a** contour fill and **b** variation with depth at different sections of the road embankment with MSW

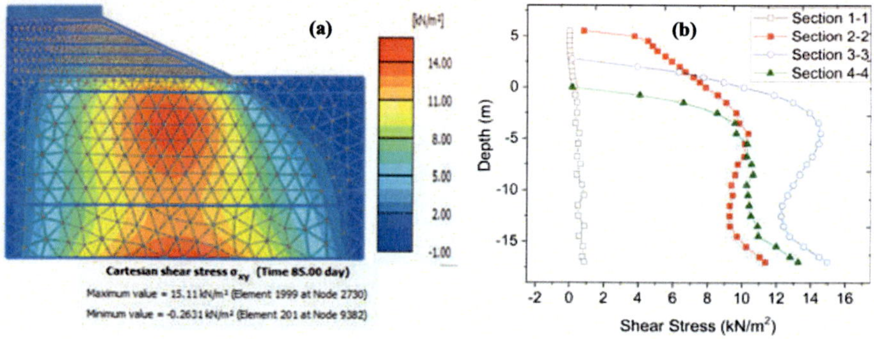

Fig. 17.6 Variation of shear stress: **a** contour fill and **b** variation with depth at different sections of the road embankment with MSW

consists of the programme designed to solve the large number of geotechnical problems commonly encountered. For this particular problem, the critical failure surface was evaluated according to Bishop, Fellenius, and Spencer limit state methods. The factor of safety (FOS) after optimization of the critical failure surface was found to be 2.67, 2.47, and 2.62 for Bishop, Fellenius, and Spencer methods, respectively, which was more than the allowable FOS of 1.25. A typical optimized failure surface is shown in Fig. 17.7. The embankments below 6 m height are considered as low embankment, and it is a common practice to adopt a factor of safety of 1.25 for the design of low embankments and 1.5 for high embankments [13].

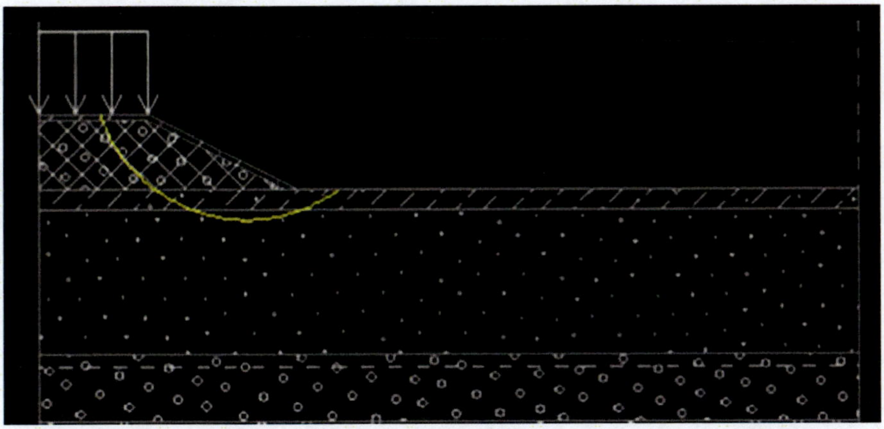

Fig. 17.7 A typical optimized critical failure surface of the road embankment with MSW

Recommendations and Limitations

The analysis was done for settlement and stability of the MSW embankment shows results within the allowable limits and even chemical and geotechnical parameters were within the safe limits. However, it is still recommended to isolate the embankment from the subsoil by providing a geomembrane with a clay liner below the MSW fill as provided in sanitary landfill sites to control the leachates percolation. This study does not include the long-term settlement of the embankment due to creep, degradation of the organic matter, unpredictable heavy traffic loads, flooding, and earthquakes. Before implementing the idea, these studies need to be done and monitoring of the embankment is required at least for 1–2 years.

Conclusions

The numerical analysis of the embankment with MSW concluded that for settlement and stability, the considered MSW can be used as a fill material in the road embankment, as the total settlement under the considered traffic loads are observed to be below the allowable limits. In addition, the factor of safety for the stability of the embankment is above the allowable limits provided by IRC: 75 (1979).

References

1. What a waste: a global review of solid waste management. Copyright © World Bank, 2012. http://hdl.handle.net/10986/17388
2. Havangi VG, Sinha AK, Parvathi GS, Chandra S (2017) Municipal solid waste in road embankment construction—a case study. J Indian Road Congr Pap 669:79–90
3. Athanasopoulos G, Grizi A, Zekkos D, Founta P, Zisimatou E (2008) Municipal solid waste as a reinforced soil: investigation using synthetic waste. ASCE-Geoinstitute Geocongress, pp 168–175. https://doi.org/10.1061/40970(309)21
4. Zekkos D, Kabalan M, Syal SM, Hambright M, Sahadewa A (2013) Geotechnical characterization of a municipal solid waste incineration ash from a Michigan monofill. Waste Manage 33(6):1442–1450. https://doi.org/10.1016/j.wasman.2013.02.009
5. Rawat P, Mohanty S (2021) Utilization of municipal solid waste as backfill material. In: Proceedings of the Indian geotechnical conference 2019: IGC-2019 volume II, 4 June 2021, vol 134. Springer, Berlin, p 49
6. Yang R, Xu Z, Chai J, Qin Y, Li Y (2016) Permeability test and slope stability analysis of municipal solid waste in Jiangcungou landfill, Shaanxi, China. J Air Waste Manag Assoc 66(7):655–662. https://doi.org/10.1080/10962247.2015.1093038
7. Hossain MS, Haque MA (2009) Stability analyses of municipal solid waste landfills with decomposition. Geotech Geol Eng 27(6):659–666. https://doi.org/10.1007/s10706-009-9265-0
8. Rajesh S, Babel K, Mishra SK (2016) Reliability-based assessment of municipal solid waste landfill slope. J Hazard Toxic Radioactive Waste 21(2):1–11
9. Durmusoglu E, Corapcioglu MY, Tuncay FK (2005) Landfill settlement with decomposition and gas generation. J Environ Eng 131(9):1311–1321. https://doi.org/10.1061/(ASCE)0733-9372(2005)131:9(1311)
10. Rawat P, Mohanty S (2021) Experimental investigation on MSW fine mixed with fibers: fiber reinforced waste. J Hazard Toxic Radioactive Waste 25(3):04021009
11. Brahma P, Mukherjee S (2010) A realistic way to obtain equivalent Young's modulus of layered soil. In: Indian geotechnical conference, Bombay, India, pp 305–308
12. Report on Geotechnical survey for preparatory survey on Ganga Rejuvenation Project. Project no. 93, Vishwa Bhumi Technologies, pp 1–186
13. IRC (1979) Guidelines for the design of high embankments, IRC 75, New Delhi, India

Chapter 18
A Numerical Study on Hydrodynamic and Liquefaction Analysis of Coastline Protected with Geotubes

A. Henitha Banumathi and S. P. Jeyapriya

Introduction

Coastline erosion is a significant problem along the shoreline due to the breaking of ocean waves, change in water levels, climate change, etc., and it is the major reason for the failure of onshore structures [1]. Conventional structures like breakwaters, seawalls, revetments, and groins are generally used to reduce shoreline erosion. These structures require a large amount of natural rocks or concrete blocks which are more expensive, and it is difficult to transport [2]. Due to the shortage of natural materials, it is suggested that it can be replaced by materials like slags, geosynthetics, gabions, etc.

Geotubes

Geosystems are the new construction systems made of geotextiles in the form of geomatresses, geotubes, geocontainers, and geocurtains for shoreline protection [2–4]. Geotextiles are typically made of synthetic fabrics made from polymeric materials like poly-ester (PET), poly-propylene (PP), poly-ethylene (PE), and poly-amide (PA). It can perform three major functions such as, (a) filtration (permeable, but soil tight), (b) reinforcement for soil against sliding, and (c) prevention of erosion of subsoil.

Geotextile tubes (geotubes) are long cylindrical tubes that are placed in the required position and filled by locally available dredged materials in the form of slurry using hydraulic pumps. While filling the tubes, the excess water will drain out from the tube, and it causes the decrease of tube height from initial, so that the geotubes

A. Henitha Banumathi (✉) · S. P. Jeyapriya
Government College of Technology, Coimbatore, Tamil Nadu, India
e-mail: henithace@gmail.com

K. Muthukkumaran et al. (eds.), *Soil Dynamics, Earthquake and Computational Geotechnical Engineering*, Lecture Notes in Civil Engineering 300,
https://doi.org/10.1007/978-981-19-6998-0_18

are needed to be filled more than once till the required height is attained [5]. If the required height is not sufficient by using the single tube, then tubes can be arranged in a stacked manner [1]. The general failure mechanisms of geotube embankment are sliding and overturning of geotubes, instability of ocean floor, scouring of the toe, and failure of geotextile material [1, 6]. Several studies have been made for the sliding and overturning failures of geotubes from physical model tests and case studies. This study mainly focuses on the failure of geotubes due to the instability of the ocean floor in terms of wave-induced liquefaction.

Liquefaction

Wave-induced liquefaction is categorized into two types: (i) due to earthquake effects and (ii) due to water pressure generated by the wave propagation. The effect of the second type of liquefaction is that, when wave propagation takes place which is a dynamic pressure acting on the soil, it results in the development of excess pore pressure and effective stresses in the soil mass. The reason is that ocean waves cause rearrangement of solid particles resulting in the reduction of volume of voids [7]. When the excess pore pressure gets exceeded the critical value, the soil liquefy leading to instability of coastal structures. Even the structure gets sunk into the liquefied soil depending upon the depth of liquefiable soils [8–10]. Hence, determination of effective stress, distribution of pore pressure, interaction between soil, and waves and the geometry of the structure are very important in the study of hydrodynamic analysis. Studies showed that, longer wave periods and waves rising to greater heights will generate pore pressure of higher magnitude on the seaward side of a breakwater and there is a decrease in pore pressure was observed in the core of the breakwater [11].

On comparing the wave-induced liquefaction with that of seismically induced liquefaction, there exists two major differences, (i) ocean wave periods are longer than earthquake shaking and (ii) ocean wave loadings are imposed at the surface of the seafloor, whereas earthquakes impose loads which are below the sea floor. Hence, wave-induced liquefaction analysis can be made by making suitable modification in the earthquake-induced liquefaction analysis.

Considering pore pressure development, the rate and amount of pore pressure build-up rely on three factors such as (i) storm characteristics, (ii) cyclic loading characteristics, and (iii) drainage and compressibility characteristics of soil profile [9, 12].

The liquefied zone is different from the shear failure zone as shear failure zone is likely to occur at both toes of breakwater, whereas in the liquefied zone, the failure will occur around the wave trough with highest upward seepage force [10, 11].

Hence the objectives of the study are to create a numerical model of a coastline having severe erosion, to analyse the factors of failures with respect to different sizes of geotube embankment and different levels of groundwater and finally to analyse

Table 18.1 Soil characteristics of the study area

Characteristics	Mandaikadu
Type of soil (texture)	Sandy
Colour	Yellowish-brown
Sand %	87.5
Gravel %	5.5
Silt and clay %	7

the wave-induced liquefaction from the change in void ratio and effective stress with respect to depth.

Location and Topography of Study Area

In this study, the coastline of Mandaikadu has been chosen for the analysis. The coastline of Mandaikadu, Kalkulam Taluk, Kanyakumari District is one of the most prone zones of beach erosion which has a fishing harbour, beach sand mining, and thick fishermen habituation along the shoreline, and it is located between 8° 10′ 30.9″ to 8° 09′ 44.1″ N Latitude and 77° 15′ 35.5″ to 77° 16′ 58.8″ E Longitude along the western coast of the Arabian Sea. Wind, wave, and coastal disturbances such as storm surges, sea-level rise, and other natural processes are the main causes of erosion in the studied region. The observed predominant wind directions of the study area were SW, SSW, and N with a 0.78% calm period. The SW monsoon is severe along the coast creating heavy erosion resulting in loss of valuable lands, roads, worship places, and houses. The soil characteristics of the study area are given in Table 18.1. Along the shore side, an elevation of (+) 6.00 m was observed at about 50 m behind the shoreline. The sea bed slope up to a water depth of (−) 10.00 m is 0.0455. The foreshore slope up to a water depth of about (−) 10.00 m is 0.0455. The bed slope up to (−) 3.00 m is 0.060 indicating that the profile has a steeper slope in the shallow depth compared to deep water. This can result in a run-up of wave height ranging from 1 to 3 m and break close to the shore. Figure 18.1 shows the cross-shore profile of the study area. The details of location and topography of the study area discussed above and Tables 18.1 and 18.3 and Figs. 18.1 and 18.2 were referred from the Environmental Impact Assessment (EIA) study report by Anti Sea Erosion Division of Public Works Department, Government of Tamil Nadu [13].

Numerical Modelling of Coastline

In this study, the modelling software PLAXIS 3D (2013) is used to evaluate the performance of coastline and geotubes for various conditions. PLAXIS is a finite element package intended for the two-dimensional and three-dimensional analysis

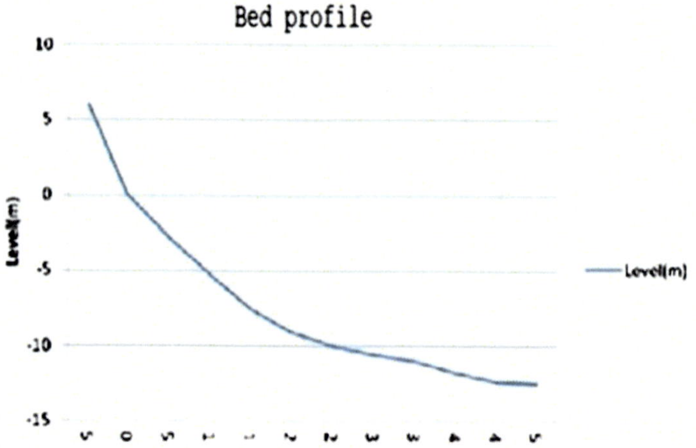

Fig. 18.1 Cross-shore profile

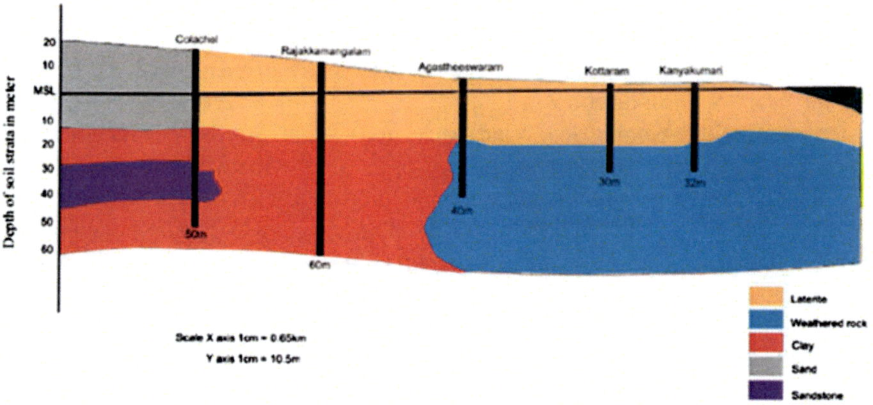

Fig. 18.2 Location of boreholes

of deformation and stability characteristics of geotechnical engineering structures. Laboratory experiments are not effective to simulate the nonlinear, time-dependent, and anisotropic behaviour of soils and/or rocks and dealing with hydrostatic and non-hydrostatic pore pressures in the soil. In the field, studies on soil and its behaviour itself is a cumbersome process in the case of tunnel projects, structures subjected to vibration and wave loading, numerical modelling simplifies the difficulties mentioned above with less cost and time. The soil properties and wave characteristics collected for the study area from various resources are given in Tables 18.2 and 18.3, respectively. Based on the data collected, a numerical model of the coastline was created having dimensions $x = 40$ m (length), $y = 10$ m (width), $z = 50$ m (depth), and the soil layers were created as per borehole data of the study area as shown in Figs. 18.2 and

18.3. Based on the cross-shore profile (Fig. 18.1) of the study area, an embankment was created on the onshore side.

The Mohr–Coulomb model was chosen to find the behaviour of soil. It is a linear elastic and perfectly plastic model and is generally used to find the first approximation of soil behaviour [14]. Numerical modelling of geotube was carried out using poly-curve option in structures mode then the shape of the curve was extruded as a tube. Totally three geotube models were created with different dimensions as shown in

Table 18.2 Properties of soil

Soil type	γ_{unsat} (kN/m^3)	γ_{sat} (kN/m^3)	k (m/day)	E (kPa)	ν	C (kPa)	Φ (deg)
Geotube fill	16.5	17.5	0.34	18,000	0.4	5	30
Topsoil	16	18	0.34	20,000	0.4	9	30
Sandy flay	15	17.5	0.86	6000	0.4	5	25
Sand	16.5	18	0.86	30,000	0.35	0	35
Kankar	19	20	0.04	360,000	0.3	35	32
Sand stone	18	19	0.34	60,000	0.32	22	30

Source Kim et al. [2]

Table 18.3 Wave characteristics of the study area

Month	Jan	Feb	Mar	Apr	May	June	July	Aug	Sept
Height of wave (m)	0.5	0.6	0.5	0.5	1.56	1.56	1.41	1.56	1.25
Time (s)	10	12	13	12	10	6	7	8	10

Source Shoreline Protection Structures, Mandaikadu, Kanyakumari District, PWD/Anti Sea Erosion Division, Govt. of Tamil Nadu

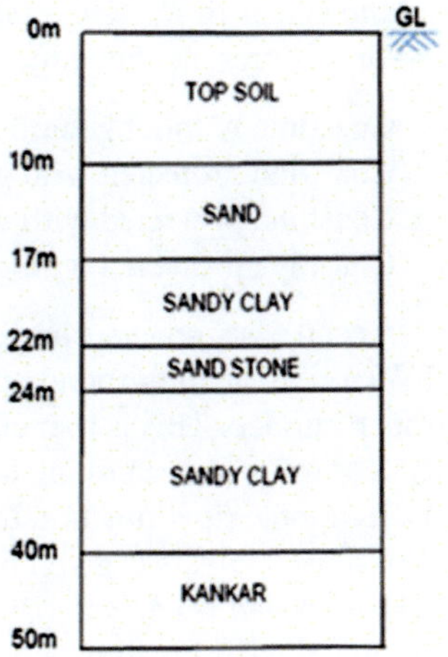

Fig. 18.3 Soil profile of study area

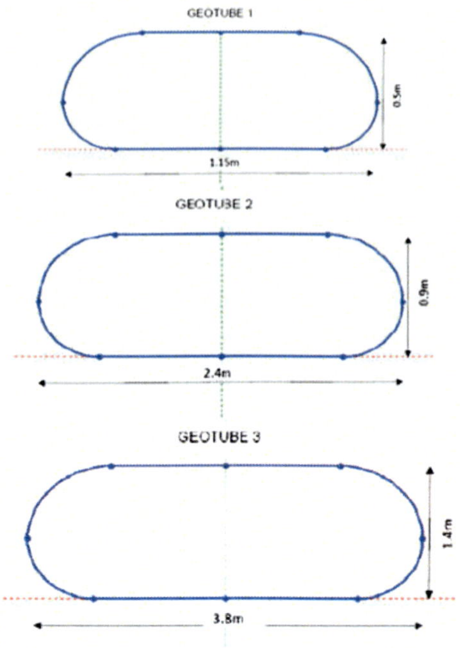

Fig. 18.4 Geotube models

Fig. 18.4, and the geotubes are assigned as a linear elastic geogrid element having axial stiffness of 600 kN/m.

Interface elements were used to simulate the exact behaviour of soil and geotube interaction. Interfaces were created for each of the geotubes and for the soil layer where the bottom of geotubes were placed. The roughness of interaction was modelled by using strength reduction factor (R_{inter}) in the material properties menu, and the values of R_{inter} was taken as 0.7 for both geotube fill and top soil [2].

For this present study, four different cases have been considered, namely

1. Coastline without geotube.
2. Coastline protected with geotube of diameter1.15 m.
3. Coastline protected with geotube of diameter 2.4 m.
4. Coastline protected with geotube of diameter 3.8 m.

In each case, analysis has been carried out for three different groundwater levels (GWL), i.e., at the ground surface, at 5 m below ground surface, at 15 m below ground surface. This is to include the variations of seepage along the shoreline. The depth of water above ground level is taken as 2 m for all the cases, and it was assigned as a harmonic flow function to simulate the ocean wave. The properties of the wave used in the analysis is given in Table 18.3.

Here, ocean wave is the dynamic load acting on the geotube embankment, and it was modelled by creating a surface water level of 2 m head and assigned as a harmonic function. For dynamic analysis, the water surface was assigned as a surface load of

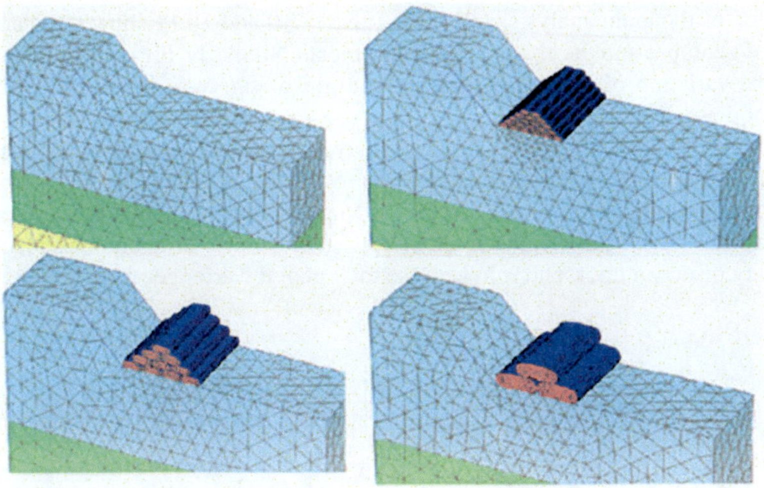

Fig. 18.5 Numerical model of the coastline

20 kN/m^2 and applied in both x and z directions as a dynamic load and activated during the dynamic analysis. After the completion of models for each case, the mesh has been generated using mesh mode. Figure 18.5 shows the numerical model of embankment along with geotubes.

Staged Construction

Staged construction is a feature to enable realistic simulation of construction, loading, and excavation processes. Starting with the initial phase, all the soil layers were activated, and K0 procedure is used for the analysis. On further phases, embankment soil, geotubes, and geotube fills were activated one by one.

Fully coupled flow deformation analysis is required when it is necessary to analyse the time-dependent development of deformations and pore pressures in saturated and partially saturated soils like drawdown of dams, dams subjected to tidal waves, dewatering of building sites, etc. [14].

Results and Discussion

From the hydrodynamic analysis, the deformed shapes of the model, total displacement of the geotube arrangement, and the global safety factors were found out for all the cases of study. Figures 18.6, 18.7 and 18.8 show the deformed shape of the model at the end of dynamic analysis, and Figs. 18.9, 18.10 and 18.11 show the displacements of geotubes and the change in displacements. Similarly, Figs. 18.12 and 18.13 show the variation of total displacement and global safety factor respectively for different diameters of geotube.

It can be observed that, the bottom most layer of geotubes are the most unstable tubes in all the cases which attains the maximum displacement. There observed a lowering of ocean floor as noted from Figs. 18.6, 18.7 and 18.8 that proves that the reason for the maximum displacement of bottom most layer of geotube is due to the instability of ocean floor, and it lowers the total geotube arrangement.

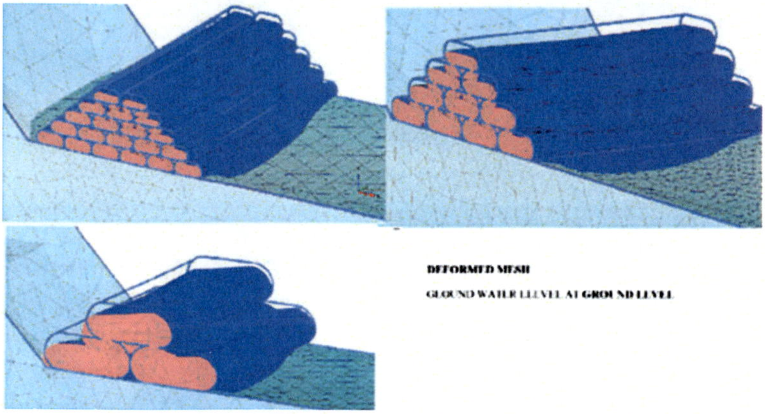

Fig. 18.6 Deformed coastline—GWL at 0 m

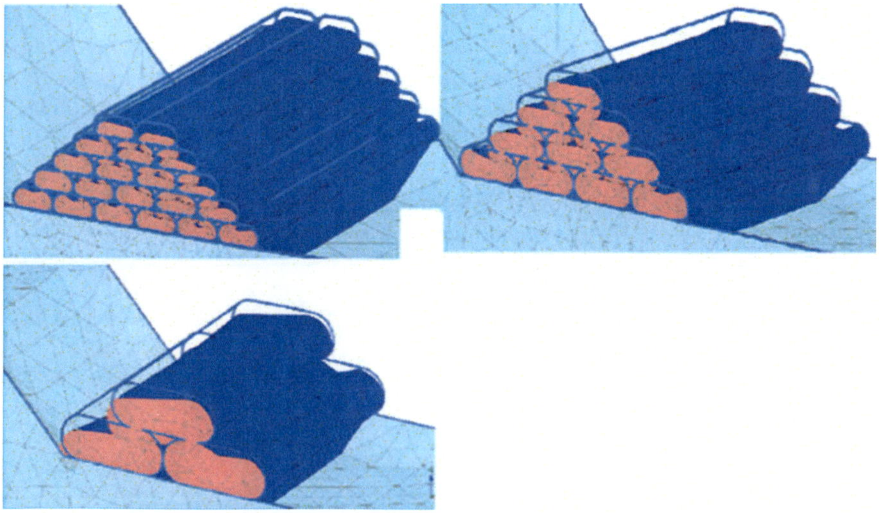

Fig. 18.7 Deformed coastline—GWL at 5 m

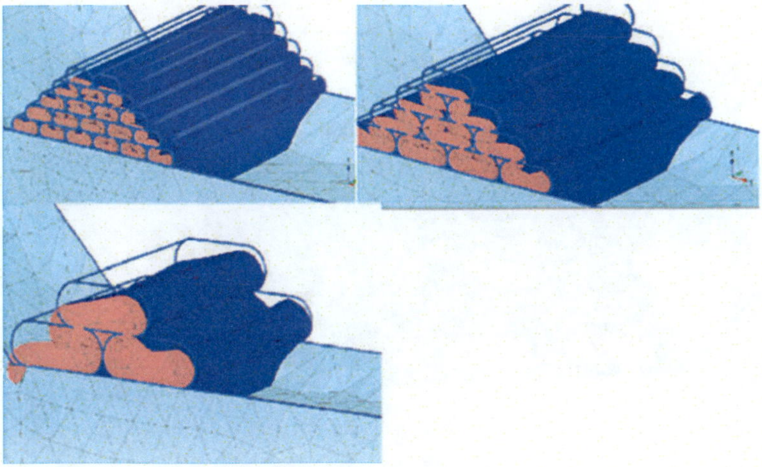

Fig. 18.8 Deformed coastline—GWL at 15 m

Fig. 18.9 Displacement of geotube of size 1.15 m

The smaller sizes of geotube arrangement like 1.15 and 2.4 m were found to be stable at the end of dynamic analysis but bigger size geotube arrangement, i.e., 3.8 m gets collapsed. It demonstrates that, lower geotube sizes may be densely packed towards the end of dynamic loading, resulting in a stable embankment. It is clear from Fig. 18.13 that, a higher safety factor for lower sizes of geotubes, and also it proved that the lowering of GWL increases the safety factor. Total displacement of geotubes showed that, the presence of geotubes reduces the total displacement when

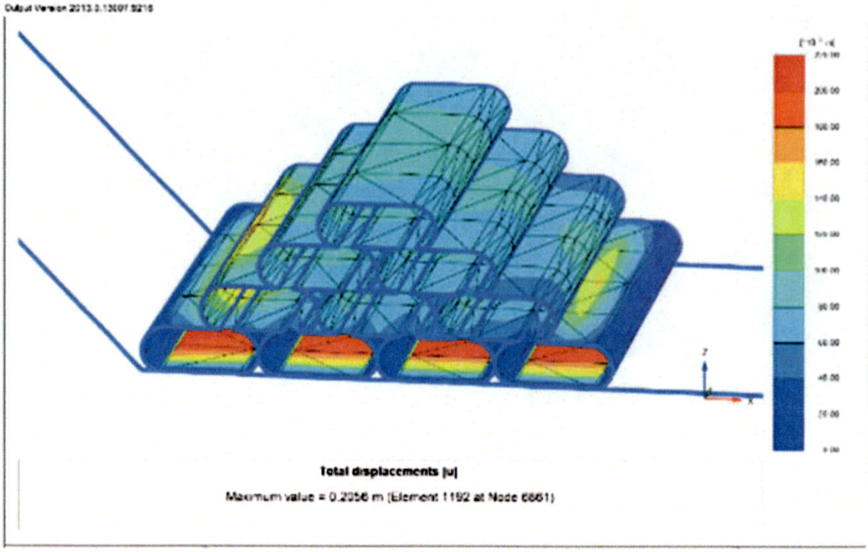

Fig. 18.10 Displacement of geotube of size 2.4 m

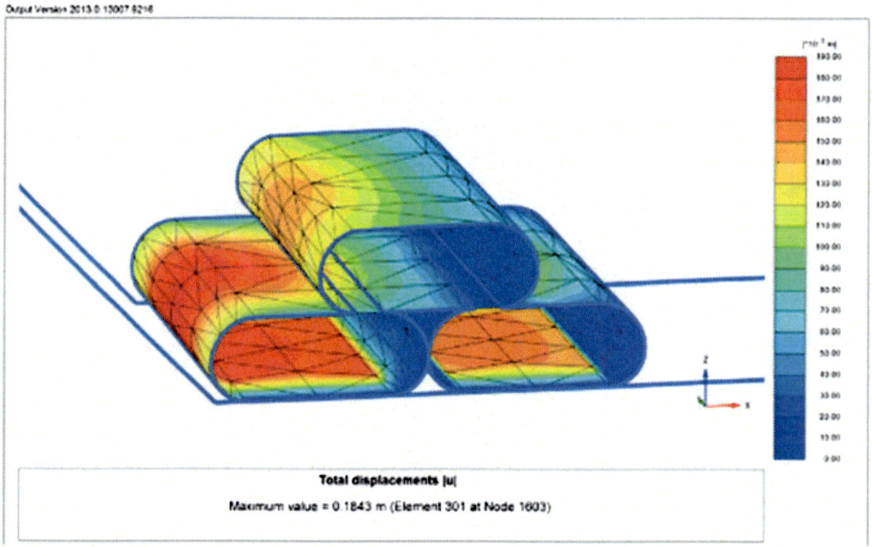

Fig. 18.11 Displacement of geotube of size 3.8 m

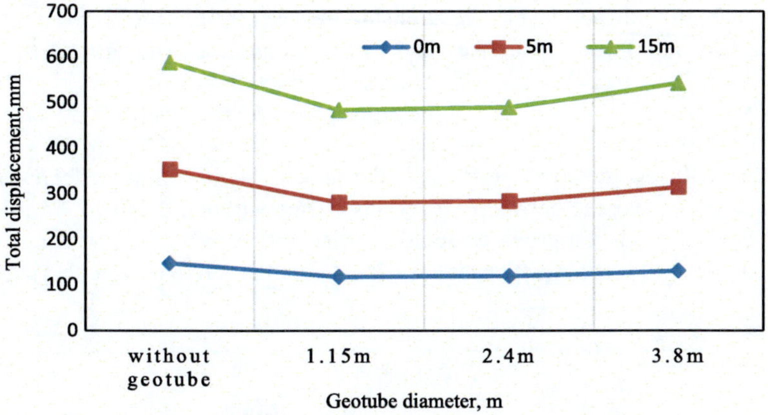

Fig. 18.12 Variation of total displacement w.r.t different diameters of geotube

compared to that of coastline without geotube as depicted in Fig. 18.12, but it shows higher displacement on lowering of GWL.

However, when GWL is much lower, the safety factor values are considerably greater, thus the geotube embankment will remain stable even if the load is increased or the strength characteristics of the bottom soil decreases.

To analyse the ocean floor instability, a comparative study was carried out for the variation of void ratio and effective stress with respect to depth for different diameters of geotubes and also with varying GWL. From the results obtained, the critical depth for liquefaction was found out. The analysis was carried out for the plane of soil at the bottom of geotube having coordinates $x = 16$–21 m, $y = 5$ m (midpoint of the width), and $z = 0$–10 m (depth of soil).

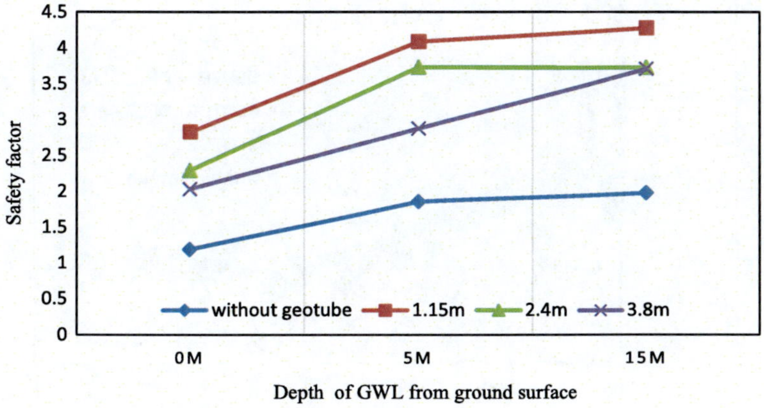

Fig. 18.13 Variation of global safety factor w.r.t depth of GWL

For each groundwater level, a relationship between the changes in void ratio vs. depth as shown in Figs. 18.14, 18.15 and 18.16 was observed and studied. In all the cases of GWL, up to a depth of 1.5 m, there is reduction in void ratio from 0.86 to 0.84 but when the GWL is nearer to the ground surface, there is more fluctuation in void ratio within the values 0.86 and 0.84.

While analysing the effective stress of the same plane of soil, the effective stress decreases on an average between 50 and 40 kN/m^2 within a depth of 0.5 m when geotube protection is not provided and thereafter it increases slowly with respect to depth. Due to the loss of effective stress, it is easy for the soil to be liquefied since the

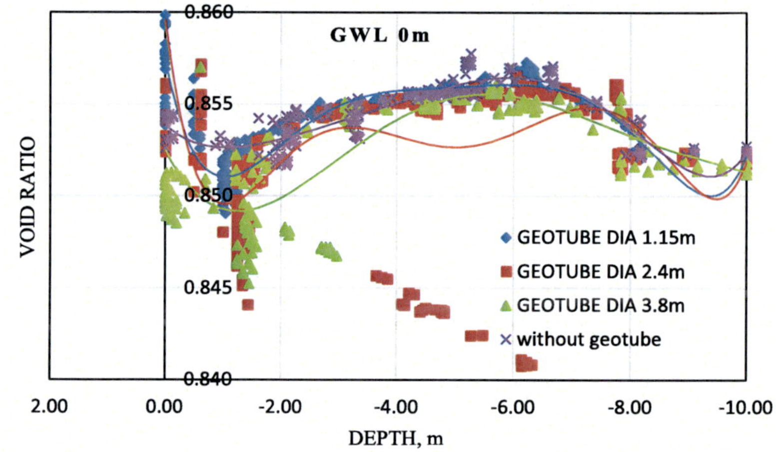

Fig. 18.14 Variation of void ratio w.r.t depth of soil when GWL 0 m

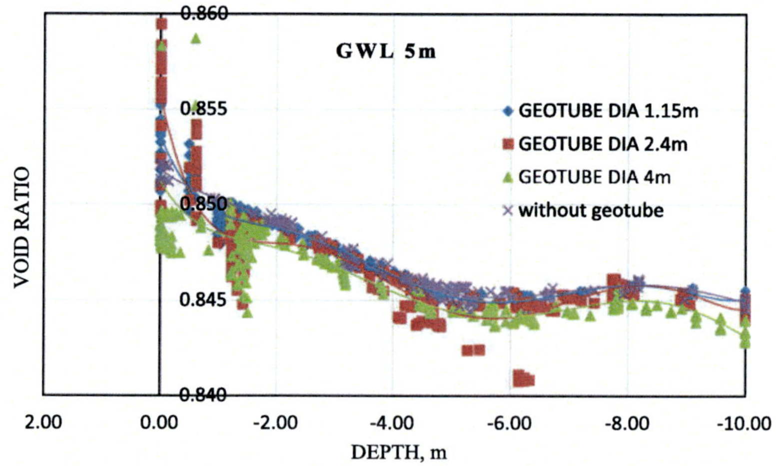

Fig. 18.15 Variation of void ratio w.r.t depth of soil when GWL 5 m

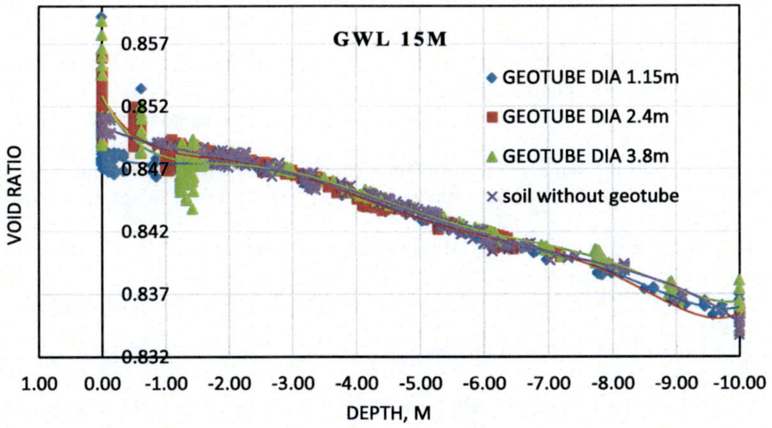

Fig. 18.16 Variation of void ratio w.r.t depth of soil when GWL 15 m

loss is within a shallow depth of 0.5 m. When there is geotube protection, the average reduction of effective stress is between 30 and 35 kN/m^2 up to a depth of 1–1.2 m. Hence, the presence of geotube increases the depth of loss of effective stress that can reduce the possibility of liquefaction of bottom soil.

Conclusions

A numerical analysis using PLAXIS 3D is performed to simulate and understand the effects of various sizes of geotube embankment with varying GWL under the action of wave loading. From the study, the following conclusions are drawn:

1. The size of the geotube has a significant effect on the stability of the geotube embankment. The smaller size of geotubes arranged in a stacked manner gives better resistance against displacement.
2. The fluctuations in groundwater level also influence the stability of geotubes. Lowering of GWL increases the total displacement of the geotube. But the safety factor values are much higher when GWL is at 15 m so the geotube embankment will be stable even on further increase of load or decrease of strength parameters of bottom soil.
3. When ocean waves propagate over a seabed, a reduction of voids ratio was observed, due to the rearrangement of the soil grains, and it decreases the effective stresses within 0.5 m and increases the possibility of liquefaction.
4. The presence of geotube embankment reduces the loss of effective stress and makes the depth of maximum loss at 1.2 m. Since the wave-induced liquefaction will occur mostly on top of the seabed, the presence of geotubes acts like a cut-off to the liquefaction.

References

1. Sherlin Prem Nishold S, Sundravadivelu Nilanjan Saha R (2018) Hydrodynamic performance of coastal geotube embankment with and without gabion boxes. Springer International Publishing AG. https://doi.org/10.1007/978-3-319-61648-3_15
2. Kim H-J, Won M-S, Jamin JC (2014) Finite-element analysis on the stability of geotextile tube—reinforced embankments under scouring. ASCE Int J Geomech. ISSN 1532-3641/06014019(13)
3. Lee SC, Hashim R, Mo KH (2018) The evaluation of geotube behaviors on muddy beach: field monitoring and numerical analysis. KSCE J Civ Eng 22(11):4185–4193
4. Pilarczyk K (1999) Geosynthetics and geosystems in hydraulic and coastal engineering, 1st edn. Rotterdam
5. Shin EC, Oh YI (2007) Coastal erosion prevention by geotextile tube technology. Geotext Geomembr 25:264–277
6. Kiran AS, Ravichandran V, Sivakholundu KM (2015) Stability analysis and design of offshore submerged breakwater constructed using sand filled geo synthetic tubes. Procedia Eng 116(2015):310–319
7. Sumer BM, Fredsøe J, Christensen S, Lind MT (1999) Sinking-floatation of pipelines and other objects in liquefied soil under waves. Coast Eng 38:53–90
8. Jeng DS, Zhang H (2005) An integrated three-dimension model of wave-induced pore pressure and effective stress in a porous seabed: II. Breaking Waves Ocean Eng 37:1950–1967
9. Bolton Seed H, Rahman MS (2008) Wave-induced pore pressure in relation to ocean floor stability of cohesion less soils. Marine Geotechnol 3(2):123–150. https://doi.org/10.1080/10641197809379798
10. Ren Y, Xu G, Xu X, Zhao T, Wang X. The initial wave-induced failure of silty seabed: liquefaction or shear failure. Science Direct. https://doi.org/10.1016/j.oceaneng.2020.106990
11. Can Ulker MB, Tatlioglu E, Aysen Lav M (2018) Dynamic response and liquefaction analysis of seabed-rubble mound breakwater system under waves. Appl Ocean Res 78(2018):75–87
12. Zhao HY, Liang ZD, Jeng D-S, Zhu JF, Guo Z, Chen WY (2018) Numerical investigation of dynamic soil response around a submerged rubble mound breakwater. Ocean Eng 156:406–423
13. Environmental Impact Assessment (EIA) report, shoreline protection structures (2011), Mandaikadu, Anti Sea Erosion Division/Public Works Department, Government of Tamil Nadu
14. PLAXIS 3D 2013 Reference manual

Chapter 19
Parametric Study for Optimum Design of Combined Piled Raft Foundation (CPRF)

E. L. Suresh Babu, G. Suneel Kumar, Suvendu Kumar Sasmal, and Rabi Narayan Behera

Introduction

Raft foundation can be used effectively when the shallow layers are capable enough to take the loads. If the loads are high as in case of high-rised buildings and layers below raft are less stiff, then the excessive settlement may occur in the structure which causes failure of structure. So, in such cases piles can be used as settlement reducers along with the raft. It is seen that use of piles also increases the load carrying capacity of the structure. Cooke [1] and Price et al. [2] observed that load is being shared by piles and raft thus increasing the load carrying capacity of the structure. Clancy et al. [3] proposed a method to analyse the piled raft, and it is called as hybrid approach including the interaction coefficients between rafts and soils. Katzenbach and Reul [4] and Katzenbach et al. [5] carried out structural model for soil-structure interaction of piled raft and proposed a piled raft coefficient, and he also proposed design procedures of piled raft for many high-rised buildings which are discussed in book foundations on bored and auger piles. Poulos [6] in his book pile foundation analysis and design discussed the analysis and design of piled raft along with the pile groups. Katzenbach et al. [7] investigated several piled raft foundations and concluded that it is more economical when compared to the pile foundation. Poulos [8] and Poulos et al. [9] analysed the pied raft in different soils and found it to be more effective in sands and stiff clays; he also presented that increasing number of piles may reduce the settlement.

Unsever et al. [10] carried out numerical analysis on 3-pile piled raft under lateral load and predicted that about 85% of load is carried by raft. Sawada and Takemura [11] proposed that settlement for pile group under combined horizontal load and moment is high compared to piled raft. Chanda et al. [12] analysed piled raft modelled

E. L. Suresh Babu · G. Suneel Kumar · S. K. Sasmal · R. N. Behera (✉)
National Institute of Technology Rourkela, Rourkela, India
e-mail: rnbehera82@gmail.com

K. Muthukkumaran et al. (eds.), *Soil Dynamics, Earthquake and Computational Geotechnical Engineering*, Lecture Notes in Civil Engineering 300,
https://doi.org/10.1007/978-981-19-6998-0_19

in finite element with vertical, horizontal loading and moment and proposed that design of piled raft considering lateral response yields in reduced number of piles. Varghese et al. [13] carried out the dynamic analysis on piled raft foundation and concluded that the load carried by piles reduced up to 40% in case of dynamic loading. Cooke [14] and Cho et al. [15] proved effective usage of piled raft in cohesive soils.

The present work deals with the analysis of the combined piled raft foundation by varying parameters like pile length, raft thickness, pie arrangement, etc., to get an idea on optimum way to design a piled raft. PAXIS 3D is used as a tool to carry out this parametric study. First, the validation is done using the experimental work as reference; then, the piled raft is modelled in PLAXIS 3D; and then, the parametric analysis is carried out on model by varying different parameters and the corresponding settlement behaviour is plotted.

Validation

Matsumoto [16] performed an experimental analysis on the piled raft, and the resulting settlement behaviour is published. This CPRF and soil is modelled in PLAXIS 3D according to Kumar et al. [17], whose model dimensions are shown in Fig. 19.1 and the observed results are compared with experimental data. The output showed the similar trend with experimental results as shown in Fig. 19.2.

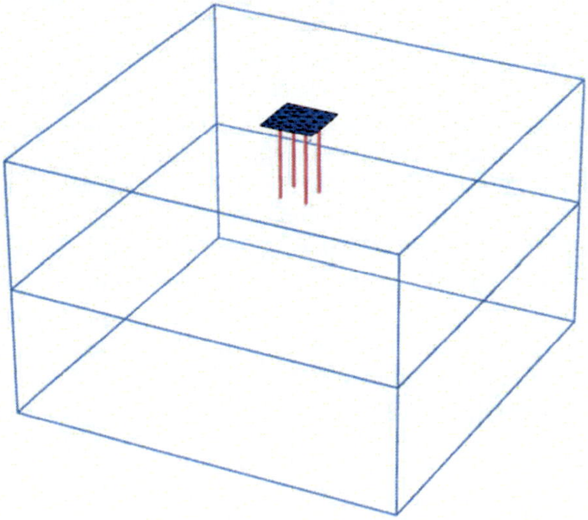

Fig. 19.1 Three-dimensional view of the modelled CPRF and soil in PLAXIS 3D

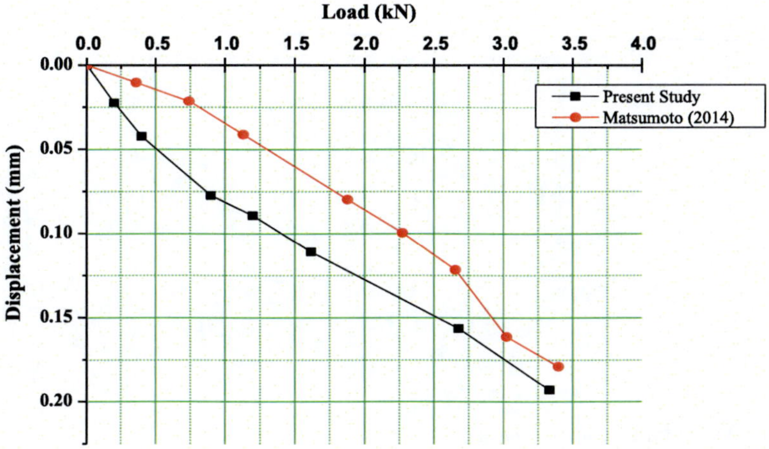

Fig. 19.2 Comparison of load-settlement curves obtained from PLAXIS 3D with experimental results

The model properties for validation are taken from Matsumoto [16] and are as following, the model is of dimension 1.5 × 1.5 × 1.5 m ($L \times B \times D$), raft is of square with dimensions 400 mm and its thickness is 40 mm, and the four piles are made of aluminium of outer diameter 40 mm and thickness is 2 mm and extends to length of 600 mm. The top 1 m layer of model is of Toyoura sand having density of 25.3 kN/m^3, angle of internal friction is 40°, maximum dry density is 15.89 kN/m^3, reference stiffness modulus is 17,000 kN/m^2, and Poisson's ratio is 0.3, and bottom layer is of brick mass with Young's modulus 6×10^6 kN/m^2, maximum dry density is 22 kN/m^3, and Poisson's ratio is 0.2. Pile and raft are of aluminium with Young's modulus of 7×10^7 kN/m^2 and Poisson's ratio of 0.3.

Parametric Analysis

A soil profile is modelled taking 40 m depth and 150 × 150 m dimensions as shown in Fig. 19.3. The raft is of 36 × 36 m^2 area, and the material used for raft, basement wall, and the piles is concrete. In this model, 16 number of piles are used with a diameter of 1.5 m. The material properties for the model are given in Table 19.1. Then, the parameters like the length of pile, thickness of the raft, and the arrangement of the piles are varied and the results are analysed in the later section.

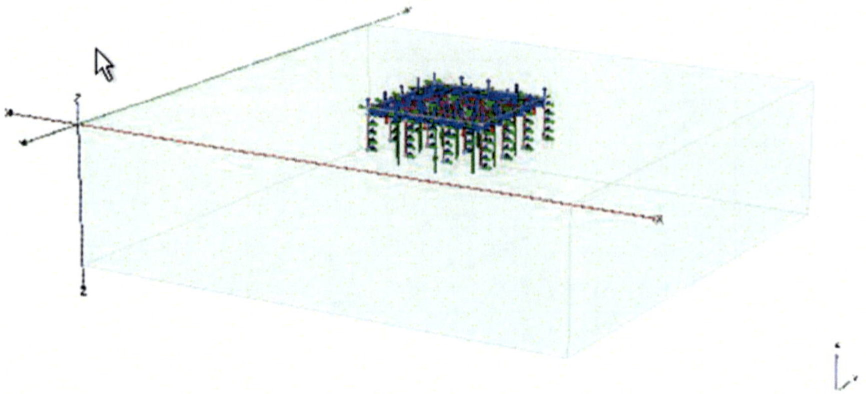

Fig. 19.3 Model for parametric analysis

Table 19.1 Material properties taken for the analysis of model piled raft

S No.	Parameter	Soil	Raft	Pile
1	Material model	Mohr–Coulomb	Linear elastic	Linear elastic
2	Unit weight (γ), kN/m^3	18	24	25
3	Young's modulus (E'), kN/m^2	15,000	30×10^6	30×10^6
4	Poisson's ratio (ν)	0.3	0.15	0.15
5	Cohesion (c'), kN/m^2	15	–	–
6	Angle of internal friction (ϕ'), degree	20	–	–
7	Permeability (k), m/day	0.65	–	–

Influence of Using Piles to Reduce Settlement

The model created is analysed with placing the raft and the piled raft individually and checked for the change in the settlement behaviour.

The variation of settlement behaviour when raft is used alone and when piled raft is used is shown in Fig. 19.4. It shows that the introducing of piles reduced the settlement in raft and increased the load carrying capacity by 30%.

Effect of Length of Pile on the Settlement Behaviour of Piled Raft

The settlement behaviour is studied with increasing pile length with increasing length to width ratio of pile. The load displacement curves are plotted as follows.

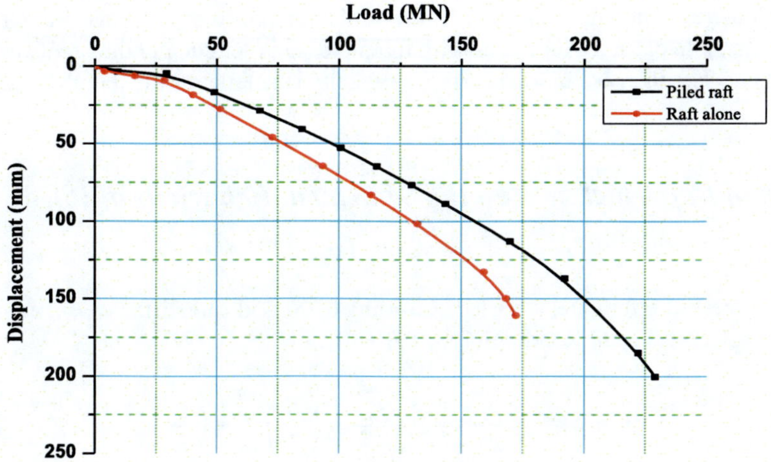

Fig. 19.4 Load displacement curves for raft alone and piled raft case

The different load-settlement curves obtained by varying pile length in clays are represented in Fig. 19.5.

It is seen that as the length of pile increases, the settlement is reduced but this is not prominent as the length exceeds some limit. The reduction in settlement after some point of increase is very less.

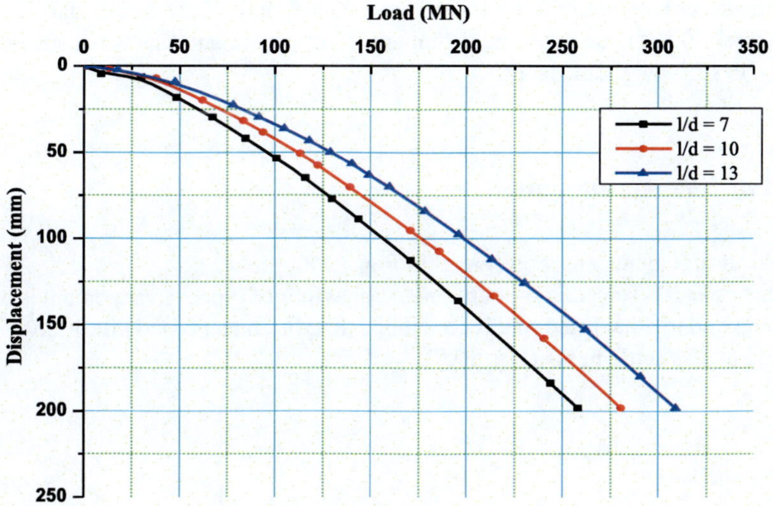

Fig. 19.5 Load-settlement curves with varying pile lengths

With increase in length of the pile, the settlement got reduced; this is due to the reason that pile capacity increases with the increase in length and thus reducing the settlement. The increase in load carrying capacity is smaller as the length increases.

Effect of Raft Thickness on the Settlement Behaviour of Piled Raft

The thickness of raft model is varied according to the soil-raft stiffness *(k_{rs})* equation given by Horikoshi and Randolph [18, 19].

$$k_{rs} = 5.57 * \frac{E_r * (1 - \mu_s^2)}{E_s * (1 - \mu_r^2)} * \left(\frac{B}{L}\right)^{0.5} * \left(\frac{t_r}{L}\right)^3 \tag{19.1}$$

In the above equation.

E_r and E_s are the Young's modulus of raft and the soil, respectively.

μ_s = Poisson's ratio of soil.

μ_r = Poisson's ratio of raft.

L, *B* are length and width of the raft.

t_r = thickness of raft.

The settlement of the piled raft is analysed with increasing raft thickness with respect to k_{rs} from 0.001 to 1 (Figs. 19.6 and 19.7).

From the graphs (Fig. 19.7), it is found that as the thickness of raft increases, the ultimate settlement of the foundation is reduced, and it is also seen that after k_{rs} value crosses 0.5, the decrease in settlement is not prominent; this is because of the increase in self-weight of the raft.

Effect of Pile Arrangement

The pile arrangements are represented in Fig. 19.8.

In this case, it is seen that the settlement behaviour of type 3 arrangement is desirable compared to other three so it is concluded that the concentrated pile arrangement can yield to less settlements (Fig. 19.9).

Conclusions

In this work, the parameters effecting the piled raft behaviour are analysed and the following conclusions are drawn:

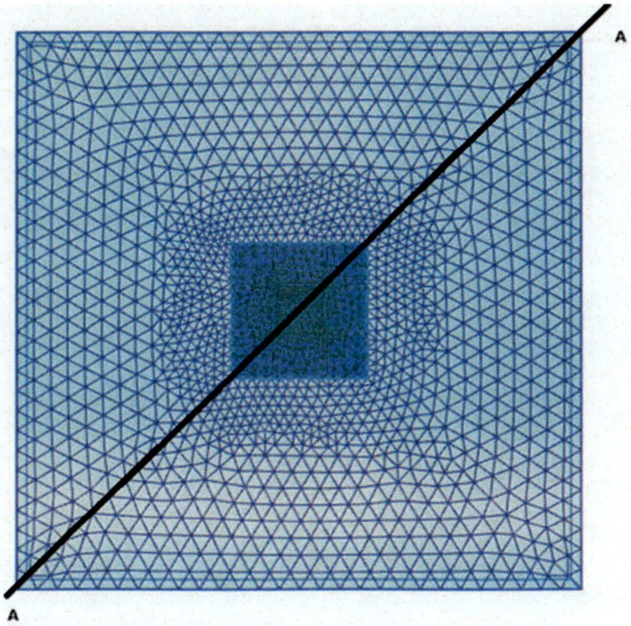

Fig. 19.6 Cross-section A-A is considered for the analysis of the settlement behaviour

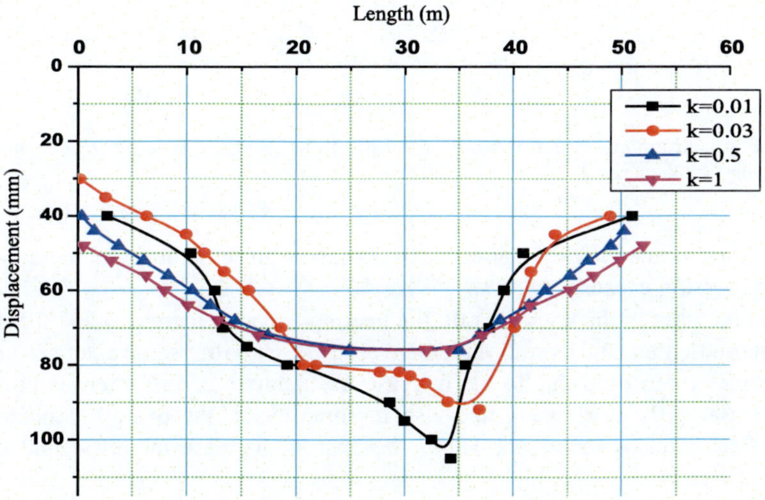

Fig. 19.7 Settlement behaviour of the raft along section A-A for different k_{rs} values

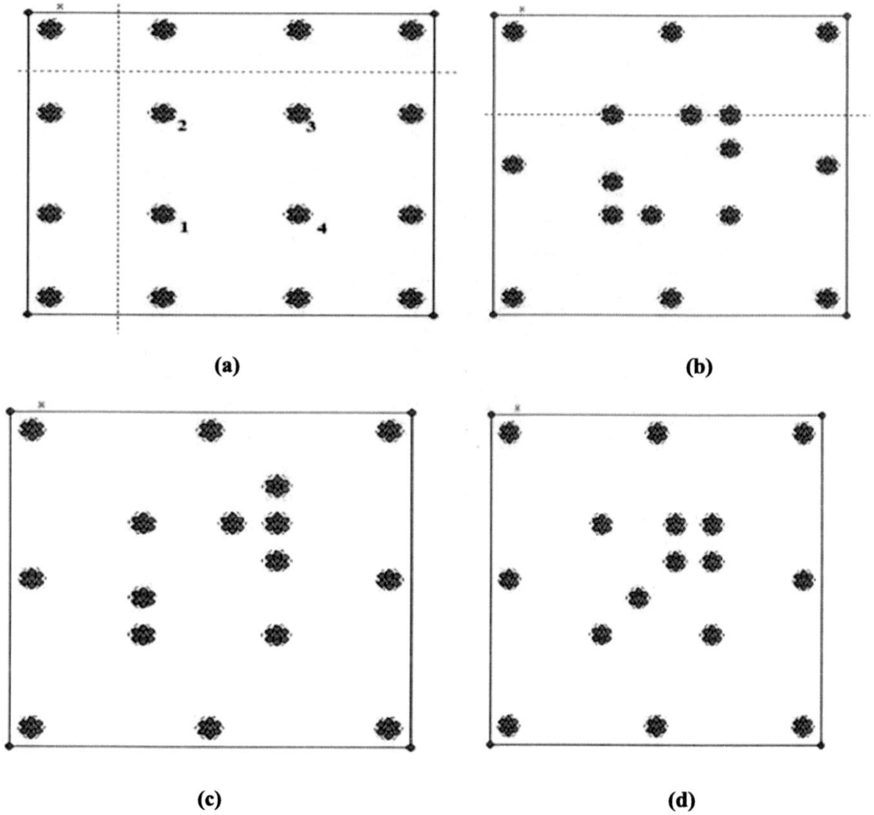

Fig. 19.8 **a** Uniform pile arrangement. **b** Type 1 pile arrangement. **c** Type 2 pile arrangement. **d** Type 3 pile arrangement

- It is noticed that providing piles for the raft reduced settlement and increased the raft's bearing capacity by 30%.
- With increase in thickness of raft, the bearing capacity increased and the differential settlement and overall settlement reduced. But after some extent as the raft becomes more rigid due to the increase in thickness of raft, there is no much decrease in the settlement, and also in some cases, the overall settlement of the foundation is increased. This is because of increase in self-weight of the foundation.
- As the length increases, there is a reduction in settlement but the rate of reduction is not of considerable extent, and also after some increase in length, there is no change in settlement behaviour.
- The concentrated pile arrangement produced less settlement while compared to the piles that are arranged uniformly, even keeping the number of piles constant.

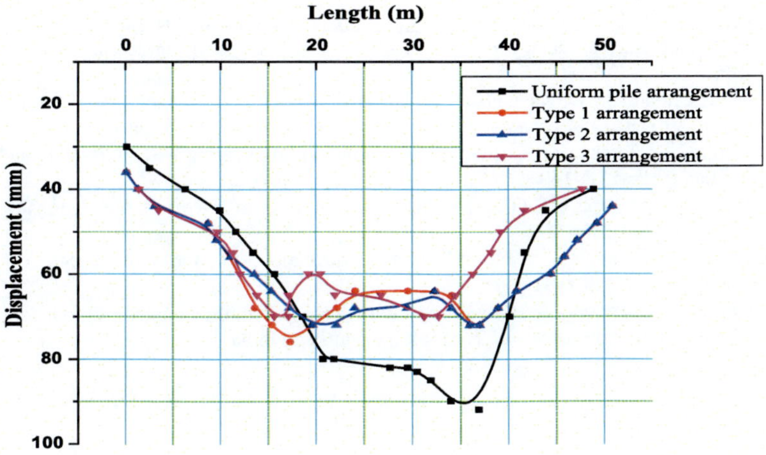

Fig. 19.9 Settlement along section A-A for various pile arrangements

References

1. Cooke RW, Sillett DF, Bryden Smith DW, Smith DWB, Gooch MV, BRE (1981) Some observations of the foundation loading and settlement of a multi-storey building on a piled raft foundation in London clay. Proc Inst Civ Eng 70(3):433–460
2. Price G, Wardle IF, BRE (1986) Queen Elizabeth Ii conference centre: monitoring of load sharing between piles and raft. Proc Inst Civ Eng 80(6):1505–1518
3. Clancy P, Randolph MF (1993) An approximate analysis procedure for piled raft foundations. Int J Numer Anal Meth Geomech 17(12):849–869
4. Katzenbach R, Reul O (1997) Theme lecture: design and performance of piled rafts. In: Proceedings of the 14th international conference on soil mechanics and foundation engineering, Hamburg, vol 4, pp 2253–2256
5. Katzenbach R, Arslan U, Moormann C (2020) Design and safety concept for piled raft foundations. In: Deep foundations on bored and auger piles. CRC Press, pp 439–448
6. Poulos HG, Davis EH (1980) Pile foundation analysis and design (No. Monograph)
7. Katzenbach R, Arslan U, Moormann C (2000) 13. Piled raft foundation projects in Germany. In: Design applications of raft foundations, vol 323
8. Poulos HG (2001) Piled raft foundations: design and applications. Geotechnique 51(2):95–113
9. Poulos HG, Small JC, Chow H (2011) Piled raft foundations for tall buildings. Geotech Eng J SEAGS & AGSSEA 42(2):78–84
10. Unsever YS, Matsumoto T, Özkan MY (2015) Numerical analyses of load tests on model foundations in dry sand. Comput Geotech 63:255–266
11. Sawada K, Takemura J (2014) Centrifuge model tests on piled raft foundation in sand subjected to lateral and moment loads. Soils Found 54(2):126–140
12. Chanda D, Saha R, Haldar S (2020) Behaviour of piled raft foundation in sand subjected to combined VMH loading. Ocean Eng 216:107596
13. Varghese R, Boominathan A, Banerjee S (2020) Stiffness and load sharing characteristics of piled raft foundations subjected to dynamic loads. Soil Dyn Earthq Eng 133:106117
14. Cooke RW (1986) Piled raft foundations on stiff clays—a contribution to design philosophy. Geotechnique 36(2):169–203
15. Cho J, Lee JH, Jeong S, Lee J (2012) The settlement behavior of piled raft in clay soils. Ocean Eng 53:153–163

16. Matsumoto T (2013) Implications for design of piled raft foundations subjected lateral loading. In: Proceeding international symposium on advances in foundation engineering, Singapore, pp 113–136
17. Kumar A, Choudhury D, Katzenbach R (2015) Behaviour of combined pile–raft foundation (CPRF) under static and pseudo-static conditions using PLAXIS3D. In: Proceedings of 6th international conference on earthquake geotechnical engineering (6ICEGE), Christchurch, New Zealand, paper ID-140
18. Horikoshi K, Randolph MF (1996) Centrifuge modelling of piled raft foundations on clay. Geotechnique 46(4):741–752
19. Horikoshi K, Randolph, MF (1998) A contribution to optimum design of piled rafts. Geotech, 48(3):301–317
20. Babu ELS (2020) Parametric study for optimum design of combined piled raft foundation. M. Tech Thesis, National Institute of Technology Rourkela, India

Chapter 20
Probabilistic Analysis of a Geosynthetic Reinforced Soil Retaining Wall Under Seismic Conditions Using Genetic Programming

Ekansh Agarwal, Ajeet Kumar Verma, Anindya Pain, and Shantanu Sarkar

Introduction

GRRW is predominantly used in the stabilization of slopes, landslide-prone areas, and embankments. The reinforced walls provide a wide variety of benefits such as improved aesthetics, economical and fast construction, to name a few. In practice, GRRW is analysed for their stability using the methods provided in the construction manuals [1, 2]. However, these methods are not always reliable because the value of the strength parameters is considered fixed in these analyses which is contrary to realistic soil conditions. Therefore, taking into consideration the uncertainty related with the strength parameters of the soil, probabilistic analysis is performed under pseudo-static conditions to examine the stability of a GRRW.

The probabilistic methodology is extensively used in stability analysis of geosynthetic reinforced structures [3–6]. The stability of the GRRW can be analysed in two ways, viz. internal and external. However, the internal stability analysis (critical tension mode [4]) is considered in the present study. If the GRRW is not failing in tension, it possibly would not fail in pullout mode. This is the prime reason pullout

E. Agarwal (✉) · A. Pain · S. Sarkar
Academy of Scientific and Innovative Research (AcSIR), Ghaziabad 201002, India
e-mail: ekansh1294@gmail.com

A. Pain
e-mail: pain_anindya@cbri.res.in

Geotechnical Engineering Group, CSIR-Central Building Research Institute, Roorkee 247667, India

A. K. Verma
Defence Institute of Advanced Technology, Deemed to be University, DRDO, Pune 411025, India

K. Muthukkumaran et al. (eds.), *Soil Dynamics, Earthquake and Computational Geotechnical Engineering*, Lecture Notes in Civil Engineering 300,
https://doi.org/10.1007/978-981-19-6998-0_20

mode is not considered in the analysis. The internal stability analysis involved calculation of total required tensile force of reinforcement (T_{total}). Different methods are available to calculate the value of T_{total}. However, the HSM is used because slices do not cut the reinforcements framing a realistic solution. The (2 N + 1) formulation of HSM is employed because it is the simplest and gives accurate results for GRRW [7].

Different probabilistic methods are available that can incorporate the HSM as a deterministic algorithm to calculate the values of performance functions. The accurate ones are FORM and MCS. However, the accuracy of the FORM decreases with the increase in stochastic parameters, and MCS suffers from curse of dimensionality. Therefore, there is a necessity of a robust and accurate formulation to probabilistically analyse the GRRW considering the variability in the soil strength parameters. This is addressed by using an evolutionary algorithm called GP which is a data-driven technique that constructs the predictive models which are capable of learning adaptively from experience and can make a sublime predictive model with least effort [8], by going through the processes of evolution, based on Darwinian theory of natural selection [9]. The results obtained from the GP are validated by performing a regression analysis and comparing the value of coefficient of determination (R^2). The use of GP in reliability-based analysis of reinforced soil structures is rare; therefore, the present study is done to showcase the capacity of GP in probabilistically analysing the same.

Deterministic Analysis for Internal Stability of GRRW

The deterministic analysis involves calculating the value of T_{total} using the (2 N + 1) formulation of HSM under pseudo-static conditions. The said formulation has been used by many researchers for deterministically analysing the reinforced soil structures [7, 10]. The formulation is very simple and efficient besides being accurate. As per the recommendations of FHWA [1], the reinforced soil is considered to be cohesionless, dry, and homogenous with log-spiral failure surface (Fig. 20.1). The unknowns along with the equilibrium equations involved in the (2 N + 1) formulation of HSM are listed in Table 20.1. The value of T_{total} is calculated by developing an algorithm in MATLAB.

T_{total} is represented in a non-dimensional form by Eq. (20.1).

$$K = \frac{T_{\text{total}}}{0.5\gamma H^2} \tag{20.1}$$

where H is the height of GRRW.

The value of K obtained from Eq. (20.1) is optimized to find out the critical slip surface. The algorithm is validated by comparing the obtained values of T_{total} with the values stated in Ling et al. [11] (Table 20.2) for $H = 5$ m, $\gamma = 18$ kN/m^3, $k_h =$ 0–0.2, $\beta = 90^\circ$, $\phi = 30^\circ$, and $k_v = 0$. Table 20.2 depicts that the values are in good

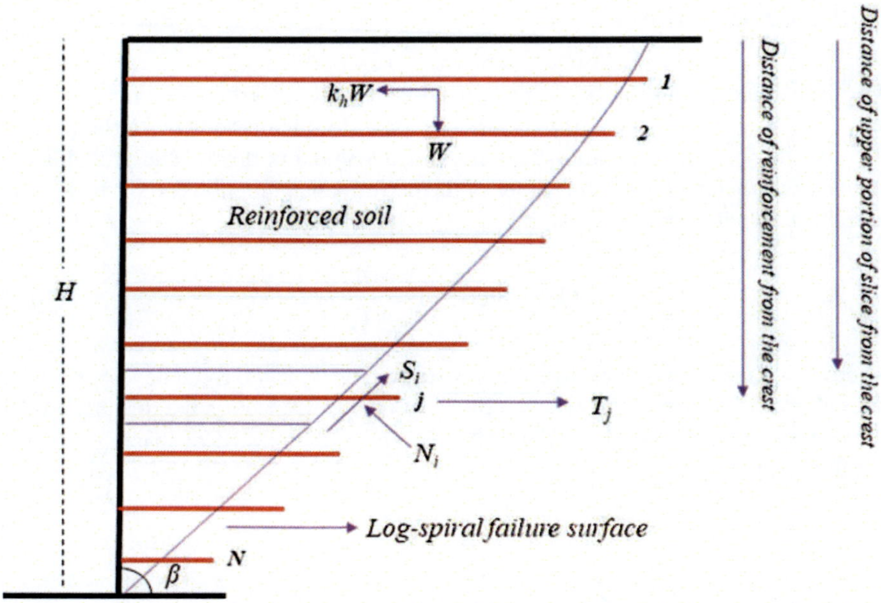

Fig. 20.1 Representational diagram of a GRRW

Table 20.1 Equilibrium equations and unknowns involved

Equations	No	Unknown	No
$\sum F_y = 0$ (for each slice)	N	Normal force on each slice (N_i)	N
$\sum F_x = 0$ (for whole wedge)	1	Shear force on each slice (S_i)	N
$\tau = \frac{\tau}{\text{FoS}}$ (for each slice)	N	Factor of safety (FoS)	1
Total	$2N + 1$		$2N + 1$

agreement with each other, confirming the accuracy of the developed deterministic formulation.

Table 20.2 Validation of the deterministic formulation

k_h	This study	Ling et al. [11]	Shahgholi et al. [10]
0	**74.98**	74	75
0.1	**89.22**	90	89
0.2	**106.48**	106	106

Probabilistic Analysis

The probabilistic analysis takes into consideration the uncertainty of random variables, and therefore, it analyses the GRRW in a more realistic way. In the present section, the proposed probabilistic method is discussed in detail. The probabilistic analysis is dependent upon the values of the performance function (FoS – 1). For the present study, this value is given by:

$$g(X) = \text{FoS} - 1 = \frac{T_{\text{ur}}}{T_j} - 1 \tag{20.2}$$

where T_{ur} is the ultimate reinforcement strength and T_j is the force developed in an individual reinforcement layer given by $T_j = (\gamma z_j)K(S_v)$. Z_j is the distance of the considered reinforcement from the wall crest, and S_v is the vertical spacing between reinforcements.

Genetic Programming

GP is a type of evolutionary algorithm. It is a data-driven technique that constructs the predictive models which are capable of learning adaptively from experience and can produce good predictive models with least computational effort by going through the processes of evolution. GP is a grey-box technique [12], and it permits the involvement of knowledge of experts in the domain of problem which results in the amalgamation of soft-computing and human intuition [13]. Implementation of GP and its modifications in the field of civil engineering [14] are relatively new. For the performance improvement of classical GP in different dimensions, many robust variants of GP have been developed [9, 15]. Models grounded on GP have been engaged to some strenuous geotechnical engineering issues with success [16–19].

In GP, the solutions are represented as computer programs commonly represented as parse tree structures (viz. genes). A typical representation of a single tree is depicted in Fig. 20.2, which denotes the predictive model in the expression: $\tanh(X1 * X2) - \cos(X3)$. These hierarchically structured trees comprising of the functional set and terminal set are user specified. Some of the functional sets and terminal sets are given in Table 20.3.

MGGP is the latest variant of GP. A single candidate solution is derived from multiple number of genes, where each gene is a tree expression of low-order nonlinear combination of the independent variables [20]. MGGP models are simpler, as they utilize a number of smaller genes (i.e. low-depth trees) as compared to the traditional GP. The root mean square error (RMSE) is considered as fitness function in this case. For more details about the procedure of implementing the GP for probabilistic analysis, readers are encouraged to refer to Gandomi and Alavi [19], Searson et al.

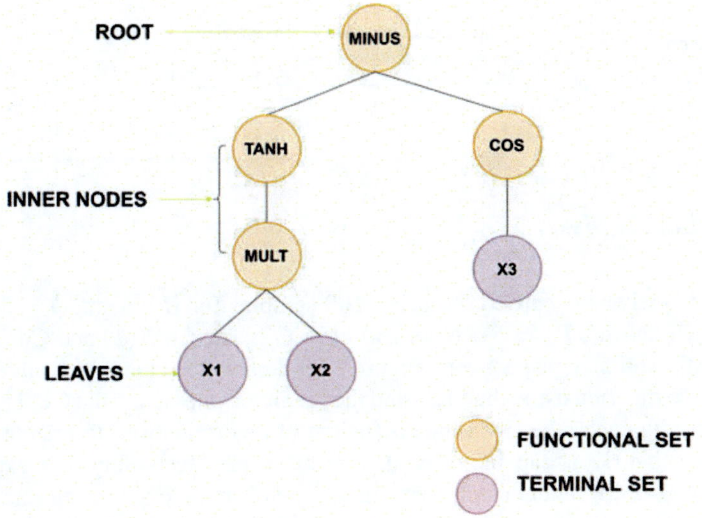

Fig. 20.2 A single tree representing a gene

Table 20.3 User-specified functional and terminal sets

Terminal sets	Inputs variables. (X_1, X_2, X_3…..)
	Functions without arguments (RAND…)
	Ephemeral random constants (ERCs) (0.45, 10….)
Functional sets	Arithmetic (+,*, -…)
	Boolean (AND, OR, ….)
	Mathematical (tanh, sin, cos, exp, sqrt…..)
	Looping (FOR, WHILE, REPEAT…)
	Conditional (IF–THEN-ELSE…)

[20], and Searson [21]. The details are omitted to avoid word similarity and to adhere to the word limit.

Results and Discussions

The statistics of the chosen random variables is provided in Table 20.4. In the present study, γ (soil unit weight) and φ (internal friction angle of the soil) are chosen as random variables. The results are evaluated for the tension mode and are verified by performing a regression analysis.

Table 20.4 Information of the stochastic variables used for this study

Stochastic variable	Properties		
	μ	CoV (%)	Distribution
ϕ	34°	5–15	Log-normal
γ	18 kN/m^3	5	Normal

Regression Analysis

Regression analysis is carried out using 100 samples for $H = 5$ m, $\phi = 34°$, $\beta = 90°$, $\gamma = 18$ kN/m^3, $T_u = 20$ kN/m, and $k_h = 0.2$. The CoV (ϕ) and CoV (γ) are fixed at 10% and 5%, respectively. Figure 20.3 depicts the calculated values of the performance function using the MCS and the predicted ones using the GP. The values are in line with each other with the coefficient of determination (R^2) being 0.9877. This proves that GP is capable of predicting the values accurately. The expression representing the GP model for the present analysis is mentioned in Eq. (20.3).

$$\begin{aligned} g(X)_{\text{predicted}} =& 3.12X2 - 0.0498X1 - 0.186\tan h(X2) - 0.128X1X2 \\ & - 0.186X1(X2^2) + 0.00595(X1^2) * X2 \\ & + 3.51 * (X2^3) - 0.0561 \end{aligned} \tag{20.3}$$

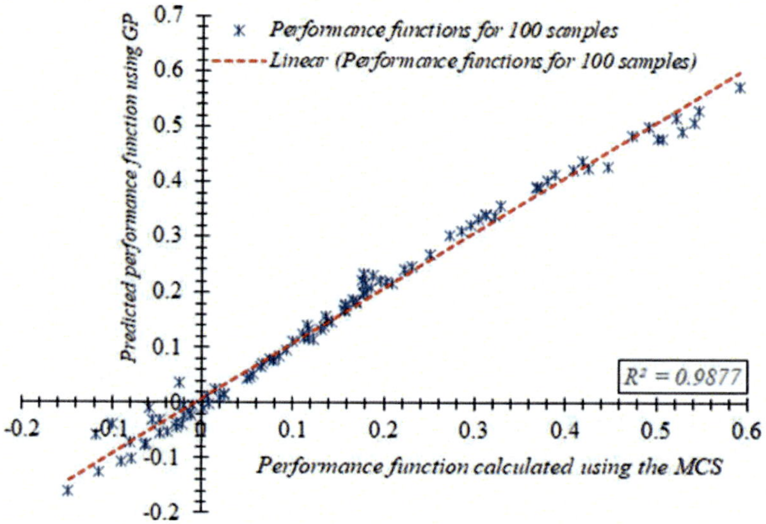

Fig. 20.3 Regression analysis

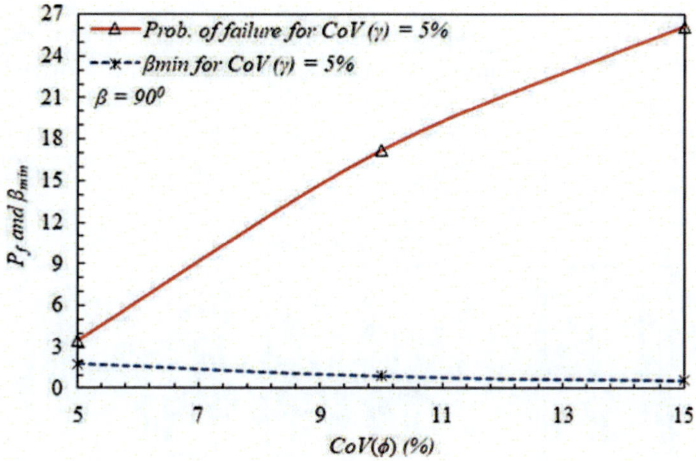

Fig. 20.4 P_f and $\beta_{\min}$ against CoV (ϕ) for $H = 5$ m, $\phi = 34°$, $\beta = 90°$, $\gamma = 18$ kN/m^3, $T_u = 20$ kN/m, and $k_h = 0.2$

Influence of Uncertainty on GRRW

This is illustrated by varying the value of CoV (ϕ) starting from 5 to 15%. This is because ϕ is more influential in comparison to γ. The analysis is performed for $H = 5$ m, $\phi = 34°$, $\beta = 90°$, $\gamma = 18$ kN/m^3, $T_u = 20$ kN/m, and $k_h = 0.2$. Figure 20.4 shows the plot of P_f and $\beta_{\min}$ against CoV (ϕ). It is seen from the figure that the P_f increases as the uncertainty (CoV (ϕ)) in the values increases. This happens because with the increase in uncertainty, the variance associated with the random variable also increases. Also, the $\beta_{\min}$ value decreases with an increase in value of CoV (ϕ).

The results depict the capacity of the proposed formulation to incorporate the uncertain nature of random variables accurately.

Results Obtained Using the GP

The GP produces many results which include the root mean square error (RMSE), value of R^2, mean squared error (MSE), performance of the model for training and testing data set, etc. These results illustrate whether the built model is predicting the correct output or not.

The values of RMSE and MSE obtained using the GP in the present study for $H = 5$ m, $\phi = 34°$, $\beta = 90°$, $\gamma = 18$ kN/m^3, $T_u = 20$ kN/m, $CoV\ (\phi) = 10\%$, $CoV\ (\gamma) = 5\%$, and $k_h = 0.2$ are presented in Table 20.5. Moreover, model performance for training and testing data is shown in Fig. 20.5.

Table 20.5 Model performance

Metric	Training values	Testing values
Root mean squared error (RMSE)	0.00088139	0.0014645
Mean squared error (MSE)	7.7684e–07	2.1449e–06

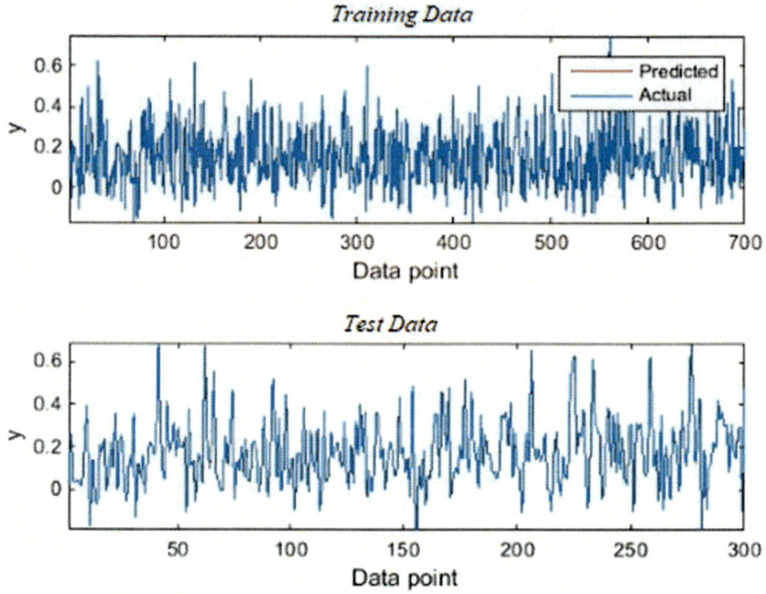

Fig. 20.5 Performance of the model for training and testing data for $H = 5$ m, $\phi = 34°$, $\beta = 90°$, $\gamma = 18$ kN/m^3, $T_u = 20$ kN/m, CoV (ϕ) = 10%, CoV (γ) = 5%, and $k_h = 0.2$

Summary and Conclusions

Probabilistic analysis of GRRW is performed in a pseudo-static framework using the evolutionary algorithm of GP. Following conclusions can be drawn from the results obtained.

- The ($2\ N + 1$) formulation of HSM is simple and effective in performing the deterministic analysis of GRRW for higher angles of slope inclination from the horizontal.
- The genetic programming is a precise evolutionary method that delivers high performance in estimating the probability of failure of GRRW. It is evident from the excellent results obtained from the regression analysis performed in the present study.

- The uncertainties in the random variables have a notable influence on the stability of GRRW. The stability decreases with increase in uncertainties associated with the random variables.
- The probabilistic analysis provides a comprehensive idea of the stability of GRRW and may provide an economical design of the GRRW.

References

1. Elias V, Christopher BR, Berg RR, Berg RR (2001) Mechanically stabilized Earth walls and reinforced soil slopes: design and construction guidelines (Updated Version) (No. FHWA-NHI-00–043). Federal Highway Administration, United States
2. AASHTO (2007) LRFD bridge design specifications. 4th ed. American Association of State Highway and Transportation Officials, Washington, D.C.
3. Basha BM, Babu GS (2010) Reliability assessment of internal stability of reinforced soil structures: a pseudo-dynamic approach. Soil Dyn Earthq Eng 30(5):336–353
4. Basha BM, Babu GS (2012) Target reliability-based optimization for internal seismic stability of reinforced soil structures. Geotechnique 62(1):55–68
5. Agarwal E, Pain A, Mukhopadhyay T, Metya S, Sarkar S (2021) Efficient computational system reliability analysis of reinforced soil-retaining structures under seismic conditions including the effect of simulated noise. In: Engineering with computers, pp1–23
6. Agarwal E, Pain A, Sarkar S (2021) Stochastic stability analysis of geosynthetic reinforced slopes subjected to harmonic base shaking. Transp Geotech 29:100562
7. Nouri H, Fakher A, Jones CJFP (2006) Development of horizontal slice method for seismic stability analysis of reinforced slopes and walls. Geotext Geomembr 24(3):175–187
8. Metenidis MF, Witczak M, Korbicz J (2004) A novel genetic programming approach to nonlinear system modelling: application to the DAMADICS benchmark problem. Eng Appl Artif Intell 17:363–370
9. Koza JR, Koza JR (1992) Genetic programming: on the programming of computers by means of natural selection, vol 1. MIT press
10. Shahgholi M, Fakher A, Jones CJFP (2001) Horizontal slice method of analysis. Geotechnique 51(10):881–885
11. Ling HI, Leshchinsky D, Perry EB (1997) Seismic design and performance of geosynthetic-reinforced soil structures. Geotechnique 47(5):933–952
12. Giustolisi O, Doglioni A, Savic DA, Webb BW (2007) A multi-model approach to analysis of environmental phenomena. Environ Model Softw 22(5):674–682
13. Rodriguez-Coayahuitl L, Morales-Reyes A, Escalante HJ (2019) A comparison among different levels of abstraction in genetic programming. In: IEEE international autumn meeting on power, electronics and computing (ROPEC), pp 1–6
14. Zhang Q, Barri K, Jiao P, Salehi H, Alavi AH (2021) Genetic programming in civil engineering: advent, applications and future trends. Artif Intell Rev 54(3):1863–1885
15. Ferreira C (2001) Gene expression programming: a new adaptive algorithm for solving problems. arXiv preprint cs/0102027
16. Javadi AA, Rezania M, Nezhad MM (2006) Evaluation of liquefaction induced lateral displacements using genetic programming. Comput Geotech 33(4–5):222–233
17. Rezania M, Javadi AA (2007) A new genetic programming model for predicting settlement of shallow foundations. Can Geotech J 44(12):1462–1473
18. Alavi AH, Aminian P, Gandomi AH, Esmaeili MA (2011) Genetic-based modeling of uplift capacity of suction caissons. Expert Syst Appl 38(10):12608–12618

19. Gandomi AH, Alavi AH (2021) A new multi-gene genetic programming approach to non-linear system modeling Part II: geotechnical and earthquake engineering problems. Neural Comput Appl 21(1):189–201
20. Searson DP, Leahy DE, Willis MJ (2010) GPTIPS: an open source genetic programming toolbox for multigene symbolic regression. In: Proceedings of the International multiconference of engineers and computer scientists, vol 1, pp 77–80
21. Searson DP (2015) GPTIPS 2: an open-source software platform for symbolic data mining. Handbook of genetic programming applications. Springer, Cham, pp 551–573

Chapter 21
Generation of Probabilistic Seismic Hazard Map of Assam State

Srijit Bandyopadhyay, Y. M. Parulekar, Aniruddha Sengupta, and J. Chattyopadhyay

Introduction

Assam state is one of the important states of North-East India. The North-East India is located in the most active tectonic region of the world. In recent past, Assam state experienced various devastating earthquakes like 1950 Assam Earthquake (Mw 8.7), 1943 NE Assam Earthquake, and 2021 Assam Earthquake (Mw 6.3). Thus, there is an increased risk of vulnerability to earthquake disasters and it is further implicated by new developments of bridges and multistoried structures built in this region. The Indian standard code [1] has described this region, as the intensity-based highest seismic zone of India, and has specified a Maximum Credible Earthquake Response spectra having return period of 2500 yrs. for this zone is PGA of 0.36 g. To mitigate the current seismic demand, there is a requirement to relook the existing seismic hazard values and generation of seismic hazard map at bedrock level.

Various researchers have conducted seismic hazard study of India. Khattri et al. [2] generated a seismic hazard map in India and its adjacent area showing peak ground acceleration (PGA) in rock level with a probability exceedance of 10% in 50 years. Parvez and Ram [3] conducted seismic hazard study in India based on PSHA as well as DSHA. Seismic hazard study of North-East India was carried out by many researchers in the past. Nath et al. [4] conducted earthquake hazard in northeast region based on seismic microzonation technique based on deterministic approach. Various researchers [5–8] performed seismic hazard study in India. In 2016, Das et al. [5] performed PSHA of North-East India considering seismic catalogue till

S. Bandyopadhyay (✉) · Y. M. Parulekar · J. Chattyopadhyay
Homi Bhabha National Institute, Mumbai 400085, India
e-mail: srijit15@gmail.com

Bhabha Atomic Research Center, Mumbai 400085, India

A. Sengupta
Indian Institute of Technology, Kharagpur 721302, India

K. Muthukkumaran et al. (eds.), *Soil Dynamics, Earthquake and Computational Geotechnical Engineering*, Lecture Notes in Civil Engineering 300,
https://doi.org/10.1007/978-981-19-6998-0_21

2010 and generated seismic hazard at the bedrock level. Later in 2020, Bahuguna and Sil [9] performed seismic hazard analysis of Assam region based on PSHA and DSHA methods. Recently in 2021, PSHA was performed by Ghione et al. [10] in India and Bhutan region based on hybrid model by combining distributed seismicity and catalogue till 2017. They found significant PGA levels of about 0.77 g at the bedrock level in North-East India due to the presence of the Himalayan Frontal Thrust. In the previous studies, seismic zonation based on earthquake mechanism was not considered yet. Here, seismic zonation is made based on fault pane mechanism.

In this present study, seismic hazard analysis of Assam state is carried out and as an outcome of the study, seismic hazard maps for Assam state is generated for different return period such as 100 years, 1000 years, 2500 years, and 1000 years. To eliminate epistemic uncertainties, logic tree analysis is performed.

Geological and Seismotectonic Background

Three different seismic plates, mainly Indian Plate, Eurasian Plate, and Sunda Plate, converge in North-East region, and these plate movement causes high seismicity in North-East region of India. A mixed kind of tectonic is observed in this area. The Indian intraplate region in south-east bounded by Himalayan Mountains in the north and Indo-Burmese Mountains in the south-east form the main part of this region. A shaded relief map of the study area is shown in Fig. 21.1. Himalayan mountain region shows that active shallow crustal tectonic and Indo-Burmese region are associated with subduction intraslab tectonic. The Assam state is divided mainly into three physiographic domains, i.e. Brahmaputra valley, Central Assam hills, and Assam Valley. Most of the Northern Assam parts are covered by Brahmaputra valley and the Mikir Hills with North Cachar hills lying in central Assam region. The Assam syntax is formed due to the convergence of Himalayan and Indo-Burma mountain ranges, and the Shillong plateau and the Mikir plateaus are located between these two boundaries. Large Bengal Basin which lies below the Assam valley has two structural regions separated by hinge zone extending in NE-SW direction from east of Calcutta. The study area is broadly subdivided into four parts such as Eastern Himalayas, Indo-Burma region, Assam valley, and Shillong-Mikir Plateau [11]. Details of these studies area are given in Angelier and Baruah [11].

Seismicity Information of the Study Area

The annual rate of earthquakes above some minimum magnitude is required for the determination of source seismicity based on recurrence relationships. The activity of any source zone can be obtained by two approaches, one from the past earthquakes in the region and the other from the local geology and tectonic information.

Fig. 21.1 Shaded relief map of Assam and adjoining region

The historical and instrumented seismic data is required for the evaluation of potential seismic sources in the study area. For the present study, the seismic events of the region with longitude 88° E–100° E and latitude 18° N–32° N are considered. Detailed documentation of seismic features of the area, like seismic faults, seismic lineaments, and earthquakes of moment magnitude, M_w greater than 3, is required to be acquired in order to carry out seismic hazard analysis. The seismic activity, tectonic features, and seismotectonic maps are obtained from the seismotectonic atlas (SEISAT, 2000) which was published by the geological survey of India. The detailed regional seismicity map of the North-East India and adjoining region is presented in Fig. 21.2.

Earthquake Catalogue

Earthquake data is collected from various agencies such as the National Earthquake Information Center (NEIC), United States Geological Survey (USGS), Indian Metrological Department (IMD), and International Seismological Center (ISC), and the earthquake events are selected. Duplicate earthquake data is eliminated, and totally, 4595 earthquake events are compiled in between latitude 20° N–31° N and longitude 86°–100° E. The earthquake data collected is generally reported data in different formats such as moment magnitude (M_w), Surface-wave magnitude (M_s), body wave magnitude (M_b), local wave magnitude, (M_l), and earthquake intensity scale. All the

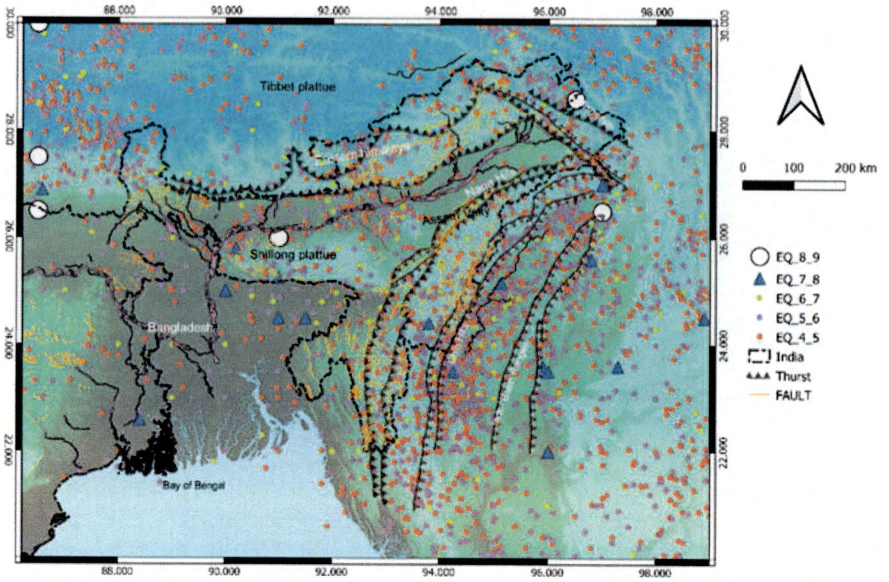

Fig. 21.2 Seismotectonic features of the study area

earthquakes of different magnitude scales are converted into moment magnitude by the equation proposed by Scordilis [12]. Details of earthquake locations are shown in Fig. 21.2. Aftershock and foreshock events are eliminated from the earthquake catalogue based on the method suggested by Reasenberg [13]. After declustering, total 3724 events remain.

The Seismic Source Models

Source models are generated for the computation of the hazard map based on seismic zones. The zonation has the purpose of dividing the seismicity into distinct source zones reflecting the tectonic conditions in order to subsequently evaluate the seismic potential of each zone. The seismicity is reasonably uniform within each area zone and hence for hazard analysis of the study area, areal-type sources are considered. The basic assumption of the area source is that earthquakes may randomly occur in anywhere of the zone and each zone has the same source properties in both time and space. Each zone has uniform seismotectonic characteristics and its size is large enough for getting a better assessment of recurrence parameters.

A detailed seismicity of the same area based on fault mechanism is explained by Angelier and Baruah [11]. They studied the seismotectonics in North-East India based on stress analysis of focal mechanism solutions of earthquakes and divided the

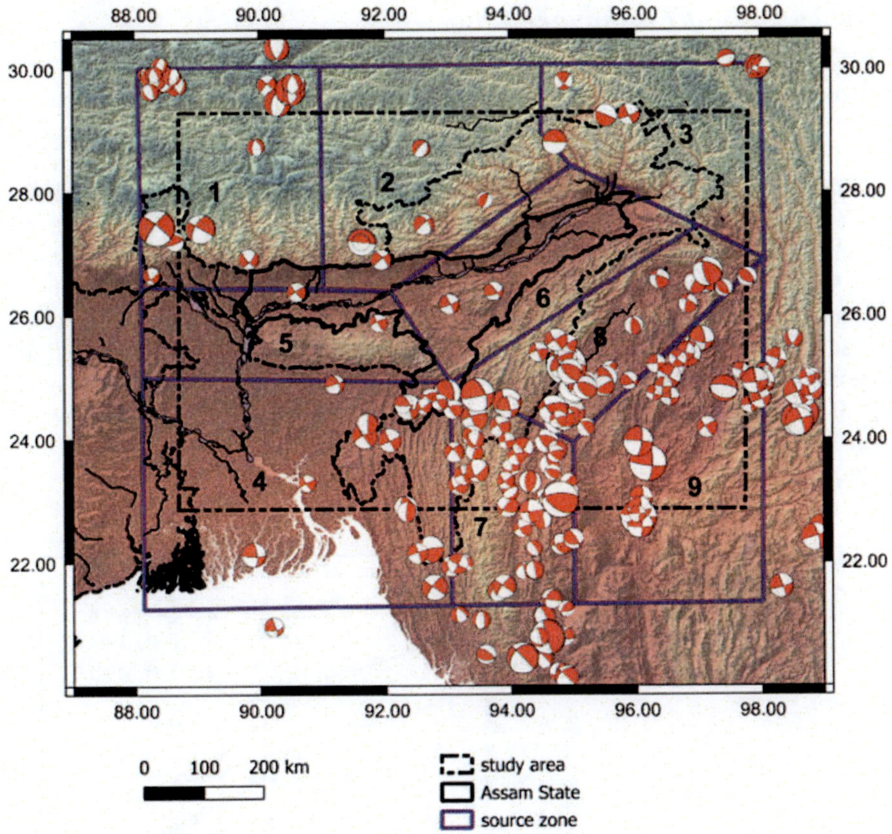

Fig. 21.3 Source zone used for PSHA analysis

study area into nine zones. In the present study, similar seismic zones are adopted for PSHA analysis. Details of the source zones are shown in Fig. 21.3.

The first three zones considered are essentially made of eastern Himalaya region and are comprised of Bhutan Himalaya, Arunachal Himalaya, and Mishmi thrust region. The next three zones consist of Tripura belt, Shillong Plateau, and Assam valley which indicate the intraplate region of Indian plate. The last three zones considered in the present work constitute the parts of Indo-Burma region included south and east Indo-Burma region and Sagaing fault region. The Indo-Burmese subduction zones are represented by Zone-7 and Zone-8 where mixed kind of fault mechanism is observed. Zone-9 represents Sagaing fault region having large number of active faults.

Completeness of Earthquake Data

Completeness of earthquake catalogue is to determine the smallest earthquake in the space and time. The declustered earthquake catalogue has to be checked for completeness in terms of magnitude and time both. ZMAP software [14] is used to determine the completeness of earthquake data and other seismic parameters. Maximum curvature method (MAXC) is used to determine the magnitude of completeness of the catalogue. It is observed the magnitude completeness of each zone varies with magnitude 4.3 to 4.7. During seismic hazard analysis threshold magnitude of each zone is considered as minimum magnitude of that respective zones. For completeness of earthquake catalogue with respect to time, method proposed by Stepp [15] is used. From Fig. 21.4, it is observed that magnitude ranges $4 \leq M_W < 5, 5 \leq M_W < 6, 6 \leq M_W < 7,$ and $7 \leq M_W < 8.6$ are complete for past 40 years, 50 years, 100 years, and 250 years. The completeness interval of different earthquake ranges of the study area is reported in Table 21.1.

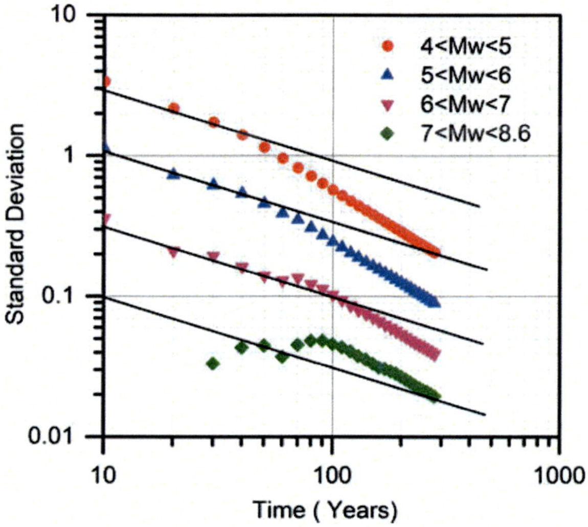

Fig. 21.4 Completeness of earthquake magnitude by year

Table 21.1 Completeness duration of different earthquake classes

Earthquake magnitude	Earthquake completeness period	Earthquake completeness year
$4 \leq M_W < 5$	1981–2021	40
$5 \leq M_W < 6$	1971–2021	50
$6 \leq M_W < 7$	1921–2021	100
$M_W > 7$	1771–2021	250

Table 21.2 Seismic zone parameters used for PSHA analysis

Zone details	EQ number	M_{obs}	b	M_c	'a' value	$N(M_0)$	M_{max}
Zone-1	762	8	0.8	4.4	3.6	1.266	8.5
Zone-2	205	7.3	0.8	4.4	3.4	0.844	7.8
Zone-3	400	8.6	0.93	4.7	4.8	2.78	8.6
Zone-4	575	7.5	0.71	4.3	2.7	0.432	8
Zone-5	575	8.1	0.71	4.3	2.7	0.432	8.6
Zone-6	575	7.2	0.71	4.3	2.7	0.432	7.7
Zone-7	846	8	1.03	4.5	5.0	2.586	8.5
Zone-8	704	7	1.03	4.5	4.9	1.729	7.5
Zone-9	207	7.5	0.8	4.4	3.3	0.636	8

Estimation of Seismic Parameters

The relationship between magnitude and the total number of earthquakes greater than a magnitude M in any given region and time period is expressed by Gutenberg–Richter (G-R) recurrence relation [16] as given Eq. 21.1

$$\log_{10} N(M) = a - bM \tag{21.1}$$

where $N(M)$ is the total number of events greater than or equal to magnitude M and 'a' and 'b' are constants. Seismic recurrence rates are calculated based on the complete earthquake catalogue. The study area is divided into a grid of 0.1°×0.1°. The seismicity parameters 'a', 'b', and 'M_c' are determined using LSM method. A 300 km radius from the centre of the grid is considered to determine the seismicity parameter of each group using Eq. 21.1. The seismicity parameter of each zone is reported in Table 21.2. The seismic activity $N(M)$ of each the zone is determined based on percentage of earthquakes with $M > M_{min}$ in each zone compared to total number of earthquakes in the region. Maximum magnitude M_{max}, is one of the source parameters which indicates the largest possible earthquake that can occur in a seismic zone. In this present study, maximum magnitude is considered as 0.5 units above the observed magnitude in each zone. The zones and maximum value of magnitude considered for each zone are given in Table 21.2.

Selection of Ground Motion Parameter

The attenuation relationship accounts for the attenuation of seismic energies of a tectonic zone. The area under consideration is divided into three categories, active tectonic region, active intraplate shallow crustal region, and subduction intraslab

Table 21.3 Combinations of GMPE used in the PSHA analysis

GMPE	GMPE used in different zones		
	Zones 1, 2, 3	Zones 4, 5, 6	Zones 7, 8, 9
Attenuation-1	Akkar and Bommer [17]	Atkinson and Boore [23]	Atkinson and Boore [20]
Attenuation-2	Chiou et. al. [18]	Nath and Thingbaijam [24]	Lin and Lee [21]
Attenuation-3	Sharma et al. [19]	–	Zaho et al. [22]

region. In recent years, different ground motion prediction relationships are developed for each of the active intraplate region, active continental crust, and subduction zones. Seismic zones 1 to 3 belong to active continental crust, and hence, attenuation relationships developed by Akkar and Bommer [17], Chiou and Youngs [18], and Sharma et. al. [19] are used. For the subduction zone region, GMPE generated by Atkinson and Boore [20], Lin and Lee [21], and Zaho et al. [22] is used. Seismic Zone-4, Zone-5, and Zone-6 belong to shallow crustal region; hence, for these zones, the attenuation relationship proposed by Atkinson and Boore [23] is used. The details of attenuation relationships are given in Table 21.3.

Logic Tree Formulation

Logic trees are applied frequently in PSHA to capture and quantify epistemic uncertainties associated with various input parameters, e.g. the description of seismic sources and choice of ground motion prediction models. The present PSHA analysis considers both seismicity parameters and GMPE in the logic tree. Figure 21.5 shows the structure of the logic tree used in this analysis. The present PSHA analysis considers both seismicity parameters (*b* values) and GMPEs in construction of the logic tree. The logic tree reflects the three different *b* values and different GMPE models. The weighting factor of '*b*' value is obtained based on statistical basis, as proposed by Grünthal and Wahlström [25] and Suckale and Grünthal [26], 0.68 weightage is assigned to the mean value of 'b', and for upper bound (mean + sigma) and lower bound (mean–sigma) branches of '*b*' value 0.16 weight is assigned, respectively. Due to the presence of less strong motion signal, the study area, weight assessment, and subsequent ranking of GMPE are not carried out. Equal weightages are assigned to each GMPE of their respective tectonic regions. The R-CRISIS software is used in PSHA analysis using various input parameters, such as geometry, source models, '*b*' values, and GMPEs.

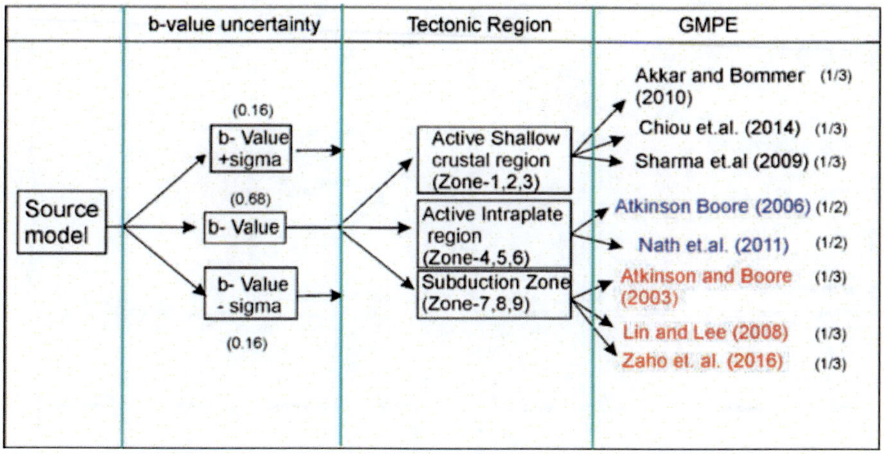

Fig. 21.5 Logic tree used in PSHA analysis

Results and Discussion

PSHA of Assam state is evaluated, and results of the analysis are presented in this section. For conducting the PSHA in Assam state, the entire study regions are divided in a grid size $0.1° \times 0.1°$ and hazards are computing in every grid point. The PGA and three different spectral accelerations at different time periods are evaluated at bedrock level with shear wave velocity of 1100 m/s. All these time period values are commonly used in building codes. The probability of exceedance value corresponding to return periods of 500 years, 1000 years, 2500 years, and 10,000 years are obtained. Logic tree approach is used, and all sources within 300 km distance from the grid point are considered for seismic hazard evaluation.

The seismic hazard map of Assam state, for PGA with different return periods of 500 yrs., 1000 yrs., 2500 yrs., and 10,000 yrs., is shown in Fig. 21.6. It is observed for 500 yrs., 1000 yrs., 2500 yrs., and 10,000 yrs. return period, the PGA value varies from 0.1 g to 0.2 g, 0.148 g to 0.28 g, 0.22 g to 0.423 g, and 0.342 g to 0.72 g, respectively. The high value is observed in front Assam due to the presence of HFT. The PGA distribution pattern is the same for all return periods. Similarly, hazard map showing acceleration for other time period of 0.2 s shown in Fig. 21.7. The maximum value of spectral acceleration is 0.39 g and 0.57 g is observed for return period 500 years and 1000 years, respectively. A similar spatial variation is observed in spectral acceleration curves for time period 1 s. and 2 s. The hazard curve showing the probability of exceedance for Guwahati for PGA and the other three time periods of the spectra are shown in Fig. 21.8, and the uniform hazard curves for Guwahati for two different return periods 2500 years and 10,000 years are shown in Fig. 21.9.

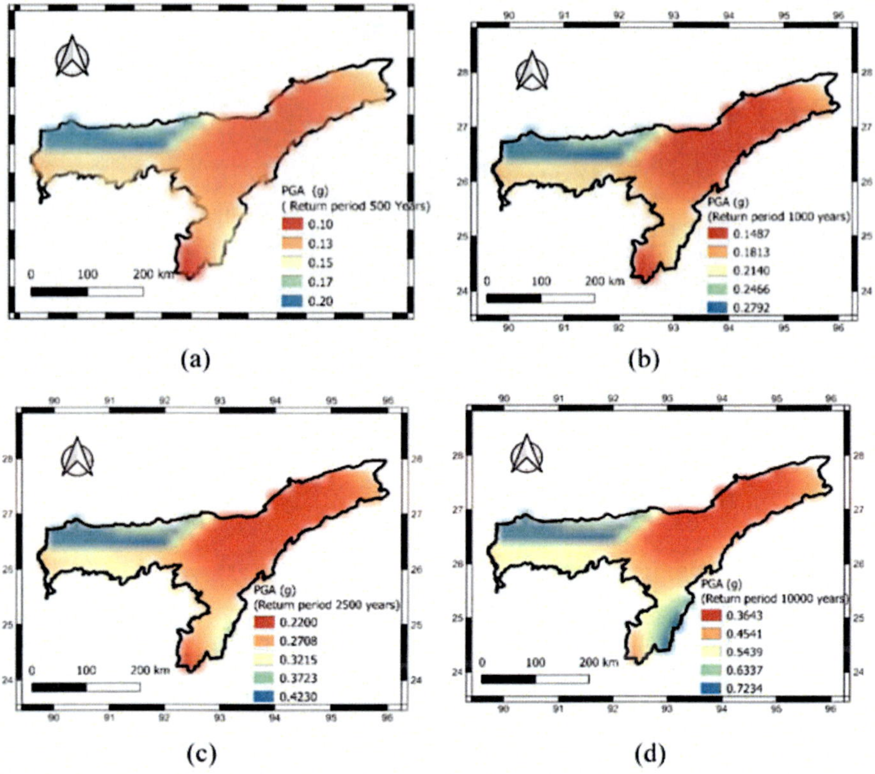

Fig. 21.6 PGA variation of Assam state for the different return periods, **a** for RP 500 years, **b** for return period 1000 years, **c** for return period 2500 years, **d** for RP 10,000 years

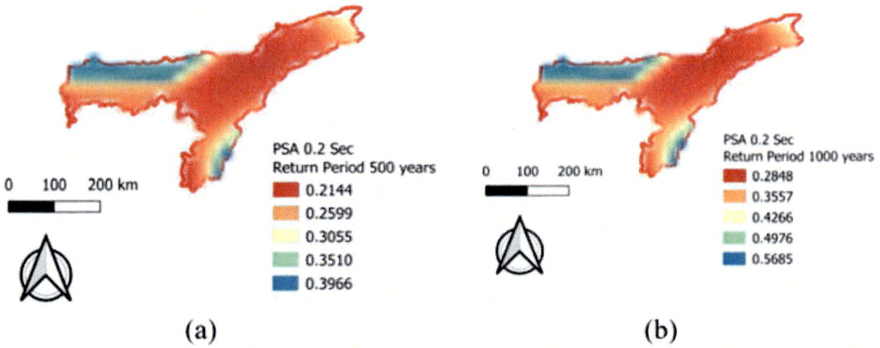

Fig. 21.7 PGA variation of Assam state for the different RPs, **a** for RP 500 years, **b** for RP 1000 years

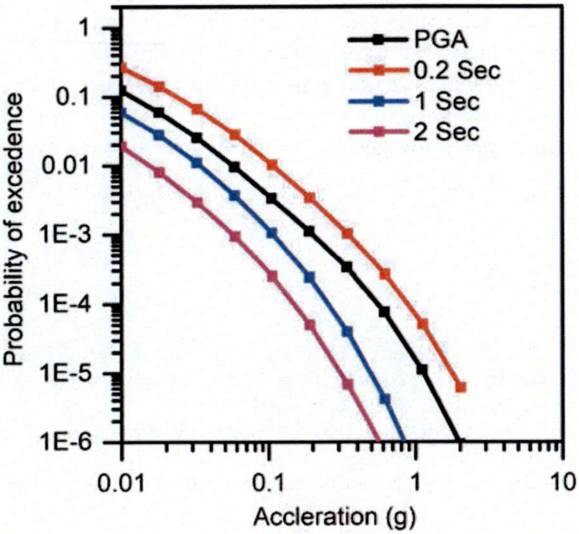

Fig. 21.8 Hazard curve of PGA and time period of 0.2 s, 1 s, and 2 s in Guwahati

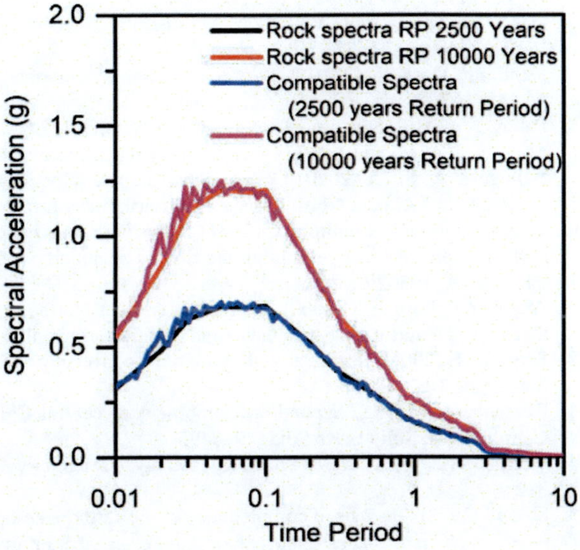

Fig. 21.9 Uniform hazard spectra of Guwahati for return period 2500 yrs and 10,000 yrs

Conclusion

- A complete earthquake catalogue having data of earthquakes up to 2021 is utilized to perform PSHA for North-East region of India. In this area, mixed kinds of tectonics are observed like active continental crust, Indian intraplate shallow crust and subduction zone. Hence, eight different GMPEs are used. The epistemic uncertainty is reduced with the help of logic tree approach, and PSHA is carried out. Hazard maps are then evaluated and uniform hazard spectra are generated for Guwahati region for 2500 yrs. and 10,000 yrs. return period. The PGA obtained for Guwahati region at the bedrock level for 2500 yrs. return period is 0.31 g and in compliance with the work done by other researchers. It is also 13.8% lesser than that suggested by IS 1893, 2016 for entire Zone-5.

References

1. BIS (2016) Criteria for earthquake resistant design of structures. Bureau of Indian Standards
2. Khattri K, Rogers A, Perkins D, Algermissen S (1984) A seismic hazard map of India and adjacent areas. Tectonophysics 108:93–134
3. Parvez IA, Ram A (1999) Probabilistic assessment of earthquake hazards in the Indian subcontinent. Pure Appl Geophys 154:23–40
4. Nath SK, Thingbaijam KKS, Raj A (2008) Earthquake hazard in Northeast India—a seismic microzonation approach with typical case studies from Sikkim Himalaya and Guwahati city. J Earth Syst Sci 117:809–831
5. Das R, Sharma M, Wason H (2016) Probabilistic seismic hazard assessment for northeast India region, Pure. Appl Geophys 173:2653–2670
6. Das S, Gupta ID, Gupta VK (2006) A probabilistic seismic hazard analysis of northeast India. Earthq Spectra 22:1–27
7. Kanth SR, Dash SK (2010) Deterministic seismic scenarios for North East India. J Seismolog 14:143–167
8. Pallav K (2010) Seismic microzonation of Imphal City and probabilistic seismic hazard assessment of Manipur state. Civil Engineering, IIT Guwahati, Assam India
9. Bahuguna A, Sil A (2020) Comprehensive seismicity, seismic sources and seismic hazard assessment of Assam, North East India. J Earthquake Eng 24:254–297
10. Ghione F, Poggi V, Lindholm C (2021) A hybrid probabilistic seismic hazard model for North-east India and Bhutan combining distributed seismicity and finite faults. Phys Chem Earth, Parts A/B/C 103029
11. Angelier J, Baruah S (2009) Seismotectonics in Northeast India: a stress analysis of focal mechanism solutions of earthquakes and its kinematic implications. Geophys J Int 178:303–326
12. Scordilis E (2006) Empirical global relations converting MS and mb to moment magnitude. J Seismolog 10:225–236
13. Reasenberg P (1985) Second-order moment of central California seismicity, 1969–1982. J Geophys Res: Solid Earth 90:5479–5495
14. Wiemer S (2001) A software package to analyze seismicity: ZMAP. Seismol Res Lett 72:373–382
15. Stepp J (1972) Analysis of completeness of the earthquake sample in the Puget Sound area and its effect on statistical estimates of earthquake hazard. In: Proceedings of the 1st international conference on microzonazion, Seattle, pp 897–910
16. Gutenberg B, Richter CF (1944) Frequency of earthquakes in California. Bull Seismol Soc Am 34:185–188
17. Akkar S, Bommer J (2010) Empirical equations for the prediction of PGA, PGV, and spectral accelerations in Europe, the Mediterranean region, and the Middle East. Seismol Res Lett 81:195–206
18. Chiou BS-J, Youngs RR (2014) Update of the Chiou and Youngs NGA model for the average horizontal component of peak ground motion and response spectra. Earthq Spectra 30:1117–1153
19. Sharma ML, Douglas J, Bungum H, Kotadia J (2009) Ground-motion prediction equations based on data from the Himalayan and Zagros regions. J Earthquake Eng 13:1191–1210

20. Atkinson GM, Boore DM (2003) Empirical ground-motion relations for subduction-zone earthquakes and their application to Cascadia and other regions. Bull Seismol Soc Am 93:1703–1729
21. Lin P-S, Lee C-T (2008) Ground-motion attenuation relationships for subduction-zone earthquakes in northeastern Taiwan. Bull Seismol Soc Am 98:220–240
22. Zhao JX, Jiang F, Shi P, Xing H, Huang H, Hou R, Zhang Y, Yu P, Lan X, Rhoades DA (2016) Ground-motion prediction equations for subduction slab earthquakes in Japan using site class and simple geometric attenuation functions. Bull Seismol Soc Am 106:1535–1551
23. Atkinson GM, Boore DM (2006) Earthquake ground-motion prediction equations for eastern North America. Bull Seismol Soc Am 96:2181–2205
24. Nath SK, Thingbaijam KKS (2011) Peak ground motion predictions in India: an appraisal for rock sites. J Seismolog 15:295–315
25. Grünthal G, Wahlström R (2006) New generation of probabilistic seismic hazard assessment for the area Cologne/Aachen considering the uncertainties of the input data. Nat Hazards 38:159–176
26. Suckale J, Grünthal G (2009) Probabilistic seismic hazard model for Vanuatu. Bull Seismol Soc Am 99:2108–2126

Chapter 22
Study on Liquefaction Evaluation of Subsoil Strata Along Visakhapatnam Coastal Area

C. N. V. Satyanarayana Reddy and A. Harika

Introduction

Earthquakes are considered as the most frequent natural disasters to occur across the globe. Consequently, its aftermath in terms of damage to the constructions is often a serious concern to the engineers. Particularly, earthquake-induced liquefaction results in loss of supporting strength of soils leading to serious damage to the structures. Initially, the term liquefy was used by Allen Hazen in describing the failure of the Calavera's Dam in California in 1918. Later, the term liquefaction was coined by Mogami and Kubo for the illustration of a phenomenon that involve soil deformations in undrained conditions [1]. Liquefaction is divided into two groups—cyclic mobility and flow liquefaction, both of which shall be considered for the liquefaction hazard assessment. Cyclic mobility occurs when the shear stresses induced during the ground motion are less than the shear strength of the soil unlike in flow liquefaction where the induced shear stress is greater than the shear strength of the soil. Although the frequency of occurrence of the flow liquefaction is rather seldom compared to cyclic mobility, its effects are very severe [1].

Liquefaction is a complex phenomenon influenced by a wide range of factors including gradation characteristics, drainage conditions, vibration characteristics, magnitude, and type of superimposed loads [2]. Casagrande opined that critical void ratio is the possible criterion for deciding the liquefaction susceptibility of the soil [2]. Sands tend to liquefy if the in-situ void ratio is greater than the critical void ratio. It was deliberated that liquefaction is a phenomenon which is associated

C. N. V. Satyanarayana Reddy · A. Harika (✉)
Department of Civil Engineering, Andhra University College of Engineering, Visakhapatnam, Andhra Pradesh 530040, India
e-mail: 319206310004@andhrauniversity.edu.in

C. N. V. Satyanarayana Reddy
e-mail: prof.cnvsreddy@andhrauniversity.edu.in

K. Muthukkumaran et al. (eds.), *Soil Dynamics, Earthquake and Computational Geotechnical Engineering*, Lecture Notes in Civil Engineering 300,
https://doi.org/10.1007/978-981-19-6998-0_22

with sands. Ishihara exposed that non-plastic coarse silts also undergo liquefaction, thereby indicating that plasticity characteristics form the criteria for liquefaction in fine grained soils as compared to gradation characteristics [1]. Clays are considered to be non-liquefiable, however, they exhibit strain softening behaviour similar to liquefied soil [1]

The ability of the soil to liquefy under a particular magnitude of ground shaking can be evaluated from the laboratory as well as field tests. The most common laboratory tests that are in use for evaluation of liquefaction potential are cyclic triaxial test and oscillatory shear box test. Although vibration table test simulates the field conditions, it is rarely used in the liquefaction studies due to the involvement of higher costs. Casagrande reported that in cyclic triaxial test, increased cyclic pore pressure is developed due to radial distribution of water content which is generated by a mechanism that is normally not active in-situ [2]. Peck reported that the laboratory tests fail to take into account several factors such as disturbance to sample, that are likely to be present in the field [2]. These drawbacks have led to the wide utilization of field data for more precise evaluation of liquefaction potential. Data obtained from SPT, CPT, and shear velocity measurements is most commonly used for liquefaction studies.

Literature Review

After realizing the importance of studying the characteristics and factors affecting the liquefaction, a lot was research was carried out in this area. A number of methods were proposed by various researchers for evaluating the liquefaction potential. Seed and Idriss [3] were popular among those researchers. They proposed a simplified method for the evaluation of liquefaction potential by taking into account the factors which are most likely to influence liquefaction. Seed et al. [4] presented a new method for evaluating the magnitude of pore water pressure generated in sands and silty sands from field tests. Iwasaki et al. [5] recommended a simplified method based on liquefaction resistance factor and liquefaction potential factor for evaluation of liquefaction susceptibility of the soils.

Arulmoli et al. [6] investigated on the applicability of electric probe in the field for evaluation of liquefaction potential at sites affected by 1906 San Francisco, Niigata, and Tangshan earthquakes. They found that the in-situ electrical measurements are in agreement with laboratory results. Kayen et al. [7] carried out the post-earthquake investigations using the standard penetration and seismic cone penetration data. They reported that both SPT and CPT methods provide a good means of assessing the soil liquefaction. Idriss and Boulanger [8] re-examined semi-empirical procedures for evaluating the liquefaction potential of saturated cohesionless soils during earthquakes and recommended the revised SPT and CTP-based liquefaction correlations for field practice. In 2008, Idriss and Boulanger [9] presented a number of potentially important correlations and recommendations associated with seismic-induced soil liquefaction.

Numerous works for the identification of liquefaction susceptible areas were done in the northern parts of India as they are seismically active compared to most parts of southern India. Research in the area of identification of liquefiable strata within south India was very limited, with few studies concentrated in Bangalore, Chennai, and parts of Andhra Pradesh. Swathi and Neelima [10] performed one-dimensional linear ground response analysis in selected areas of Visakhapatnam and reported that most locations in north and central parts of the city are prone to liquefaction. Sai and Rao [11] evaluated the liquefaction potential index for the new capital region of Andhra Pradesh and concluded that most of the study locations in the capital region are not susceptible to liquefaction under earthquakes of low magnitude.

A research conducted by University of Hyderabad in collaboration with National Institute of oceanography and Oil and Natural Gas Commission (ONGC) revealed the presence of a 300 km long fault line in the Krishna Godavari basin and few kilometres away from the coastline [12]. Though the fault line is inactive, but the movements within the interiors of the earth may make it active in future with its activity mostly concentrated in the north-eastern parts of Andhra Pradesh. It was believed that this fault line has the potential to trigger major hazardous events associated with seismicity, along the coastline of Visakhapatnam. As liquefaction is associated with the seismic activity, it is necessary to identify the liquefaction-prone areas along the coastline to mitigate or to minimize its deleterious effects.

Liquefaction Analysis of the Study Areas

The study is carried out for the selected areas along the coastline of Visakhapatnam. The 135 km long coastline of Visakhapatnam has substantial extents of silty sands and fine sands with ground water table close to the ground surface. These soils in their saturated state tend to liquefy when subjected to a short time dynamic impulse.

Three study areas along the Visakhapatnam coast, two at DGNP and one at the port area, were considered for the liquefaction analysis. The subsoil at all the three study areas comprised of considerable thicknesses of liquefiable soils. Site 1 is characterized by the presence of 10 m thick deposit of fine sand sandwiched between clayey gravel and clay. SDR and fractured rocks are encountered at deeper depths of 20.0 m. Site 2 consisted of 4.5 m extent of fine to medium sand underlain by very soft to soft clay with hard rock available at 28 m. Site 3 consisted of 3 m of silty sand followed by soft marine clay, SDR, and hard fractured rock. The ground water table at locations 1 and 3 is encountered at the ground surface, while at site 2, it is present at a depth of 1.2 m below the ground level. The recorded standard penetration resistances of the fine sand/ Silty sand at all the three study areas are observed to be less than 15. The typical borelog corresponding to the study areas is presented in Tables 22.1, 22.2 and 22.3, and the properties of the liquefiable layers at the study areas are enlisted in Table 22.4.

The liquefaction potential of the subsoil strata at the study areas is assessed utilizing the data obtained from the standard penetration tests conducted at the sites

Table 22.1 Borelog of study area 1 (SA_1)

Depth (m)	Type of soil	Sample		Blow count for the penetration of the split spoon sampler through			N	CR (%)	RQD (%)
		Depth (m)	Type	0–15 cm	15–30 cm	30–45 cm			
0.0–3.0	Clayey gravel	0.0–1.5	DS						
		1.5–1.95	SPT	08	08	11	19		
3.0–13.5	Silty fine sand	3.0–3.45	SPT	02	02	03	05		
		4.5–4.95	SPT	04	03	04	07		
		6.0–6.45	SPT	06	07	07	14		
		7.5–7.95	SPT	05	07	05	12		
		9.0–9.45	SPT	03	05	07	12		
		10.5–10.95	SPT	04	04	07	11		
		12.0–12.45	SPT	06	07	07	14		
13.5–15.0	Medium sand	13.5–13.65	SPT	03	16	25	41		
15.0–22.5	Medium to stiff clay	15.0–15.45	SPT	03	03	03	06		
		16.5–16.95	SPT	02	02	04	06		
		18.0–18.45	SPT	03	02	04	06		
		19.5–19.95	SPT	03	03	04	07		
		21.0–21.45	SPT	03	04	05	09		
22.5–23.5	Very stiff clay	22.5–22.95	SPT	07	11	13	24		
23.5–24.2	Soft Disintegrated rock (SDR)	23.5–23.82	SPT	20	25	R	> 100		
		23.82–24.2	DS						

(continued)

Table 22.1 (continued)

Depth (m)	Type of soil	Sample		Blow count for the penetration of the split spoon sampler through			N	CR (%)	RQD (%)
		Depth (m)	Type	0–15 cm	15–30 cm	30–45 cm			
24.2–27.0	Highly fractured rock	24.20–25.5	RCS					30	08
		25.5–27.0	RCS					18	10
27.0–29.1	Fractured Hard rock	27.0–28.5	RCS					46	22
		28.5–29.1	RCS					54	16

Table 22.2 Borelog of study area 2 (SA_2)

Depth (m)	Type of soil	Sample		Blow count for the penetration of the split spoon sampler through			N	CR (%)	RQD (%)
		Depth (m)	Type	0–15 cm	15–30 cm	30–45 cm			
0.0–2.5	Filled up soil	1.5–1.95	SPT	07	07	09	16		
2.5–7.0	Fine to medium sand	3.0–3.45	SPT	04	03	06	09		
		4.5–4.95	SPT	04	04	06	10		
		6.0–6.45	SPT	05	07	08	15		
7.0–12.0	Very soft to soft clay	7.5–7.95	SPT	00	00	00	00		
		9.0–9.45	SPT	00	00	01	01		
		10.5–10.95	SPT	01	02	02	04		
12.0–16.5	Medium clay	12.0–12.45	SPT	02	03	02	05		
		13.5–13.65	SPT	03	03	04	07		
		15.0–15.45	SPT	03	03	05	08		
16.5–24.0	Stiff to very stiff clay	16.5–16.95	SPT	05	05	06	11		
		18.0–18.45	SPT	03	05	06	11		
		19.5–19.95	SPT	04	06	06	12		
		21.0–21.45	SPT	07	07	09	16		
		22.5–22.95	SPT	07	09	09	18		
24.0–26.0	Clayey sand with pebbles	24.0–24.45	SPT	21	36	40	76		
		25.5–25.95	SPT	29	40	45	85		
26.0–27.0	Soft disintegrated rock	27.0–27.01	SPT	R	–	–	>100		
27.0–28.5	Soft rock	27.01–28.5	RCS					49	12
28.5–31.75	Hard rock	28.5–30.0	RCS					60	17
		30.0–31.75	RCS					70	28

Table 22.3 Borelog of study area 3 (SA_3)

Depth (m)	Type of soil	Sample		Blow count for the penetration of the split spoon sampler through			N	CR (%)	RQD (%)
		Depth (m)	Type	0–15 cm	15–30 cm	30–45 cm			
0.0–2.0	Filled up soil	0.0–2.0	DS						
2.0–5.0	Silty sand	2.0–2.45	SPT	02	04	04	08		
		2.45–3.5	DS						
		3.5–3.95	SPT	02	04	06	10		
		3.95–5.0	DS						
5.0–8.0	Soft marine clay	5.0–5.45	SPT	01	01	01	02		
		5.45–6.5	DS						
		6.5–6.95	SPT	01	01	01	02		
		7.0–8.0	DS						
8.0–11.5	Soft disintegrated rock	8.0–8.02	SPT	R	–	–	>100		
		8.02–9.5	DS						
		9.5–9.52	SPT	R	–	–	>100		
		9.52–11.0	DS						
		11.0–11.02	SPT	R	–	–	>100		
		11.02–11.5	DS						
11.5–14.0	Fractured hard rock	11.5–12.5	RCS					59	00
		12.5–14.0	RCS					40	09

Table 22.4 Geotechnical properties of the liquefiable layers at the study areas

Property	Value		
	Site 1	Site 2	Site 3
Grain size distribution			
Gravel (%)	01	00	00
Sand (%)	96	95	85
Fines (%)	03	05	15
Bulk density (g/cc)	1.83	1.98	2.03
Natural moisture content (%)	9.2	8.5	15.2
IS classification	SP	SP	SM

at various depths. The susceptibility of the subsoil to liquefaction under a particular magnitude of dynamic loading is expressed in terms of a factors of safety (FOS) defined as the ratio of cyclic resistance ratio (CRR) to cyclic stress ratio (CSR).

$$\mathrm{FOS} = \frac{\mathrm{CRR}}{\mathrm{CSR}} \tag{22.1}$$

The factors of safety against liquefaction are evaluated using method given in annexure F of IS 1893 Part 1 2016 [13] and Idriss and Boulanger [9] method. At present, Visakhapatnam is located in zone II (Z_2) of seismic mapping of IS 1893-2016. As the seismic zoning is dynamic, it is preferable to evaluate the liquefaction susceptibility for zone III (Z_3) seismic activity considering the possible chances for updation. Also, extension of study for liquefaction vulnerability for the subsoil strata for zones IV and V will be useful to provide sufficient information for similar sites located in such zones. Visakhapatnam and its surrounding areas have a history of experiencing earthquakes of magnitude in between 4.5 and 6, and also the north and north-eastern parts of India frequently experience earthquakes of magnitudes in the same range. With this point of view, the factors of safety are computed for earthquake magnitudes of 5, 5.5, 6.0, and 6.5.

Liquefaction Potential from IS 1893 Part 1 (2016) Method

Cyclic stress ratio (CSR) is a function of the horizontal acceleration induced by an earthquake at the site. IS 1893 recommends the use of zone factors as values for peak ground acceleration in the computation of CSR. The cyclic resistance ratio (CRR) is calculated based on the standard penetration resistance values obtained after correcting the recorded values in the field for overburden. The CRR for a magnitude of 7.5 corresponding to different N values and percentage of fines is obtained from Fig. 22.1. The cyclic resistance ratio for magnitudes other than 7.5 is determined by multiplying the CRR at 7.5 by magnitude scaling factor. (MSF). The following

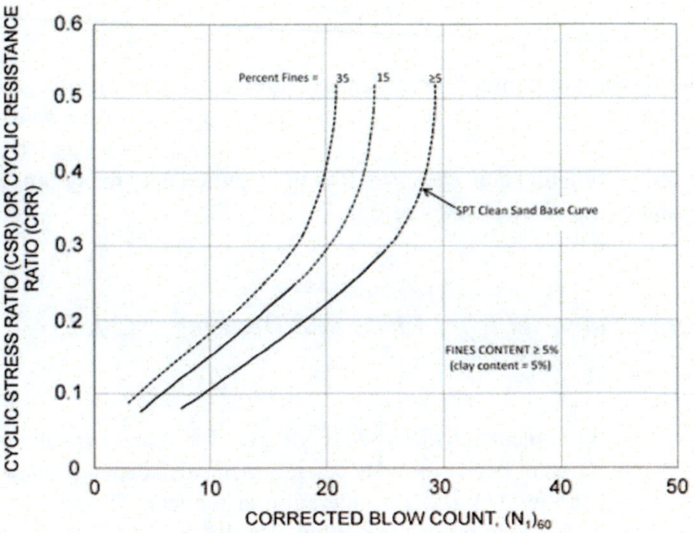

Fig. 22.1 Relation between CRR and $(N_1)_{60}$ for an earthquake of magnitude 7.5 (IS 1893 Part 1 2016)

equations are proposed by IS 1893 Part 1(2016) [13] for the computation of CRR and CSR.

$$CSR = 0.65 * (a_{max}/g) * (\sigma v/\sigma v') * r_d \quad (22.2)$$

where

$$\begin{aligned} r_d &= \text{Stress reduction factor} \\ &= 1 - 0.00765Z \text{ for } 0 < Z < 9.75 \\ &= 1.174\text{–}0.0267Z \text{ for } 9.75 < Z < 23 \end{aligned}$$

$$(N_1)_{60} = C_N C_{60} N$$
$$C_N = (P_a/\sigma v')^{0.5} \leq 1.7 \quad (22.3)$$

IS recommends to take the value of C_{60} as 1; $\sigma v'$ is the effective overburden pressure. Corresponding to the value of $(N_1)_{60}$, the value of $CRR_{7.5}$ is obtained.

$$CRR = CRR_{7.5} * MSF * K_\sigma K_\alpha \quad (22.4)$$

where

$$MSF = \text{Magnitude scaling factor}$$

$$= 10^{2.24}/M_{\mathrm{w}}^{2.56}$$

K_σ is the correction for high overburden pressure, required for depths greater than 15 m. $K_\sigma = (\sigma v'/P_a)^{(f-1)}$, f is a factor which depends on the relative density of the soil.

K_α is the correction for static shear stresses and required only for sloping grounds. For plain grounds, its value is taken as one.

Liquefaction Potential from Idriss and Boulanger (2008) Method

Unlike in the method suggested by IS 1893, CSR from Idriss and Boulanger method [9] is not only a function of peak ground acceleration but also depends on the magnitude of the earthquake inducing this acceleration at the site. The stress reduction factor varies with the magnitude of the earthquake and so is CSR. For the present study, in the evaluation of CSR from Idriss and Boulanger method, corresponding zone factors are used in lieu of peak ground acceleration values. A separate equation is proposed for calculating the CRR corresponding to a magnitude of 7.5. Multiplying this value with the magnitude scaling factor gives the CRR for other magnitudes of earthquake. The following equations proposed by Idriss and Boulanger are used for the evaluation of factors of safety.

$$\mathrm{CSR} = 0.65 * (a_{\max}/g) * \left(\sigma v/\sigma v'\right) * r_{\mathrm{d}} \tag{22.5}$$

where

$$r_{\mathrm{d}} = \mathrm{Exp}[\alpha(Z) + \beta(Z)M]$$
$$\alpha(Z) = -1.012 - 1.126\sin\left[\frac{z}{11.73} + 5.133\right]$$
$$\beta(Z) = 0.106 + 0.118\sin\left[\frac{z}{11.73} + 5.142\right]$$

$$(N_1)_{60\mathrm{cs}} = C_{\mathrm{N}}C_{60}(N_1)_{60}$$
$$C_{\mathrm{N}} = \left(P_{\mathrm{a}}/\sigma v'\right)^{\mathrm{m}} \le 1.7 \tag{22.6}$$

where $m = 0.784–0.0768\sqrt{N_{1_{60}}}$

$$\mathrm{CRR}_{M=7.5,\sigma v'=1\ \mathrm{atm}} = \exp\left[\frac{(N_1)_{60cs}}{14.1} + \left(\frac{(N_1)_{60cs}}{126}\right)^2 - \left(\frac{(N_1)_{60cs}}{23.6}\right)^3 + \left(\frac{(N_1)_{60cs}}{25.4}\right)^4 - 2.8\right]$$
$$\mathrm{MSF} = 6.9\exp(-\mathrm{M}/4) - 0.058 \le 1.8 \tag{22.7}$$

where $(N_1)_{60\mathrm{cs}}$ is the corrected standard penetration resistance.

If the factors of safety from both IS 1893 Part 1 2016 method and Idriss and Boulanger method are less than 1, then the soil is considered to liquefy under the particular magnitude of earthquake. Using the factors of safety obtained, the extents of liquefiable strata in the study areas are established. The factors of safety against liquefaction for the sub soil profile at the study areas calculated from the above two methods are given in Tables 22.5 and 22.6.

Evaluation of Liquefaction Potential Index (LPI) Using **Iwasaki et al.** *[14] Method*

The factor of safety against liquefaction gives a fair idea about the liquefaction susceptibility of the sub soil at a site. But the damage potential or the severity of liquefaction cannot be interpreted with factor of safety alone. Hence, liquefaction potential index is evaluated for assessing the probable hazard associated with liquefaction. Iwasaki et al. in 1978 [14] proposed the concept of liquefaction potential index to ascertain the ability of the liquefaction to cause damage to the foundations of the structure. He also put forward the degree of damage due to liquefaction based on the liquefaction potential index. According to Iwasaki et al., if the LPI is less than 0, the potential of the liquefaction is low. Sites with LPI in between 0–5 and 5–15 are rated to have low and high liquefaction potential respectively. Liquefaction potential is very high for sites with LPI greater than 15 [5]. The following formula proposed by Iwasaki et al. was used for evaluating the liquefaction potential indices of the study areas.

$$\mathrm{LPI} = \int_0^{20} F(Z)W(Z)\mathrm{d}z \tag{22.8}$$

where

$$\begin{aligned} F(Z) &= 1 - \mathrm{FOS} && \text{for FOS} < 1 \\ F(Z) &= 0 && \text{for FOS} > 1 \\ W(Z) &= 10\text{–}0.5Z && \text{for } Z < 20\,\mathrm{m} \\ W(Z) &= 0 && \text{for } Z > 20\,\mathrm{m} \end{aligned}$$

The LPI values for the three study areas (SA_1, SA_2, and SA_3) corresponding to the factors of safety determined from IS 1893 method and Idriss and Boulanger method are presented in Tables 22.7 and 22.8.

Table 22.5 Factors of safety against liquefaction using IS 1893 Part 1 2016 method

Study area	Depth (m)	$(N_1)_{60}$	Factor of safety															
			5.0				5.5				6.0				6.5			
			Z_2	Z_3	Z_4	Z_5	Z_2	Z_3	Z_4	Z_5	Z_2	Z_3	Z_4	Z_5	Z_2	Z_3	Z_4	Z_5
SA1	3.0	8	1.93	1.20	0.80	0.53	1.51	0.94	0.63	0.42	1.21	0.75	0.50	0.34	0.98	0.63	0.42	0.28
	4.5	10	2.43	1.52	1.01	0.67	1.91	1.19	0.79	0.53	1.53	0.95	0.63	0.42	1.24	0.77	0.51	0.34
	6.0	18	4.55	2.84	1.89	1.26	3.56	2.22	1.48	0.98	2.85	1.78	1.18	0.79	2.32	1.45	0.96	0.64
	7.5	14	3.49	2.18	1.45	0.97	2.73	1.70	1.14	0.76	2.19	1.37	0.91	0.61	1.78	1.11	0.74	0.49
	9.0	12	2.98	1.86	1.24	0.83	2.34	1.46	0.97	0.65	1.87	1.17	0.78	0.52	1.53	0.95	0.64	0.42
	10.5	10	2.62	1.64	1.09	0.73	2.05	1.28	0.85	0.57	1.64	1.03	0.68	0.45	1.34	0.84	0.56	0.37
	12.0	12	3.26	2.04	1.36	0.90	2.56	1.60	1.06	0.71	2.04	1.27	0.85	0.56	1.66	1.04	0.69	0.46
SA2	3.0	14	4.66	2.92	1.94	1.29	3.65	2.28	1.52	1.01	2.92	1.82	1.21	0.81	2.38	1.49	0.99	0.66
	4.5	13	3.89	2.41	1.61	1.07	3.05	1.91	1.27	0.84	2.44	1.52	1.02	0.67	1.99	1.24	0.83	0.55
	6.0	18	4.96	3.10	2.07	1.38	4.12	2.57	1.71	1.14	3.29	2.06	1.37	0.91	2.68	1.67	1.12	0.74
SA3	2.0	13	3.11	1.95	1.30	0.86	2.44	1.53	1.02	0.68	1.95	1.22	0.81	0.54	1.59	0.99	0.66	0.44
	3.5	16	4.15	2.59	1.73	1.15	3.25	2.03	1.35	0.92	2.60	1.62	1.08	0.72	2.11	1.32	0.88	0.59

Table 22.6 Factors of safety against liquefaction using Idriss & Boulanger (2008) method

Study area	Depth (m)	$(N_1)_{60cs}$	Factor of safety															
			5.0				5.5				6.0				6.5			
			Z_2	Z_3	Z_4	Z_5	Z_2	Z_3	Z_4	Z_5	Z_2	Z_3	Z_4	Z_5	Z_2	Z_3	Z_4	Z_5
SA1	3.0	08	2.49	1.55	1.03	0.69	2.19	1.37	0.91	0.61	1.82	1.13	0.76	0.50	1.51	0.94	0.63	0.42
	4.5	11	2.98	1.86	1.24	0.82	2.62	1.63	1.09	0.72	2.17	1.35	0.90	0.60	1.80	1.12	0.75	0.50
	6.0	18	4.38	2.74	1.82	1.22	3.86	2.41	1.60	1.07	3.20	2.00	1.33	0.88	2.65	1.65	1.10	0.73
	7.5	14	3.53	2.21	1.47	0.98	3.11	1.94	1.29	0.86	2.58	1.61	1.07	0.71	2.14	1.33	0.89	0.59
	9.0	12	3.16	1.97	1.32	0.87	2.77	1.73	1.15	0.77	2.31	1.44	0.96	0.64	1.91	1.19	0.79	0.53
	10.5	10	2.82	1.76	1.17	0.78	2.48	1.55	1.03	0.68	2.06	1.28	0.86	0.57	1.71	1.07	0.71	0.47
	12.0	12	3.17	1.98	1.32	0.88	2.78	1.73	1.16	0.77	2.31	1.44	0.96	0.64	1.91	1.19	0.79	0.53
SA2	3.0	14	4.88	3.05	2.03	1.35	4.31	2.69	1.79	1.19	3.57	2.23	1.48	0.99	2.95	1.84	1.23	0.82
	4.5	14	4.35	2.72	1.81	1.21	3.84	2.40	1.60	1.06	3.18	1.98	1.32	0.88	2.63	1.64	1.09	0.73
	6.0	18	5.06	3.16	2.11	1.40	4.47	2.79	1.86	1.24	3.70	2.31	1.54	1.02	3.06	1.91	1.27	0.85
SA3	2.0	13	3.41	2.13	1.42	0.94	3.00	1.87	1.25	0.83	2.49	1.55	1.03	0.69	2.06	1.28	0.85	0.57
	3.5	17	4.24	2.65	1.76	1.17	3.73	2.33	1.55	1.03	3.10	1.93	1.29	0.86	2.56	1.60	1.06	0.71

Table 22.7 Liquefaction potential indices corresponding to factors of safety obtained from IS 1893 Part 1 2016 method

Study area	Depth (m)	$(N_1)_{60}$	Liquefaction potential index															
			5.0				5.5				6.0				6.5			
			Z_2	Z_3	Z_4	Z_5	Z_2	Z_3	Z_4	Z_5	Z_2	Z_3	Z_4	Z_5	Z_2	Z_3	Z_4	Z_5
SA1	3.0	8	00	00	34	80	00	10	62	98	00	42	85	112	03	63	98	122
	4.5	10	00	00	00	51	00	00	32	73	00	08	57	90	00	35	76	102
	6.0	18	00	00	00	00	00	00	00	03	00	00	00	29	00	00	06	50
	7.5	14	00	00	00	04	00	00	00	30	00	00	11	48	00	00	32	64
	9.0	12	00	00	00	18	00	00	03	38	00	00	24	53	00	05	40	64
	10.5	10	00	00	00	25	00	00	14	43	00	00	30	52	00	15	48	60
	12.0	12	00	00	00	08	00	00	00	23	00	00	12	35	00	00	25	43
SA2	3.0	14	00	00	00	00	00	00	00	00	00	00	00	32	00	00	02	58
	4.5	13	00	00	00	00	00	00	00	25	00	00	00	51	00	00	26	70
	6.0	18	00	00	00	00	00	00	00	00	00	00	00	14	00	00	00	36
SA3	2.0	13	00	00	00	25	00	00	00	57	00	00	34	83	00	02	61	101
	3.5	16	00	00	00	00	00	00	00	13	00	00	00	46	00	00	20	67

Table 22.8 Liquefaction potential indices corresponding to factors of safety obtained from Idriss & Boulanger (2008) method

Study area	Depth (m)	$(N_1)_{60cs}$	Liquefaction potential index															
			5.0				5.5				6.0				6.5			
			Z_2	Z_3	Z_4	Z_5	Z_2	Z_3	Z_4	Z_5	Z_2	Z_3	Z_4	Z_5	Z_2	Z_3	Z_4	Z_5
SA1	3.0	08	00	00	00	53	00	00	15	66	00	00	41	85	00	10	63	98
	4.5	11	00	00	00	28	00	00	00	43	00	00	15	62	00	00	39	77
	6.0	18	00	00	00	00	00	00	00	00	00	00	00	17	00	00	00	38
	7.5	14	00	00	00	02	00	00	00	17	00	00	00	41	00	00	14	51
	9.0	12	00	00	00	14	00	00	00	25	00	00	04	40	00	00	23	52
	10.5	10	00	00	00	21	00	00	00	30	00	00	13	41	00	00	27	50
	12.0	12	00	00	00	10	00	00	00	00	00	00	03	34	00	00	17	37
SA2	3.0	14	00	00	00	00	00	00	00	00	00	00	00	02	00	00	00	31
	4.5	14	00	00	00	00	00	00	00	00	00	00	00	19	00	00	00	42
	6.0	18	00	00	00	00	00	00	00	00	00	00	00	00	00	00	00	21
SA3	2.0	19	00	00	00	11	00	00	00	31	00	00	00	56	00	00	27	77
	3.5	17	00	00	00	00	00	00	00	00	00	00	00	23	00	00	00	48

Results and Discussion

Tables 22.5 and 22.6 indicate that the factors of safety obtained from the Idriss and Boulanger method are comparatively higher than those determined from IS 1893 Part 1 2016 method. Further, the FOS against liquefaction under the considered magnitudes of the earthquake for study areas 2 and 3 obtained from both IS method and Idriss and Boulanger method corresponding to zone 2 are more than one signifying that the subsoil at these areas is not susceptible to liquefaction under the considered magnitudes of the earthquake. While for study area one, the FOS corresponding to zone 2 against magnitudes 5, 5.5, and 6 are more than one from both the methods. The top layer (i.e. at 3 m) of the subsoil at site 1 yielded FOS less than 1 when evaluated from IS method for a magnitude of 6.5. The FOS determined from IS method for sites 1 and 3 corresponding to zone 5 are less than 1.2 for all the considered magnitudes of the earthquake, while for site 2, the FOS are more than one for magnitudes 5 and 5.5 and less than one for magnitudes 6 and 6.5. Few pockets of liquefiable layers are identified in subsoil profiles corresponding to seismic zones 3 and 4 for earthquake magnitudes of 6 and 6.5. Tables 22.7 and 22.8 give that the liquefaction potential indices of the three study areas in zone 2 are zero, suggesting that the potential to liquefaction is very low for the considered magnitudes of earthquakes. The LPI values of the three study areas corresponding to zone 5 for a magnitude of 6.5 are between 36 and 122. It indicates that the liquefaction potential of study area soil profiles if present in seismic zone 5 is very high.

Conclusions

Based on the interpretation of the factors of safety and liquefaction potential indices evaluated for the three study areas under consideration, the following conclusions were made:

i. Factors of safety against liquefaction from Idriss and Boulanger (2008) method are higher compared to IS 1893 Part 1 2016 method and is therefore the more conservative of the two methods.
ii. The three study areas considered along the coastline of Visakhapatnam are not prone to liquefaction under the present seismic zoning 2 for earthquakes of magnitudes up to 6.5.
iii. Subsoil profiles prevailing at the study areas 1 and 3 if present in seismic zone 5 are prone to liquefaction even under a magnitude of 5.0. While, sites in seismic zone 5 similar to those encountered at study area 2 liquefy under a magnitude of 5.5.
iv. Subsoil profiles at study area 1 if present in seismic zone 4 will be at a risk of liquefaction under earthquakes of magnitudes 5.5, 6, and 6.5
v. The liquefaction potential of the similar subsoil profiles prevailing at the study areas is very high for all considered magnitudes of earthquakes corresponding to seismic zone 5.

References

1. Kramer SL (2013) Geotechnical earthquake engineering, 2nd edn. Pearson Education Inc, New Delhi
2. Prakash S (1981) Soil dynamics. McGraw Hill Book Company, USA
3. Seed HB, Idriss IM (1971) Simplified procedure for evaluating soil liquefaction potential. J Soil Mech Foundations Division 97(9)
4. Seed HB, Idriss IM, Arango I (1983) Evaluation of liquefaction potential using field performance data. J Geotechnical Eng 109(3)
5. Iwasaki T, Tokida K, Tatsuoka F (1981) Soil liquefaction potential evaluation with use of the simplified procedure. In: Proceedings of the international conferences on recent advances in geotechnical earthquake engineering and soil dynamics, vol 12
6. Arulmoli K, Arulanandan K, Seed HB (1985) New method for the evaluation of liquefaction potential. J Geotechnical Eng 111(1)
7. Kayen RE, Mitchell JK, Seed RB, Lodge A, Nishio SY, Coutinho R (1992) Evaluation of SPT-, CPT-, and shear wave-based methods for liquefaction potential assessment using Loma Prieta data. In: Hamada M, O'Rourke TD (eds) Proceedings of the 4th Japan-US workshop on earthquake resistant design of lifeline facilities and countermeasures for soil liquefaction
8. Idriss IM, Boulanger RW (2006) Semi-empirical procedures for evaluating liquefaction potential during earthquakes. Soil Dyn Earthquake Eng 26(2–4):115–130
9. Idriss IM, Boulanger RW (2008) Soil liquefaction during earthquake. EERI Publication, Monograph MNO-12, Earthquake Engineering Research Institute, Oakland
10. Swathi PP, Neelima S (2018) Ground response analysis and liquefaction hazard assessment for Visakhapatnam City. Innov Infrastructure Solutions 3:12

11. Bandaru US, Rao S (2020) G.V. R. Seismic liquefaction potential assessment of Andhra Pradesh Capital Region. J Earth Syst Sci 129:SP-144
12. Geological fault line may trigger tsunami, earthquakes in Vizag: Study (yovizag.com), https://www.yovizag.com/lncs. Last accesses 15 Aug 2021
13. IS 1893 Part 1 (2016) Criteria for earthquake resistant design of structures. Bureau of Indian Standard, New Delhi
14. Iwasaki T, Tokida K, Tatsuoka F (1978) A practical method for assessing soil liquefaction potential based on case studies at various sites in Japan. In: Proceedings of the 2nd international conference on microzonation for safer construction research and application, pp 885–896

Chapter 23
Seismic Performance Evaluation of E-Waste-Georeinforced Embankment and Pavement

K. Muthukkumaran, Yagnya Prasad Pradhan, and Rima Das

Introduction

E-wastes are the waste of electrical/electronic goods those have reached end of their lives and mostly consisting of 80% of plastics and 20% of other materials. Plastics presents in the E-waste are categorized into different types such as poly-propylene, poly-ethylene, and poly-styrene. E-wastes were crushed to a size ranging from 4.75 mm to 425 microns, and the optimum percentage of replacement of E-waste, i.e. 2%, is mixed with soil to find the dynamic parameter of soil mixed with E-waste [1–3]. Along with China, India is the largest importer of electronic devices from developed countries like US, UK, and Japan. There is no proper way to dispose such a waste. So, it is important to find a proper way to implement the use of E-waste into soil without affecting the environment.

A geogrid is geosynthetic material used to reinforce soils and similar materials. Geogrids can be used as a reinforcement between various layers in embankment as well as pavement. Due to its lower density, light weight pavement can be constructed using geogrids [4–7].

Several researchers have used E-waste as soil stabilizer in many parts of the world. Researchers have performed cyclic triaxial test on sand mixed with E-waste to evaluate the improvement of strength properties of soil [8] (Ayse Edincliler et al. 2015; Keeran Kumar et al. 2019). Karthi et al. [9] have performed different test to identify the shear strength of soil by mixed with E-waste, geogrids, geotextile, and coir to make it useful for engineering practice. Some studies have performed several test to find out the suitability of geotextiles as strengthening material in pavement

K. Muthukkumaran · Y. P. Pradhan · R. Das (✉)
Department of Civil Engineering, National Institute of Technology, Tiruchirappalli 620015, India
e-mail: 403120005@nitt.edu

K. Muthukkumaran
e-mail: kmk@nitt.edu

K. Muthukkumaran et al. (eds.), *Soil Dynamics, Earthquake and Computational Geotechnical Engineering*, Lecture Notes in Civil Engineering 300,
https://doi.org/10.1007/978-981-19-6998-0_23

construction and soil stabilization of highways, dams, bridges, and railway structures (Ayse Edincliler et al. 2008; Sugandini et al. 2017). Some numerical modelling study was also conducted to determine the optimum tensile strength of geotextile as the reinforcement in road embankment (Danial Tjandra et al. 2015). Most of the studies are on cohesive soil, and very few number of studies have been done on cohesionless soils. Dynamic properties of c-ϕ soils mixed with waste material have not been determined. In this study, attempt has been made to conduct cyclic tests of soil mixed with E-waste and reinforced with geogrid in pavement and embankment under both saturated and unsaturated condition to evaluate the suitability of E-waste-georeinforcement in pavement and embankment for seismic load.

Application of Geogrid in Embankment and Pavement

Due to the cyclic loading behaviour of earthquake, pavement layers are greatly affected as it starts from the inner pavement layers and reflected towards the upper layers of the pavement which results into different types of deformations to the pavement. So, it is planned to use the geogrids in pavement and embankment as shown (Figs. 23.1 and 23.2).

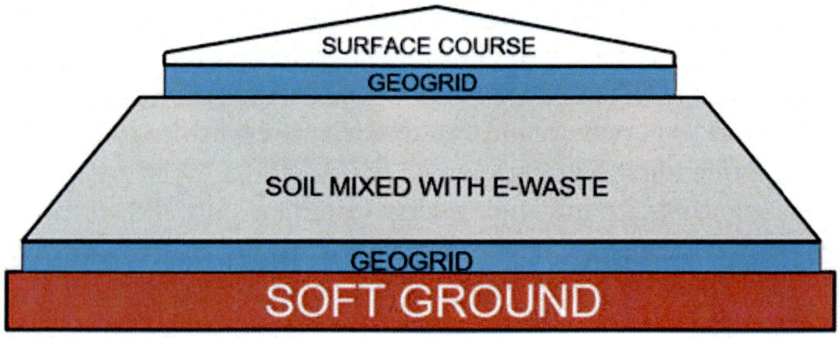

Fig. 23.1 Geogrid reinforced embankment

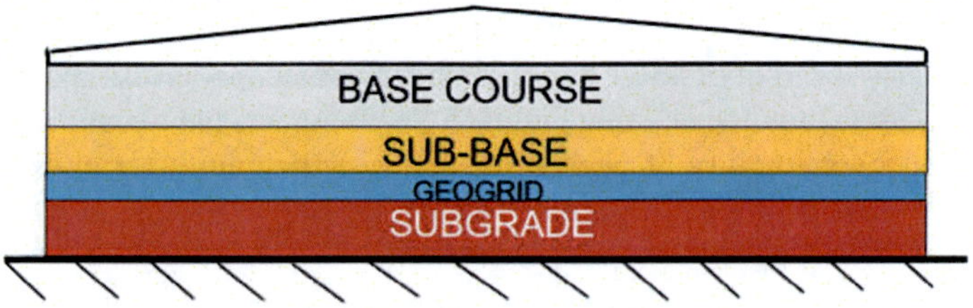

Fig. 23.2 Geogrid reinforced pavement

Table 23.1 Soil parameter

Parameters	Values
Specific gravity	2.69
Gravel (%)	4
Sand (%)	66
Fines (%)	31
Liquid limit (%)	28
Plastic limit (%)	16
Plasticity index (%)	12
Classification based on plasticity chart	SC
Optimum moisture content (%)	10.8
Maximum dry density (kN/m^3)	20.1

Testing Methods

Soil

Bulk quantities of soil have been collected from excavation site at chemical engineering department NIT—Tiruchirappalli. The collected soil samples were characterized in the laboratory. Tests were conducted to determine physical and engineering properties of soil samples as per respective IS codes. 50% of soil was found to have passed the 4.75 mm size sieve. So, it is coarse grained soil (sand). 31% of soil was found to have passed the 75micron size sieve (less than 50%). So, it is not fine grained soil. Plasticity index is 12%. Percentage of fines is more than 12%, and it is found to be non-plastic soil. Hence, the soil is classified as SC (silty clay) (as per IS 1498:1970). The results' index on collected soil was presented in Table 23.1 (Fig. 23.3).

Cyclic Triaxial Test

The test involves subjecting a cylinder soil sample to cyclic axial load at both undrained and drained condition. Specimens shall have a minimum diameter of 51 mm with a height-to-diameter ratio between 2.0 and 2.5. The input parameters taken are for the Indian earthquake data (Wikipedia) taken for zone-iii and zone-iv.

a. Frequency: 0.25 Hz, 0.5 Hz, 1 Hz
b. Confining pressure: 1 kg/cm^2, 1.5 kg/cm^2
c. Amplitude: 2 mm
d. Number of cycles: 30

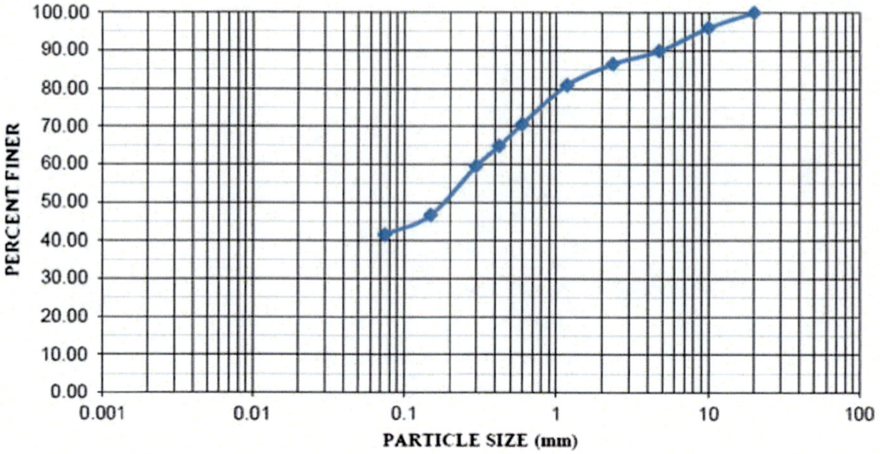

Fig. 23.3 Cumulative % finer versus sieve size for soil (SC)

By taking the above parameters as input, the setup will give dynamic parameters such as shear modulus, damping ratio, and Young's modulus as outcomes. Then, modulus of subgrade reaction is calculated by Eq. (23.1)

$$K = 1.13\frac{E}{\left(1-\mu^2\right)}\frac{1}{\sqrt{A}} \tag{23.1}$$

where E = Young's modulus; A = Area of load plunger; μ = Poisson's ratio of soil (0.3–0.35).

Cyclic Triaxial Test Results

By applying same confining pressure (1 and 1.5 kg/cm^2) and amplitude of 2 mm with different frequency of magnitude 0.25, 0.5, and 1 Hz, testing has been performed under both unsaturated and saturated condition on soil mixed with E-waste (Figs. 23.4 and 23.5; Table 23.2).

Comparison Study

Comparison of dynamic parameters with frequency for virgin soil as well as modified soil, i.e. soil mixed with E-waste in various combination, has been done (Figs. 23.6 and 23.7).

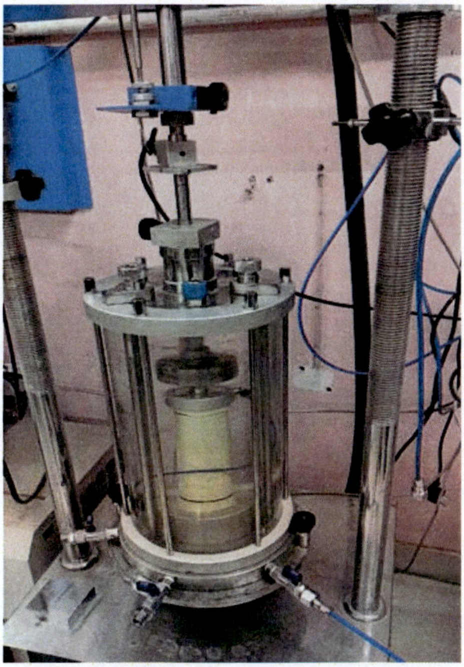

Fig. 23.4 Testing on sample

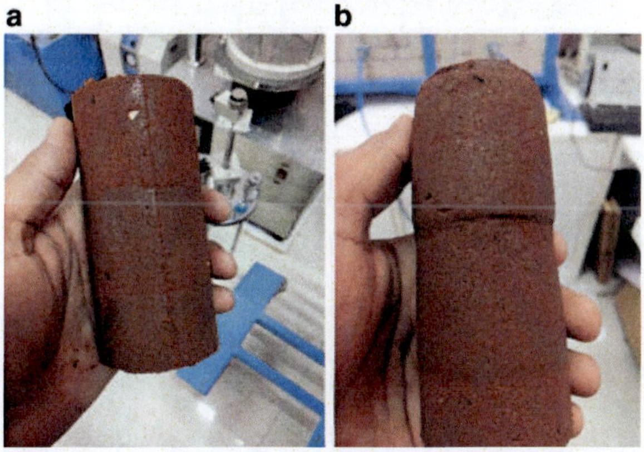

Fig. 23.5 Triaxial sample: **a** Before testing and **b** after testing

Bender Element Test

Bender element helps in measuring the maximum shear modulus (G_{max}) of soil samples to evaluate their stiffness. G_{max} (typically with shear strain levels of about 0.001%) is used to predict soil behaviour or soil-structure interaction during earthquakes, explosion, or machine and traffic vibrations in low-strain dynamic analysis.

Table 23.2 Triaxial test results at different conditions

	Trial 1 (unsaturated condition)			Trial 1 (saturated condition)			Trial 2 (unsaturated condition)		
CP (kg/cm^2)	1	1	1	1	1	1	1.5	1.5	1.5
Amplitude (mm)	2	2	2	2	2	2	2	2	2
Frequency (Hz)	0.25	0.50	1.00	0.25	0.50	1.00	0.25	0.50	1.00
Young's modulus (kg/cm^2)	83–72	69–64	54–38	62–49	51–37	39–23	86–73	72–66	63–49
Shear modulus (kg/cm^2)	35–29	27–25	21–15	28–21	25–20	19–12	37–30	29–25	22–18
Damping ratio (%)	16–12	21–23	25–22	12–10	19–15	22–17	13–9	19–17	21–19
Modulus of subgrade reaction (kg/cm^3)	22.19–19.23	18.58–17.29	14.61–10.19	16.75–13.25	13.77–9.91	10.39–6.22	24.28–20.8	20.47–18.8	17.4–13.24
	Trail 3 (unsaturated condition)			Trail 4 (Unsaturated Condition)					
CP (kg/cm^2)	1	1	1	1.5	1.5	1.5			
Amplitude (mm)	2	2	2	2	2	2			
Frequency (Hz)	0.25	0.50	1.00	0.25	0.50	1.00			
Young's modulus (kg/cm^2)	69–58	55–42	43–29	73–61	57–50	48–36			
Shear modulus (kg/cm^2)	29–23	22–20	16–10	33–29	28–23	17–15			
Damping ratio (%)	13–9	17–15	24–19	11–9	14–11	18–17			
Modulus of subgrade reaction (kg/cm^3)	18.56–15.6	14.8–11.3	11.57–7.8	19.64–16.41	15.33–13.45	12.91–9.68			

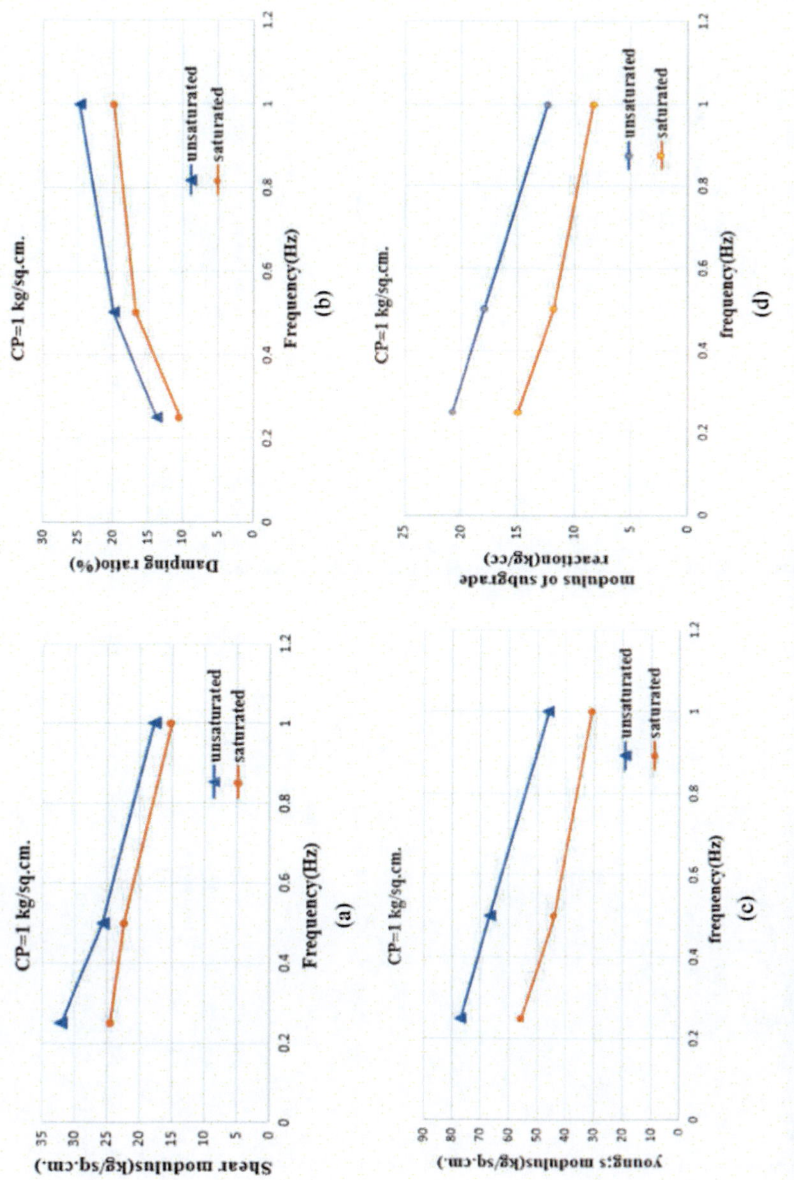

Fig. 23.6 Dynamic output parameters at confining pressure 1 kg/cm^2 under saturated and unsaturated condition for **a** shear modulus, **b** damping ratio, **c** Young's modulus, and **d** Modulus of subgrade reaction

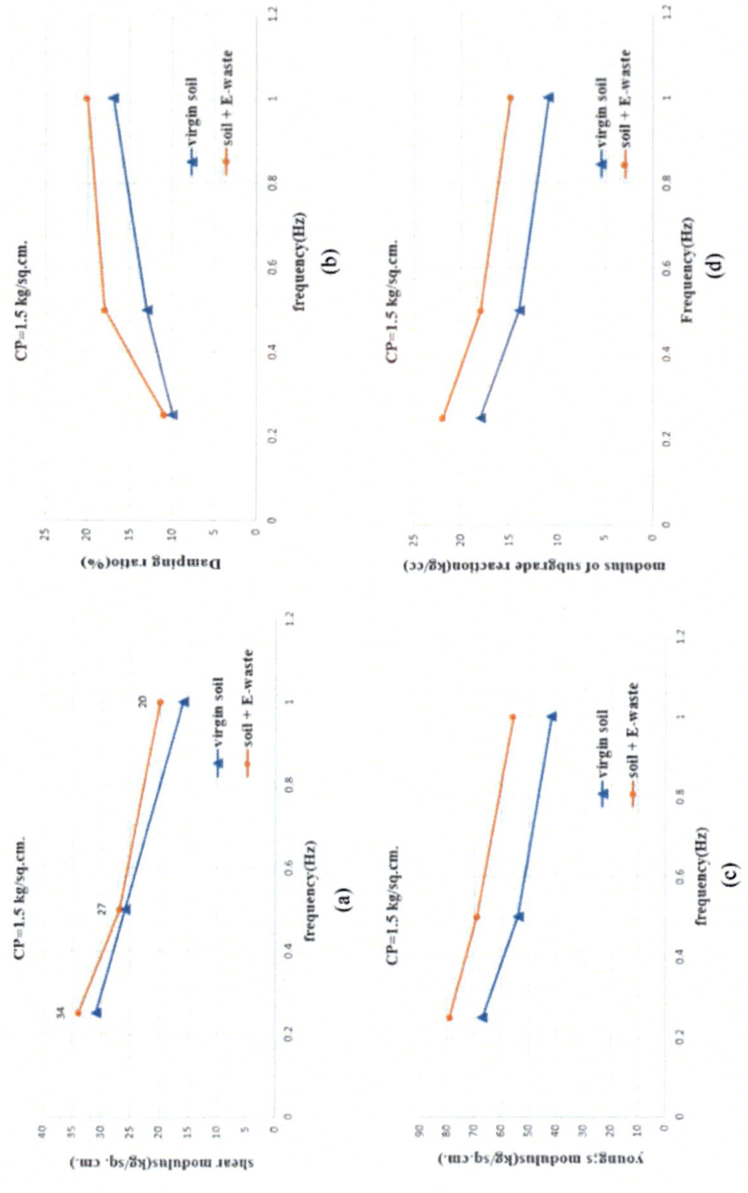

Fig. 23.7 Dynamic output parameters at confining pressure 1.5 kg/cm^2 for virgin and soil mixed with E-waste for **a** shear modulus, **b** damping ratio, **c** Young's modulus, and **d** modulus of subgrade reaction

Fig. 23.8 Bender element testing system

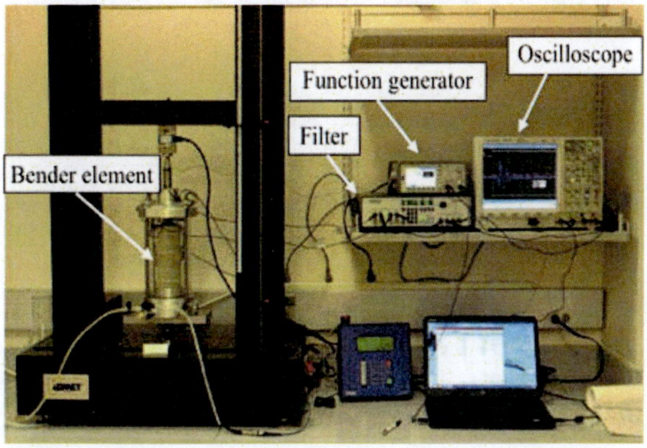

Fig. 23.9 Bender element setup

$$G_{max} = \rho \cdot (V_s)^2 \tag{23.2}$$

where ρ is the mass density of the soil sample and V_s is shear wave velocity (Figs. 23.8 and 23.9).

Bender Element Test Results

A sample of 50 mm diameter with length-to-diameter ratio of 2 to 2.5 is prepared. To fix the sample, it is drilled 15 mm in both top and bottom of the sample. To apply the confining pressure, the sample is confined with rubber membrane. Then, the same procedure is repeated as triaxial test. Sample is tested with a constant frequency of 1 and 1.5 kHz and varying confining pressure (Fig. 23.10; Table 23.3).

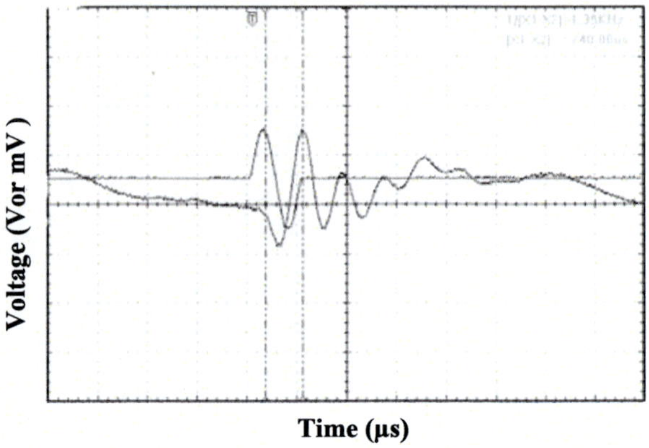

Fig. 23.10 Element signals during triaxial testing

Table 23.3 Bender element test with frequency 1 and 1.5 kHz

Frequency (kHz)	Confining pressure (kPa)	Velocity of shear wave (m/s)	G_{max} (Pa)	E (Pa)	k (N/m^3)
1	50	68.55	5.591×10^6	14.762×10^6	432.39×10^6
1	100	73.28	6.392×10^6	16.869×10^6	479.26×10^6
1	105	74.56	6.620×10^6	17.477×10^6	496.53×10^6
1	200	77.27	7.104×10^6	18.744×10^6	532.53×10^6
1.5	50	71.63	6.092×10^6	16.083×10^6	466.93×10^6
1.5	100	77.36	7.106×10^6	18.760×10^6	534.98×10^6
1.5	150	79.59	7.532×10^6	19.884×10^6	564.92×10^6
1.5	200	83.32	8.251×10^6	21.783×10^6	618.87×10^6

Comparison Study

See Fig. 23.11.

Conclusion

In this present study, the dynamic parameters of SC soil and modified soil have been studied. Under unsaturated condition, dynamic parameters such as Young's modulus, shear modulus, damping ratio, and modulus of subgrade reaction were evaluated and found to be 86–38 kg/cm^2, 35–15 kg/cm^2, 25–12%, and 22.19–10.19 kg/cm^3, respectively. Similarly, the dynamic parameters under saturated condition were evaluated and found to be 62–23 kg/cm^2 for Young's modulus, 28–12 kg/cm^2 for shear modulus, 22–10% damping ratio, and 16.75–6.22 kg/cm^3 for modulus of subgrade

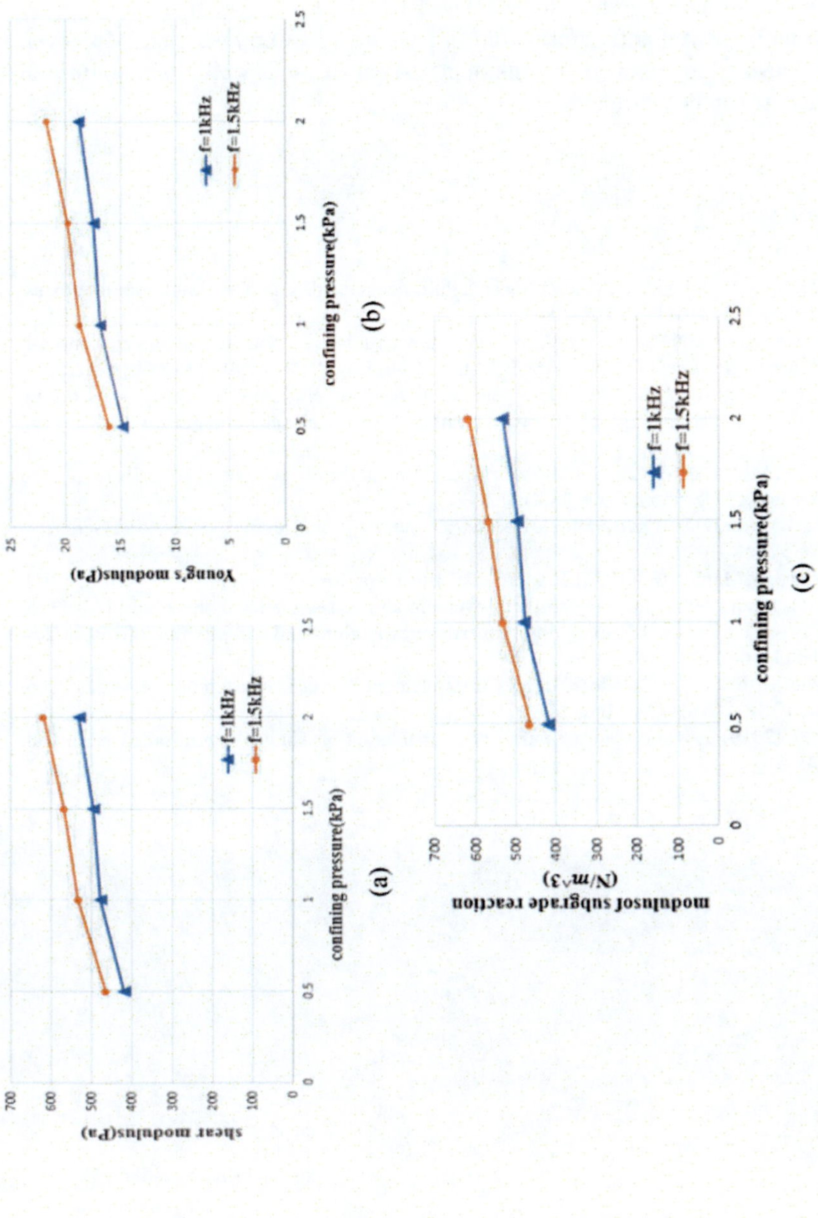

Fig. 23.11 Graphs between confining pressure and **a** Shear modulus, **b** Young's modulus, and **c** modulus of subgrade reaction with different frequency

reaction. Shear modulus is more influenced by frequency compared to damping ratio. Shear modulus and damping ratio which are the key parameters for seismic evaluation have been improved by using E-waste with virgin soil. Young's modulus and modulus of subgrade reaction which plays an important role in rigid and flexible pavement design also significantly increased which shows that stiffness of pavement can be improved and deformation in pavement can be restricted. Moreover, we can achieve an economical embankment and pavement as well as environmental hazardousness can be restricted.

References

1. Frost, DeJong JT, Recalde M (2002) Shear failure behavior of granular-continuum interfaces. Eng Fracture Mech 69(17):2029–2048
2. Mohanty S, Ranjan N (2014) Cyclic behavior and liquefaction potential of Indian pond ash located in seismic zones iii and iv. ASCE J Mater Civil Eng. ISSN 0899-1561/06014012(5)
3. Ozaydin K, Erguvanh A (1980) The generation of pore pressures in clayey soils during earthquakes. In: Proceedings of 7th world conference on earthquake engineering, vol 3, pp 326–330
4. Noorzad R, Mirmordi SH (2010) Laboratory evaluation of the behavior of a geotextile reinforced clay. Geotextiles Geomembranes 28(4):386–392
5. Palmeria EM (2009) Soil-geosynthetic interaction: modelling and analysis. Geotextile Geomembrane 25(5):368–390
6. Humphrey D, Manion W (1992) Properties of tire chips for lightweight fill. In: Borden RH, Holtz RO, Juran I (eds) Grouting, soil improvement and geosynthetics, issue 2, pp 1344–1355
7. Edil T, Bosscher P (1994) Engineering properties of tire chips and soil mixtures. Geotech Test J 17(4):453–464
8. Chaugule M, Deore S, Gaeade K (2017) Improvement of black cotton soil properties using E-waste. IOSR J Mech Civil Eng 14(3):76–81
9. Karthi M (2019) E-waste-soil-geo-reinforcement interaction study for reinforced earth wall application

Chapter 24
Effects of Soil Properties and Input Motion on Liquefaction Zones in Backfill and Subsoil of Quay Walls

K. Pushpa, S. K. Prasad, and P. Nanjundaswamy

Quay Wall

They are the essential part of a structure which has to overcome the presence of water and has a direct impact on the infrastructural development of a nation. It becomes essential to conserve them from the catastrophic effects resulting from strong earthquakes. At times when there are strong earthquakes, there is total failure of these structures even though there has been a lot of research in this area. The stability of any structure mainly depends on the efficient design methodology adopted. Analogous to this condition, the result of the behaviour of soil when it comes in contact with a structure on reducing into liquid state during earthquakes is given by Rollins and Seed [1].

The behaviour of the adjoining soil and the retaining wall is dependent on its properties, stiffness of the soil and the wall, condition of wall, geometry of retaining wall, subsoil stability, ground motion characteristics, and a few others. The destruction caused to these structures has substantial effects on strength, stability, and displacement in the form of sliding, overturning, and settlement of backfill soils and overall failure. It becomes essential to make a thorough research and revisit the design aspects of these structures.

K. Pushpa (✉)
Construction Technology and Management, J.S.S. Science and Technology University, Mysuru 570006, India
e-mail: pushpa_k@sjce.ac.in

S. K. Prasad
Department of Civil Engineering, Vidyavardhaka College of Engineering, Mysuru 570006, India
e-mail: skprasad@vvce.ac.in

P. Nanjundaswamy
Department of Civil Engineering, J.S.S. Science and Technology University, Mysuru 570006, India
e-mail: pnswamy@sjce.ac.in

K. Muthukkumaran et al. (eds.), *Soil Dynamics, Earthquake and Computational Geotechnical Engineering*, Lecture Notes in Civil Engineering 300,
https://doi.org/10.1007/978-981-19-6998-0_24

The properties of the backfill and subsoil have an impact on the zone of liquefaction as well as the amount of displacement of the quay wall. A number of case histories of seismic events were studied to develop some expressions to find out the probability of liquefaction by Liao et al. [2]. The changes in the geotechnical parameters, viz. density, phi, and cohesion, significantly impact the performance of quay walls.

Analysis of Quay Walls

The extent of liquefaction depends mainly on the physical properties of soil, and the qualitative assessment of these parameters is the major step towards understanding liquefaction as mentioned by Seed et al. [3]. The effects of soil parameters, viz. cohesion, phi, and unit weight, are studied using an analytical model through GeoStudio. The soil near water retaining structures is mainly sand, and the level of water is usually high in such soils. At higher magnitudes of earthquakes, higher is the chance of increase in the pressure in the pores. But the loosening of soil at ports occurs even at moderate magnitudes. This will result in reduction in the stability and shear strength of the soil, thereby inducing failure of ground and settlement of soil, increasing the lateral earth pressure and reduction in passive resistance.

Major assumptions made in the present study are:

1. The failure of quay wall is mainly due to sliding displacement.
2. The wall is embedded into the ground to a very small depth.
3. Before liquefaction, the backfill and subsoil act as a rigid-plastic material and as a fluid of higher density than that of water after liquefaction.

Quake/w gives us the displacement values under varying ground motion. A quay wall is drawn in GeoStudio software. A retaining wall laid on subsoil and a backfill on one side and sea water on the other. The boundaries are fixed in both directions at the bottom and fixed against vertical movement on either side. The water exerts a dynamic pressure on the wall and the water in the subsoil adds to the pressure on the other side of the retaining wall. This hydrodynamic pressure can be calculated as per Westergarrd [4]. The stresses applied on the quay wall are the water pressure, the pressure due to backfill soil and acceleration due to the ground motion. The minimum input acceleration is 0.1g m/s^2, and the maximum is 1g m/s^2.

Effect of Input Acceleration on the Sliding and Tilting Displacements and Shear Stress on the Quay Wall

Sliding and rotation models coupled together represent displacement of the water retaining wall. This effect on the retaining wall is analysed using GeoStudio. Quake/w employs the use of integrating the acceleration with time details twice to calculate the resulting displacement of the retaining wall. The inputs for the analysis are taken as cohesionless soil with unit weight 16 kN/m^3, phi is 30°, damping ratio 0.6 and coefficient of lateral earth pressure are 0.5 for the backfill soil. The unit weight of cohesionless subsoil is 20 kN/m^3 and phi is 40°. The quay wall has a unit weight of 24 kN/m^3 and dimensions 6 m by 8 m.

Cohesionless soil tends to reduce to a liquid faster when compared to dense sand. However, dense sand is less susceptible and the same is observed in the analytical model. Figure 24.1 shows the change in displacement with changes in input acceleration. It can be seen that the effect of liquefaction is more pronounced when both subsoil and backfill soil liquefy than when they individually liquefy. Further, the displacement is large when the subsoil liquefies compared to liquefaction of only backfill soil. The permanent displacement of retaining structures is not restricted to walls that move along their base as a result of inertial forces exerted during earthquake shaking. For some retaining wall systems, permanent displacements may also result from the rotation of a retaining wall about a point along its wall-to-subsoil interface. Many a times, tilt is coupled with linear movement. The tilting displacement is found to be very sensitive to input ground motion properties.

The amount of movement of the retaining wall depends on the extent of gain in momentum of the ground movement. Table 24.1 gives a variation of tilting angle with the increase in acceleration of earthquake motion.

Figures 24.2, 24.3, and 24.4 show the influence of input acceleration of ground motion on the tilting failure. The amount of tilt of shows a linear variation in all the three cases.

Figure 24.5 indicates the change in shear with input acceleration in all the three cases.

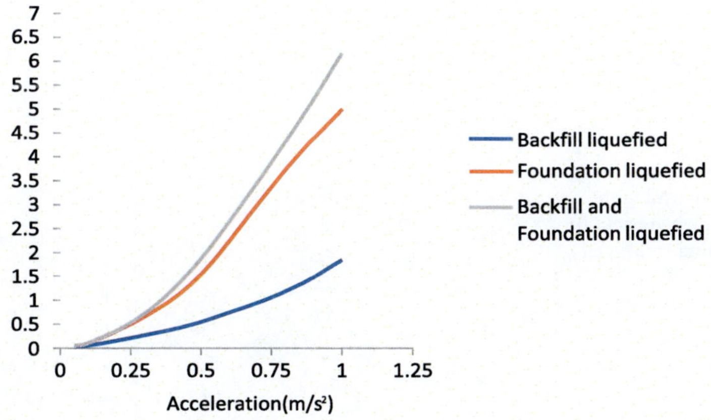

Fig. 24.1 Effect of input acceleration on liquefaction

Table 24.1 Influence of input ground motion on the tilting failure

Acceleration (g, m/s^2)	Angle of tilt (∘)		
	Backfill liquefied	Foundation liquefied	Backfill and foundation liquefied
0.05	0.167	0.0057	0.005
0.1	0.296	0.1094	0.176
0.3	1.096	0.7843	1.022
0.5	0.210	6.140	8.75
0.8	7.960	18.204	17.62
1.0	4.37	22.956	27.00

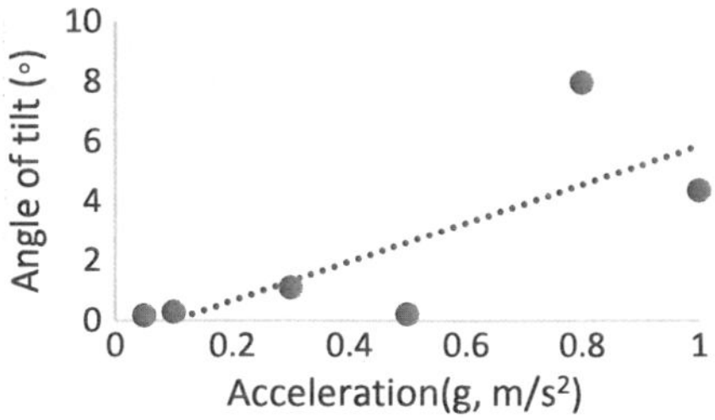

Fig. 24.2 Variation of angle of tilt of quay wall with acceleration when backfill soil liquefies

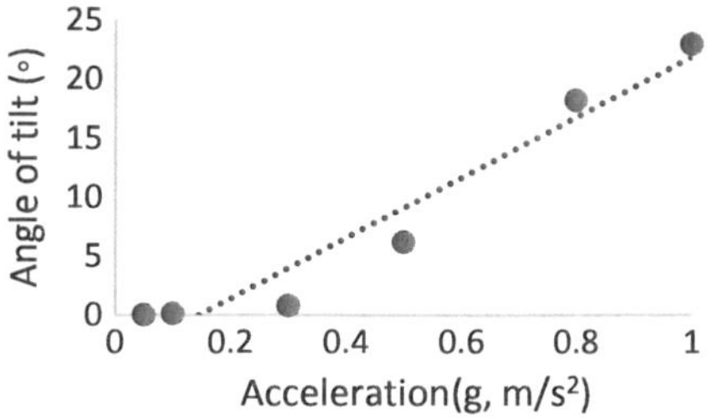

Fig. 24.3 Variation of angle of tilt of quay wall with acceleration when foundation soil liquefies

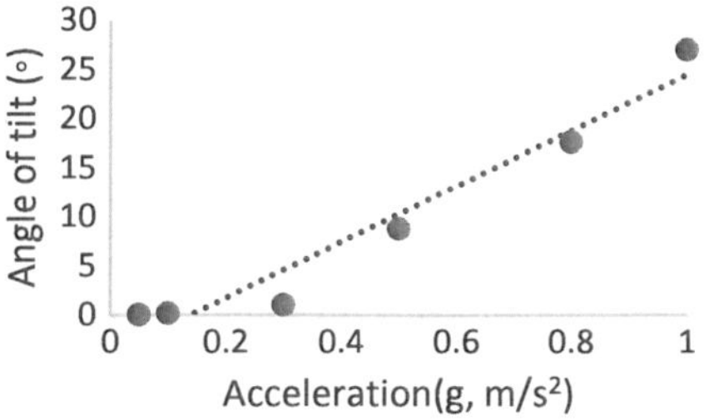

Fig. 24.4 Variation of angle of tilt of quay wall with acceleration when backfill and foundation soil liquefy

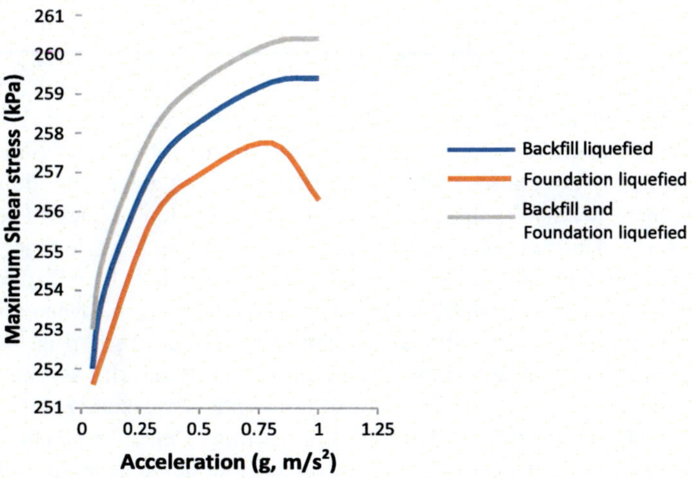

Fig. 24.5 Influence of input acceleration of ground motion on the maximum shear stress on the quay wall

Effect of Soil Properties on Liquefaction Zones of Backfill and Subsoil of Quay Walls

The various parameters which influence the liquefaction zones in backfill and subsoil studied here are cohesion of soil, phi, and unit weight of soil with changes in input acceleration of ground motion. Here, five cases of varying unit weight and angle of internal friction are studied for loose and dense sand as shown in Table 24.2. Table 24.2 gives the effect of acceleration on the maximum linear displacement of the retaining wall for change in soil parameters. It can be seen that the maximum horizontal displacement is more when the backfill liquefies. On the other hand, when subsoil liquefies, the quay wall experiences punching motion in to the subsoil and hence the horizontal displacement is reduced.

Table 24.2 Properties of backfill and subsoil for different cases

	Backfill soil		Subsoil	
Property	Unit weight (kN/m^3)	Angle of internal friction (°)	Unit weight(kN/m^3)	Angle of internal friction (°)
Case 1	15	20	20	40
Case 2	17.5	30	20	40
Case 3	20	40	20	40
Case 4	20	40	17.5	30
Case 5	20	40	15	20

Figure 24.6 indicates the changes due to input ground motion on the displacement of quay walls when the soil properties are varying from loose to dense in backfill and subsoil. It shows that there is an increase in the linear movement of walls with increase in the input acceleration and decrease in the displacement with increase in the denseness of backfill and subsoil. The influence of the soil properties mentioned above can be clearly visualized after analysing the model in Quake/w for liquefaction and displacement results. Figure 24.7 shows the liquefaction of backfill soil when the backfill soil has lesser unit weight and cohesion as in case 1, while Figs. 24.8 and 24.9 show the liquefaction of backfill soil and subsoil when both backfill and subsoil are loose. Figure 24.9 shows slight liquefaction in the backfill soil when it is dense. The liquefaction is shown by yellow regions in backfill and subsoil (Fig. 24.10).

The present study is useful to identify the influence of backfill soil and subsoil on the performance of quay wall in terms of horizontal displacement and mode of failure such as sliding, tilting, and both sliding and tilting. Depending on the situation of backfill and subsoil liquefactions and the input motion, the changes in modes of failure are observed. When only backfill starts flowing, the horizontal displacement of quay wall is more pronounced. On the other hand, when the subsoil liquefies,

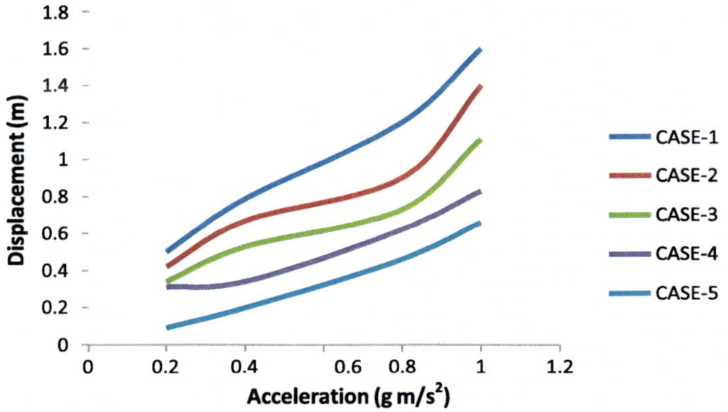

Fig. 24.6 Influence of input acceleration of ground motion on the displacement of quay wall

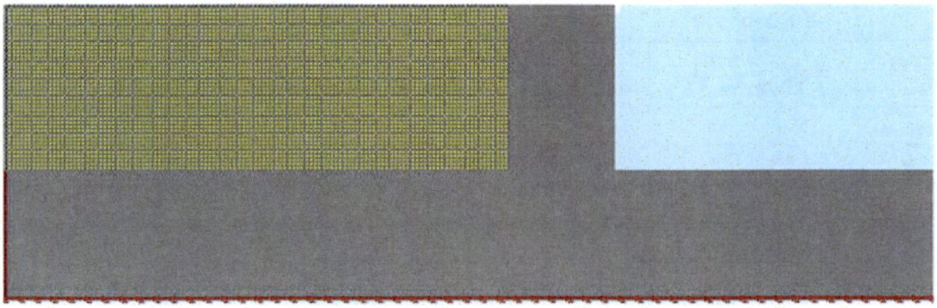

Fig. 24.7 Liquefaction of loose backfill soil

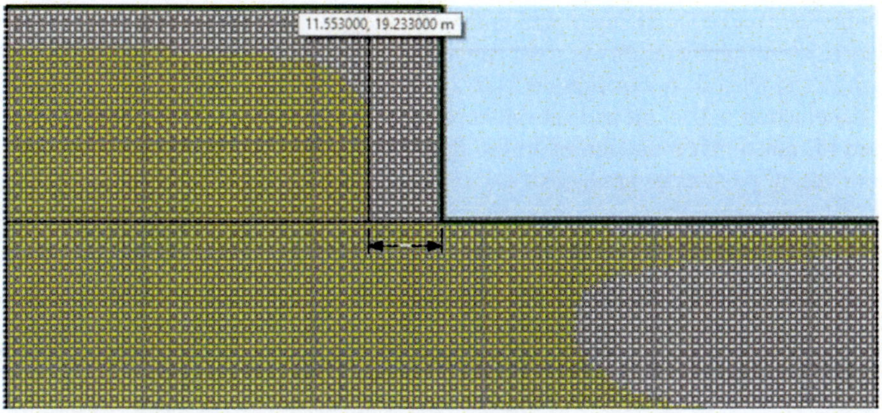

Fig. 24.8 Partial Liquefaction of loose backfill and subsoil

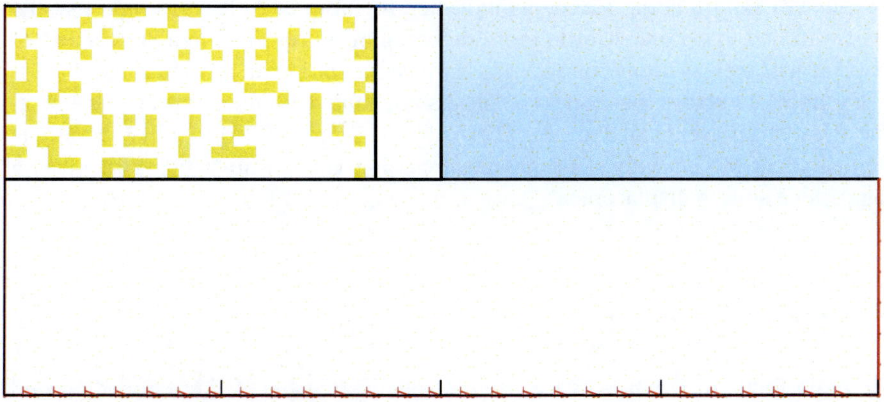

Fig. 24.9 Liquefaction of dense backfill soil

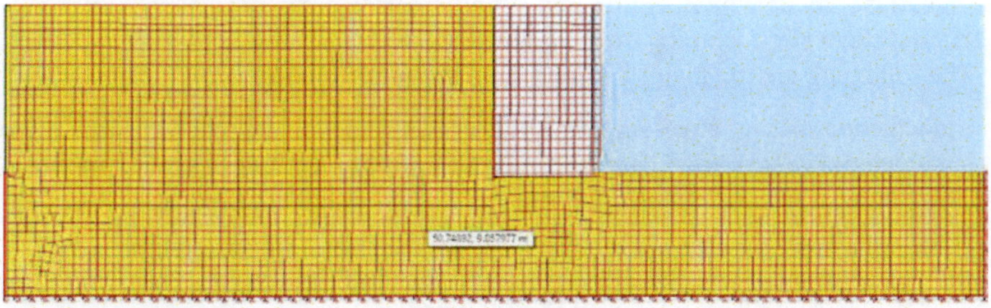

Fig. 24.10 Complete Liquefaction of loose backfill and subsoil

tilting mode will become more predominant. Tilting failure also can be catastrophic. Hence, subsoil vulnerability is most dangerous, and extreme situation is when both backfill and subsoil liquefy. Figures 24.7, 24.8, and 24.9 show the locations and extent of liquefaction in backfill and subsoil depending on their vulnerability. Kramer and Seed [5] noted that reduction of soil to liquid is induced by small ground motions when the initial stresses are large.

Numerical Validation

Model test was carried out using manual shaking table [6] for a quay wall system comprising of subsoil which has high unit weight, backfill soil is cohesionless and quay wall is medium heavy. The same was modelled in GeoStudio, and the input data as in the physical model test was provided. The results were comparable with the mode of failure being sliding as indicated in Table 24.3. It was found that the quay wall had experienced little more deformation in physical model in terms of movement as well as tilt as compared to the observations from analytical model test.

Figure 24.11 shows the model test on shaking table with failure as displacement. The effect on adjacent soil and subsoil can also be observed. Figure 24.12 represents the analytical model with same parameters using GeoStudio which also gives a comparable sliding displacement.

Concluding Remarks

The main focus of the present work is to identify the liquefaction zones created during earthquake due to the changes in the soil properties. Such a study will assist in performance-based design of quay walls. The present study also aims to analyse the situations under which sliding and tilting displacements occur and assess the maximum shear stress coming on the quay wall.

The following are the major findings from the present work.

- Liquefaction has a huge impact on the linear movement of retaining wall in situations where backfill soil and subsoil liquefy as a result of ground motion.
- The sliding displacement of quay walls is highly sensitive to input acceleration of ground motion.
- The rotational or tilting failure of quay walls is observed to be more dominant when the subsoil under quay wall system liquefies.
- The zone of liquefaction is concentrated in the region which has low density. When backfill is of loose density, liquefaction zone will be more pronounced in backfill region. On the other hand, when subsoil is of loose density, liquefaction is concentrated in subsoil region. When both backfill and subsoil are loose, liquefaction is distributed throughout.

Table 24.3 Details of model tests on shaking table and analytical model tests using GeoStudio

Condition			Model tests on shaking table [6]				Analytical model tests using GeoStudio			
Foundation soil type	Backfill type	Wall weight	Predominant mode	H (mm)	V (mm)	R (deg)	Predominant mode	H (mm)	V (mm)	R (deg)
Dense	Loose	Medium	SL	+ 80	−10	−2	SL	+ 66	−5	0

SL—Sliding, R—Rotation (+), S—Sinking and BR—Backward rotation (−)
H—Horizontal displacement, V—Vertical settlement and
R—Rotation with respect to initial position
+ve = Downward settlement and rotation, −ve = Uplift and backward rotation

Fig. 24.11 Model test on shaking table [6]

Fig. 24.12 Analytical model tests using GeoStudio

- It is possible to identify the modes of failure, such as sliding, tilting, or over turning depending on the vulnerability of backfill and subsoil in addition to the magnitude of ground shaking.
- The displacement results from the GeoStudio analysis are comparable with that obtained from shake table tests for similar soil and quay wall properties.

References

1. Rollins KM, Seed HB (1990) J Geotech Engr Div ASCE 116(2):165–185
2. Liao SSC, Veneziano D, Whitman RV (1988) Regression models for evaluating liquefaction probability. J Geotech Eng 114(4):389–409
3. Seed RB, Cetin KO, Moss RE, Kammerer AM, Wu J, Pestana JM, Riemer MF, Sancio RB, Bray JD, Kayen RE, Faris A (2003) Recent advances in soil liquefaction engineering: a unified and consistent framework. In: Proceedings of the 26th annual ASCE Los Angeles geotechnical spring seminar, ASCE
4. Westergaard M (1931) Water pressure on dams during earthquakes. Trans ASCE. Paper No. 1835, pp 418–433
5. Kramer SL, Seed HB (1988) Initiation of Soil Liquefaction under Stat- ic Loading Conditions. J Geotech Eng 114:412–430

6. Nanjundaswamy P (2008) A study on seismic response of quay walls. Doctoral thesis submitted to Kuvempu University, India
7. Geo-Slope (1992) User's manual, Alberta, Canada: Geo Slope Office, Geo-Slope International, Calgary
8. Martin GR, Liam Finn WO, Seed HB (1975) Fundamentals of liquefaction under cyclic loading. J Geotech Eng Div ASCE 101: 423–438. No. GTS, Proc. Paper 11284, May

Chapter 25
Effect of Fine Content on Resisting Liquefaction Potential of Sandy Soil

Nishigandha Rathod and Seema Shringeri

Introduction

When saturated, loose granular deposits are dynamically stressed; they undergo a quick increase in pore pressure and a temporary reduction in effective normal stress, which can cause the soil to lose almost all of its shear strength and behave like a fluid, a phenomenon known as soil liquefaction. Buildings settle or tip due to liquefaction, and the ground structure fails. Because of the high frequency of liquefaction during earthquakes and the potential for damage, liquefaction has become a major problem in geotechnical earthquake engineering.

Analytical approaches for forecasting ground motions owing to dynamic loading necessitate a thorough understanding of soil deposits stress–strain and energy dissipation characteristics. A variety of laboratory and field approaches are available to measure the dynamic properties of soil. The choice of a technique, on the other hand, is dependent on the specific problem to be solved. Overall densification is produced by cyclic shearing of a sample of cohesion-less soil under constant confining pressure. In the presence of confinement, the soil particles move relative to each other in response to the undrained shearing, and the sample tends to compact after each cycle. Typically, soil with a higher particles concentration is thought to have greater liquefaction resistances. Silty soil samples have distinct deformation characteristics and pore pressure generation than clean sand samples. The liquefaction potential of fines-containing soils is mostly determined by the percentage of fines present and the plasticity of fines, and hence, a clear understanding of the behaviour of silty sands undergoing earthquake loading is required to study.

N. Rathod (✉) · S. Shringeri
Basaveshwar Engineering College, Bagalkot, Karnataka, India
e-mail: nisharathod2019@gmail.com

K. Muthukkumaran et al. (eds.), *Soil Dynamics, Earthquake and Computational Geotechnical Engineering*, Lecture Notes in Civil Engineering 300,
https://doi.org/10.1007/978-981-19-6998-0_25

Literature Review

In terms of literature, there has been a lot of disagreement in determining the effect of fine contents on sand liquefaction qualities. According to several writers, sand liquefaction resistance increases as fine concentration increases. Chang et al. [1] discovered that adding fine content up to 60% fines to sandy soil significantly increases cyclic resistance. When particles up to 60% of the cyclic resistance is mixed with soil, the results show that about half of the cyclic resistance improves. The outcomes are observable and measurable. With the addition of particles to sand, some authors have predicted the opposite behaviour. When sand is mixed with 30% fines, the cyclic resistance of the sand is reduced to roughly 60%, according to Shen et al. [2], Tronsco and Verdugo [3].

Materials and Methodology

Material used: In the present work, sand used is collected from the Malaprabha River near Kamatagi village Bagalkot.

Sand

The specific gravity of sand 2.69, it has a maximum grain size 1 mm, minimum grain size of 0.15 mm; the relative density of sand is 52%; the coefficient of uniformity of sand is 4.66, and the coefficient of curvature is 0.74. The coefficient of curvature is less than 1, and coefficient of uniformity is lesser than 6 as per the standard; based on results, it is classified as poorly graded sand.

Fine Content

Silt is taken from the portion of the regionally accessible sand which was passed through I.S sieve 75μ and is classified as non-plastic silt. Different percentage of silt is added to establish the impact of fines content on liquefaction behaviour such as 0, 10, 20, 30, and 40% (by weight) to the clean sand at relative density (Dr = 40%), and specific gravity of silt is 2.07.

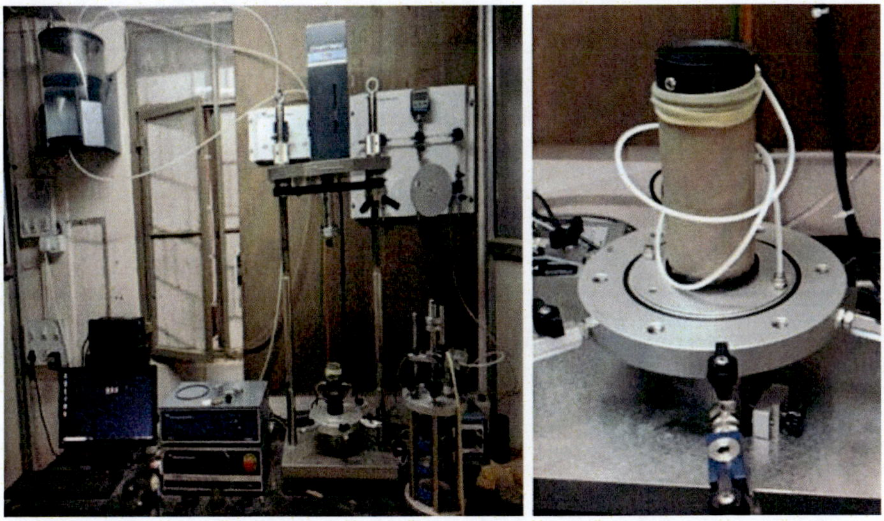

Fig. 25.1 Cyclic triaxial test apparatus and sample preparation

Digital Cyclic Triaxial System

See Fig. 25.1.

Methodology

In this current study, to know the index properties of sand, all the basic tests are conducted on sample. Sand mixed with silt content in range of 0–20% is used for testing. Consolidated undrained tests are conducted on the sample to study the liquefaction potential.

Test Setup

Make sure compressor is ON maintain air pressure above 8 bar on the compressor dial gauge; water supply is ON, fill the de-airing tank up to 80%, de-air the water by turning ON vacuum pump attached and after de-airing (indicated by stabilizing of bubble emerging from water) turn ON vacuum release value B1.

Sample Preparation

Specimens of 100 mm in height and 50 mm in diameter were used in this study. The specimens were prepared using spilt mould by compaction of the soil sample in 3 layers.

Saturation Stage

The specimens were saturated by applying back pressure and compilation of the saturation process by ensuring B check (Skempton's pore water pressure parameter B greater than 0.95 or 95%).

Isotropic Consolidation Stage

The cell pressure and back pressure target values will be the value that of in the end of saturation stage which are to be entered in the respective column and then start consolidation when all the pressure reached that target, consolidation process will start and wait until either change in the volume ceases or degree of consolidation reaches above 95% which indicates that the consolidation of the sample is over.

Cyclic Shear Stage

The test will conducted at frequency of 1 Hz. However, failure can be characterized with respect to peak to peak strain (2%, 5% etc.) that a sample be subjected to cyclic loading, the initial liquefaction standard was taken on as the noticed patterns regarding important variables like development of excess of pore water pressure.

Results and Discussions

For the case of clean washed sand, when the deviator stress is applied, at the beginning, there was increment in the excess pore water pressure with increase in axial strain; later, it starts to reduce and reach negative value. When the sample is loaded, the sand grains try to roll over or split; since the testing condition is undrained test and particle size of sand is more than 75 μm, the sand particles unable to roll over or split so easily. Thus, enormous negative pore water pressure (suction pressure) will be developed in the space between the particles. For the case of 10% silt, the existence of silt in the sample due to consolidation the silt particles slide inside the void between the sand grains and are in fully confined state due to which regional least silt density and regional maximum void ratio of silt will be produced; this will continue till limiting fine content is reached. Consequently, for the 10% of silt content, the silt grains are not able to bear the sand particles structure but create blockage in order to develop excess pore pressure. Hence, for sample with 10% of silt, the negative excess pore pressure generated is relatively less. Now, for 20% of silt content, the silt particles begin to support the sand grain. The silt particles will occupy almost all the voids present, which results in less permeability, and hence, the excess positive pore water pressure developed will be more, which is the indication that at 20% of silt content; the silt particles turn to active contributor in the internal force chain as they fill intergranual voids. For further increment of silt, i.e. 30 and 40%, the sand grain skeleton is practically not stable without the silt particles.

Effect of Fines Content on Pore Water Pressure Response

On clean sand, a combination of sand and fine content, and the following loading conditions, stress-controlled basic cyclic triaxial tests are performed. The loading conditions are relative density (Dr = 40%), frequency (f) = 1 Hz, and deviator stress (d) = 60 kPa; the variation in pore water pressure as a function of the number of loading cycles is shown below.

As seen in Fig. 25.2. It shows that at a deviator stress of 60 kPa, pore water pressure increases slowly at first and then gradually, as in the case of 10%, 30%, and 40% silt added to clean sand; however, in the case of 20% silt content, pore water pressure increased quickly with less number of cycles and then became stable after some time, and thus, resistance will be less at that phase which indicates that 20% of silt content mixture is the limiting fine content.

The individual nature of sand and mixtures of sand and fines has been noticed the last pattern has appeared to interpret the change in pore pressure with varied amount of fines exhibited, as well as the number of cycles limited to100. From Fig. 25.3, it was observed that by adding fines content, amount of pore water pressure built up at

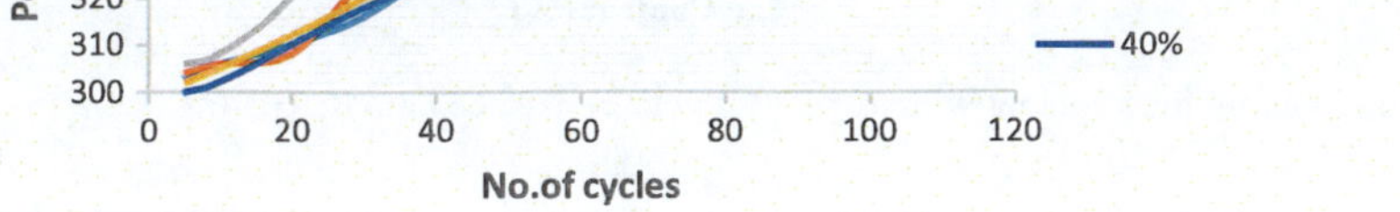

Fig. 25.2 Impact of silt contents on the pore water pressure

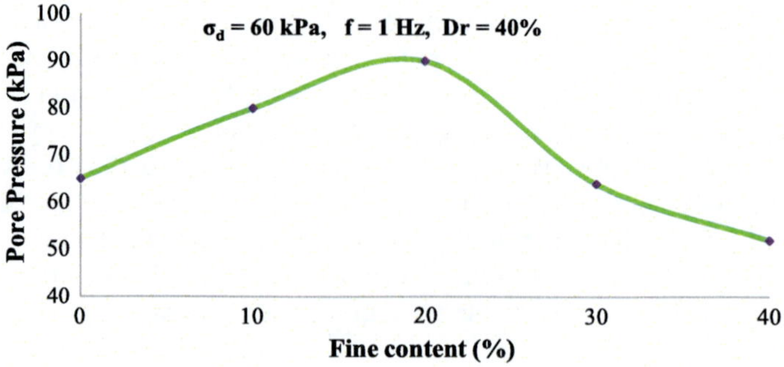

Fig. 25.3 Limiting fines content with respect to pore water pressure

first rise till 20%; afterwards, this pattern observed was reversed after 20% of fines added in other words pore pressure is decreased.

Limiting Fines Content with Respect to Cyclic Resistance Ratio (CRR)

Figure 25.4 shows the impact of fines on CRR. The cyclic stress ratio (CSR) required to commence the first stage of liquefaction in the number of loading cycles is defined as the CRR. Cyclic resistance is described by cyclic stress. The cyclic resistances response is usually opposite to the pore water pressure response in the soil.

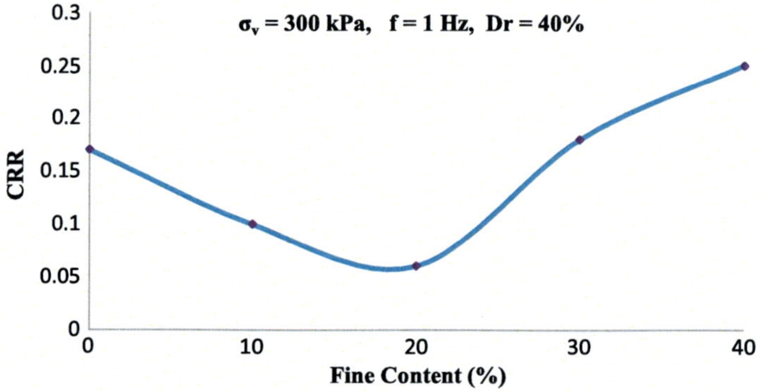

Fig. 25.4 Limiting fines content with respect to cyclic resistance ratio (CRR)

Shear Strain of Sample

The effect of fine content on the relationship between shear strain (%) and the number of cycles for initial liquefaction was studied by conducting a series of cyclic triaxial tests by using stress-controlled techniques.

Figure 25.5 shows the number of cycles required to fail sample by adding a silt content of 20%. The development of the shear strain is quick because of voids present in the sample, and the sample reaches the standard axial strain value 5% is often used for the liquefaction studies.

Figure 25.6 shows the development of shear strain in a slow rate for the sample treated with 20% of silt content, but for 40% silt content which will help to fill the voids present in the matrix and strengthen the sample; the shear strain developed in the sample slowly reaches the standard axial strain of 5%. As the sample is dense, the soil does not exhibit complete loss of stiffness or strength and shows some amount of residual strength even though the pore pressure induced in the sample eventually reaches the applied confining pressure. The development of 100% pore pressure ratio can be observed at about 39 cycles of uniform strain application. In Fig. 25.6, the graph shows the quick growth of double amplitude of axial strain for sample with for 10% of silt content; the axial strain increases to about 3.6% during the first loading cycle; in the second cycle, it was about 4.4%, and later, in 6th cycle, it arrives at 7.3% and reaches initial liquefaction. Now, for sample with 20% fines (i.e. @ 20% silt), in the first loading cycle, the double amplitude of axial strain (εc) turns out to be around 5.4%; in the second cycle, it was 8.2%, and later in 4th cycle, it arrives at liquefaction stage. For the sample with fines of 30% of silt, the double amplitude axial strain found to be 2.7% during the first cycle; in the second cycle, it was about 3.1%, and later, it arrives to 6% in 23rd cycle and initial liquefaction begins. Lastly, for the sample with 40% of colloidal silt content, the increment in the double amplitude axial strain at the time of cyclic loading was generally much slower in rate, especially in beginning stages. The initial liquefaction stage was not achieved, and in the first cycle, the double amplitude axial strain (εc) turns out to be

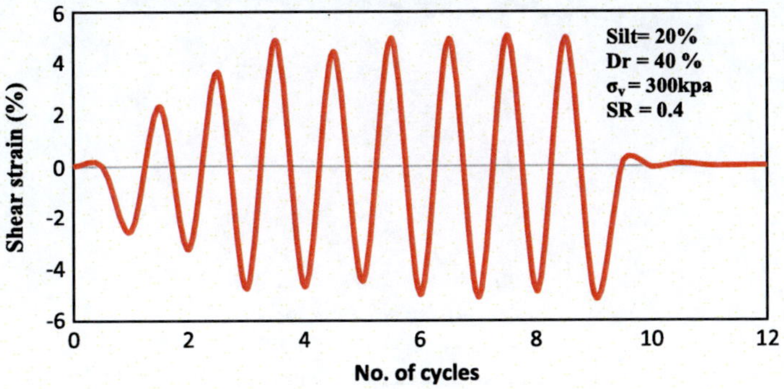

Fig. 25.5 Shear strain of sample with silt content 20%

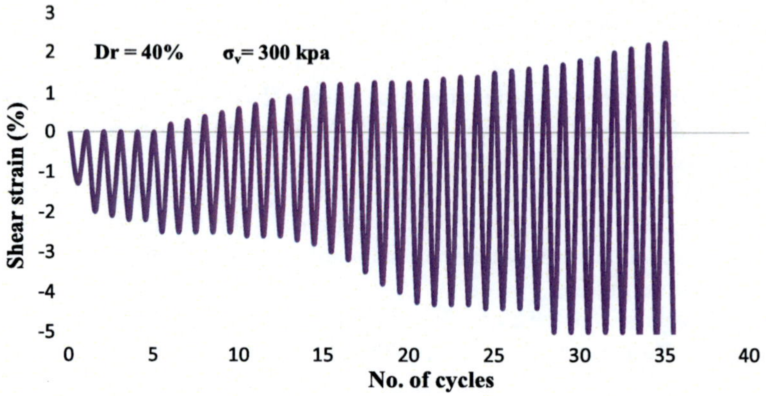

Fig. 25.6 Shear strain of sample with silt content 40%

around 1.6%; in the second cycle, it was 1.8%, and it reaches 2.2% at around 30th cycle. From the above results, we can observe for 20% fines the development of axial strain was quick, and after the amount of silt beyond the limiting fine content the negligible amount of axial strains it might conclude that after limiting silt content adding of a small amount of silt proven advantageous.

Deviator Stress

Figure 25.7 shows the variation of deviator stress with respect to number of cycles; the pore water pressure progressively builds up causing deviator stress path progressively increases and became constant.

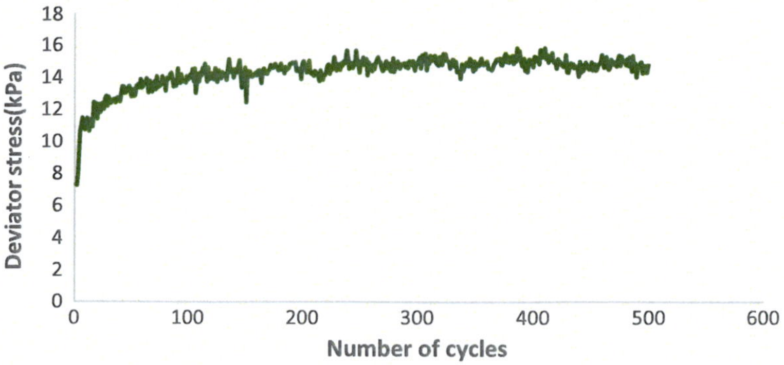

Fig. 25.7 Sand deviator stress versus loading cycles 20% fines

Conclusions

Based on the experimental investigation the results and observation as follows:

- For clean sand sample, when stress is applied, there was increment in excess pore water pressure at beginning; later, it starts to reduce and reach negative value because the development of negative pore pressure in the space between the particles.
- During the initial stage, observations made was the amount of pore pressure generated was comparably high for 20% fines content and reduces for 30% and 40%. In other words, it can also be concluded that the cyclic resistance behaviour is exactly inverse to response of pore water pressure.
- The negligible shear strains development which is the beneficial of using silt. For the sample with 20% silt, the axial strains reaches within a limit of 8 cycles whereas for 40% of silt, it reaches within 58 cycles.
- It is concluded that deviator stress is directly proportional to pore water pressure generation.

References

1. Chang NY, Yeh ST, Kaufman LP (1982) Liquefaction potential of clean sand and silty sand. In: Proceedings, 3rd international earthquake microzonation conference, Seattle, vol 2, pp 1017–1032
2. Shen CK, Vrymoed JL, Uyeno CK (1977) The effects of fine content on liquefaction of sands. In: Proceedings of the 9th international conference on soil mechanics and foundation engineering, Tokyo, Japan, vol 2, pp 381–385
3. Transco JH, Verdugo R (1985) Silt content and dynamic behavior of tailing sand. In: Proceedings 25th international conference on soil mechanics and foundation engineering, San Francisco, USA, pp 1311–1314

Chapter 26
Regional Rupture-Based Seismic Hazard Analysis of Tripura State—NE India

P. Anbazhagan, Arindam Das, and G. Silas Abraham

Introduction

The north-eastern state of India, Tripura, is located between two of the most tectonically active plate boundaries, the Indian-Sunda plate boundary in the east and the Indian-Eurasian plate boundary in the north. The Indian plate is moving under the Eurasian plate as well as the Sunda plate at the boundaries. In the past few decades, many destructive earthquakes were recorded in the region. The Shillong earthquake of 1897, Bihar-Nepal earthquake of 1934, and Assam earthquake of 1950 are some of the mass devastating earthquakes. Recently in 2017, a 5.6 magnitude earthquake at Manu induced liquefaction and also damaged more than 6000 houses in many parts of the state [5, 16]. The rapidly growing population resulting in industrialization and urbanization have intensified the hazard. Hence, currently used construction practices need an upgradation to account for the regional seismicity to mitigate the hazard of the region. The design PGA given in IS 1893 [25] is 0.36 g for the entire state of Tripura. However, the Indian standard code provides the zonation map considering only the past seismic activity of the region and does not consider other factors, like the rupture mechanism and the tectonic features of the area. Therefore, seismic hazard analysis for the region should be carried out in detail, considering all the seismotectonic features, rupture mechanisms, and past seismicity. This study provides the spatial distribution of the bedrock Peak Ground Acceleration for the state of Tripura by DSHA. The seismic hazard of the state was analysed using rupture-based approach (considering the rupture characteristic of the region) and the conventional approach (considering the observed magnitudes from the past seismic activities) both to understand better the influence of rupture characteristics in the seismic hazard analysis for the region.

P. Anbazhagan (✉) · A. Das · G. Silas Abraham
Department of Civil Engineering, Indian Institute of Science, Bangalore 560012, India
e-mail: anbazhagan@iisc.ac.in

K. Muthukkumaran et al. (eds.), *Soil Dynamics, Earthquake and Computational Geotechnical Engineering*, Lecture Notes in Civil Engineering 300,
https://doi.org/10.1007/978-981-19-6998-0_26

Study Area and Its Tectonic Framework: Tripura

Tripura, one of the hilly states of North-East India, is bordered on three sides by Bangladesh. The capital of Tripura is Agartala which is situated near the Bangladesh border in the north-western part of the state. The third-smallest state in the country, Tripura, has an area of 10,486 km^2. In 2011, the state had 36,71,032 residents, constituting 0.3 per cent of the country's population. There are few seismically active regions of the country, and Tripura is one of them. As per the IS 1893 part-1 (2016), the whole state of Tripura is in seismic zone V. The map of Tripura is presented in Fig. 26.1. The red coloured stars in the figure represent the earthquakes whose felt intensity was IV in Tripura and were considered for choosing the seismic study area.

Hills, plains, and valleys characterize the topography of Tripura. The geology of Tripura shows a variety of sedimentary rocks that dates back to the Oligocene period as per Tripura state pollution control board.

The whole North-East India has a high seismicity. The tectonic environment of the region is very complex. Thingbaijam et al. [42] pointed out that the seismic activities of the region include intraplate activities, subduction, and thrusting. Main Boundary Thrust (MBT), Dauki fault, Sagaing fault, Main Central Thrust (MCT), Main Frontal Thrust (MFT), etc., are some of the major active seismic sources in the region. In the past, these faults had significantly ruptured causing some of the major devastating earthquakes.

Various researchers have studied the regional seismotectonics and found: Indian and the Eurasian plates are converging in the north resulting in the formation of the

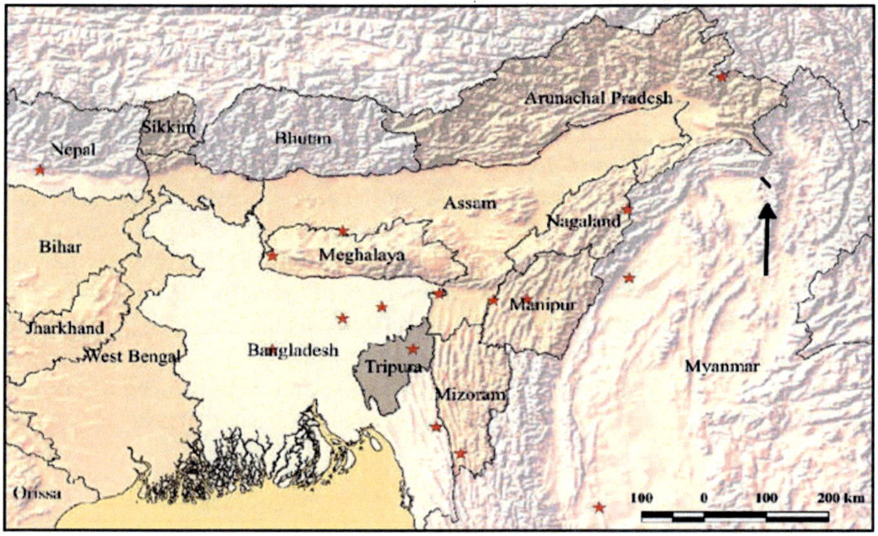

Fig. 26.1 Geographic location of Tripura

Himalayan range and due to subduction activity, in the east, of the Indian plate and the Sunda plate, the Indo-Burmese Ranges are created.

The tectonics of the study area include Himalayan thrust zone, Bengal basin and Shillong plateau, and the Indo-Burmese subduction zone which are in the northern side of the state, eastern side of the state, and western side of the state, respectively, as per Anbazhagan and Balakumar [5].

Himalayan Belt

The seismicity of the Himalayan belt trending from east to west is well explained by the converging of the Indian plate under the Eurasian plate, leading to the creation of the mountain arc of Himalayas. The Main Central Thrust (MCT), Main Frontal Thrust (MFT), and Main Boundary Thrust (MBT) are the major thrust faults in the Himalayan Arc [27]. These faults delineate the boundary of convergence between the two plates. The dipping of the Indian plate below the Eurasian plate is occurring at an approximate rate of 4.5 cm/year [11]. Moreover, Bilham [11] suggested the probable existence of many faults in the region waiting for potential reactivation. The range of Himalaya is not equally seismogenic, indicating the presence of non-uniform seismic slip. From the observations of paleoseismic, microseismic, and modern Global Positioning System, Srivastava et al. [39] located around ten seismic gaps in the Himalayan ranges. The seismic gaps are the Kashmir gap, Kangra gap, Jammu gap, Eastern Himachal Pradesh gap, Western Nepal gap, Sikkim Bhutan gap, Central Nepal-Bihar gap, Shillong gap, and Arunachal gap as mentioned by the above study. But among these gaps, the Shillong, Arunachal, and Sikkim Bhutan gap are the gaps of concern for this present study.

Indo-Burma Ranges (IBR)

The Indo-Burma ranges located in the eastern part of the study area (SA) is the intraplate boundary between the Indian and Burmese plates. The north-western part of the Sunda plate is formed by the Indo-Burma ranges and the Sagaing fault. The Indo-Burma ranges are almost trending from north to south and meet the Himalayan Arc at the Eastern Himalayan Syntaxis. A study by Kundu and Gahalaut [28] reported that in the Indo-Burma wedge, the focal mechanism involves strike-slip or thrust faults, while the Sagaing fault exhibits the focal mechanism of strike-slip. The Indo-Burma wedge shows much deeper earthquake events as compared to the Sagaing fault. Recently, Steckler et al. [40] studied the GPS data along the plate boundaries and reported that the Indo-Burma region is still having active subduction. They discovered the accumulation of convergence exceeding 5.5 m in the Indo-Burma wedge over the past 400 years, suggesting that an 8.2–9 M_W mega-thrust earthquake is likely to occur in that region in future. Wang et al. [44] divided this region into five major regimes depending on tectonics and seismicity, namely the Indo-Burma subduction zone, Naga zone, Dhaka zone, Shan zone, and Sagaing zone. The empirical relations provided by different researches indicate that the seismic sources in this area have capacity to produce an earthquake with $M_W = 7$. However, incomplete rupture of the faults will result in earthquakes with smaller magnitude.

Shillong Plateau

Shillong plateau can be regarded as the projection of the Indian shield in the north-eastern direction which is surrounded by the Himalayan Arc, Dauki fault, and Assam Syntaxis in the north, south, and northeast, respectively. This plateau, which is also known as Shillong Mikir Massif, had risen to an elevation of about 1000 m in the geological past [27]. This region shows intraplate earthquakes such as the Shillong earthquake of 1897.

Bengal Basin

The Bengal basin, in the south of the Dauki fault, is one of the world's largest sedimentary basins which is created by the flowing rivers in that region. This basin possesses its own complex tectonic features, which can be categorized into four major zones, namely (i) Eocene Hinge Zone, the tectonic feature trending in NE-SW, (ii) north-western shelf zone, (iii) southeast deep basin, and (iv) western shear zone [30]. Most of the regions in this basin face the threat of seismic hazards. Due to interplate activities, the seismic activity in the fold belt of Tripura is higher than that in the Eocene Hinge Zone. The Bengal basin is characterized by thick sediment deposits and less crustal thickness. The sediment deposits can intensify the ground motion by 3–4 times, as they did during the Shillong earthquake, 1897. There is a high probability of liquefaction in that region due to the presence of wet sediments [41]. Rao et al. [33] investigated the cause of the magnitude M_W 6 earthquake that occurred in the Bay of Bengal, on May 21, 2014. The reason behind the earthquake might be the reactivation of a strike-slip fault zone of the Precambrian era in the lithosphere of the ocean.

Preparation of Seismotectonic Map

Seismotectonic map should be comprised of all the faults, lineaments, seismic gaps, and past seismic events of the seismic study area (SSA) which can affect the study area (SA). The seismotectonic map is essential to get an overall view of the seismic status of the SSA and to forecast future seismic hazards of the SA [8]. The earthquake catalogue and all the seismic sources of the SSA are compiled, to prepare the seismotectonic map, from different sources.

Selection of Seismic Study Area (SSA)

As suggested by Anbazhagan et al. [6], the past damage intensity maps, i.e. isoseismal maps of the region, should be studied for selecting the SSA for any hazard analysis. The selection of SSA should be done in such a way that it captures all the past earthquake events and potential seismic sources that might cause damage to the study area. Generally, for damage to the structures and other facilities, intensity of IV or V on the Modified Mercalli Intensity (MMI) scale is considered to be sufficient. So, around 16 earthquakes were considered in this study which had caused damages of intensity IV on MMI scale at the state capital of Tripura, Agartala. The epicentres of each of the 16 earthquakes considered for the study are shown in Fig. 26.1 as red star marks.

From the observation, it is clear that earthquakes such as the Bihar-Nepal earthquake of 1934 and the Assam earthquake of 1950 (even from further distances) damaged Agartala significantly. So, such earthquakes should be included in the assessing the seismic hazard of Tripura.

A circle of radius 700 km with its centre at Nunachara (23.805, 91.720) was chosen as the SSA to capture all the potential damaging earthquakes. The centre of the circle was considered as Nunachara since it was the centre of Tripura approximately.

Seismic Sources

Seismic sources that were not ruptured for a long period of time are more likely to rupture according to the elastic rebound theory. Considering the serious implications of the seismic gaps, this study takes into consideration of the seismic gaps as well as faults and lineaments. The faults and lineaments of India are mapped in the SEISAT [19] which was published by the Geological Survey of India (GSI). But it did not provide any information about the tectonic features and the seismic gaps around Bangladesh, Myanmar, and Tibet. Therefore, those details were taken from different studies like Yin and Harrison [46], Wang et al. [44], and Hoque and Khan [23]. ORIGIN PRO software was used for digitization of the sheets and is included in the seismotectonic map provided here. In total, 414 linear seismic sources with four seismic gaps along with the Himalayan ranges and two seismic gaps along the Sagaing fault were identified and considered for the study. The seismotectonic map prepared is presented here in Fig. 26.2.

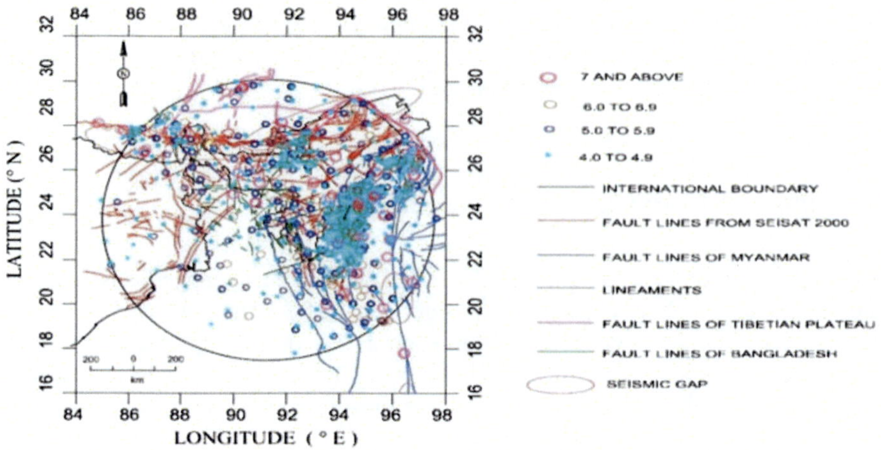

Fig. 26.2 Seismotectonic map of Tripura

Earthquake Catalogue

An earthquake catalogue was compiled for the seismic study area (SSA) in this study. So, all the earthquake events in the seismic study area (SSA) were collected from different reporting agencies like European Mediterranean Seismological Centre (EMSC), United States Geological Survey (USGS), Incorporated Research Institutions for Seismology (IRIS), and International Seismological Centre (ISC). Details of the earthquakes like date of event, time of event, depth of epicentre, magnitude, the reporting agency, magnitude scale, etc., are present in the compiled catalogue. The earthquake catalogue consisted of instrumented data of the last 121 years from 1900 to 2021.

In Fig. 26.3, a histogram of earthquake data is presented which shows the distribution of the number of earthquake events with each decade starting from 1900 to 2021.

Earthquake events compiled in this study consisted of earthquakes in different magnitude scales like M_L, M_b, $M_{s,}$ and M_w. Moment magnitude (M_w) represents the energy released during any earthquake event. As M_w does not get saturated even for larger values, it is widely used for seismic hazard studies [22]. But compared to M_w, other magnitude scales like M_s, M_L, and M_s saturate at higher values and do not represent the actual released energy. So, there was a need of homogenization, i.e. converting the earthquake events that were recorded in different magnitude scales to a single magnitude scale. In this study, all the recorded events were converted to moment magnitude (M_w) using equations provided by Anbazhagan and Balakumar [5] developed for the study region (Tripura and surrounding region). There are several other conversion equations available like Sitharam and Sil [38] developed for Tripura and Mizoram, Das et al. [14] developed for the North-East India region. Beyond the magnitude ranges, Das et al. [14] was used for conversion.

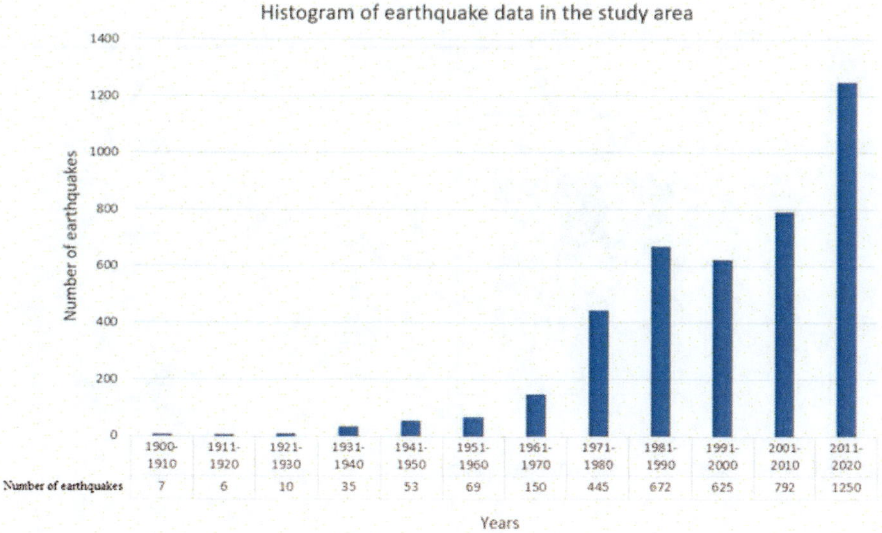

Fig. 26.3 Histogram of earthquake data

Table 26.1 Summary of the constants of the magnitude scaling equation given above

Equations	Intercept (α)	Slope (β)	R^2 value	Magnitude range
m_b to M_w	−1.5725	1.3296	0.64	4.7 to 6.6
M_L to M_w	−0.3573	1.1327	0.73	3 to 4.9
M_s to M_w	2.3172	0.6189	0.79	3 to 6.1
M_s to M_w	−0.1401	1.0238	0.68	6.1 to 6.8

The magnitude scaling equation used in this study is given below:

Anbazhagan and Balakumar [5]: General form of the equation developed for Tripura and its surrounding region: $M1 = \alpha + \beta M2$. The values of the constants in the equation are given in Table 26.1.

Uhrhammer [43], Reasenberg [34], and Gardner and Knopoff [18] are some of the widely used algorithms. In this study, Reasenberg [34] algorithm was used for declustering. The declustered catalogue contained 4114 events, which were identified as main shocks. To prepare the seismotectonic map of the SSA, superimposition of the declustered events of the catalogue was done on the fault map.

Seismicity Parameters

An earthquake catalogue is complete for a minimum magnitude, and that minimum magnitude is called magnitude of completeness. Using the Gutenberg–Richter relation given below, estimation of M_c was done.

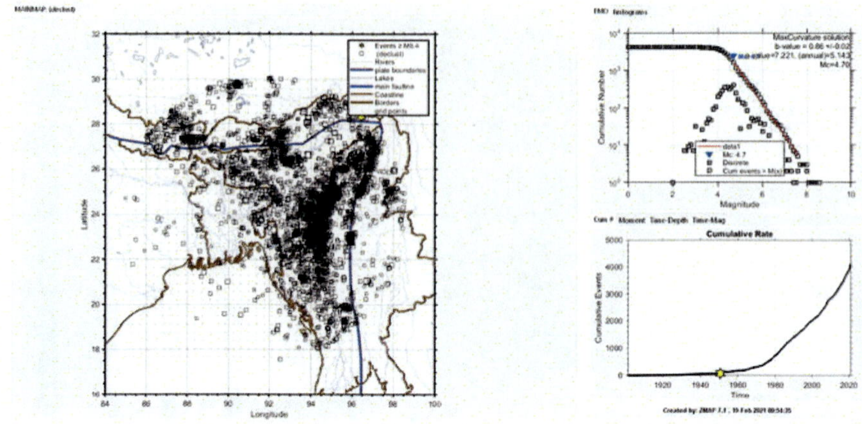

Fig. 26.4 Declustered map obtained from Zmap software

$$\log N = a + bM \tag{26.1}$$

where a is intercept, b is slope, and N is number of earthquakes of magnitude greater than M.

The magnitude of completeness (M_c) was found to be 4.7. And the seismicity parameters were found to be $a = 5.143$ and $b = 0.86$. All of these were found using Zmap software. The declustered map obtained from Zmap is presented in Fig. 26.4.

We have extrapolated the a and b value up to a magnitude of 4 due to lack of data even though the catalogue is complete for 4.7 magnitude.

Maximum Magnitude ($M_{\max}$) Estimation

$M_{\max}$ is defined as the maximum magnitude of earthquake that can be expected from a fault. Estimation of the $M_{\max}$ is a fundamental part of any hazard study. The present study makes use of three methods: the observed magnitude on each seismic source, the incremental method, and the rupture-based method which Anbazhagan et al. [4] proposed for the calculation of $M_{\max}$.

The methods used for estimating the maximum magnitude ($M_{\max}$) are described below:

Observed magnitude method: $M_{\max}$ was estimated considering the maximum observed magnitude on each fault. From the past observed seismic events record, it was found that out of 414 faults in the region, only 270 faults had an observed magnitude on it. So those 270 faults were considered in this method for hazard estimation.

Incremental method: M_{max} was estimated by adding 0.5 magnitude unit (to consider for the high seismicity of the SSA) to the maximum observed magnitude for each of the faults.

Anbazhagan et al. [4] suggested a method to estimate M_{max} of each fault by considering the regional rupture characteristics. In this method, the total length of the fault and ruptured length of the fault are considered. The subsurface rupture length (RLD) is calculated by using the widely popular relation of Wells and Coppersmith [45]. Relation given by Wells and Coppersmith is empirical which relates the subsurface rupture length (RLD) with the moment magnitude (M_W) and is given below.

$$\log(RLD) = 0.59M_W - 2.44 \tag{26.2}$$

As per Wells and Coppersmith [45], the above equation is valid for all types of earthquakes, all types of faults, and for different magnitude and distance ranges. As per Anbazhagan et al. [4], the trend between Total Fault length (TFL) and Percentage Fault Rupture (PFR) is unique and remains same for different extents of seismic study area (SSA). The plot between TFL and PFR for the radius of SSA (700 km) obtained in the present study is given below in Fig. 26.5.

In this study, the TFL was divided into three distance bins based on the trend observed and was fixed as 0–30 km, 30–160 km, and > 160 km. Corresponding to each bin, the maximum PFR was found to be 0.94, 0.70, and 0.12 respectively. Then RLD for each fault, whichever was falling in that distance bin, was increased. Now, using the same Well and Coppersmith [45] equation, the M_{max} was determined for each of the faults with their new increased RLDs. M_{max} calculated from the observed method and the incremental method was used for the conventional analysis. For the rupture-based analysis, M_{max} calculated from the regional rupture characteristics approach given by Anbazhagan et al. [4] was adopted.

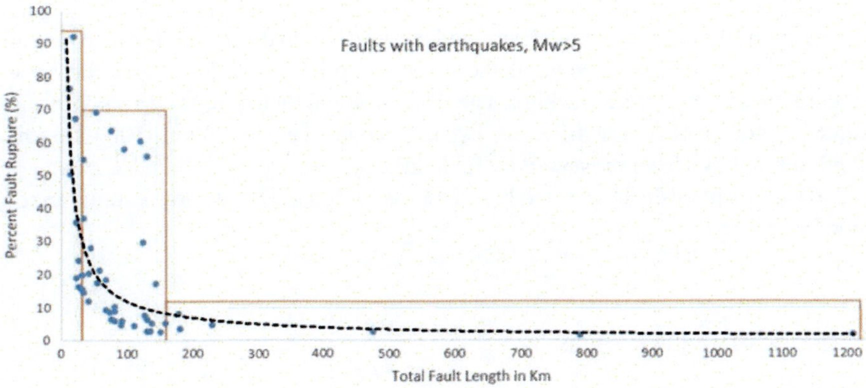

Fig. 26.5 Plot between Percentage Fault Rupture (PFR) versus Total Fault Length (TFL)

Ground Motion Prediction Equation (GMPE)

The ground motion parameters like Peak Ground Acceleration (PGA), Peak Ground Velocity (PGV), etc., at a site can be estimated using empirical equations called the Ground Motion Prediction Equations (GMPEs). Region-specific GMPEs must be selected for any site as most of the GMPEs depend on tectonic setting of the region. In this study, the SSA cannot be represented by a single GMPE as the seismic study area has a very complex tectonic framework. Therefore, based on the tectonics features and fault system, the SSA was divided into three zones.

Zone 1: Himalayan Region: Himalayan thrust zone, which is on the northern part of SSA, is demarcated by the collision of the Indian and Eurasian plates. It is believed that the tectonics in the region are due to thrusting. There is no active subduction going on in the region.

Zone 2: Indo-Burma Ranges (IBR): Indo-Burma subduction zone, which is in the eastern part of the SSA, is demarcated by the collision of Indian and Sunda plate. There are evidences that the tectonics in the region are due to subduction.

Zone 3: Bengal Basin and Shillong Plateau: This region has moderate seismicity which can be seen as the continuation of the Indo-Gangetic basin. Southern and south-western parts of the SSA are bounded by this region. No plate boundaries are present in this region.

Because of this complex setting in its tectonic features, a set of GMPEs were collected for the above-mentioned zones from the compilation of GMPEs given by Douglas [17]. The GMPEs compiled were considered to be suitable for the particular tectonic environment if they had been proposed either for the same region or similar tectonic regime worldwide.

In this study, Loglikelihood (LLH) and Data Support Index (DSI) values of each of the GMPEs are calculated using past 14 earthquakes of the region. Only the GMPEs with lowest LLH values and positive DSI values are finally selected from the set of the GMPEs.

The magnitude of the earthquake ranged between 4 and 8, and the distance ranged between 100 and 500 km. Beyond the distance and magnitude ranges, the same equations are assumed to be working. The DSI value and their weights, in parenthesis, assigned to the GMPEs for different distance ranges, i.e. 0–100 km, 100–300 km, and >300 km, have been presented in Table 26.2.

The final list of GMPEs selected based on the LLH and DSI values is summarized in the Table 26.3.

Seismic Hazard Analysis

The deterministic seismic hazard analysis (DSHA) of the study area was done using conventional methods and regional rupture-based method. In DSHA, a maximum earthquake is to be selected for each seismic source and PGA at the bedrock level for

Table 26.2 DSI values of GMPEs for different distance ranges

SL No	GMPEs	References	DSI <100 km	DSI 100–300 km	DSI >300 km
1	KT-18	Bajaj and Anbazhagan [10]	−92.6179	−36.491	−99.849
2	ANBU-13	Anbazhagan et al. [8]	−100	−100	−100
3	Kano06	Kanno et al. [26]	−100	−99.983	−100
4	ASK-14	Abrahamson et al. [1]	−97.5931	−100	−100
5	IDR-14	Idriss [24]	−96.5158	−99.872	−100
6	Zhao-06	Zhao et al. [48]	−99.125	−99.44	−100
7	SHAR-09	Sharma et al. [36]	−100	−98.93	−100
8	AMB-05	Ambraseys et al. [3]	**212.7589** ***(0.17)***	−100	−100
9	AKBO-10	Akkar and Bommer [2]	−100	−100	−100
10	ndmaH-10	NDMA [32]	−100	−100	−100
11	ndmaNE-10	NDMA [32]	−99.3728	−100	−100
12	CB-14	Campbell and Bozorgnia [12]	−10.1821	−99.886	−100
13	CY-14	Chiou and Youngs [13]	−92.4182	−96.395	−100
14	ABIF-03	Atkinson and Boore [9]	−99.8057	−98.122	−99.8057
15	ABIS-03	Atkinson and Boore [9]	−82.8113	−99.734	−82.8113
16	GT-10	Gupta [21]	−100	−99.909	−100
17	Youngs_IS97_SZ	Youngs et al. [47]	**931.6013** ***(0.55)***	**583.4183** ***(0.42)***	**801.1039** ***(1.00)***
18	Youngs_If97_SZ	Youngs et al. [47]	**422.9016** ***(0.28)***	**55.67283** ***(0.10)***	−1.25491
19	GREG-02	Gregor et al. [20]	−100	−100	−100
20	LILEIf-08	Lin and Lee [29]	−99.6573	−100	−100
21	LILEIs-08	Lin and Lee [29]	−97.1759	−99.999	−100
22	MAK-18	Mahani and Kao (2018)	−99.9882	−99.974	−100
23	SI-16	Singh et al. [37]	−99.9985	**689.645** ***(0.48)***	−100
24	NATH-12	Nath [31]	−100	−100	−100

the region due to an earthquake magnitude $M_{\max}$ on each source at the closest source to site distance is calculated. In this study, a constant average focal depth of 15 km was considered. For carrying out conventional seismic hazard analysis, the $M_{\max}$ for each source was calculated by two methods: (a) maximum observed magnitude from the past seismic events and (b) incremental method, i.e. the observed maximum magnitude for each source is incremented by adding a 0.5 moment magnitude unit to it. The incremental method was employed because of its repeated usage in the

Table 26.3 Final list of selected GMPEs

GMPE	Abbreviation of the equation	M_w range	Distance range	Standard form of equation	Coefficients value for Zero periods
Singh et al. [37]	SI-16	4 to 8.5	≤ 300	$lnln(Y_{br}) = c_1 + c_2(M_w - 6) + c_3(M_w - 6)^2 - lnR - c_4R$ where $R = \sqrt{R_e^2 + h^2}$ where Y_{br} is the Peak Ground Acceleration at bedrock level, R_e is the epicentral distance, and h is the focal depth	$c_1 = 2.0282$ $c_2 = 0.8569$ $c_3 = -0.0472$ $c_4 = 0.0091$ where c_1, c_2, c_3 and c_4 are regression coefficients
Youngs et al. [47]	Youngs-97	≥ 5	10–500	$lnln(Y) = 0.2418 + 1.414M + c_1 + c_2(10 - M)^3 + c_3lnln(r_{rup} + 1.7818e^{0.554M}) + 0.00607H + 0.3846Z_T$ where Y is Peak Spectral Acceleration in g	$c_1 = 0$ $c_2 = 0$ $c_3 = -2.552$ $c_4 = 1.45$ $c_5 = -0.1$ where c_1, c_2, c_3, c_4 and c_5 are regression coefficients
Ambraseys et al. [3]	AMB-05	≥ 5	≤ 100	$lnln(Y) = a_1 + a_2M + (a_3 + a_4M)ln\sqrt{(d^2 + a_5^2}$ where Y is spectral acceleration in m/s^2	$a_1 = 0.835$ $a_2 = -0.083$ $a_3 = -2.489$ $a_4 = 0.206$ $a_5 = 5.60$ where a_1, a_2, a_3, a_4 and a_5 are regression coefficients

seismological community. The study area was divided into grids of 0.1° × 0.1° (11.1 km × 11.1 km), and at the centre of each grid, PGA values were estimated.

Considering the regional rupture characteristics of all the seismic sources in the region, the rupture-based seismic hazard analysis was carried out. In this method, the M_{max} was estimated from the rupture-based method. This approach for M_{max} estimation could model the future worst-case scenario earthquake. The GMPEs were selected using the statistical procedure (based on LLH and DSI values) along with the weights. Similar grid size and interpolation technique were used to map the PGA values.

Results and Discussion

The prediction of Peak Ground Acceleration (PGA) plays a significant role in mitigating hazards caused due to earthquakes by improving the construction methods.

However, the estimated value must be realistic, and it should not be under-predicting or over-conservative. An under-predicted hazard value can cause severe damages in case of any earthquake in the region that may come in future, and if the estimated value is over-conservative, then it will be unfeasible in the sense of economy as it is same as designing any structure for such a ground motion that might ever not come in that region.

Conventional Approach

The PGA distribution map obtained from the conventional DSHA analysis has been presented in Fig. 26.6a, b. From Fig. 26.6a, it can be observed that the northern part of Tripura is more vulnerable to severe ground shaking as compared to the other parts of the state, whereas Fig. 26.6b indicates that both north and north-eastern parts of the state are vulnerable to severe ground shaking when compared to the other parts. The west and south-eastern parts of the state are liable to comparatively lesser ground shaking. The reason for this observation could be the influence of potential faults such as MBT, MCT, MFT, and Dauki fault, which are present to the north of the state.

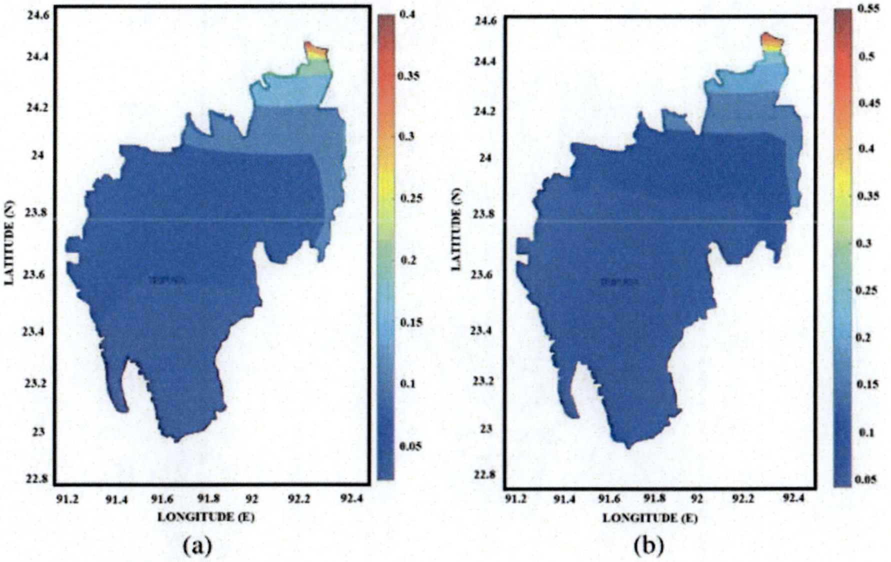

Fig. 26.6 PGA distribution map of Tripura obtained by conventional method: **a** observed magnitude approach and **b** incremental method

Rupture-Based Method

The PGA distribution map obtained from the rupture-based DSHA analysis is presented in Fig. 26.7. It can be noted from Fig. 26.7 that the north-eastern parts of the state are vulnerable to higher levels of shaking. Slightly deviating from the results of conventional analysis, the rupture-based procedure predicts a little higher levels of ground shaking for most of the southern parts of the state.

The comparison of the results of the present study with other reported literatures for this region is shown in Table 26.4. The obtained PGA values of Agartala city are matching well with the results of the study by Das et al. [15] and Sitharam and Sil [38], but do not match well with the results of the study by Sharma and Malik [34]. The reason behind this difference in results is because of the different seismic sources and GMPEs used in the studies. Das et al. [15] has used GMPE developed by them while Sharma and Malik [35] only used GMPE developed by Youngs et al. [47].

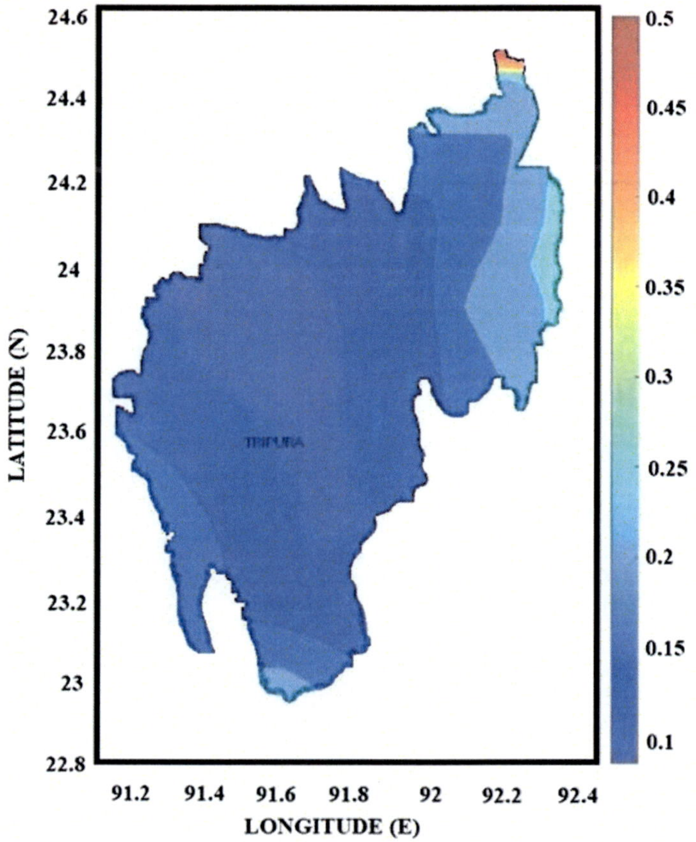

Fig. 26.7 PGA distribution map of Tripura obtained by rupture-based method

Sitharam and Sil [38] used two equations developed by Gupta [21] and Atkinson and Boore [9]. In this study, a set of GMPEs are used in combination, which were selected based on proper statistical analysis. In addition to this, the range of PGA value that the state can experience obtained by using different methods of analysis is given in Table 26.5. According to IS 1893 [25], Tripura state falls in seismic zone V which implies the highest PGA value of 0.36 g. But the present study shows that there is a spatial variation of hazard even within the small area of the state. The reason behind this variation is because at each grid points, all sources and events are considered separately.

In the present work, it was observed that the incremental method assigns larger magnitude values to such faults which have already ruptured a lot without considering their actual fault length and its potential to rupture. However, the rupture-based procedure for M_{max} calculation takes into account the rupture behaviour of the seismic sources in terms of the percentage of rupture. Since the PFR cannot take values beyond 100%, the rupture-based procedure will never produce unrealistic estimates of the maximum magnitude and thereby seismic hazard. 100% PFR would indicate the seismic source had ruptured its entire fault length. This discussion shows that the conventional technique was over-conservative in predicting hazard, whereas the rupture-based approach optimized the hazard analysis, especially when smaller active faults were considered. The rupture-based procedure could prove to be a robust tool for maximum magnitude calculation and a reliable seismic hazard study.

Table 26.4 Comparison of results with other studies

Study	PGA value (g)	Methodology of Seismic Hazard Analysis	Location	Study Area
Present study	0.10–0.22	Conventional Deterministic	Agartala	Tripura
Present study	0.14–0.20	Rupture-based Deterministic	Agartala	Tripura
Sitharam and Sil [38]	0.29	Deterministic	Agartala	Mizoram and Tripura
Sitharam and Sil [38]	0.11–0.20	Probabilistic	Agartala	Mizoram and Tripura
Das et al. [15]	0.18–0.22	Probabilistic	Agartala	North-East
Sharma and Malik [35]	0.33	Probabilistic	Agartala	North-East

Table 26.5 Summary of results obtained using different approaches

Method	Maximum observed magnitude	Incremental method	Rupture based
DSHA	0.05–0.43 g	0.05–0.57 g	0.10–0.50 g

Conclusion

Several past earthquakes, including the recent 2017 Manu earthquake of 5.7 magnitude, had shown high levels of damages, which highlight the requirement of detailed seismic hazard analysis of Tripura [7]. However, there is no region-specific seismic hazard map for the state of Tripura. In the present study, the hazard map (seismic) of Tripura was prepared by the rupture-based and conventional approach of hazard analysis. This study shows that the PGA value at bedrock level varies between 0.1 and 0.50 g for rupture-based analysis, whereas the results from conventional seismic hazard analysis show that the PGA at bedrock level varies between 0.05 and 0.57 g throughout the state. These values differ from the bedrock PGA of 0.36 g given by IS1893:2016 for the entire state. The northern part of the state is observed to be more vulnerable to hazard as compared to the rest of the state. The southern part of the state has the lowest seismic hazard potential. The capital city Agartala shows moderate hazard values. This deviation of the estimated PGA values and its spatial variation from the Indian standard code states the necessity of region-specific hazard studies. It is evident from the recent earthquake in Manu that the state of Tripura is liable to liquefaction damages as well. Therefore, it is advisable to carry out the seismic microzonation of the state to facilitate urban planning in the state. The seismic hazard analysis carried out in this study can be used for further studies such as mapping of surface ground motion, site response analysis, and liquefaction potential mapping of the state.

References

1. Abrahamson NA et al (2014) Summary of the ASK14 ground motion relation for active crustal regions. Earthq Spectra 30(3):1025–1055
2. Akkar S, Bommer JJ (2010) Empirical equations for the prediction of PGA, PGV, and spectral accelerations in Europe, the Mediterranean region, and the Middle East. Seismol Res Lett 81(2):195–206
3. Ambraseys NN et al (2005) Equations for the estimation of strong ground motions from shallow crustal earthquakes using data from Europe and the Middle East: horizontal peak ground acceleration and spectral acceleration. Bull Earthq Eng 3(1):1–53
4. Anbazhagan P et al (2015) Maximum magnitude estimation considering the regional rupture character. J Seismolog 19(3):695–719
5. Anbazhagan P, Balakumar A (2019) Seismic magnitude conversion and its effect on seismic hazard analysis. J Seismolog 23(4):623–647
6. Anbazhagan P et al (2013) Ground motion prediction equation considering combined dataset of recorded and simulated ground motions. Soil Dyn Earthq Eng 53:92–108
7. Anbazhagan P et al (2019) Reconnaissance report on geotechnical effects and structural damage caused by the 3 January 2017 Tripura earthquake, India. Nat Hazards 98(2):425–450
8. Anbazhagan P et al (2013) Estimation of design basis earthquake using region-specific Mmax, for the NPP site at Kalpakkam, Tamil Nadu, India. Nuclear Eng Des 259:41–64
9. Atkinson GM, Boore DM (2003) Empirical ground-motion relations for subduction-zone earthquakes and their application to Cascadia and other regions. Bull Seismol Soc Am 93(4):1703–1729

10. Bajaj K, Anbazhagan P (2018) Determination of GMPE functional form for an active region with limited strong motion data: application to the Himalayan region. J Seismolog 22(1):161–185
11. Bilham R (2004) Earthquakes in India and the Himalaya: tectonics, geodesy and history. Ann Geophys 47
12. Campbell KW, Bozorgnia Y (2014) NGA-West2 ground motion model for the average horizontal components of PGA, PGV, and 5% damped linear acceleration response spectra. Earthq Spectra 30(3):1087–1115
13. Chiou BS-J, Youngs RR (2014) Update of the Chiou and Youngs NGA model for the average horizontal component of peak ground motion and response spectra. Earthq Spectra 30(3):1117–1153
14. Das R, Mukhopadhyay S (2020) Regional variation of coda wave attenuation in Northeast India: an understanding of the physical state of the medium. Phys Earth Planet Inter 299:106404
15. Das S et al (2006) A probabilistic seismic hazard analysis of northeast India. Earthq Spectra 22(1):1–27
16. Debbarma J et al (2017) Preliminary observations from the 3 January 2017, MW 5.6 Manu, Tripura (India) earthquake. J Asian Earth Sci 148:173–180
17. Douglas J (2020) Ground motion prediction equations 1964–2018. http://www.gmpe.org.uk. Last accessed 2021/02/12
18. Gardner J, Knopoff L (1974) Is the sequence of earthquakes in Southern California, with aftershocks removed, Poissonian? Bull Seismol Soc Am 64(5):1363–1367
19. Dasgupta S, Narula PL, Acharyya SK, Banerjee J (2000) Geological Survey of India. Seismotectonic atlas of India and its environs, Report by Geological Survey of India
20. Gregor NJ et al (2002) Ground-motion attenuation relationships for Cascadia subduction zone megathrust earthquakes based on a stochastic finite-fault model. Bull Seismol Soc Am 92(5):1923–1932
21. Gupta I (2010) Response spectral attenuation relations for in-slab earthquakes in Indo-Burmese subduction zone. Soil Dyn Earthq Eng 30(5):368–377
22. Hanks TC, Kanamori H (1979) A moment magnitude scale. J Geophys Res: Solid Earth 84(B5):2348–2350
23. Hoque M, Khan A (2001) Seismicity and seismotectonic regionalization of the Bengal Basin, Bangladesh. Bangladesh J Geol 20:77–86
24. Idriss I (2014) An NGA-West2 empirical model for estimating the horizontal spectral values generated by shallow crustal earthquakes. Earthq Spectra 30(3):1155–1177
25. IS 1893-Part 1 (2016) Criteria for earthquake resistant design of structures—general provisions and buildings. Bureau of Indian Standards, New Delhi
26. Kanno T et al (2006) A new attenuation relation for strong ground motion in Japan based on recorded data. Bull Seismol Soc Am 96(3):879–897
27. Kayal J (2008) Microearthquake seismology and seismotectonics of South Asia, 1st edn. Springer Science & Business Media, India
28. Kundu B, Gahalaut V (2012) Earthquake occurrence processes in the Indo-Burmese wedge and Sagaing fault region. Tectonophysics 524:135–146
29. Lin P-S, Lee C-T (2008) Ground-motion attenuation relationships for subduction-zone earthquakes in northeastern Taiwan. Bull Seismol Soc Am 98(1):220–240
30. Nath S (2017) Seismic hazard, vulnerability and risk microzonation Atlas of Kolkata. Open File Report, Geoscience Division, Ministry of Earth Sciences, Government of India
31. Nath SK et al (2012) Ground-motion predictions in Shillong region, northeast India. J Seismolog 16(3):475–488
32. NDMA (2010) Development of probabilistic seismic hazard Map of India. Technical report by National Disaster Management Authority, Government of India
33. Rao GS et al (2015) A seismotectonic study of the 21 May 2014 Bay of Bengal intraplate earthquake: evidence of onshore-offshore tectonic linkage and fracture zone reactivation in the northern Bay of Bengal. Nat Hazards 78(2):895–913

34. Reasenberg P (1985) Second-order moment of central California seismicity, 1969–1982. J Geophys Res: Solid Earth 90(B7):5479–5495
35. Sharma M, Malik S (2006) Probabilistic seismic hazard analysis and estimation of spectral strong ground motion on bed rock in north east India. In: 4th international conference on earthquake engineering, Taipei, Taiwan, pp 12–13
36. Sharma ML et al (2009) Ground-motion prediction equations based on data from the Himalayan and Zagros regions. J Earthquake Eng 13(8):1191–1210
37. Singh N et al (2016) A new ground-motion prediction model for Northeastern India (NEI) Crustal Earthquakes. Bull Seismol Soc Am 106(3):1282–1297
38. Sitharam T, Sil A (2014) Comprehensive seismic hazard assessment of Tripura and Mizoram states. J Earth Syst Sci 123(4):837–857
39. Srivastava H et al (2015) Discriminatory characteristics of seismic gaps in Himalaya. Geomat Nat Haz Risk 6(3):224–242
40. Steckler MS et al (2016) Locked and loading megathrust linked to active subduction beneath the Indo-Burman Ranges. Nat Geosci 9(8):615–618
41. Stone R (2011) A Bengali recipe for disaster. Science 332(6035):1256–1258
42. Thingbaijam KKS et al (2008) Recent seismicity in northeast India and its adjoining region. J Seismolog 12(1):107–123
43. Uhrhammer R (1986) Characteristics of northern and central California seismicity. Earthq Notes 57(1):21
44. Wang Y et al (2014) Active tectonics and earthquake potential of the Myanmar region. J Geophys Res: Solid Earth 119(4):3767–3822
45. Wells DL, Coppersmith KJ (1994) New empirical relationships among magnitude, rupture length, rupture width, rupture area, and surface displacement. Bull Seismol Soc Am 84(4):974–1002
46. Yin A, Harrison TM (2000) Geologic evolution of the Himalayan-Tibetan orogen. Annu Rev Earth Planet Sci 28(1):211–280
47. Youngs R et al (1997) Strong ground motion attenuation relationships for subduction zone earthquakes. Seismol Res Lett 68(1):58–73
48. Zhao JX et al (2006) Attenuation relations of strong ground motion in Japan using site classification based on predominant period. Bull Seismol Soc Am 96(3):898–913

Chapter 27
An Energy-Based Approach Towards Liquefaction Potential Analysis: Agartala City

Abhileen Chatterjee and Sima Ghosh

Introduction

Soil Liquefaction can be described as a natural phenomenon occurring when the induced shear strain increases the pore water pressure in the soil resulting in loss of shear strength and bearing capacity of the soil. Liquefaction failures are primarily categorized into two broad heads—(i) Flow Failures and (ii) Cyclic Mobility. Flow failures occur in response to static shear stresses coming from mostly overlying superstructures on loose sands. When the static shear stress exceeds the shear strength of the underlying saturated cohesionless soil layer, the soil starts to flow, causing settlements and failure of the infrastructure. On the contrary, cyclic mobility occurs due to repeated loadings such as earthquakes and is considered to be more common than flow failures.

Seed and Idriss [17] mentioned that soil response is a function of increasing pore water pressure due to cyclic loading. Bozorgnia and Bertero [4] supports the conventional fact that generation of pore pressure in saturated soils is firmly associated with strain amplitude than the stress amplitude. Due to parallel complexities determining strain amplitudes in soil, a theory was developed by Nemat-Nasser and Shokooh [14], which relates densification of drained soils and pore pressure generation in undrained cases to the dissipated energy. Arias [3] studied ground shaking and characterized response based on a parameter involving seismic energy evaluated from recorded ground acceleration, velocity, or displacement time histories. The gross energy consumed by a group of oscillators with single degrees of freedom (SDOF) was represented as a ground motion parameter known as the Arias Intensity. Kayen and Mitchell [10] devised a methodology to evaluate the liquefaction potential of a site employing Arias Intensity. An aspect of the superiority of that method over

A. Chatterjee (✉) · S. Ghosh
Department of Civil Engineering, National Institute of Technology Agartala, Tripura, India
e-mail: abhileen96@rnd.iitg.ac.in

K. Muthukkumaran et al. (eds.), *Soil Dynamics, Earthquake and Computational Geotechnical Engineering*, Lecture Notes in Civil Engineering 300,
https://doi.org/10.1007/978-981-19-6998-0_27

the stress-based approach is the obliteration of the Magnitude Scaling Factor (MSF) because Arias Intensity itself consists of the amplitude, duration, and frequency content of the earthquake motion.

In this paper, an earthquake has been considered to characterize the liquefaction loading, recorded on 3rd January 2017, at 09:09:0.5 UTC [1]. A moderate moment magnitude of $M_W = 5.6$ was recorded as per GCMT data, and a Modified Mercalli Intensity (MMI) of IV was reported at various cities in Tripura, including Agartala. The epicentre was located near the Tripura-Mizoram border possessing coordinates of 23.98°N and 92.03°E, and the hypocentral depth was estimated to be 24.5 km.

Characterizing the Liquefaction Loading

Arias Intensity is calculated from two orthogonal accelerograms and is mathematically given as below:

$$I_h = \frac{\pi}{2g}\left[\int_0^{t_0} a_x^2(t)\mathrm{d}t + \int_0^{t_0} a_y^2(t)\mathrm{d}t\right] \tag{27.1}$$

Here t_0 is the strong motion duration. This paper considers the x- and y- components of the acceleration time history of the scenario earthquake as given in Fig. 27.1 [1]. The acceleration time histories adopted for this study is seen to be irregular ground motion signals depicting fluctuation of ground acceleration with temporal variation. For the ease of calculation of Arias Intensity for the concerned earthquake event, a smooth equivalent dynamic load has been considered from the given time history.

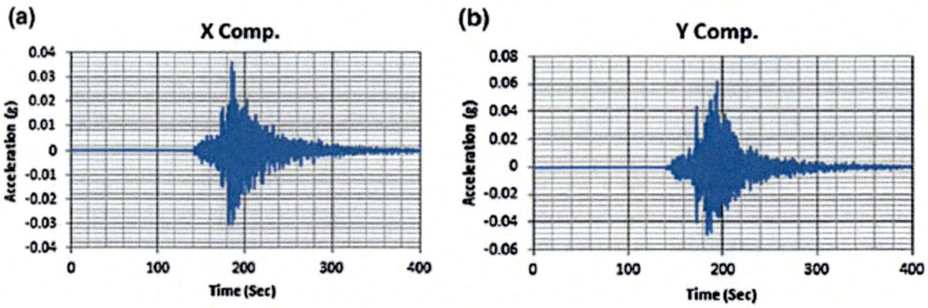

Fig. 27.1 The plot of acceleration—time history of the scenario earthquake, **a** *X*—Component Plot, **b** *Y*—component Plot [1]

Equivalent Dynamic Load

For ease of calculations, the paper assumes that similar deformations are experienced due to a uniformly varying load from the irregular earthquake loading. Seed and Idriss [16] claimed that the mean acceleration is almost similar to 0.65 times the Peak Ground Acceleration (PGA). The significant cycle (NS) count varies with the earthquake magnitude. Lee and Chan [12] gave a procedure for the irregular acceleration time history to be transformed into an equivalent no. of cyclic shear stress cycles having Maximum Magnitude $= k - M_W$ (where k is a constant and $k < 1$). The conversion parameters for each cycle have been adopted from the chart produced by Seed et al. (1975). The number of equivalent cycles for the x- and y- components of the accelerogams are evaluated to be 7.2 and 4.3, respectively.

Calculation of Arias Intensity at Surface

In this context, to evaluate the equations of motion for each of the acceleration component peak ground acceleration plays an important role. Mog et al. [1] specifies that the ground motion records are acquired from the North East Institute of Science and Technology (NEIST), India which is located at an approximate distance of 297 km from the estimated epicentre. Since Fig. 27.1b depicts the maximum value of peak ground acceleration to be around 0.064 g at a distance of 297 km, it is evident that studies specific to target regions are necessary for design of buildings and disaster management. Youd et al. [22] while answering some of the questions regarding selection of peak ground acceleration noted that $a_{\max}$ can be evaluated from accelerograms. It has also been suggested to take up a group of earthquake records with compatible magnitudes.

In Neo-deterministic seismic hazard analyses, earthquakes of magnitudes greater than 5 are considered from synthetic records developed from geophysical and geotechnical aspects of the site. That said, the study takes up the Design Ground Acceleration (DGA) formulated by Parvez et al. [15] using Neo-deterministic analysis approach based on the natural frequencies of infrastructure. It has been found that the DGA values in the proximity of Tripura region varies from 0.36 to 0.6 g. DGA being analogous to peak ground acceleration with an added advantage of frequency content of neighbouring infrastructures, is safe to use for this study. Mog et al. [1] estimates the maximum spectral acceleration to be 0.25 g at a distance of 297 km and considering the frequency parameter designated for the infrastructures in Agartala area, a peak ground acceleration value of 0.4 g has been adopted for this study.

Employing Eq. (27.1) the Arias Intensity at the surface has been calculated using the acceleration parameters from the earthquake records. The integration has been performed in Matlab to safeguard any human error using the traditional trapezoidal rule. The number of steps for the calculation has been kept to be 1000 for precise results and the integration over total ground shaking duration of 3 s [1] gives us the surface Arias Intensity ($z = 0$) to be 1.25 m/s. Analogous to that in stress-based approach, this methodology takes into account a factor that addresses the influence of depth on the Arias intensity. It is called the Burial Reduction Factor (r_b) which when multiplied with the Arias intensity at the surface, produces the equivalent Arias Intensity generated due to the earthquake of a given magnitude at the desired depth where liquefaction potential is to be evaluated. Mathematically, r_b is given as in Eq. (27.2) below.

$$r_b = \exp\left[\frac{35}{(M_W)^2} \times \sin(-0.09 \times z)\right] \tag{27.2}$$

Here z is the critical depth of liquefaction and the term $(-0.09 \times z)$ is in radians. Now, the Arias intensity at the depth z, also known as the demand produced during the earthquake event is denoted by $I_{h,\text{eq}} = I_h(z) = r_b \times I_h(z = 0)$.

Characterizing the Resistance to Liquefaction

Liquefaction Resistance is generally demonstrated in terms of similar parameters as used for the loading to be characterized. In the case of the traditional stress-based approach, the liquefaction resistance is expressed in terms of Cyclic Resistance Ratio (CRR), which can be illustrated as the cyclic stress ratio that initiates liquefaction. In this approach, the resistance to liquefaction is calculated in terms of requisite Arias Intensity for the initiation of the phenomenon. This paper attempts to determine the liquefaction resistance using Standard Penetration Test (SPT) data from different subsurface layers at the target region. A sample BH- 02 (Table 27.1) at Gandhi Ghat, Agartala [23.8274° N, 91.2758° E], has been taken up to account for the subsurface calculations, as shown later in this paper.

Figure 27.2a–c show the variation of SPT N-count, fine content of each soil layer at the target site and the corrected SPT $(N_1)_{60,CS}$ variation with incremental depth below the ground level.

Susceptibility Criteria

Soils demonstrating sand-like, clay-like or intermediate behaviour were categorized by Boulanger and Idriss [5] based on behavioural patterns of fine-grained soils under monotonic and cyclic undrained loading on the basis of consistency limits. Behavioural consistency for samples of cyclic test data from that site was calculated taking into account the collations of stress–strain loops and ratios of excess pore water pressure. Andrews and Martin [2] proposed liquefaction susceptibility criteria of clayey and silty soils based on their liquid limits and clay content, i.e. the weight of grains finer than 0.002 mm. Three general possibilities are proposed considering

Table 27.1 Soil Profile of a sample borehole (BH-02) at Gandhi Ghat within Agartala city

Thickness (m)	Layers	γ_b (kN/m^3)	LL	PL	G	w_n	N-values
0.8	Filled up brown sand	17.7	25.50	21.47	2.66	29.54	2
3.3	Dark grey soft sandy clayey silt	19.0	-	-	2.66	28.45	15
2.10	Light brownish fine to medium sand	18.5	32.50	21.07	2.65	42.68	12
4.3	Light grey soft to medium silty clay / clayey silt	15.6	-	-	2.40	108.66	6
4.85	Blackish grey soft decomposed wood grass mixed up with silty clay	18.8	42.50	25.67	1.88	44.15	5
4.65	Light grey firm soft silty clay	17.7	25.50	21.47	2.66	29.54	2

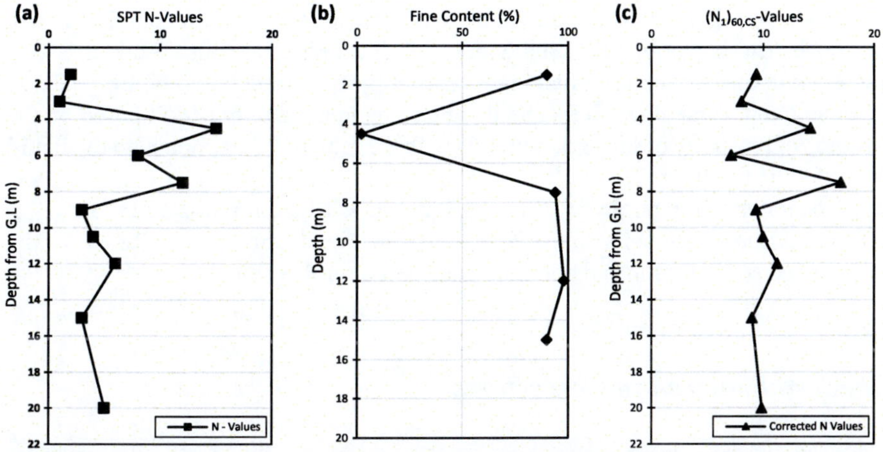

Fig. 27.2 **a** SPT N-values of each layer with increasing depth from the ground level, **b** fine content in percentage for each layer, **c** corrected SPT $(N_1)_{60,CS}$ values varying with depth

the grain size criterion: susceptible to liquefaction, not susceptible and cases where further studies are required to draw solid conclusions. Skempton [19] launched a parameter known as the Activity (A) of the soil given as the ratio of plasticity index (PI) to the percentage of clay-size particles (2 μm). It is seen that both activities, along with particle size, are insufficient to predict soil behaviour from field observations on ground failure.

Boulanger and Idriss [6] proposed behaviour of fine-grained soil under categories of clay-like, intermediate, and sand-like behaviours. An A-line was plotted on experimental investigations, which depicted soils behaving differently under shear loadings

Table 27.2 Layer wise liquefaction susceptibility

Depth (m)	Liquefaction susceptibility
1.5	No
3	Yes
4.5	Yes
6	Yes
7.5	No
9	No
10.5	No
12	–
15	–
20	No

considering liquid limit and plasticity index. In general practice, soils having PI $\geq$ 7 can be safely considered to behave as clays. CL-ML soils with PI values ranging from 3–6 show intermediate behaviour but may have greater cyclic strength than non-plastic fine-grained soil at the same tip resistance. Any soil not complying with the above criteria can be classified under sand-like behaving soils. While it is also suggested to discard the use of Chinese susceptibility criteria possibly, Bray [7], on the other hand, came up with factors to determine liquefiability of soil based on the variation of plasticity of fine-grained soils with the ratio of water content to the liquid limit of soils (wc/LL). Three classifications – susceptible (wc/LL > 0.85 and PI < 12), moderately susceptible (wc/LL > 0.8 and PI < 18), and non-susceptible (PI > 18) were proposed for soils. Table 27.2 shows the liquefaction susceptibility for the fine-grained soil layers in the borehole considered for this study.

Standard Penetration Corrections

Bozorgnia and Bertero [4] tells us that with increasing density of soil the requisite threshold energy to trigger liquefaction, increases. Although the correlation between excess pore water pressure generated and energy are soil specific, the level of liquefaction triggering Arias intensity is evaluated by empirical relation to penetration resistance. From case studies with liquefaction and non-liquefaction, Kayen and Mitchell [10] provided a chart depicting the variation of threshold Arias intensity as a function of $(N_1)_{60}$, as shown in Fig. 27.3. Singnar and Sil [18] showed that the correction for measured SPT values in the field are needed to be corrected based on a certain array of factors, such as, overburden pressure (C_N), hammer efficiency (C_E), borehole diameter (C_B), rod length (C_R), and sampler with or without liner (C_S). The observed SPT values are corrected for approximately 100 kPa of overburden pressure, and is given in Eq. (27.3).

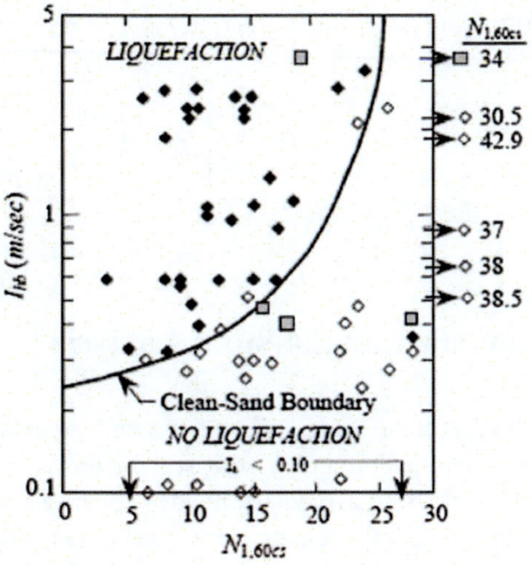

Fig. 27.3 Liquefaction chart based on Arias Intensity as given by Kayen and Mitchell [10]

$$(N_1)_{60} = N_m \times C_N \times C_E \times C_B \times C_R \times C_S \tag{27.3}$$

In this case, N_m stands for the observed SPT N–count from field investigation and $(N_1)_{60}$ denotes the preliminary corrected Singnar and Sil [18], and Das et al. [8] adopted the relations to evaluate the overburden correction factor C_N as given by Liao and Whitman [13]. This paper adopts the following expression to calculate the overburden correction factor demonstrated by Kayen et al. [11] in Eq. (27.4).

$$C_N = \frac{2.2}{\left(1.2 + \frac{\sigma'_{vo}}{P_a}\right)} \tag{27.4}$$

Here, σ'_{v0} represents the effective overburden stress at the soil strata where the SPT value is being considered and P_a is the atmospheric pressure taken as a value of reference.

Following Das et al. [8], the other correction factors C_E, C_B, and C_S are assigned 0.75, 1.05, and 1.0, respectively. Although the correction factor for rod length (C_R) adopts the value of 0.75 since the possible depth of liquefaction susceptibility at the concerned site is less than 3 m. Since sands having high silt content are generally observed to have a lower standard penetration tip resistance than clean sands with equivalent capacity of liquefaction prevention, the correction for fine content is provided as:

$$(N_1)_{60,CS} = (N_1)_{60,SS} + \Delta N_1 \tag{27.5}$$

$(N_1)_{60,CS}$ = SPT resistance value equivalent to clean sand, $(N_1)_{60,SS}$ = SPT resistance of silty sand and ΔN_1 = correction factor for content of fine particles (FC). The value of ΔN_1 is determined by the fine content of the soil as per the following conditions given in Eq. (27.6).

$$\Delta N_1 = \begin{cases} 0 & \text{for} \quad FC \leq 5\% \\ \frac{7}{30}(FC - 5) & \text{for} \ 5\% < FC < 35\% \\ 7 & \text{for} \quad FC \geq 35\% \end{cases} \quad (27.6)$$

Liquefaction Potential Analysis

The factor of safety (F_S) of the soil against liquefaction is defined as the ratio of capacity to demand of the soil in terms of energy. In this work, capacity is the threshold Arias intensity needed to trigger the liquefaction, and demand holds the value of Arias intensity developed during the scenario earthquake. The chart below (Fig. 27.3) by Kayen and Mitchell [10] gives the triggering intensity of liquefaction ($I_{hb,eq}$) as derived from various case studies based on corrected $(N_1)_{60,CS}$ values. The curved line represents the clean sand boundary for $I_{hb,eq}$ values as a function of corresponding resistances to penetration of the potentially liquefiable soil in the subsurface profile.

Results and Discussion

In this paper, evaluation of Liquefaction Potential Index (LPI) is attempted using the Arias intensity approach. Iwasaki et al. [9] categorized liquefaction damage by the severity of liquefaction. He devised the parameter called LPI to address the severity of liquefaction of an area based on the Factor of Safety (F_S) against the geohazard. Moreover, Iwasaki et al. [9] ruled out the category of moderate liquefaction in his study and considered only the possibility of low and high liquefaction. Sonmez [20] modified the LPI and the liquefaction potential categories to overcome the short-comings given above. Due to involvement of geological and seismological criteria for liquefaction analyses, the uncertainties accompanying the fines content and the SPT values cannot be ignored. While Ulusay and Kuru [21] proposed $F_S = 1.2$ as the triggering value for marginally liquefiable and non-liquefied soils, $1.25 \leq F_S \leq 1.5$ are considered acceptable by Seed and Idriss [18]. Hence, Sonmez [20] took up 1.2 as the least value for non-liquefied category and devised the following LPI evaluation procedure:

$$\text{LPI} = \int_0^{20} W(z) F(z) \mathrm{d}z \quad (27.7)$$

Proposed by Iwasaki [9], $W(z)$ is expressed as a function of depth = z and given as:

$$W(z) = 10 - 0.5z \tag{27.7a}$$

The term $F(z)$ is expressed in terms of the factor of safety at the desired depth z.

$$F(z) = \begin{cases} 1 - F_S & \text{for} \quad F_S \leq 0.95 \\ 2 \times 10^6 \times e^{-18.427F_S} & \text{for } 0.95 < F_S < 1.2 \\ 0 & \text{for} \quad F_S \geq 1.2 \end{cases} \tag{27.7b}$$

Since the GWT is located at a depth of approximately 2.15 m below the ground level, the soil above the level of commencement of the water table is considered non-liquefiable. As per Boulanger [6], in this study, it was found that the plasticity index for the dark grey soft sandy clayey silt fell within the range of 4 to 5 and when reduced by a point or two, the CL-ML soil converts to ML and it behaves like a sand. For a further sense of validation of the results, the layer was tested according to the susceptibility criteria proposed by Bray [7], and according to the parameters the soil falls within the ‘susceptible’ zone. Hence, for the remaining depth within the water table the soil has been considered to be liquefiable.

Sonmez [20] considered a constant factor safety for the total soil column up to an excavated maximum depth of 20 m. In this paper, the factor of safety has been calculated for each soil stratum. F_S values are calculated for the thickness of each soil layer that has been regarded for this study and subsequently liquefaction potential has been calculated individually which layer sums up to an average LPI value for the entire borehole. That said, the limits in the integral given in Eq. (27.7) vary from layer to layer. The categorization of LPI as provided by Sonmez [20] is given in Table 27.3.

Hence, the liquefaction potentials for all the layers in the borehole cross-section has been given in Table 27.4. From depth 10.5 to 20 m, the soil layers are found to be non-susceptible to liquefaction as mentioned earlier. Hence the LPI for the last two strata are ignored in this study since they have no contribution towards the liquefiability of the soil profile. The light grey soft to medium clayey silt as per Andrews and Martin [2] is not susceptible because it has a clay content of 32%, whereas, the liquid limit is 32.5%. Cross-verified with criteria proposed by Boulanger

Table 27.3 Modified liquefaction potential classification proposed by Sonmez [20]

LPI	Factor of Safety (F_S)	Liquefaction potential
0	$F_S \geq 1.2$	Not liquefied
$0 < LPI \leq 2$	$1.2 > F_S \geq 1$	Low
$2 < LPI \leq 5$	$1 > F_S \geq 0.95$	Moderate
$5 < LPI \leq 15$	$0.95 > F_S \geq 0.85$	High
$LPI > 15$	$F_S < 0.85$	Very high

Table 27.4 Classification of BH-02 sample layers based on the liquefaction potentials

Depth	r_b	$I_{h,\mathrm{eq}}$	$I_{h,1}$	F_S	LPI	Average LPI
1.5	0.86	1.08	0.55	0.51	0	4.16
3	0.74	0.93	0.5	0.54	8.53	
4.5	0.64	0.81	0.7	0.87	1.88	
6	0.56	0.70	0.4	0.57	6.21	
7.5	0.50	0.62	0.8	1.29	–	
9	0.45	0.56	0.55	0.99	–	
10.5	0.40	0.51	0.6	1.19	–	
12	0.37	0.47	0.6	1.28	–	
15	0.34	0.42	0.52	1.24	–	
20	0.34	0.42	0.55	1.30	–	

[6], the soil has a plasticity index higher than 7 and hence confidently depicts a clay-like behaviour. Both of the tests show non-susceptibility to liquefaction and hence is not considered in this study. The last layer of light grey firm clayey silt in the borehole cross-section exhibits a clay-like behaviour as well supported by its Atterberg parameters and hence is ignored.

From Table 27.4 it is evident that the borehole (BH - 02) represents a moderately liquefiable soil profile taking into account all the soil strata at the location. Figure 27.4a exhibits the curves showing the fluctuation of F_S with increasing depth from the ground level and (b) shows Liquefaction Potential Indices of different layers of the borehole.

Conclusions

In this deterministic approach, going by the liquefaction potential classifications proposed by various researchers, it is not ultimate to believe that if the soil exceeds a specific range of values, then liquefaction may be moderate, high or low as the case may be. The liquefaction of soil is determined by the frequency of earthquake loading and even the duration for which the ground shakes. In the context of the present study, considering an earthquake event of a fixed magnitude having a particular frequency and duration content, the area under target is 'moderately' liquefiable. For higher magnitude earthquakes, the same region may be highly liquefied with a considerable duration of ground shaking. Nevertheless, this study proves the efficacy of an energy-based approach towards a deterministic evaluation of liquefaction, and the results are congruent since slight liquefaction phenomena are reportedly sighted at this place after the seismic event. Hence, it can be safely noted that Arias intensity approach shines as a liquefaction potential analysis mechanism using the acceleration time history of an earthquake without the hassle of scaling the magnitudes as done for

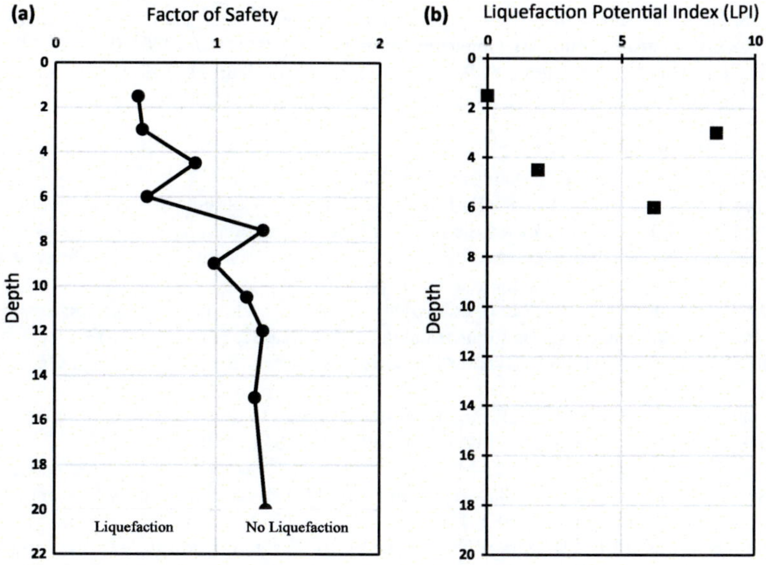

Fig. 27.4 **a** Variation of Factor of safety (F_S) with depth considering the entire cross-section of borehole, **b** Liquefaction Potential Index (LPI) values for the liquefiable layers

the stress-based approaches. A probabilistic evaluation can further be carried out to study liquefaction's recurrence rate at a previously liquefied location.

Acknowledgements This paper is a part of the author's M. Tech dissertation project. Earnest gratitude is extended to his supervisor and co-author, Dr. Sima Ghosh, for her constant motivation and support.

References

1. Anbazhagan P, Mog K, Nanjunda Rao KS, Siddharth Prabhu N, Agarwal A et al (2019) Reconnaissance report on geotechnical effects and structural damage caused by the 3 January 2017 Tripura earthquake, India. Nat Hazards 425–450. https://doi.org/10.1007/s11069-019-03699-w
2. Andrews DA, Martin GR (2000) Criteria for liquefaction of silty soils. In: 12th world conference on earthquake engineering. NZ Soc. for EQ Engrg, Upper Hutt
3. Arias A (1970) A measure of earthquake intensity. In: Press M (ed) Seismic design for nuclear power plants, pp 438–483
4. Bertero VV, Bozorgnia Y, Bolt BA, Campbell KW, Bertero RD (2004) Earthquake engineering from engineering seismology to performance-based engineering. Taylor & Francis, New York
5. Boulanger RW, Idriss IM (2004) Evaluating the potential for liquefaction or cyclic failure of silts and clays. University of California, Davis, California
6. Boulanger RW, Idriss IM (2006) Liquefaction susceptibility criteria for silts and clays. J Geotech Geoenviron Eng 132:1413–1426
7. Bray JD, Sancio RB (2006) Assessment of the liquefaction susceptibility of fine-grained soils. J Geotech Geoenviron Eng 132:1165–1177

8. Das S, Ghosh S, Kayal JR (2018) Liquefaction potential of Agartala City in Northeast India using a GIS platform. Bull Eng Geol Env. https://doi.org/10.1007/s10064-018-1287-5
9. Iwasaki T, Tokida K, Tatsuoka F, Watanabe S, Yasuda S, Sato H (1982) Microzonation for soil liquefaction potential using simplified methods simplified methods. In: 3rd international conference on microzonation, vol 3, Seattle, pp 1319–1330
10. Kayen RE, Mitchell JK (1997) Assessment of liquefaction potential during earthquakes by Arias Intensity. J Geotech Geoenviron Eng 1162–1174
11. Kayen RE, Mitchell JK, Seed RB, Lodge A, Nishio S, Coutinho R (1992) Evaluation of SPt-, CPT-, and shear wave based methods for liquefaction potential Assessment using Loma Prieta Data. In: 4th Japan-US workshop on earthquake-resistant design of lifeline fac. and countermeasures for soil liquefaction, vol 1, pp 177–204
12. Lee KL, Chan K (1972) Number of equivalent significant cycles in strong motion earthquakes. In: International conference on microzonation, vol II. Seattle, WA, pp 609–627
13. Liao S, Whitman RV (1986) Overburden correction factors for SPT in sand. J Geotech Eng 112:373–377
14. Nemat-Nasser S, Shokooh A (1979) A unified approach to densification and liquefaction of cohesionless sand in cyclic shearing. Can Geotech J 16:649–668
15. Parvez IA, Magrin A, Vaccari F, Mir RR, Peresan A, Panza GF (2017) Neo-deterministic seismic hazard scenarios for India—a preventive tool for disaster mitigation. J Seismol. https://doi.org/10.1007/s10950-017-9682-0
16. Seed HB, Idriss IM (1971) Simplified procedure for evaluating soil liquefaction potential. J SoilMech Found Div 1249–1274
17. Seed HB, Idriss IM (1982) Ground motions and soil liquefaction during earthquakes. Earthquake Engineering Research Institute
18. Singnar L, Sil A (2017) Assessment of liquefaction potential of Guwahati city based on geotechnical standard penetration test (SPT) data. In: Disaster advances, pp 10–21
19. Skempton AW (1953) The colloidal activity of clay. In: Proceedings of the 3rd international conference on soil mechanics and foundation engineering, vol 1, pp 57–61
20. Sonmez H (2003) Modification of the liquefaction potential index and liquefaction susceptibility mapping for a liquefaction-prone area (Inegol, Turkey). Environ Geol 44:862–871. https://doi.org/10.1007/s00254-003-0831-0
21. Ulusuay R, Kuru T (2004) 1998 Adana-Ceyhan (Turkey) earthquake and a preliminary microzonation based on liquefaction potential for Ceyhan town. Nat Hazards 32(1):59–88. https://doi.org/10.1023/B:NHAZ.0000026790.71304.32
22. Youd TL, Idriss IM (2001) Liquefaction resistance of soils: summary report from the 1996 Nceer and 1998 Nceer/Nsf workshops on evaluation of liquefaction resistance of soils. J Geotech Geoenviron Eng 127:297–313

Chapter 28
Determination of Site Specific Response Spectra for Site Class D and E for Guwahati City, NE India Region

Niranjan Borah, Joy Kumar Mondal, and Abhishek Kumar

Introduction

While designing a structure as EQ resistant, particularly in a seismically active region, the design response spectrum for the specific location is a crucial input. This design response spectrum is mostly acquired from seismic hazard analyses (SHA) which is mainly classified into two types: (a) deterministic seismic hazard analysis (DSHA) and (b) probabilistic seismic hazard analysis (PSHA). The North East India (NE India) region is seismically very active due to its location near the Indian-Eurasian plate boundary and the Indian-Burmese plate boundary. Many studies, including Sharma and Malik [1], Kanth et al. [2], NDMA [3], Sil et al. [4], Sitharam and Sil [5], Bahuguna and Sil [6], and Baro et al. [7] performed DSHA and PSHA for specific areas of NE India as well as the entire region to get the design response spectrum. Various GMPEs have been employed for these objectives, including Anbazhagan et al. [8], Atkinson and Boore [9], Gupta [10], Jain et al. [11], Nath et al. [12], NDMA [3], and Toro [13]. In the majority of studies, these GMPEs are solely used to determine the Peak Horizontal Acceleration (PHA) at the bedrock level. The most of these GMPEs, however, determine ground motion at the surface level for Site Class A (SC A) or SC B. Determining the hazard level for other SCs, such as SCs D and E, is equally essential since these types of SCs are also prevalent in the region. In this current work, DSHA for the city of Guwahati in NE India is carried out

N. Borah (✉) · J. K. Mondal · A. Kumar
IIT Guwahati, Guwahati, Assam, India
e-mail: niran521152@iitg.ac.in

J. K. Mondal
e-mail: joy.mondal@iitg.ac.in

A. Kumar
e-mail: abhiak@iitg.ac.in

K. Muthukkumaran et al. (eds.), *Soil Dynamics, Earthquake and Computational Geotechnical Engineering*, Lecture Notes in Civil Engineering 300,
https://doi.org/10.1007/978-981-19-6998-0_28

corresponding to SC A condition. In addition, site specific response spectra for SCs D and E are also developed using site response analysis for Guwahati.

Study Area and Its Past Earthquake Scenario

The present study area of Guwahati is located in NE India at the bank of the Brahmaputra River. Centre of Guwahati lies at 26.14° N and 91.74° E. It is home to Assam's capital, Dispur, and serves as the entrance to NE India. As a result, Guwahati is a vital city for both Assam and the rest of NE India. Guwahati, which is located in seismic zone V (as per IS 1893:2016 [14]), has been impacted by several earthquakes (EQs) in the past. The Assam EQ of 1897 occurred relatively close to the Guwahati city, which had a magnitude of 8.1 Mw. The Cachar EQ of 1869, with a magnitude of 7.5 Mw, also caused considerable damages in terms of several landslides, collapse of numerous masonry buildings and road damage in the area [15–18]. Occurrence of frequent low magnitude EQs as well as past great and major EQs makes the region as one of the most seismically active regions in the world.

Figure 28.1 shows past EQs that occurred within 500 km radius of Guwahati city centre. The information for these EQs is gathered from the NDMA [3], the US Geological Survey [19], International Seismological Centre [20], and Indian Meteorological Department [21]. As the magnitudes of these EQs are reported in various units (mb, ML, and Ms), they are all converted to Mw using the correlations given by Sitharam and Sil [5] to develop EQ catalogue for this work;

$$\mathrm{Mw} = 0.862\,\mathrm{mb} + 1.034 \tag{28.1}$$

$$\mathrm{Mw} = 0.673\,\mathrm{ML} + 1.730 \tag{28.2}$$

$$\mathrm{Mw} = 0.625\,\mathrm{Ms} + 2.350 \tag{28.3}$$

A total 3777 numbers of EQ events (non-repeating) are found to be within 500 km radius from Guwahati city centre (Fig. 28.1). Out of these, the numbers of events having magnitude Mw < 4, 4 ≤ Mw < 5, 5 ≤ Mw < 6, 6 ≤ Mw < 7, 7 ≤ Mw < 8, and 8 ≤ Mw are 183, 2916, 513, 139, 22, and 4, respectively. These EQs include the most recent Dhekiajuli EQ (MW = 6.4) occurred in April 2021. For DSHA, all of these EQs are considered for $M_{\max}$ calculation of the sources without declustering.

Deterministic Seismic Hazard Analysis for the Area

DSHA is based on the worst-case EQ scenario, where ground motion level at a specific site is determined by considering the EQ scenario that gives the highest

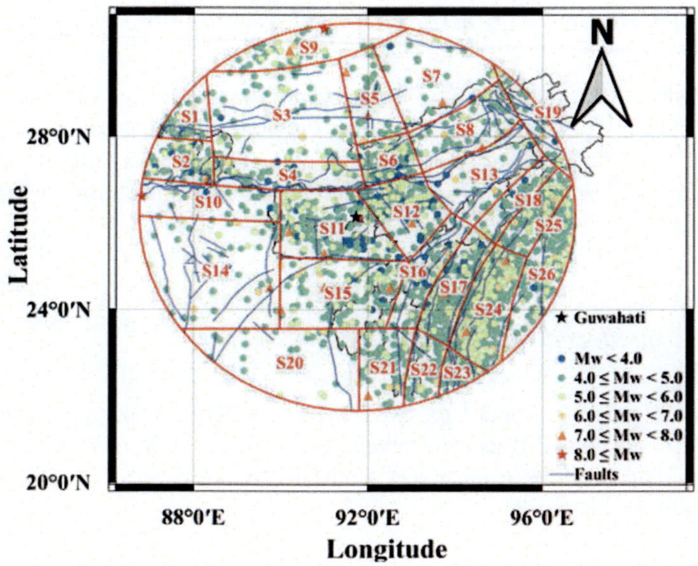

Fig. 28.1 Past EQs within 500 km of the Guwahati city centre and source zones and faults within 500 km of the Guwahati City centre

level of hazard out of all seismic sources. This hazard level is calculated based on information about EQ sources and ground motion attenuation. The whole DSHA method can be broken into four stages; (1) creation of a seismotectonic map and identification of all sources capable of causing substantial EQ in and around the study location. (2) Determining the shortest distance between each detected source and the location of interest. (3) Determining the largest probable EQ that may produce the highest shaking for each of the detected sources. (4) The overall maximum level of ground motion, which may be calculated by comparing the maximum level of ground motions from all identified sources at the location of interest.

Seismic Sources Around the Study Area

Faults considered for the current study are depicted in Fig. 28.1. The fault information is acquired from Bhukosh [22], Kundu and Gahalaut [23], and Chen [24]. Figure 28.1 also illustrates the areal sources that were considered for the study. These areal sources are considered in the light of previous research by Gupta [25] and Kolathayar and Sitharam [26]. The DSHA is performed taking into account a total of 26 areal sources and 74 significant faults.

Maximum Potential Magnitude Considered for the Analyses

It is critical in DSHA to understand the maximum magnitude of EQ that a source may produce. This value can be higher than the largest known magnitude that has occurred on that source. The maximum magnitude of the sources (Mp) is calculated in this study by increasing 0.5 Mw value in the observed maximum magnitude of the sources [27].

GMPEs Considered for the Analyses

For DSHA, the present work utilizes two GMPEs, namely Anbazhagan et al. [8] and NDMA [3]. These two GMPEs were developed for SC A condition in the Himalayan and NE Himalayan regions, respectively. Both GMPEs are given identical weightages (of 0.5) in the current work.

DSHA Results

Because multiple source models and GMPEs are used in this study, the logic tree shown in Fig. 28.2 is taken into account. This logic tree depicts the weightages assigned to the source models and GMPEs in order to get the final outcome. For all conditions (areal and linear sources separately), GMPEs are used to calculate the Sa value for all sources while taking the shortest distance to the site location and Mp values into account. Further, for each time period, the maximum value of all sources' Sa values is used as the final Sa value. Finally, the weightage displayed within bracket in the logic tree (Fig. 28.2) is used to construct final response spectrum (Table 28.1) for the site. The PGA observed for the location is 0.86 g. It is to be noted here, the worst scenario in case of areal source is 8.1 Mw EQ at 33 km hypocentral distance and, in case of areal source, it is 8.1 Mw at 63.5 km hypocentral distance. Considering the present PGA value, one more thing is needed to be highlighted here that the PGA values obtained for Guwahati region by earlier DSHA studies Bahuguna and Sil [6], Parvez et al. [28], and Nath et al. [12] are 1.23, 0.6–1.325 g, and 1.27 g (maximum within the city), respectively.

Consideration of Local Site Effects

For the present study, equivalent linear ground response analyses (ELGRA) are carried out to quantify LSE numerically. For this purpose, 20 boreholes belonging to SC D, and 20 boreholes belonging to SC E (refer to Kumar et al. [29]), that were

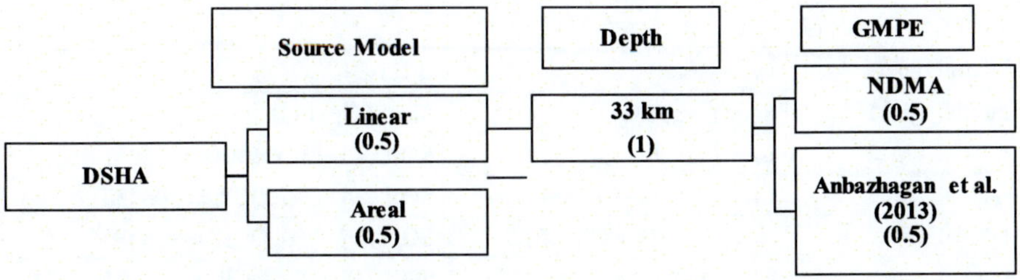

Fig. 28.2 Logic tree considered for DSHA

Table 28.1 Sa (g) of the response spectrum obtained from DSHA

Time period (s)	Sa (g)
0	0.86
0.1	1.69
0.2	1.15
0.3	0.94
0.4	0.79
0.5	0.74
0.6	0.67
0.8	0.60
1	0.48
1.2	0.43
1.5	0.39
2	0.26

drilled for Mass Rapid Transit System (MRTS) project are collected from Guwahati Metropolitan Development Authority (GMDA). For the present study, 30 globally recorded ground motions originally selected by Kumar et al. [30] are considered as input for ELGRA. According to Kumar et al. [30], these ground motions were selected taking into consideration all possible variations in ground motion properties (amplitude, frequency content, and duration). Details of these ground motions are available at 'Table 28.1' in Kumar et al. [30].

Dynamic Soil Properties Considered

Another important input parameter for ELGRA is the dynamic soil properties for the different existing soil layers. Dynamic soil properties consist of low shear strain (γ) corresponding shear modulus ($G_{\max}$) and damping ratio (β) and dynamic soil property curves (DSPCs). For the present study, however, only SPT-N values are available from the boreholes. Therefore, a correlation developed by Kumar et al.

Table 28.2 Details of DSPCs used

Material type	G/G_{max} curve	β curve
Material_type1	Sand (avg) [33]	Sand (avg) [33]
Material_type2	Clay PI = 1–10, [35]	SOIL, PI = 0, OCR = 1–8 [35]
Material_type3	Clay PI = 10–20, [35]	SOIL, PI = 15, OCR = 1–8 [35]
Material_type4	clay PI = 20 = 40, [35]	SOIL, PI = 30, OCR = 1–8 [35]
Material_type5	Clay (upper) [36]	SOIL, PI = 50, OCR = 1–8 [35]
Material_type6	Rock [34]	Rock [34]

[29] for Guwahati city that correlates SPT-N with shear wave velocity (V_s) is used to calculate V_s values of different soil layers. It should be highlighted here that the correlation developed by Kumar et al. [29] can be used for both clay and sandy types of soils. Further, the G_{max} values are calculated using the formulae $G_{max} = \rho V_s^2$. The low γ corresponding β values, on the other hand, is taken as 1% for all the soil layers. DSPCs consist of modulus degradation (G/G_{max}) and β curves. Ideally, DSPCs for the soil layers present at the site should be used for ELGRA. However, like regional bedrock motions, regional DSPCs also are not available for most of the times. Therefore, DSPCs that are developed considering soils from other part of the globe are employed for carrying out GRA. These DSPCs can be selected based on soil type, confining stress, plasticity index (PI), over consolidation ratio, and many other parameters. For the present study, from the 40 boreholes considered, it is observed that the subsoil lithology for Guwahati consists mainly of clay layers with low to high compressibility and silty sand layers. Similar types of soil were also reported by Raghukanth et al. [16], Halder et al. [31], and Kumar and Halder [32] for Guwahati city. Thus, based on the soil types, and PI, 6 different sets of DSPCs are considered for the present study. The sand and silty sand layers are modelled as material_type1 and average G/G_{max} curve for sand developed by Seed and Idriss [33] is assigned to them. The β curves are taken as average sand developed by Seed and Idriss [33] for modelling of sand layers. Further, the clay layers are modelled as material_type2-material_type5 in the ascending order based on their PI value. Further, the bedrock layer is modelled with DSPCs for rock as per Schnabel [34]. A summary of the different DSPCs used for the analysis is presented in Table 28.2.

ELGRA

For the present work, all the 30 ground motions mentioned earlier are applied as SC A motion for all the 40 boreholes. Therefore, total 40 × 30 = 1200 ELGRA are performed. For the purpose, ELGRA code developed by Kumar and Mondal [37] has been utilized. It has to be highlighted here that all the 40 borelogs considered for the present study are not available till 30 m depth. Some of the boreholes were terminated before 30 m depth as SPT-N value reached 100 in those boreholes. In such cases, additional soil layers each having 2 m thickness and soil properties similar to the bottom most soil layer present in the original borehole are added till 30 m depth. Further, the underlying bedrock layer is modelled with V_s value of 1500 m/s. As the ground motions are applied as bedrock motion, above considered 30 motions are first transferred to the bedrock level by performing deconvolution, following the procedure mentioned in Annie et al. [38] later, these deconvoluted motions are applied at the bedrock level to perform ELGRA. While performing ELGRA, boreholes corresponding to SC D and SC E are dealt separately. Results are obtained in the form of 5% damped response spectra at the ground surface and at bedrock. Additionally, 5% damped response spectra for the input motions (considering them to be rock outcrop motion) are also calculated.

Results

It is to be highlighted here, determination of spectral acceleration (SA) at the ground surface by means of response spectra is useful for EQ resistant design. However, in standard building codes, such information regarding SA is provided in the form of site coefficients [32]. According to Hwang et al. [39], site coefficient is defined as the ratio of SA obtained at the ground surface and SA obtained at the bedrock for a particular time period. Therefore, it is evident that site coefficient obtained at 0 s time period corresponds to amplification factor between bedrock PHA and ground surface PGA. In the present study, however, ratio of SA obtained at the ground surface and SA obtained at SC A (rock outcrop) for a particular time period is determined and further used for consideration. For the purpose of the present study, the above obtained ratio is termed as site coefficient subsequently. Further, site coefficients for each of the 30 globally recorded ground motions mentioned earlier are determined separately till 2 s time period, for all 40 boreholes. It has to be highlighted here, above calculated site coefficients belong to both the SCs (SC D and SC E) considered in the present study. Thus, they need to be separated based on their respective SCs. After accomplishing the above step, site coefficients for each of the 2 SCs are obtained overall for 20 boreholes. Further, the site coefficients for each of the 30 ground motions are sorted from 20 boreholes within an SC based on the PGA values of the outcrop motions. Finally, the site coefficients obtained collectively from 20 boreholes for each of the 30 ground motions are averaged to evaluate average site coefficients for the particular outcrop PGA value. Figure 28.3a, b show the average site coefficients for SC D and SC E for a few selected ground motions. In Fig. 28.3, the entities along the legends indicate the outcrop PGA value. It can be observed from Fig. 28.3 that the site coefficients are different for different PGAs. Another, observation that can be made from Fig. 28.3 is as the outcrop PGA increases the amplification factor decreases regardless of SCs. The last observation is consistent with the findings of Kumar and Halder [32] and Kumar et al. [30, 40].

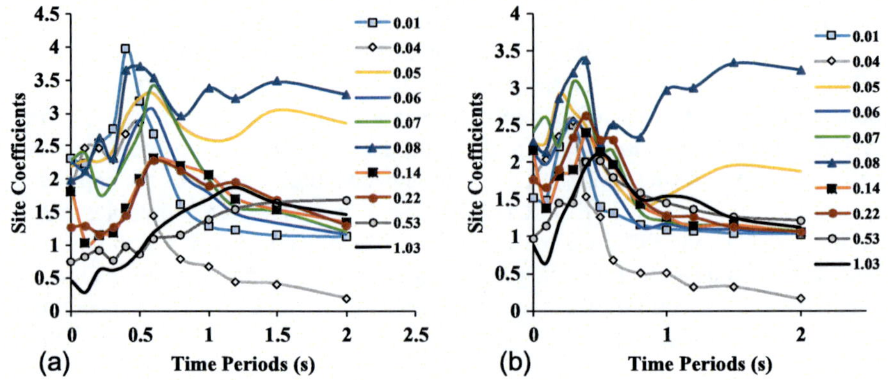

Fig. 28.3 Average site coefficients for SC D and SC E for different PGAs (units in g)

Site Specific Response Spectra for Guwahati City

The response spectrum obtained in Sect. 28.3 is for SC A condition (as listed in Table 28.1). Further, this response spectrum can be utilized in combination with LSE to determine response spectra for SC D and SC E conditions. Two approaches are used for this purpose, which will be discussed in subsequent subsections. In the first approach, LSE is estimated by considering globally recorded earthquakes. On the other hand, in the second approach, the outcome from the first approach is validated by considering spectrum compatible ground motions.

Site Specific Response Spectra from ELGRA of Global EQs

As mentioned earlier, site coefficients are dependent on PGA of outcrop motion. Therefore, for further assessment, it is imperative to consider them based on the seismic hazard value. Since for the present study, PGA from DSHA is obtained as 0.86 g at the rock outcrop (or SC A), site coefficients should be selected in a way that conform to the mentioned PGA value. However, in present study, none of the 30 globally recorded ground motion mentioned earlier has a PGA value close to 0.86 g. For this reason, average of the two sets of average site coefficients (see Fig. 28.3a, b) obtained for outcrop PGA 1.03 and 0.53 g (for ground motion 10 and 20 as per Kumar et al. [30]) is utilized to calculate the response spectrum of SC D and response spectrum of SC E. The resulting average site coefficients based on (ground motion 10 and 20) for SC D and SC E are tabulated for few time periods in Table 28.3. The final response spectra for SC D and SC E are obtained by multiplying these site coefficients (Table 28.3) with the Sa values of response spectrum of SC A (Table 28.1).

Table 28.3 Average site coefficients for SC D and SC E

Time periods (s)	SC D	SC E
0	0.94	0.60
0.1	0.90	0.56
0.2	1.32	0.77
0.3	1.50	0.69
0.4	1.98	0.84
0.5	2.08	0.89
0.6	1.90	1.14
0.8	1.55	1.32
1	1.49	1.54
1.2	1.43	1.71
1.5	1.26	1.64
2	1.17	1.58

Site Specific Response Spectra from ELGRA of Spectrum Compatible Ground Motions

In an another approach, which is followed widely, spectrum compatible ground motions are used for site response analysis to obtain response spectra for SC D and SC E [41]. To accomplish this, 10 numbers of spectrum compatible ground motions are generated considering the response spectrum based on DSHA. SeismoArtif [42] software is utilized for this purpose. The site response analyses are carried out utilizing the same method described above considering the spectrum compatible ground motions as input. The acquired site specific response spectra for each SC from site response analyses are averaged to get the final site specific response spectra for that particular SC (Fig. 28.4).

Figure 28.4a compares of response spectra for SC D obtained by following the above two mentioned methods. Same way, Fig. 28.4b shows comparison of response spectra for SC D. It is observed from Fig. 28.4 that both approaches give comparable results. This suggests that the first approach mentioned in this study can be utilized to get an overall idea of the response spectra on different SCs when response spectrum on SC A is known.

Conclusion

An effort is made in the present work to develop site specific response spectrum for Guwahati city on the basis of DSHA for SC A as well as Site SC D and SC E conditions. To begin with, by considering different source models (linear and areal) and available GMPEs, response spectrum for SC A is obtained by utilizing logic

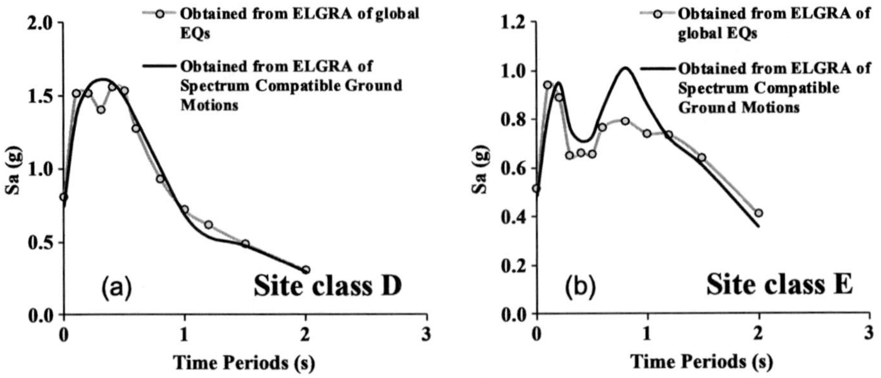

Fig. 28.4 Response spectra for **a** SC D and **b** SC E

tree. In addition, the response spectra for SC D and SC E are then produced using response spectrum obtained based on DSHA and LSE. LSE is taken into account by considering two separate approaches, considering globally recorded ground motions and, spectrum compatible ground motions for the Guwahati city. While using globally recorded ground motions, average site coefficients for SC D and SC E are determined based on different outcrop PGA values. Further, these average site coefficients can be used with DSHA value to develop site specific response spectrum. This way even without considering spectrum compatible ground motions, LSE can be taken into account in seismic hazard analysis. The above approach is validated by considering spectrum compatible ground motions for the Guwahati city in the second approach. Results from the both the approaches are found comparable. The approaches adopted in this present study can also be used for other locations in other region and also by considering PSHA. Utilization of PSHA along with a larger dataset of boreholes can significantly improve outcomes from the present study.

References

1. Sharma ML, Malik H (2006) Probabilistic seismic hazard analysis and estimation of spectral strong ground motion on bed rock in north east India. In: 4th international conference on earthquake engineering, Taipei, Taiwan
2. Raghu Kanth STG, Singh KD, Pallav K (2009) Deterministic seismic scenarios for Imphal city. Pure Appl Geophys 166:641–672
3. NDMA (2010) Development of probabilistic seismic hazard map of India. Technical Report
4. Sil A, Sitharam TG, Kolathayar S (2013) Probabilistic seismic hazard analysis of Tripura and Mizoram states. Nat Hazards 68(2):1089–1108
5. Sitharam TG, Sil A (2014) Comprehensive seismic hazard assessment of Tripura and Mizoram states. J Earth Syst Sci 123(4):837–857
6. Bahuguna A, Sil A (2018) Comprehensive seismicity, seismic sources and seismic hazard assessment of Assam, north east comprehensive seismicity, seismic sources and seismic. J Earthq Eng

7. Baro O, Kumar A, Ismail-zadeh A (2018) Seismic hazard assessment of the Shillong Plateau, India. Geomatics Nat Hazards Risk 9(1):841–861
8. Anbazhagan P, Kumar A, Sitharam TG (2013) Ground motion prediction equation considering combined dataset of recorded and simulated ground motions. Soil Dyn Earthq Eng 53:92–108
9. Atkinson GM, Boore DM (2003) Empirical ground-motion relations for subduction-zone earthquakes and their application to Cascadia and other regions. Bull Seismol Soc Am 93(4):1703–1729
10. Gupta ID (2010) Response spectral attenuation relations for in-slab earthquakes in Indo-Burmese subduction zone. Soil Dyn Earthq Eng 30:368–377
11. Jain KS, Roshan AD, Arlekar JN, Basu PC (2000) Empirical attenuation relationships for the Himalayan earthquakes based on Indian strong motion data. In: Proceedings of the sixth international conference on seismic zonation
12. Nath SK, Raj A, Thingbaijam KKS, Kumar A (2009) Ground motion synthesis and seismic scenario in Guwahati city—a stochastic approach. Seismol Res Lett 80(2):233–242
13. Toro GR (2002) Modification of the Toro et al. (1997) Attenuation equations for large magnitudes and short distances. Risk Eng Inc. (2002)
14. BIS. IS-1893 2016 part 1: Criteria for Earthquake Resistant Design of Structures sixth revision (2016)
15. Baro O, Kumar A (2015) A review on the tectonic setting and seismic activity of the Shillong plateau in the light of past studies. Disaster Adv 8(7):34–45
16. Raghukanth S, Dixit J, Dash S (2011) Ground motion for scenario earthquakes at Guwahati city. Acta Geod Geophys Hungarica [Internet] 46(3):326–346
17. Kayal JR (2008) Microearthquake seismology and seismotectonics of South Asia. Springer, India
18. Oldham RD (1899) Report on the great earthquake of 12 June 1897. Mem Geol Soc India (29):379
19. USGS. Search earthquake catalog. https://earthquake.usgs.gov/earthquakes/search/. Accessed on 1 May 2019
20. ISC. Event CatalogueSearch. http://www.isc.ac.uk/iscbulletin/search/catalogue/. Accessed on 1 May 2019
21. IMD. Indian Meteorological Department. http://www.imd.gov.in/pages/earthquake_prelim.php. Accessed on 1 May 2019
22. Bhukosh. Bhukosh. https://bhukosh.gsi.gov.in/Bhukosh/MapViewer.aspx
23. Kundu B, Gahalaut VK (2013) Tectonic geodesy revealing geodynamic complexity of the Indo-Burmese arc region, North East India. Curr Sci 104(7):920–933
24. Chen B (2020) Stress-induced trend: the clustering feature of coal mine disasters and earthquakes in China. Int J Coal Sci Technol 7(4):676–692
25. Gupta ID (2006) Delineation of probable seismic sources in India and neighbourhood by a comprehensive analysis of seismotectonic characteristics of the region. Soil Dyn Earthq Eng 26(8):766–790
26. Kolathayar S, Sitharam TG (2012) Characterization of regional seismic source zones in and around India. Seismol Res Lett 83(1):77–85
27. Gupta ID (2002) The state of the art in seismic hazard analysis. ISET J Earthq Technol 39(4):311–346
28. Parvez IA, Vaccari F, Panza GF (2003) A deterministic seismic hazard map of India and adjacent areas. Geophys J Int 155(2):489–508
29. Kumar A, Harinarayan NH, Verma V (2018) Seismic site classification and empirical correlation between standard penetration test N value and shear wave velocity for Guwahati based on thorough subsoil investigation. Pure Appl Geophys 175(8):2721–2738
30. Kumar A, Baro O, Harinarayan NH (2016) Obtaining the surface PGA from site response analyses based on globally recorded ground motions and matching with the codal values. Nat Hazards 81(1):543–572
31. Halder S, Mondal JK, Kumar A (2019) Empirical correlations for strain dependent linear 1D ground response analysis for Guwahati, India based on detailed analyses. In: 7ICEGE. Roma

32. Kumar A, Suman H (2020) Design response spectra and site coefficients for various seismic site classes of Guwahati, India, based on extensive ground response analyses. Geotech Geol Eng [Internet] 38(6):6255–6280
33. Seed H, Idriss I (1970) Soil moduli and damping factors for dynamic response analysis. University of California, Berkeley
34. Schnabel PB (1973) Effects of local geology and distance from source on earthquake ground motions. University of California, Berkeley
35. Sun JI, Golesorkhi R, Seed HB (1988) Dynamic moduli and damping ratios for cohesive soils. Report no EERC 88-15. University of California Berkeley
36. Seed HB, Sun JH (1989) Implication of site effects in the Mexico City earthquake of September 19, 1985 for earthquake-resistance-design criteria in the San Francisco Bay area of California. Report No. UCB/EERC-89/03
37. Kumar A, Mondal JK (2017) Newly developed MATLAB based code for equivalent linear site response analysis. Geotech Geol Eng 35(5):2303–2325
38. Annie OL, Jonathan P, Kwok AOL, Asce M, Stewart JP, Asce M et al (2007) UCLA use of exact solutions of wave propagation problems to guide implementation of nonlinear seismic ground
39. Hwang HHM, Lin H, Huo JR (1997) Site coefficients for design of buildings in Eastern United States. Soil Dyn Earthq Eng 6:29–40
40. Kumar A, Harinarayan NH, Baro O (2017) Nonlinear soil response to ground motions during different earthquakes in Nepal, to arrive at surface response spectra. Nat Hazards 87:13–33
41. Nguyen VQ, Aaqib M, Nguyen DD, Luat NV, Park D (2020) A site-specific response analysis: a case study in Hanoi, Vietnam. Appl Sci 10(11)
42. SeismoArtif. Artificial earthquake accelerograms (2013). Available from https://seismosoft.com/products/seismoartif/

Chapter 29
Performance Evaluation of Single and Multiple Trenches in Screening Ground Vibrations

Nitish Jauhari and Amarnath Hegde

Introduction

The ground-borne vibrations emanated from various sources have become a major concern. These vibrations induce detrimental effects on nearby structures and health of the people working in the vicinity. The effects of ground vibrations have produced an increasing alarm in recent years and became a relevant area of research.

Vibration propagation in soils takes place in the form of surface waves. About two-thirds of the energy released due to vibration travels in the form of Rayleigh waves [1]. The effect of ground-borne vibrations can be diminished by placing a barrier in the propagation path of the waves. These barriers can be open trenches, infilled trenches, wave-impeding blocks, sheet piles, etc. The infilled wave barrier system creates a solid-to-solid or solid-to-fluid interface. At the solid-to-solid interface, both the P-waves and S-waves can travel, while in the solid-to-fluid interface, only P-waves can travel. The open wave barrier system creates a solid-to-void interface. These interfaces restrict the movement of both P-waves and S-waves further [2]. Thus, reflection, diffraction, and scattering of Rayleigh waves by the wave barriers play a major role in screening ground-borne vibrations.

Active and passive vibration isolations are the two scenarios to screen the wave energy emanated from the source. In active isolation, the barrier is constructed near the source of vibration. However, In passive isolation, the barrier is placed far from the source but near a structure that has to be protected [3]. Figures 1a, b show the schematic representations of active and passive vibration isolation barrier systems.

N. Jauhari (✉) · A. Hegde
Department of Civil and Environmental Engineering, Indian Institute of Technology Patna, Bihar 801106, India
e-mail: nitish_2021ce15@iitp.ac.in

A. Hegde
e-mail: ahegde@iitp.ac.in

K. Muthukkumaran et al. (eds.), *Soil Dynamics, Earthquake and Computational Geotechnical Engineering*, Lecture Notes in Civil Engineering 300,
https://doi.org/10.1007/978-981-19-6998-0_29

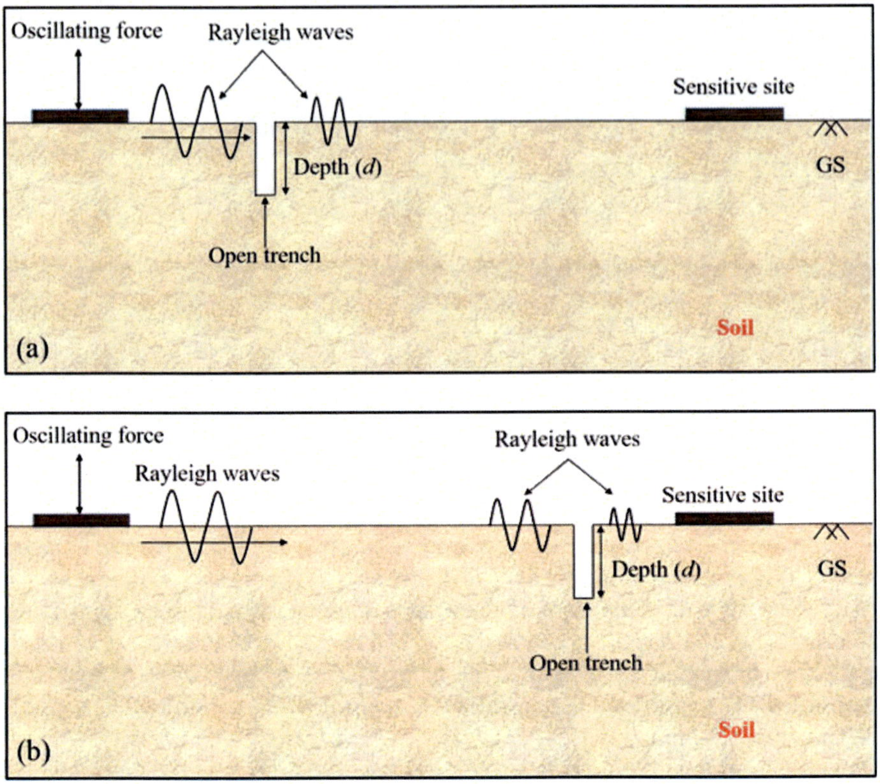

Fig. 29.1 Schematic of isolation systems: **a** active isolation; **b** passive isolation

Several studies were carried out by considering the single open trenches as a screening measure [4–12]. It has been conclusively shown that the trench depth is the most influential parameter used to screen the vibration amplitude. The effect of the width was found insignificant as compared to depth [2, 13]. However, the trench depth required to screen vibrations is typically 0.8–1.5 times Rayleigh wavelength (L_R). On account of the significant trench depth and the associated construction difficulties, open trenches are not recommended widely. Therefore, the core interest of the present study is to explore the possibilities of using dual open trenches to reduce the trench depth. Numerical studies have been performed using PLAXIS 2D to study the performance of the dual trenches. The depth and spacing between the trenches have been varied in order to arrive at optimum values for the effective mitigation of the vertical and horizontal vibrations. The performance of the dual open trenches has been compared with single open trench wherever applicable.

Materials and Methods

Problem Formulation

A sinusoidal harmonic load of the magnitude (P_o) of 1 kN and frequency (f) of 31 Hz was applied in dynamic analysis for a time duration of 0.5 seconds. The amplitudes of vertical and horizontal components of velocities were measured under the above-stated vertical excitation. The frequency of 31 Hz was selected to get an integer value of Rayleigh wavelength (L_R). The two rectangular open trenches of depth d and width w were placed at a distance l_1 and l_2 from the source of excitation. The geometrical parameters of trench, viz. depth, width, and locations, were normalized by Rayleigh wavelength (L_R) as $d = D \times L_R$, $w = W \times L_R$, $l_1 = L_1 \times L_R$, and $l_2 = L_2 \times L_R$, to prevent the dependency of results on load frequency. Here, D and W represent the normalized depth and normalized width of the trench, whereas L_1 and L_2 are the normalized distance of the first and second trench from the source. A schematic of the dual trench barrier system is shown in Fig. 29.2. To study the effect of spacing of dual trenches, the location of the first trench was varied from 3.5 to 4.5 L_R. However, the position of the second trench, i.e. l_2, was fixed at a distance of 5 L_R. To examine the vibration isolation of a single trench system, the trench was positioned at 5 L_R from the source. A set of normalized benchmark parameters ($W = 0.2, D = 0.4, L_1 = 4, L_2 = 5$) were defined for the barrier system. The parametric study was performed as per Table 29.1.

The vibration screening of the barrier system was evaluated in the context of amplitude reduction ratio (ARR) as defined by Woods [2],

$$\text{ARR} = \frac{\text{Velocity amplitude of ground surface with a barrier}}{\text{Velocity amplitude of ground surface without a barrier}} \tag{29.1}$$

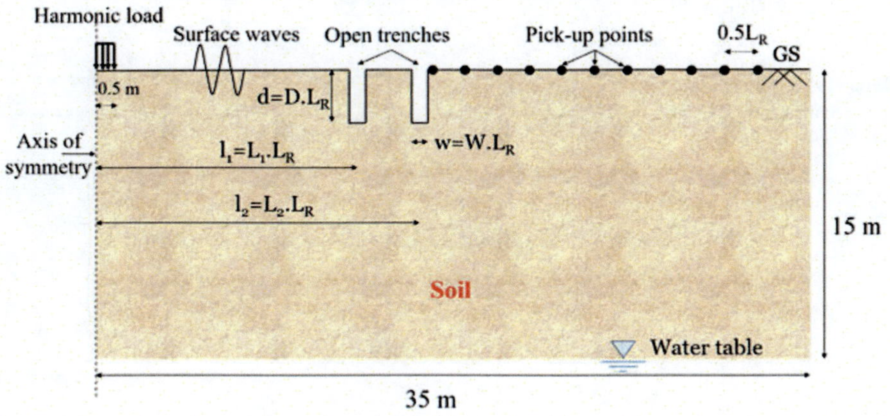

Fig. 29.2 Schematic of a dual trench barrier system

Table 29.1 Numerical parametric study

Normalized barrier width, W	0.2
Normalized depth of single trench, D	0.2, 0.3, 0.4, 0.5, 0.6, 0.7, 0.8, 0.9, 1.0
Normalized depths of dual trenches, D	0.2, 0.3, 0.4, 0.5, 0.6
Normalized location of first trench, L_1	3.50, 3.75, 4.00, 4.25, 4.50

The amplitude reduction ratios were calculated for the vertical (V_V) and horizontal (V_h) velocity components. The ARR values were calculated beyond the second barrier and up to a distance of 10 L_R at intervals of 0.5 L_R. Duvall and Fogelson [14] recommended studying the vibration effects in terms of velocity rather than displacement or acceleration. To assess the effectiveness of the barrier system, an average value of amplitude reduction ratio over the barrier influence zone was calculated and reported as average ARR. The criteria for selecting the optimum trench depth is when it reduces the amplitudes of vibration by at least 75% [2, 3]. It represents that average ARR should be lower or equal to 0.25.

Material Properties

In this study, the soil medium was considered as linear elastic, homogeneous, and isotropic. The input parameters for the soil media were selected similar to Saikia and Das [10]. The density (ρ), elasticity modulus (E), and Poisson's ratio (ν) of elastic half-space were considered 1800 kg/m^3, 46,000 kN/m^2 and 0.25, respectively. The soil medium was assumed to have a material damping ratio of 5%. The ground motion parameters were estimated in Table 29.3 from the chosen elastic parameters and listed in Table 29.2.

Table 29.2 Soil properties (after Saikia and Das [10])

Parameters	Value
Young's modulus, E (kN/m^2)	46,000
Unit weight, γ_d (kN/m^3)	17.65
Poisson's ratio, ν	0.25
Shear wave velocity, V_S (m/s)	101.3
Rayleigh wave velocity, V_R (m/s)	93
Rayleigh wavelength, L_R (m)	3.0

Table 29.3 Calculation of ground motion parameters

Parameter	Expression	Value
V_S	$(\sqrt{E/\rho})/(\sqrt{2(1+\nu)})$	101.1 m/s
V_R	$V_s((0.87+1.12\nu)/(1+\nu))$	93.02 m/s
L_R	V_R/f	3.0 m

Numerical Modelling

To understand the vibration isolation efficacy of dual open trenches, a 2D axisymmetric numerical model was developed in the finite element package PLAXIS 2D. The model consisted of 15 noded triangular elements. Previous studies have suggested a soil domain of 10 L_R from the vibration source for wave barrier analysis [15, 16]. Therefore, the model dimensions in this study were kept as 35 m × 15 m.

The mesh was discretized using very fine elements. Furthermore, the local refinements were applied along the surface and trenches for better accuracy. A discretized 2D numerical model is shown in Fig. 29.3. The Rayleigh α and Rayleigh β parameters were used for defining the damping in the numerical model. The values of Rayleigh α and Rayleigh β were presumed as 0.9 and 0.000488 to satisfy the relationship, $\alpha+\beta\omega^2 = 2\xi\omega$. In the relationship, ξ and ω represent material damping and angular frequency of excitation, respectively [10]. The vertical sides of the model were vertically restrained ($u_x = 0$) while the bottom boundary was completely restrained ($u_x = u_y = 0$). To avoid the undue reflections from the boundary edges, absorbent boundaries were assigned at the bottom and the right edges of the model.

A steady-state vertical harmonic load of magnitude 1 kN and frequency of 31 Hz was applied on the soil media over a width of 0.5 m. The time interval for dynamic analysis was considered as 0.5 s, which is adequate to study the wave propagation

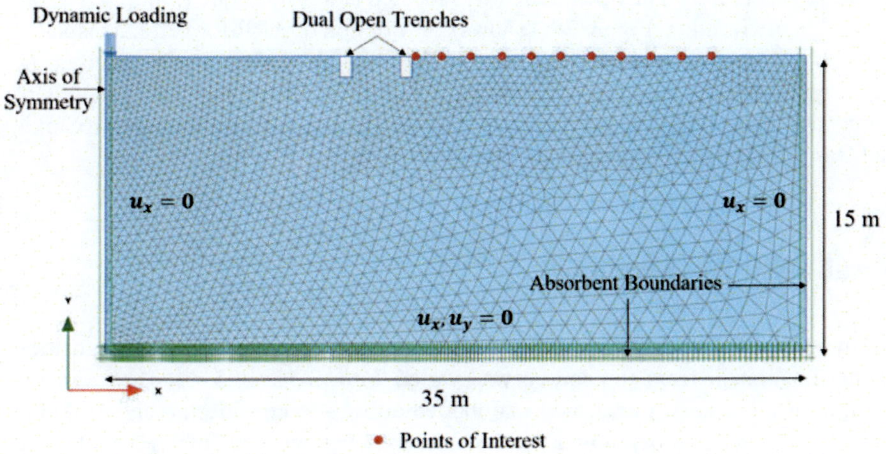

Fig. 29.3 2D numerical model of a dual trench barrier system in PLAXIS

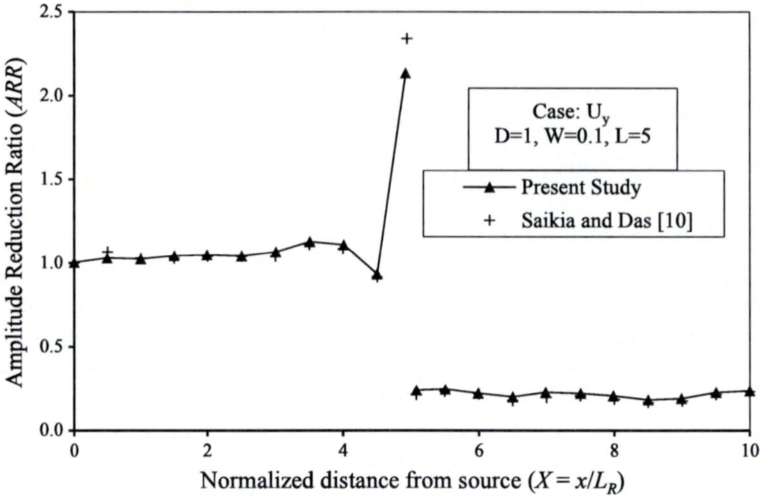

Fig. 29.4 Finite element model validation

in the zone of interest. The footing carrying the vibration load was not included in the model as it does not affect the study results as suggested by Kattis et al. [17].

Model Validation

The model was validated using the result by Saikia and Das [10]. The result concerning an open trench of a depth of 3 m (1 L_R) and a width of 0.3 m (0.1 L_R) were considered. The open trench was located at a distance of 15 m (5 L_R) from the source of vibration. Figure 29.4 shows the variation of ARR for vertical displacement (U_y) with the normalized distance (x/L_R) for both models. In the abscissa, L_R and x represent Rayleigh wavelength and distance of the points from the axis of symmetry, respectively. A good agreement between the numerical results of Saikia and Das [10] and the present model was observed.

Results and Discussion

To investigate the effect of spacing between two trenches, the normalized position of the first trench from the source was varied from 3.5 to 4.5 L_R. However, the position of the second trench was maintained at 5 L_R to consider passive isolation case. The normalized depth and width of the trenches were maintained as 0.4 and 0.2. Figure 29.5 shows a significant reduction in the average ARR for the horizontal

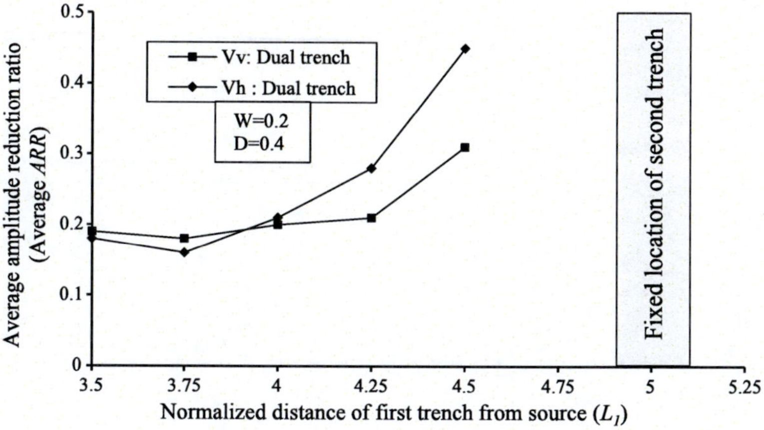

Fig. 29.5 Effect of trench location on vibration screening

vibration with the increase in spacing between two trenches. The average ARR for horizontal vibration was reduced from 0.45 to 0.16 when the first trench was repositioned from 4.5 to 3.75 L_R. The average ARR for the vertical vibration was reduced from 0.31 to 0.18 for the similar reposition of the trench. However, an insignificant reduction was observed in vertical attenuation when the trench position varied up to 4.25 L_R.

To compare the vibration attenuation of single and dual trenches, the variation of ARR with the normalized distance beyond the second trench was considered, as shown in Figs. 29.6a, b. The benchmark parameters ($W = 0.2$, $D = 0.4$, $L_1 = 4$, $L_2 = 5$) were used to perform the analysis for the dual trench system. For a single trench system, the normalized width and depth were maintained the same as those of dual trench, while the normalized location of the trench (L) was considered as five. From the Figs. 29.6a, b, ARR values were found significantly reduced in dual trenches in comparison with the single trench. This may be attributed due to the reflection of waves from the soil–barrier interface. Furthermore, the waves transmitted from the base of the first trench suppress within the space between two trenches. From Fig. 29.6a, attenuation rate of vertical vibration was relatively high in dual trench system up to 1.5 L_R.

Figure 29.7 shows the effect of normalized depth on the average ARR for single and dual trenches. It was observed that the dual trenches screen the vertical and horizontal vibration effectively. Furthermore, it can be seen that there was a minimal difference in vibration screening when the normalized depth of dual trenches was increased beyond 0.5. On the contrary, average ARR continues to decrease with the increase in normalized depth in single trenches. The results also revealed that the single trenches are more effective in screening vertical vibrations than horizontal vibrations. The single trench reflects the vertical vibrations deep into the half-space. This leads to a change in the propagation mode of waves from surface waves to body waves. On the contrary, the horizontal vibrations do not reflect into the deeper

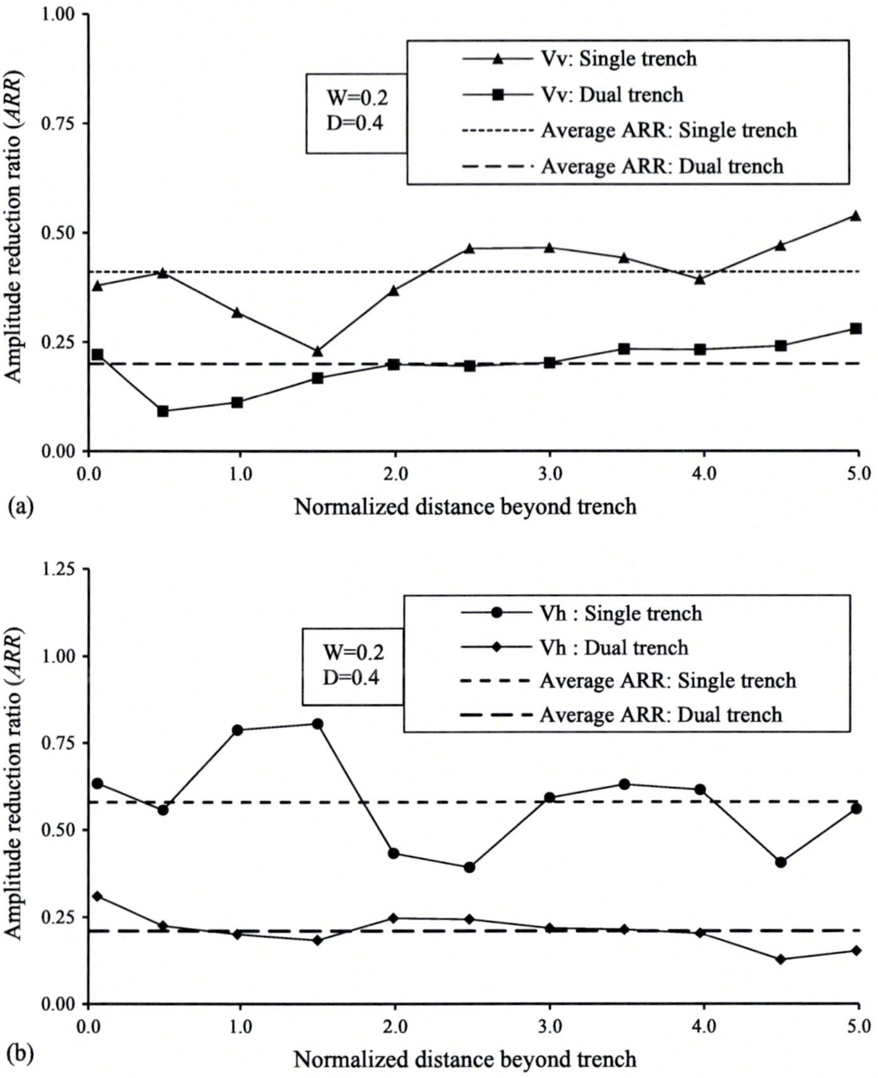

Fig. 29.6 Variation of ARR for single and dual trench: **a** vertical component of velocity; **b** horizontal component of velocity

space resulting in less attenuation. Table 29.4 shows the optimum normalized depths required to screen 75% vibrations for single and dual trenches.

Figure 29.8a, b show the comparison of vibration effectiveness for single and dual trenches. A dual trench was found to require much lesser depth to attain a desired degree of isolation than a single trench. The average ARR criteria equal to or lower than 0.25 was met when the normalized depth of the dual trenches was 0.4. The average ARR was calculated as 0.20 and 0.21 for vertical and horizontal vibrations,

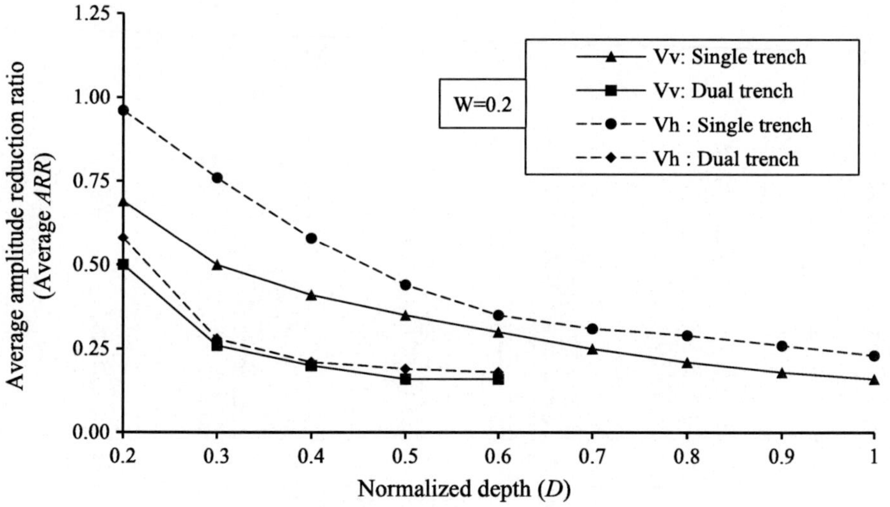

Fig. 29.7 Effect of normalized depth on average ARR for single and dual trenches

Table 29.4 Optimum normalized depths to screen vibrations

Vibrations	Single trench	Dual trench
Vertical vibrations	0.7	0.4
Horizontal vibrations	1.0	0.4

respectively. This indicates that the amplitudes were reduced by 80 and 79% for vertical and horizontal vibration.

Conclusions

A 2D axisymmetric finite element analysis was performed to interpret the effectiveness of single and dual open trenches as vibration isolators. The effects of normalized depth and spacing of trenches on vibration attenuation were studied extensively in the study. The comparison of single and dual trenches was carried out based on the depth required to attain a particular degree of isolation. The following conclusions can be drawn from the parametric study.

The effectiveness of single trenches was found more in screening vertical vibrations compared to horizontal vibrations. To attain the criteria of average ARR equal to or lower than 0.25, optimum depths for screening vertical and horizontal vibrations were found to be 0.7 L_R and 1.0 L_R, respectively.

The dual open trenches performed exceptionally well in screening ground vibrations. The spacing between two trenches was found to influence the isolation effectiveness. A spacing of 1.25 L_R reduces vertical and horizontal vibrations by 82%

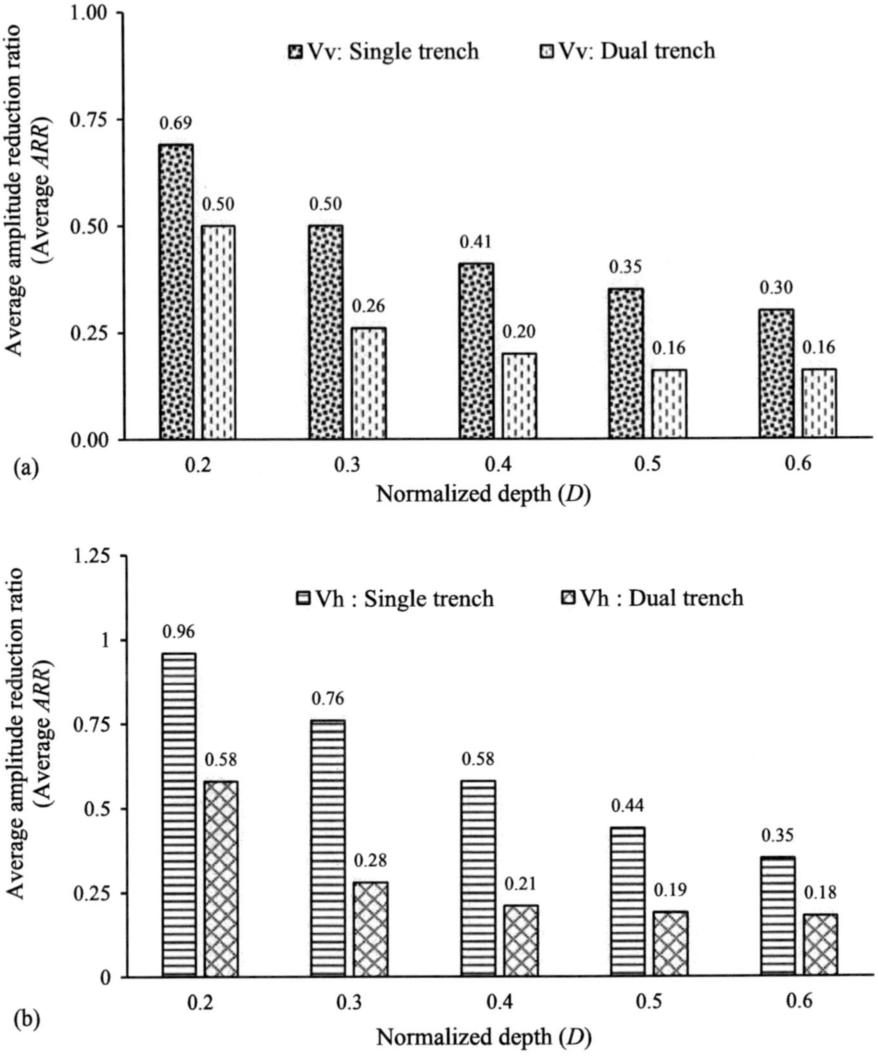

Fig. 29.8 Comparison of average ARR for single and dual trenches: **a** vertical vibrations; **b** horizontal vibrations

and 84%, respectively. Also, the minimum depth required to reduce the vertical and horizontal vibrations by 75% was reported as 0.4 L_R when the trenches were 1 L_R apart. The dual open trenches effectively isolated both the vertical and horizontal vibrations. The average ARR for vertical and horizontal vibrations corresponding to depth 0.4 L_R was 0.20 and 0.21, respectively. It implies that the vertical and horizontal amplitudes of vibrations can be reduced by 80% and 79%, respectively.

It was observed from the study that the depth required for the dual trenches was much smaller than the single trenches to achieve the same degree of vibration screening. The findings of this study support the idea of using dual trenches in place of single trenches under challenging conditions.

References

1. Sanchez-Sesma F, Weaver R, Kawase H, Matsushima S, Luzon F, Campillo M (2011) Energy partitions among elastic waves for dynamic surface loads in a semi-infinite solid. Bull Seismol Soc Am 101(4):1704–1709
2. Woods RD (1968) Screening of surface wave in soils. J Soil Mech Found Div 94(4):951–979
3. Richart FE, Hall JR, Woods RD (1970) Vibrations of soils and foundations. Prentice-Hall
4. Beskos DE, Dasgupta B, Vardoulakis IG (1986) Vibration isolation using open or filled trenches. Comput Mech 1(1):43–63
5. Ahmad S, Al-Hussaini TM (1991) Simplified design for vibration screening by open and infilled trenches. J Geotech Eng 117(1):67–88
6. Ahmad S, Al-Hussaini T, Fishman KL (1996) Investigation on active isolation of machine foundations by open trenches. J Geotech Eng 122(6):454–461
7. Di Mino G, Giunta M, Di Liberto CM (2009) Assessing the open trenches in screening railway ground-borne vibrations by means of artificial neural network. Adv Acoust Vib 2009:1–12
8. Alzawi A, Hesham El Naggar M (2011) Full scale experimental study on vibration scattering using open and in-filled (GeoFoam) wave barriers. Soil Dyn Earthquake Eng 31(3):306–317
9. Bo Q, Limam A, Djeran-Maigre I (2014) Numerical study of wave barrier and its optimization design. Finite Elem Anal Des 84:1–13
10. Saikia A, Das UK (2014) Analysis and design of open trench barriers in screening steady-state surface vibrations. Earthq Eng Eng Vib 13(3):545–554
11. Ulgen D, Toygar O (2015) Screening effectiveness of open and in-filled wave barriers: a full-scale experimental study. Constr Build Mater 86:12–20
12. Majumder M, Ghosh P (2020) Screening of train-induced vibration with open trench—a numerical study. In: Prashant A, Sachan A, Desai CS (eds) Advances in computer methods and geomechanics. Springer, Singapore, pp 453–466
13. Segol G, Abel JF, Lee PCY (1978) Amplitude reduction of surface waves by trenches. J Eng Mech Div 104(3):621–641
14. Duvall WI, Fogelson DE (1962) Review of criteria for estimating damage to residences from blasting vibrations. United States Department of the Interior, Bureau of Mines, Washington, DC
15. Yang Y-B, Hung H-H (1997) A parametric study of wave barriers for reduction of train-induced vibrations. Int J Numer Meth Eng 40(20):3729–3747
16. Naggar MHE, Chehab AG (2005) Vibration barriers for shock-producing equipment. Can Geotech J 42(1):297–306
17. Kattis SE, Polyzos D, Beskos DE (1999) Vibration isolation by a row of piles using a 3-D frequency domain BEM. Int J Numer Meth Eng 46(5):713–728

Chapter 30
Nonlinear Analysis of Soil Pile Interaction Under Seismic Loads

K. Siri Reddy, G. V. S. K. Harini, and Thanikella Vijayasri

Introduction

Piles are long slender columns, adopted as a deep foundation to bear massive superstructures like towers, multi-storeyed buildings, dams, bridges, etc., when the founding soil is not strong enough, which results in settlement and bearing capacity problems. Apart from carrying the vertical compressive loads, piles are designed to resist the dynamic lateral loads which are common in the offshore structures, retaining walls, and the structures in the earthquake prone regions. A typical dynamic load on a soil-pile system is earthquake vibrations, and analysis of performance of the pile foundation is important in the seismic assessment and design of existing and new structures. During earthquakes, the pore water pressures developed in clay soils are not dissipated quickly. Sensitivity of clays causes liquefaction. When the soil adjacent to pile liquefies, it loses its strength and stiffness, and the piles behave as unsupported long slender columns and simply buckle when subjected to the axial loads. Hence, the evaluation of the performance of a structure supported on pile foundations is of greater importance in the seismic design and assessment of new and existing structures.

In the early days, Timber Piles were driven into the ground by hand or holes were dug and filled with sand and stones. Steel piles have been used since 1800 and concrete piles since about 1900. Various case studies and literature work developed over the past three decades have thrown sufficient light over many aspects of pile foundations, soil-pile interactions, dynamic characteristics of pile foundations, influence of many critical parameters on their performance. These studies have clearly shown the suitability of different types of pile foundations, considering safety and economic aspects, over a wide range of soil conditions. A paper on Soil-Pile-Structure

K. Siri Reddy · G. V. S. K. Harini · T. Vijayasri (✉)
MVSR Engineering College, Nadergul, Hyderabad, Telangana 500097, India
e-mail: Vijayasri.Thanikella@jacobs.com

K. Muthukkumaran et al. (eds.), *Soil Dynamics, Earthquake and Computational Geotechnical Engineering*, Lecture Notes in Civil Engineering 300,
https://doi.org/10.1007/978-981-19-6998-0_30

Interaction of Offshore Structures during an Earthquake was studied by Kagawa [1]. Boulanger and Curras studied the Seismic Soil-Pile-Structure Interaction Experiments and Analyses [2] in 1999. Lok et al., analysed Seismic Soil-pile-superstructure Interaction using Numerical Modelling [3]. Bhattacharya and Bolton, worked in the area of Buckling of Pile During Earthquake Liquefaction [4].

Based on the above literature studies, the present study focuses on a robust nonlinear FEM soil-pile interaction code that has been developed by OpenSees considering a single uniform cross-section pile and understanding its efficiency during seismic events in relevance to the region's seismicity. Response analysis of soil-pile interaction by using forced vibration theory is performed. Proper emphasis is given to the dynamic characterization of the soil structure interaction by evaluating natural period, displacement and acceleration amplification profiles, pore water pressure, and the potential of liquefaction.

Problem Definition

The idea behind conducting this study is to understand the efficiency of pile foundations around the areas with high water tables (areas neighbouring the rivers or offshores) in presence of dynamic loads. Hence, we choose to work in the Musi River area, located in Hyderabad, which falls under seismic zone II. The soil profile near Musi River is found to be layered, containing gravels on the surface, silts, sensitive clays, and soft clays underlying the dense sands. The water table is known to be located at 2 m depth from the ground surface, indicating the low suitability of the site for construction of heavy structures. The bedrock layer is found at the depth of 30 m, so it was decided to use the combined piles (constructed to gain the bearing capacity from the interfacial skin and end bearing of the pile-soil system) rather than the end bearing piles. The pictorial representation of the soil profile is shown in Fig. 30.1.

The two types of problems chosen in our numerical study are the fixed boundary and viscous boundary value problems. The three-dimensional soil and soil-pile domains have been modelled using 3D FEM using OpenSees software. The soil domain model of size 20 m $\times$ 20 m $\times$ 30 m. The base of the soil profile is assumed to be on a hard stratum. The element of pile size is considered as 1 m $\times$ 1 m $\times$ 30 m. (Taken from an approximate equation using grid spacing equals to it is the ratio of minimum shear wave velocity to the maximum frequency). The properties of sensitive clays have been determined, and the results are mentioned below in the table.

Numerical Modelling

The modelling characteristics of the Finite Element Method are described in this section. For single pile configurations, OpenSees software is used to describe the soil

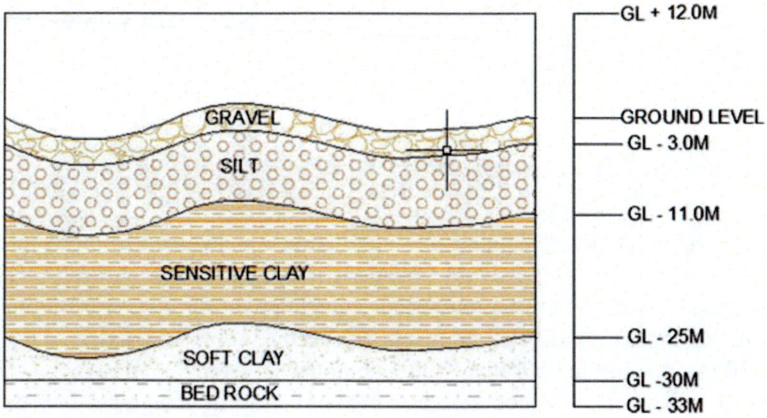

Fig. 30.1 Soil profile at Musi River area

and pile elements, the elements at their interface, the type of boundary conditions, and to perform the dynamic analysis. The present work focuses on adopting the 3D numerical method or the direct method to model the detailed three-dimensional soil-pile systems. Numerical studies are the most useful techniques in solving engineering problems in the field of design and research in geotechnical engineering as a result of rapid advancements in computers. The relationship between stresses and strains for the materials incorporated in the analysis is what is important to deal with the SSI problems. The advantage of the numerical study or the analysis of soil problems is that the behaviour of soil, their preexisting stress states, mechanism involved in their failure, the creep and plastic deformations can be understood well. Constitutive models represent the stress–strain relationships through the use of a collection of mathematical equations that simulate the behaviour of the soil. The level of the study can be extended by incorporating hydraulic processes into the 3D soil system. The Finite Element Method is used in this research. The Finite Element Method is used to describe the geometrical structure and elemental connections, as well as properties of constituent material properties, in a realistic manner. The dynamic characterization of soil-pile interaction is carried out under influence of harmonic and earthquake vibrations using 3D FEM. The study of this system is hindered by the complications involved in solving governing equations of motion of the structures, which includes foundation interaction. The equation showing motion of the complete system is as follows,

$$[M_G]\{\Delta\ddot{u}\}_{nG} + [C_G]\{\Delta\dot{u}\}_{nG} + [K_G]\{\Delta u\}_{nG} = \{\Delta F\}_G$$

where M, C, and K indicate the mass matrix, damping matrix, and stiffness matrix of the structure. Δu, $\Delta\dot{u}$, and $\Delta\ddot{u}$ denote the displacements, velocities, and accelerations vector, and ΔF is load vector. In order to take the nonlinear behaviour into account, it is essential to deal with the incremental equations rather than with the original equations.

Finite Element Mesh Generation

A Finite Element Model of the soil-pile interaction was evolved in the OpenSees. OpenSees software has elements and material models that can be used to model a soil structure interaction situation successfully. For meshing the soil domain in the OpenSees model, bbar brick elements are used to represent a size of 20 m × 20 m × 30 m, and Nonlinear Beam Column elements are employed to represent pile columns line elements. The concrete pile is assumed to be integrated into the soil domain exactly at the centre. In total, 6000 elements in soil, each element sized 2 m × 2 m × 0.5 m and 60 elements in the pile were created. There are 7381 nodes in soil mesh for which pore water fixities are defined in accordance with the location of the water table situated near Musi River area. The nodes of these elements have 3 translational DOF. The size of each element has been established such that, it is constricted to 1/8th to 1/10th of the shortest wavelength in the soil medium. To simulate the behaviour of cohesive material in an undrained condition, Fluid Solid Porous Material is used. The Fluid Solid Porous Material enables us to couple the responses of solid and fluid phase of the saturated soil mass. In the present study, the solid phase is modelled using a pressure independent-multi-yield model, whereas for silt and gravel which 4.75 mm size is modelled using pressure-dependent multi-yield material.

Soil-Pile Connectivity

The EqualDOF command is used to enable good connection at the soil-pile interface, making pile nodes as the master nodes and soil nodes as the slave nodes in order to simulate the stress transfer mechanism from the piles to surrounding soil as shown in Fig. 30.2. The connection between the pile and the soil was modelled using EqualDOF to simulate perfect bonding. The above procedure has been followed using fixed boundary conditions as well as viscous boundary conditions. To simulate the concrete behaviour of pile, concrete material has been used in this study. Elastic section of the pile has been considered. Experiments are conducted to determine the physical and engineering properties of soil and as well as concrete material are incorporated into FEM code. The schematic diagrams of soil and soil-pile domains are depicted in Figs. 30.3 and 30.4 using different boundary conditions.

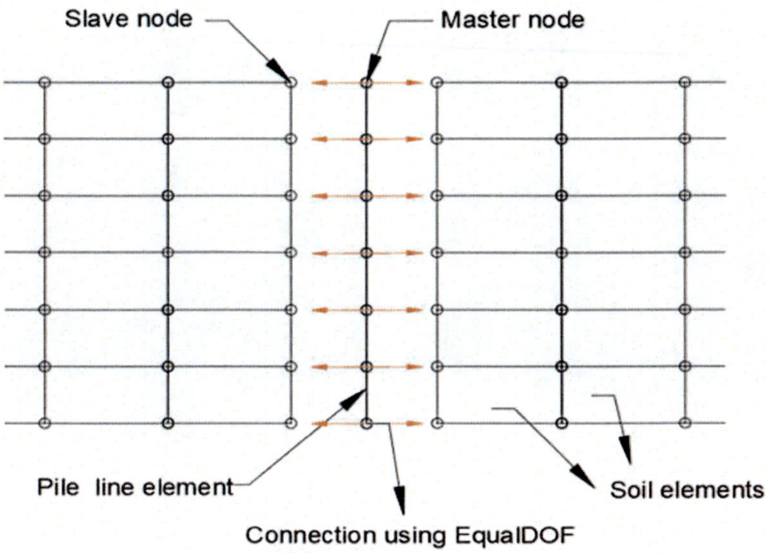

Fig. 30.2 Schematic diagram of soil-pile connectivity

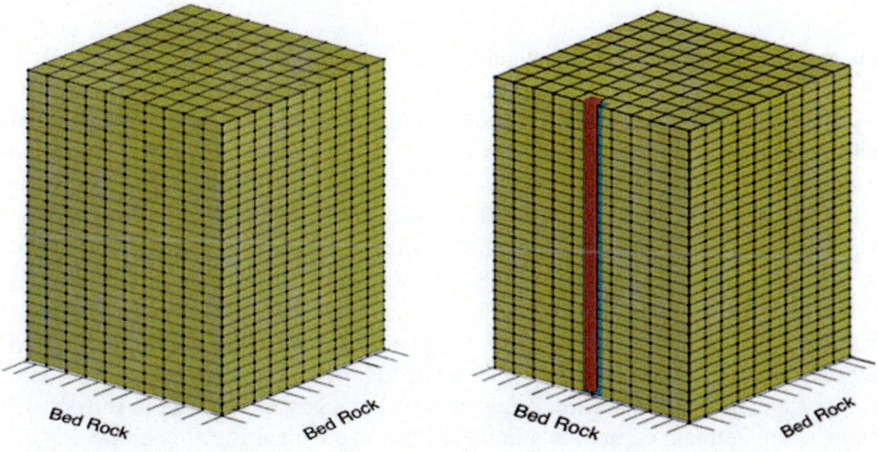

Fig. 30.3 Schematic diagram of mesh model of **a** soil domain and **b** soil-pile domain for fixed boundary condition

Boundary Conditions

The dynamic analysis of the soil-pile domain was performed for two different boundary conditions, namely the fixed boundary and the viscous boundary conditions. The first one being the 'fixed boundary conditions', where the soil domain is assumed to be laid on the hard strata/rock bed, i.e., all the nodes at the base are

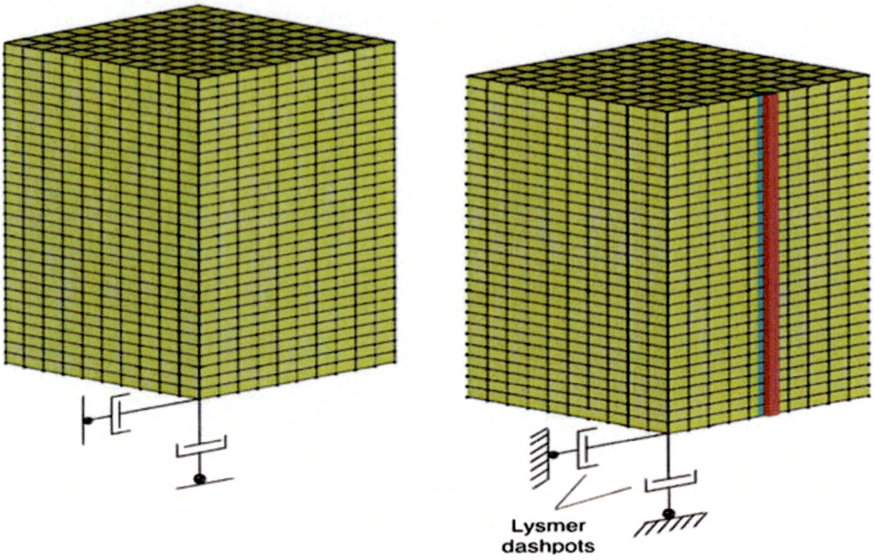

Fig. 30.4 Schematic diagram of mesh model of **a** soil domain and **b** soil-pile domain for viscous boundary condition

constrained from undergoing translations in X, Y, and Z directions as in Fig. 30.3. When an earthquake originates from the soil domain considered, the seismic waves propagate in all directions to touch all the boundaries of the domain and get reflected back into the same medium, which is not in a good agreement with real ground scenarios.

The second boundary condition is the 'viscous boundary conditions', where the base of the domain is connected with infinitesimal lysmer dashpots oriented in normal and tangential directions to generate the real ground wave transmission conditions as shown in Fig. 30.4. Dashpots are the simple mechanical devices used to create friction and dissipation of energy for propagation of the waves, as the result of which they undergo attenuation as they travel. Reflection of the seismic waves back into the domain is barred. Therefore, using viscous boundary conditions, to restrict the infinite computational domain to a finite domain one with artificial boundaries.

For modelling the viscous boundary conditions, a set of 3 dashpot nodes were created at the location of original bottom soil nodes. Proper connection is established between soil and dashpot nodes using equalDOF command. Zero Length element command is used to connect the dashpot nodes, and then viscous material is assigned to these elements. The energy loss/wave attenuation within the soil because of microstructural mechanisms such interparticle sliding, friction, pore fluid viscosity, and structure rearrangement is achieved by the viscous damping material.

On considering a soil system composed of a layer of gravel (4.75 mm sieve) underlain by a layer of silt and clays, an advanced nonlinear robust FET model using multi-yield surfaces of layered soils is developed. The soil is discretized in a finite domain, whose nature is assumed to be nonlinear. Until the defined yield stress is reached, the behaviour of the pile members is regarded as linearly elastic. In the OpenSees code, nodes situated above the water table are constrained from experiencing the pore water pressure. The time required to dissipate the excess pore pressure is dependent on the soil profile and its characteristics, which may take from minutes to hours. The seismic analysis of soil has to be carried out in undrained conditions as the earthquake duration is very less when compared with the time taken for dissipation of excess pore water pressure.

Results and Discussions

Some experiments have been performed in the laboratory in order to evaluate the physical and engineering properties of soil found near the area of work, the results of which were incorporated into the FEM code to simulate the 3D soil-pile system efficiently. Having performed this study, the results of experimental work and numerical study are presented in the following subsections.

Laboratory Test Results

Experimental work has been performed to determine the physical and engineering properties of soil and cement, strength of the concrete, and to evaluate the effect of earthquake loads on performance of piles. The physical and engineering properties of the red soil are determined by conducting experiments such as specific gravity as per IS 2386 Part-3 (1963), Standard Proctor Test as per IS-2720-Part-7-1980, Atterberg's Tests as per IS: 2720 (part-5) 1985, Hydrometer Analysis as per IS: 2720 (Part 4)-1985 in the laboratory according. The test results from each experiment are listed in Table 30.1.

Table 30.1 Properties of red soil

Test	Result
Specific gravity	2.66
Standard proctor test	The maximum dry density 1.6215 gm/cc, achieved at 20% OMC
Liquid limit	26%
Plastic limit	11%
Gravel percentage (>4.75 mm)	Nil
Fine sand (0.075–0.425 mm)	25.9%
Silt percentage (0.002–0.075)	18.85%
Clay percentage (<0.002 mm)	35.49%

Table 30.2 Cement parameters

Property	Result
Cement properties	
Specific gravity of cement	2.2
Normal consistency of cement	33%
Initial setting time of cement	32 min
Fineness of cement	2%
Concrete properties	
Slump test	0 cm
Compressive strength	34 MPa

The specific gravity of red soil obtained is 2.66, which indicates soil is porous and lies within the range of 2.68–2.80 confirming with the Indian standards. According to IS: 2720 (part-5) 1985, for low compressible soils, the liquid limit < 35%. The liquid limit obtained for the red soil is 26%, so this red soil is recognized as low compressible soil. From the gradation curve of red soil, it is observed that the soil sample predominantly consists of clay particles (35.49%).

Experiments were conducted on cement to determine the specific gravity as per IS 2720-Part 3, Normal consistency as per IS: 269-1989 and IS: 4031-1988 (part 4), initial setting time as per IS: 269-1989 and IS: 4031-1988 (part 5), and fineness of the cement as per IS: 269-1989 and IS: 4031-1999 (part 2) using pozzolana Portland cement of grade 53 in the laboratory. The result of each experiment is mentioned in Table 30.2 (Fig. 30.5).

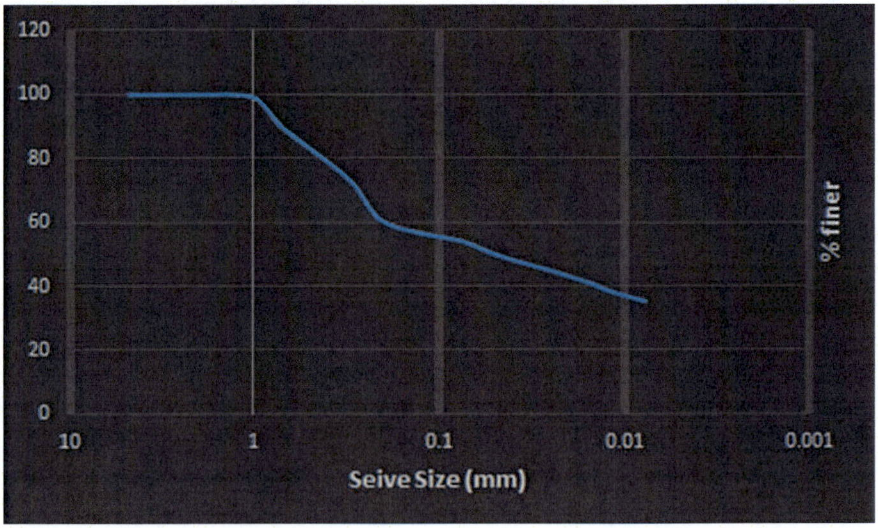

Fig. 30.5 Gradation curve of red soil

Table 30.3 Concrete mix design (IS 10262-2019)

Materials	Quantity
Cement	480 kg
Water	192 lit
Fine aggregate	560 kg
Coarse aggregate (20 mm)	542 kg
Coarse aggregate (10 mm)	542 g

The cement is used as pozzolana Portland cement. The normal consistency of cement must be between 25 and 35% as per Indian standards, and the obtained value is 33% which is in good agreement with Indian standards. The obtained result of initial setting time of the cement was 32 min (should not be less than 30 min is in agreement with IS: 10262 (2009)). The obtained fineness value of cement is 2% which is not exceeding 10% by weight as per IS recommendation.

Concrete (grade 35) mix is designed according to IS: 10262-2019. For the concrete mix, pozzolana Portland cement of 53 grade, water, sand as fine aggregate, and coarse aggregate of sizes 10 and 20 mm are being used for nominal mix design. The mix was prepared in the laboratory using a mixer. The quantity of each material used is listed in Table 30.3.

The compression test on the concrete cylinder of dimensions 15 cm × 30 cm is performed at a pace rate of 1.1 KN/s in the laboratory according to the ASTM C39 test method using a digital testing machine. Slump test is performed to determine the workability of the concrete after preparing the mix in the laboratory as per Code 1199-1959 using a slump cone. Figure 30.6 shows the stress (MPa) versus strain (%) profile of concrete cylinder which is obtained from the compression test for 28 day compressive strength. The compressive strength achieved by the concrete cylinder is observed to be 40 MPa with a strain of 0.025%.

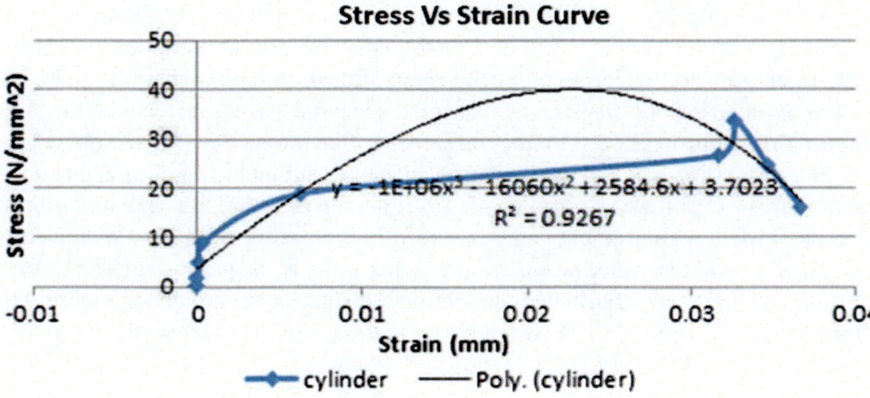

Fig. 30.6 Stress versus strain for concrete cylinder

Numerical Study Results

Dynamic characterization of soil-pile interaction was done by performing seismic response analysis or time history analysis. The following presents the results from seismic analysis.

Seismic Analysis

A typical earthquake motion (Chamba) from local earthquakes was chosen for performing seismic analysis or dynamic time history analysis. The ground motion data was extracted from COSMOS website. As per IS 1893:2002 code, the soil profile was chosen in the seismic zone 2, which has a seismic intensity of 0.1 g, where g represents gravitational acceleration. The amplitude of Chamba earthquake motion was scaled down to obtain a seismic intensity of 0.1 g by using Newmark time integration scheme.

Effect of Pile on Soil Domain

The results of the 3D Finite Element Analysis include the variation of displacements, accelerations, and pore water pressures along the depth of the soil domain. These results are compared with those of soil-pile domains to highlight the significant alterations in the results occurring on the inclusion of concrete piles as shown in the figures above. The graphs have been plotted considering both the fixed boundary condition and the viscous boundary condition.

Displacement Results

On the application of the static and dynamic loading, the soil nodes undergo vertical and horizontal displacements. The comparison graphs have been plotted for the soil domain before and after inclusion of the concrete pile to show the trends in displacements at varying depths along the central line of the soil domain (Fig. 30.7).

It can be noted that the displacement of soil is greater when compared with the pile system. This can be explained by the difference of the interactions of pile elements with the surrounding soil and the increased stiffness with inclusion of pile at centre of soil domain. In soil and soil-pile systems, it is indicated that the peak horizontal and peak vertical displacements are significantly increasing from base of the soil domain to the ground surface because there is flexibility at the ground surface where the strains are more. The horizontal and vertical displacement are understood to be reduced by 25.55% and 80.24%, respectively at the depth of 7 m from the base.

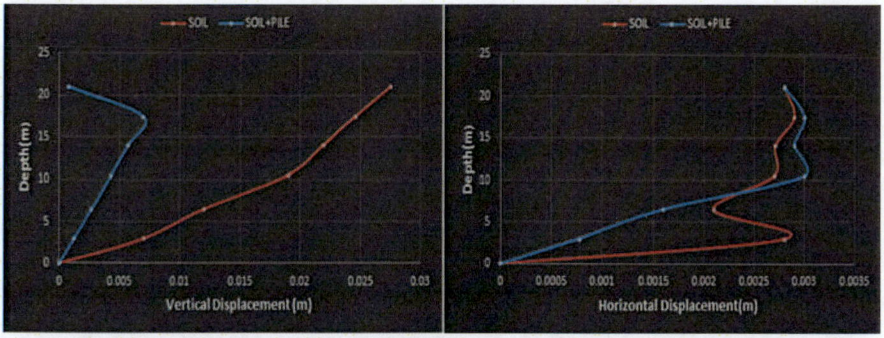

Fig. 30.7 Effect of pile on variation of **a** vertical displacement profile and **b** horizontal displacement profile along the depth using fixed boundary condition

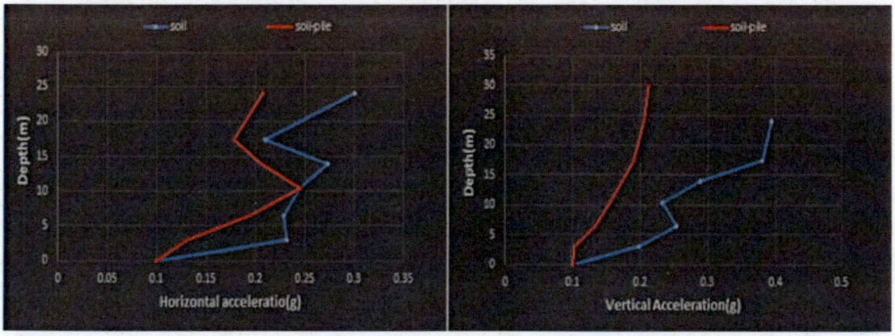

Fig. 30.8 Effect of pile on variation of **a** vertical acceleration profile and **b** horizontal acceleration profile along the depth using fixed boundary condition

Acceleration Results

Comparison of variation of the vertical and horizontal accelerations with the depth for soil and soil + pile domain are presented in Fig. 30.8. From the figure, it is observed that accelerations amplified drastically from the base (bedrock) to the ground surface, as the soil depth increased from 0 m (base) to 30 m. The horizontal and vertical accelerations are known to have diminished by 30% and 89.5%, respectively, at the ground surface, whereas at 3 m depth, reduction of 42.5 and 45% has been observed from the base, with the inclusion of concrete piles.

Figure 30.9 shows the acceleration time history profile at three depths of 10, 17, and 30 m along the central line of the soil domain. It is clearly noted that the acceleration response increases more significantly from the base of the soil domain to the ground surface (30 m). This can be explained by more flexibility of soil particles at ground level, resulting in greater displacements and accelerations, whereas the bottom nodes are fixed to hard strata. The peak bedrock acceleration of 0.1 g rises

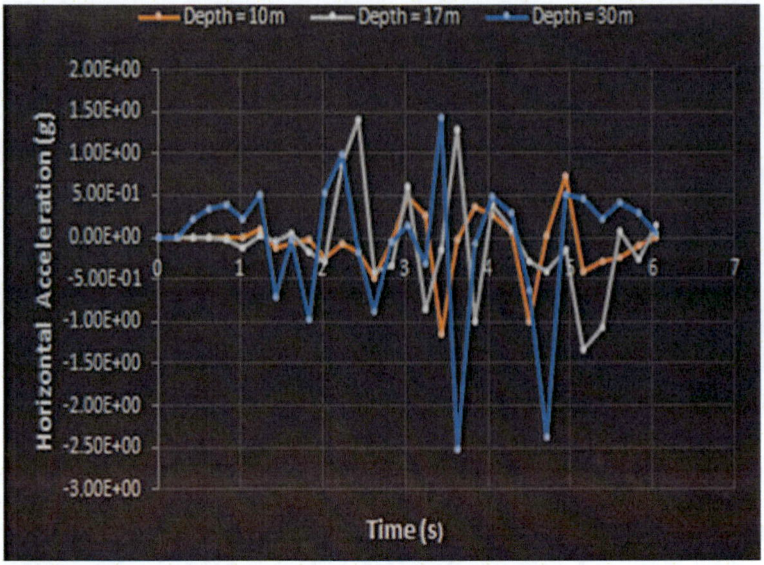

Fig. 30.9 Time history profile of horizontal accelerations of soil domain at various depths

to 0.244 g, 0.209 g, and 1.49 g at the depths of 10 m, 17 m, and 30 m, respectively, indicating an amplification factor of 2.44, 2.09, and 14.9.

Pore Water Pressure

Figure 30.10 explains that the pore water pressure increases from the ground surface to the base of the domain. The peak values of 56.7 and 34.2 kPa were observed in soil and soil + pile domain at around 25 m from base.

Effect of Boundary Conditions

In this part of the study, the dynamic analysis of the soil domain was performed for two different boundary conditions, namely the fixed boundary and the viscous boundary conditions. The results, i.e., displacement amplitudes versus depth and acceleration amplitudes versus depth were presented graphically for both the conditions as shown in the figures below.

Displacement Results

It is clearly evident from the comparison graphs (Fig. 30.11) that the displacements in the soil domain with viscous boundary conditions are relatively lower than that of

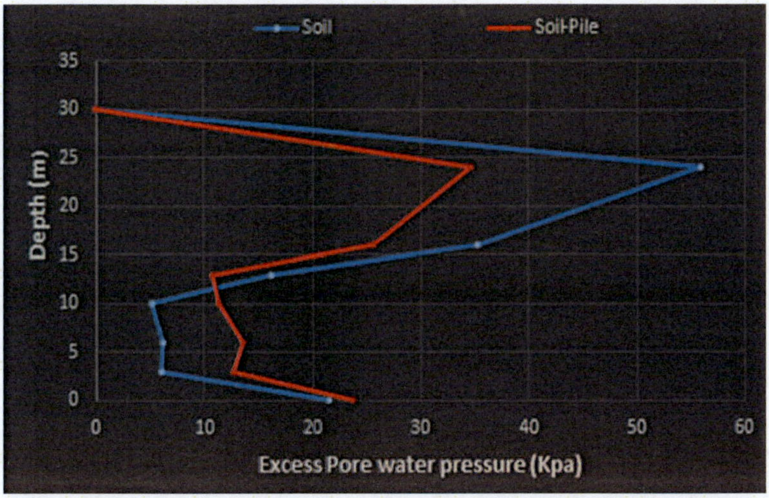

Fig. 30.10 Variation of pore water pressure with depth for fixed condition (soil vs soil-pile)

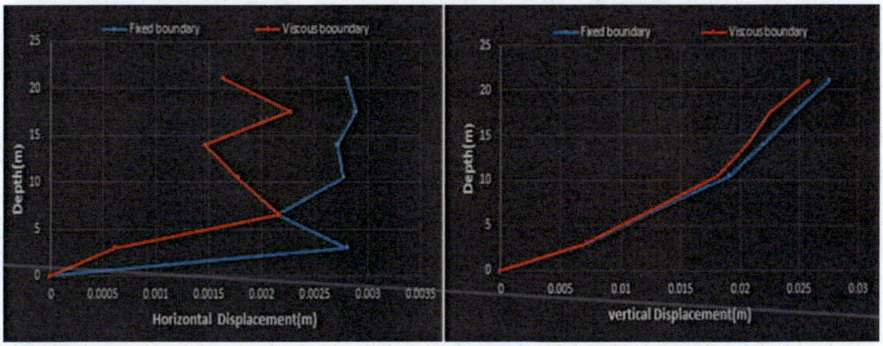

Fig. 30.11 Effect of viscous dashpots on **a** horizontal displacement profile and **b** vertical displacement profile of soil domain

soil domain with fixed boundary. The horizontal displacement amplitude and vertical displacement amplitude at the ground surface reduced by nearly 41.58% and 9.2%, respectively. From the results, it can be concluded that viscous boundary condition replicates the actual field situation gives accurate results, whereas fixed boundary condition is not acceptable for the unbounded domains and gives most inaccurate results.

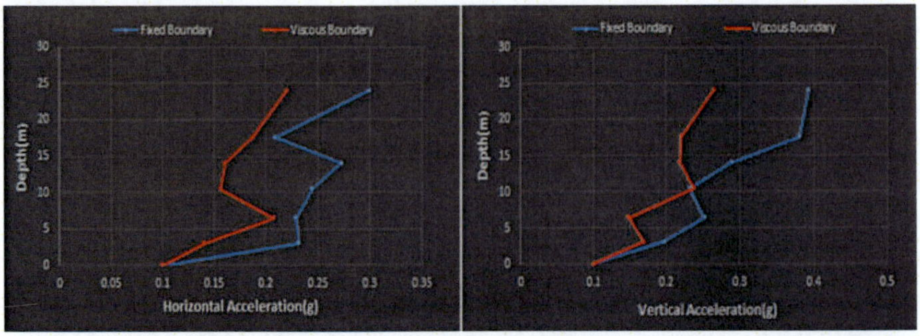

Fig. 30.12 Effect of viscous dashpots on **a** horizontal acceleration profile and **b** vertical acceleration profile along the depth of soil domain

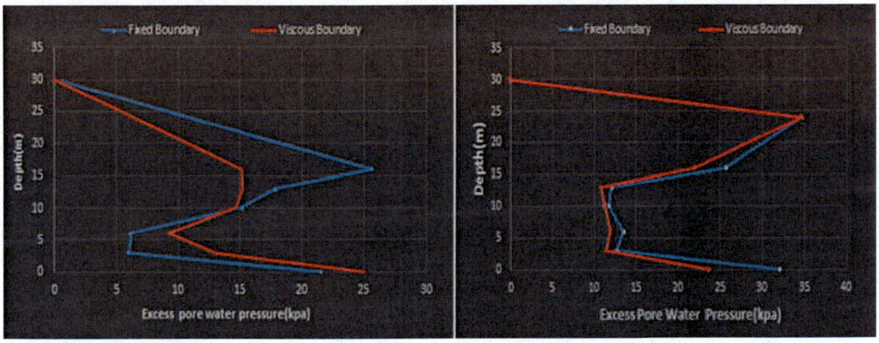

Fig. 30.13 Variation of pore water pressure in **a** soil domain and **b** soil-pile domain in both fixed and viscous boundary conditions

Acceleration Results

As explained previously, the dashpots help in attenuating the seismic waves in all the directions, thereby replicating the real ground conditions. Hence, lesser accelerations are experienced by soil particles in viscous boundary conditions. The reduction of 26.67 and 33.58% is noticed in the horizontal acceleration and vertical acceleration at 25 m from the base, with the introduction of dashpots to the soil domain (Fig. 30.12).

Pore Water Pressure Results

Pore water pressures obtained in case of viscous boundary conditions are relatively lesser than those obtained in fixed boundary cases. This can be explained by the fact that there is attenuation of seismic waves in viscous cases. The pore water pressures are noted to be increasing from ground surface to the base of the domain (Fig. 30.13).

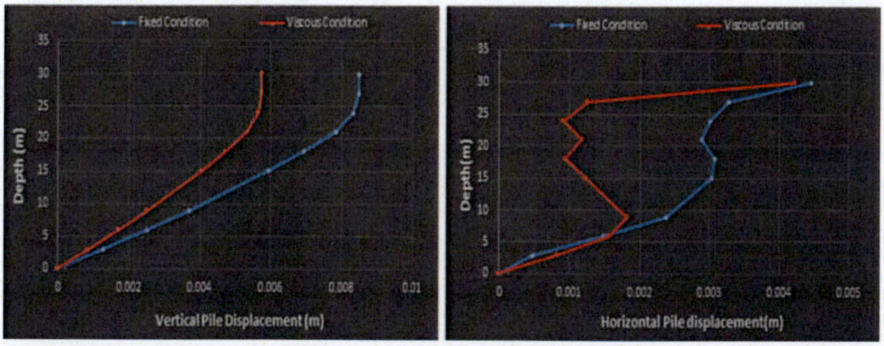

Fig. 30.14 Effect of viscous dashpots on **a** horizontal pile displacement profile and **b** vertical displacement profile along the length of pile

On calculating the pore pressure ratio Ru (where Ru = excess pore water pressure/initial effective pressure) values at various depths along the central line of the domain, it is found that these values lie between 0.11 and 0.91 in soil domain, 0.075 and 0.68 in soil + pile domain for fixed case, whereas for viscous case, these ratios lie between 0.089 and 0.467 in soil domain, 0.064 and 0.433 in soil + pile domain, indicating that there is no susceptibility of liquefaction occurring anywhere in the domain.

Pile Displacements

The pile is experiencing peak horizontal and vertical displacements at the ground surface, i.e., at 30 m owing to the reason that enough compaction of soil is not achieved near the ground surface. Figure 30.14a, b show that lesser displacements occur in case of viscous boundary conditions. There is a reduction of 5.26 and 32.22% in horizontal and vertical displacements of pile at ground surface, respectively, in case of viscous boundary condition.

Shear Force and Bending Moment Results in the Pile

Figures 30.15a, b show calculated peak shear forces and bending moment profiles for various depths along the span of the pile inserted at the centre of the soil domain under input acceleration of 0.1 g. It was identified that the both shear force and bending-moment profiles showed peak values at around 25 m from the base, when chamba earthquake was applied at input excitations. Figure 30.15b shows the bending moment profile of pile. As the seismic intensity rises, gradual increase in the bending moment is noticed. The maximum bending moment is observed at the boundary of the coarse sand (Gravel) and the silt layers. The bending moment is also seen to be escalating from the bedrock to the soft clay layer. It is clearly shown that the bedrock

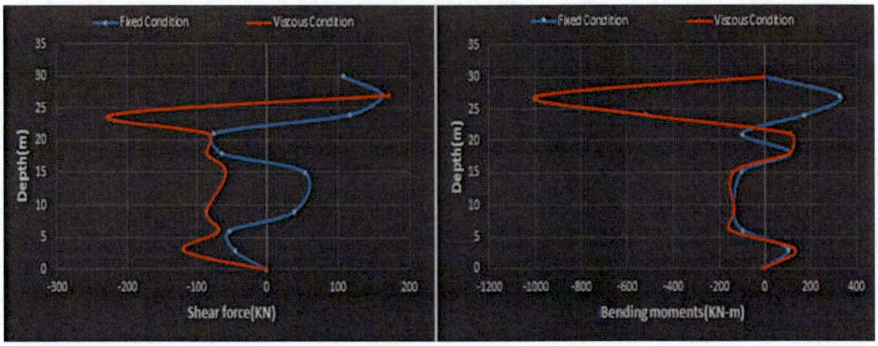

Fig. 30.15 Variation of **a** shear forces and **b** bending moments occurring in the pile for fixed and viscous boundary conditions

has a strong embedded effect on the pile. From the results obtained, the seismic design of the pile foundation in the soil domain is proven to be appropriate for the seismic fortification intensity of seismic zone II to V.

Conclusion

The paper presents the results of 3D nonlinear finite element analysis of soil-pile interaction subjected to earthquake vibrations. Assessment and performance of 3D nonlinear dynamic analysis of soil-pile interaction using different absorbing boundary conditions based on advanced multi-surface plasticity models has been studied. To verify the applicability of the viscous and fixed boundary dynamic analyses was performed on soil-pile interaction, both boundaries were subjected to a challenging numerical test and they both performed very well. A notable amplification in acceleration response was observed between the base and surface of the soil domain. Also, from the displacement profile, it is found that viscous boundary condition gives accurate results, whereas for fixed boundary condition gives most inaccurate results due to the domain being suddenly fixed at the end so energy waves will be reflected back at boundary. It may conclude that viscous boundary condition replicates the actual field condition. With the inclusion of concrete piles in the soil, the vertical displacement and acceleration amplitudes have reduced by 97% and 82.38%, respectively. The acceleration amplitudes have depreciated 60.16%, respectively, at the ground surface. The excess pore water pressures are noted to be increasing from the ground surface to the bottom of the soil domain. The shear force and bending moment profiles of the pile show that peak values are obtained at around 25 m from the base.

References

1. Kagawa T (1980) Soil-pile-structure interaction of offshore structures during an earthquake. In: 12th annual OTe in Houston, Tex., May S.a, 1980. OCT 3280
2. Boulanger RW, Curras CJ, Kutter BL, Wilson DW, Abghari A (1999) Seismic soil-pile-structure interaction experiments and analyses. J Geotech Geoenviron Eng 1999(125):750–759
3. Lok TMH, Pestana JM, Seed RB (2000) Numerical modeling of seismic soil-pile-superstructure interaction. University of California, Berkeley, p 0914
4. Tabeshi A, Poulos HG (2000) A simple method for the seismic analysis of piles and its comparison with the results of centrifuge tests. In: 12 WEEE 2000
5. Cai YX, Gould PL, Desai CS (2000) Nonlinear analysis of 3D seismic interaction of soil–pile–structure systems and application. Eng Struct 22:191–199
6. Bhattacharya S, Bolton M (2004) Buckling of pile during earthquake liquefaction. In: 13th world conference on earthquake engineering, Vancouver, B.C., Canada, 1–6 August 2004, Paper No. 95
7. Chu D, Truman KZ (2004) Effects of pile foundation configuration in seismic soil-pile-structure interaction. In: 13th world conference on earthquake engineering, Vancouver, B.C., Canada, 1–6 August 2004, Paper No. 95
8. Pitilakis K, Gazepis C, Anastasiadis A (2004) Design response spectra and soil classification for seismic code provision. In: 13th world conference on earthquake engineering, Vancouver, B.C., Canada, 1–6 August 2004, Paper No. 2904
9. Wotherspoon LM (2006) Three dimensional pile finite element modelling using OpenSees. In: 2006 NZSEE Conference Paper Number 26
10. Xu R, Fatahi B (2007) Effects of pile group configuration on the seismic response of buildings considering soil-pile-structure interaction. In: 2018 proceedings of GeoShanghai 2018 international conference: advances in soil dynamics and foundation engineering, pp 279–287
11. Deendayal, Muthukkumaran K, Sitharam TG (2019) Experimental investigation on behaviour of a laterally loaded single pile located on sloping ground. Int J Geotech Eng
12. Rathod D, Muthukkumaran K, Sitharam TG (2020) Analysis of laterally loaded group of piles located on sloping ground. Int J Geomech
13. Rathod D, Krishnanunni KT, Nigitha D (2020) A review on conventional and innovative pile system for offshore wind turbines. In: Geotechnical and geological engineering. Springer
14. Rathod D, Nigitha D, Krishnanunni KT (2021) Experimental investigation on the behaviour of monopile under asymmetric two-way cyclic lateral loads. Int J Geomech. ASCE
15. IS:2720 (part 5) 1985: Determination of liquid and plastic limit
16. IS-2720-PART-7-1980: Determination of water content-dry density relation using heavy compaction
17. IS: 2386 (Part-3): 1963, Methods of test for aggregates for concrete (Determination of Specific Gravity of aggregates)
18. IS: 2720 (Part 4)-1985 Methods of tests for soils
19. IS: 272O (Part17)-1986 Methods of tests for soils
20. IS: 2386 (Part III)-1963 Specific gravity, density, voids, absorption, and bulking
21. IS: 2720 (Part 15)-1986 Methods of tests for soils
22. IS: 4031 (Part 4)-1988 Determination of consistency of standard cement paste.
23. IS: 4031 (Part 5)-1988 Methods of physical tests for hydraulic cement
24. IS: 10262-2019 Concrete mix proportioning—guidelines (Second Revision)

Chapter 31
Transient Response of Shallow Foundation Under Eccentric Static and Cyclic Load

Suvendu Kumar Sasmal and Rabi Narayan Behera

Introduction

The analysis and design of shallow foundations are complex tasks due to the complicated behaviour of the underlying soil. The analysis becomes even more complex when the line of load application does not pass through the centreline of the footing. In other words, the presence of load eccentricity largely influences the behaviour of the shallow foundation. Although consideration of load eccentricity makes the analysis complex, it is also true that more practical situation is simulated with the consideration of the load eccentricity. This is attributed to the fact that, in real-time conditions, it is a rare event that the resultant of amount of load applied on the foundation will pass through the centreline of the footing. Due to its applied nature, this problem has attracted researchers from many disciplines.

The study of foundation behaviour under the influence of eccentric static load is well established. The governing equations for eccentrically loaded foundations were given by Meyerhof [13]. Georgiadis and Butterfield [8] investigated the effect of static load eccentricity on the response of footing resting on sand. Bransby [3] studied the failure surface of a footing under eccentric loading. Fedorovskii [5] presented exact solutions for estimating the bearing capacity of foundation under the influence of eccentric load. Patra et al. [15] presented experimental analysis of shallow foundation considering the influence of load eccentricity. Lu et al. [12] presented an elasto-plastic model for numerical analysis of foundation under eccentric load and compared their results with the results obtained from experimental study. Sargazi and Hosseininia

S. K. Sasmal · R. N. Behera (✉)
Department of Civil Engineering, National Institute of Technology Rourkela, Rourkela, Odisha, India
e-mail: rnbehera82@gmail.com

S. K. Sasmal
e-mail: suvendukumarsasmal@gmail.com

K. Muthukkumaran et al. (eds.), *Soil Dynamics, Earthquake and Computational Geotechnical Engineering*, Lecture Notes in Civil Engineering 300,
https://doi.org/10.1007/978-981-19-6998-0_31

[18] presented a numerical analysis for observing the influence of eccentricity of load on the response of ring footing.

Although the influence of load eccentricity on foundation response in static condition is well known, the effect of load eccentricity in cyclic condition is not fully explored yet. This study is an attempt to extend the concept to cyclic analysis domain. In this work, the settlement of a shallow strip footing is observed under the influence of eccentrically applied allowable static load and one cycle of rectangular pulse load. From rigorous numerical analysis, a dataset is generated, and this dataset is analysed with the help of Adaptive Neuro-Fuzzy Interface System (ANFIS). Lastly, sensitivity analysis is carried on the trained ANFIS model to observe the importance of parameters.

Beam on Nonlinear Winkler Foundation (BNWF) Model

A shallow foundation is numerically modelled using BNWF model with the help of Open System for Earthquake Engineering Simulation (OpenSEES). The BNWF model was successfully implemented for the analysis of shallow foundation response problem by Harden et al. [10], Gajan et al. [6]. Figure 31.1 represents the cross-section of the strip footing and the eccentric static and cyclic load (one cycle). The soil structure interface is represented by a set of springs (Fig. 31.2), viz. Qzsimple springs (to observe vertical response), Pysimple springs (to observe passive resistance), and Tzsimple springs (to observe the sliding resistance). The footing (0.5 m × 0.1 m × 0.03 m ($L \times B \times t$)) is divided into 100 parts and using 101 nodes. The nodes are joined using one-dimensional elastic elements. The capacities of the springs are calculated using the expressions of Meyerhof [13]. The stiffnesses of the mechanistic springs are calculated using expressions mentioned in Gazetas [7], [Eqs. (31.1)–(31.6)].

$$K_V = \frac{\mathrm{GL}}{1-\nu}\left[0.73 + 1.54\left(\frac{B}{L}\right)^{0.75}\right] \tag{31.1}$$

$$K_H = \frac{\mathrm{GL}}{2-\nu}\left[2 + 2.5\left(\frac{B}{L}\right)^{0.85}\right] \tag{31.2}$$

$$K_{H'} = \frac{\mathrm{GL}}{2-\nu}\left[2 + 2.5\left(\frac{B}{L}\right)^{0.85}\right] - \frac{\mathrm{GL}}{0.75-\nu}\left[0.1\left(1 - \frac{B}{L}\right)\right] \tag{31.3}$$

where K_V is the surface stiffness (vertical translation), K_H is surface stiffness (horizontal translation towards long side), $K_{H'}$ is surface stiffness (horizontal translation towards short side), G represents the shear modulus of the soil.

For embedded footing, embedment factor, used in Gazetas [7], is multiplied to the corresponding surface stiffness. The embedded factors are calculated using Eqs. (31.4)–(31.6). In this study, the embedment ratio (D_f/B) is taken as 0, 0.5, and 1.

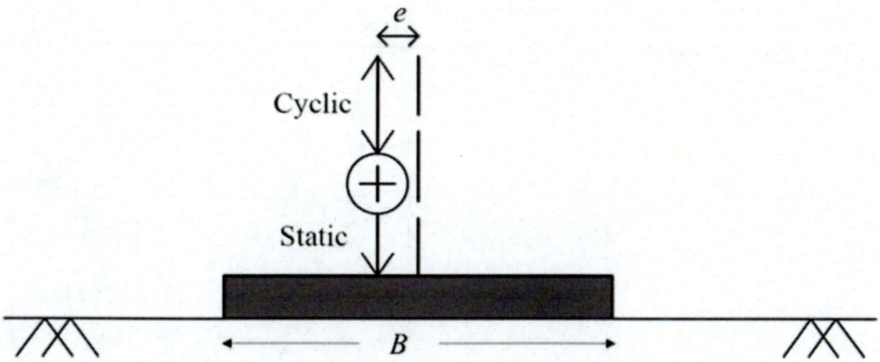

Fig. 31.1 Footing and loading condition

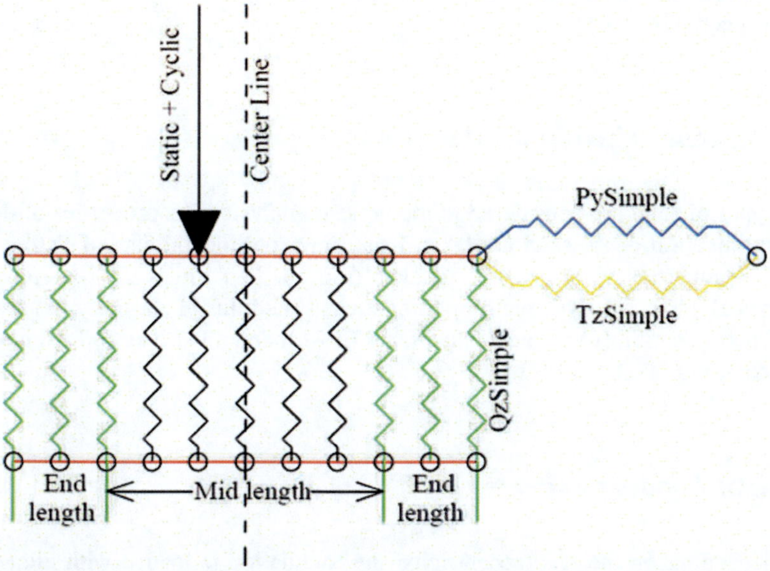

Fig. 31.2 Spring distribution

$$e_{V'} = \left[1 + 0.095\frac{D_f}{B}\left(1 + 1.3\frac{B}{L}\right)\right]\left[1 + 0.2\left(\frac{2L + 2B}{\mathrm{LB}}H\right)^{0.67}\right] \tag{31.4}$$

$$e_H = \left[1 + 0.15\left(\frac{2D_f}{B}\right)^{0.5}\right]\left[1 + 0.52\left(\frac{(D_f - \frac{H}{2})16(L + B)H}{\mathrm{BL}^2}\right)^{0.4}\right] \tag{31.5}$$

$$e_{H'} = \left[1 + 0.15\left(\frac{2D_f}{L}\right)^{0.5}\right]\left[1 + 0.52\left(\frac{(D_f - \frac{H}{2})16(L + B)H}{\mathrm{LB}^2}\right)^{0.4}\right] \tag{31.6}$$

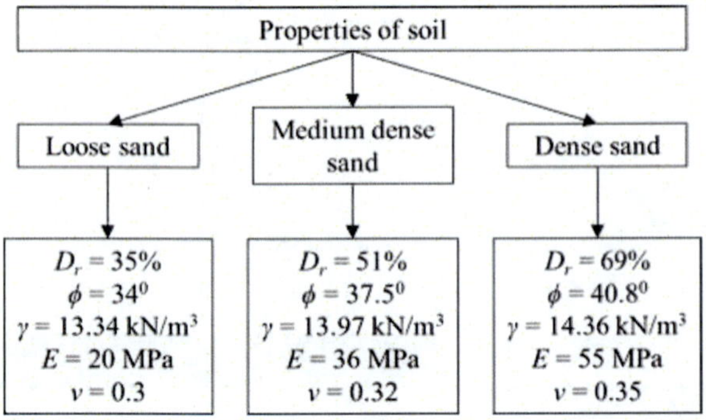

Fig. 31.3 Soil properties used for numerical modelling

where e_V, e_H and $e_{H'}$ represent the stiffness embedment factors for vertical translation, horizontal translation (towards long side), and horizontal translation (towards short side), respectively.

Soil Properties Used in BNWF Model

Three types of sand are considered in the present analysis. The properties of dense sand and medium dense sand are taken from the experimental data of Patra et al. [15]. The properties of loose sand are taken from the experimentally reported data of Sahu et al. [17]. The dynamic properties such as modulus of elasticity (*E*) as well as the Poisson's ratio (ν) are calculated from EPRI manual [4]. The soil parameters are listed in Fig. 31.3.

Details of Loading Pattern

First, depending on the load eccentricity, the foundation is loaded with allowable static load which is calculated as; the ultimate load on the foundation divided by a factor of safety (*FS*). The *FS* values are taken as 2, 2.5, 3, and 3.5. Then, the applied static load is kept constant, and vertical pulse of one cycle is applied, the intensity/amplitude of which is 5, 10, and 13% of the ultimate load carrying capacity of the foundation. Both static and cyclic load are applied at an eccentricity depending of the eccentricity ratio (*e*/*B*). The *e*/*B* values are taken as 0, 0.05, 0.10, and 0.15.

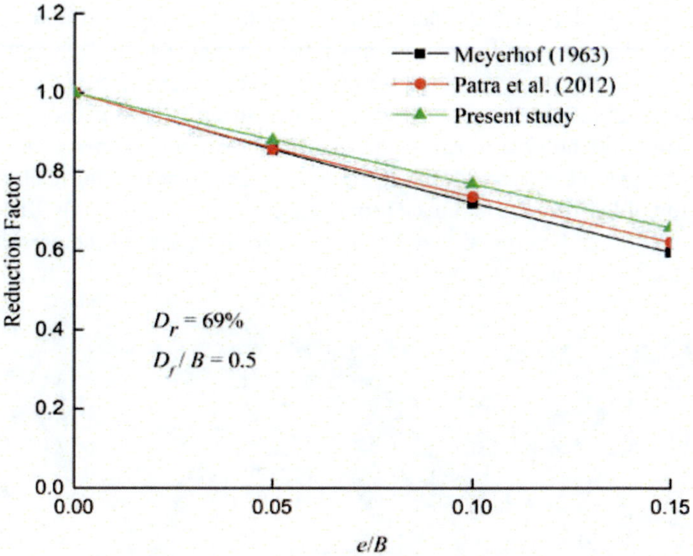

Fig. 31.4 Comparison of present result with existing solutions

Model Validation

The static response of the foundation is compared with existing analytical and experimental analysis. The reduction factor obtained from the present BNWF model is compared with those reported in Mayerhof [13] and Patra et al. [15] which is shown in Fig. 31.4. It is found that the result obtained from the present analysis has good agreement with the reported outcomes.

Adaptive Neuro-Fuzzy Interface System (ANFIS)

ANFIS is one of the key soft computing techniques to establish a relationship between a set of inputs and the output. In the last decade, ANFIS has been used by a number of researches in geotechnical engineering. ANFIS was used for predicting swelling potential of clay [22], blast-induced fly rock distance [21], prediction of deformation modulus of rock mass [1], ultimate bearing capacity prediction of rectangular footing [20], migration of leachate [11], prediction of Unconfined compressive strength (UCS) of soil [16], transient settlement of strip footing under centric static and rectangular pulse load [19].

Normalization is done within [0, 1] using Eq. (31.7)

$$X_{\text{normalized}} = \frac{X - X_{\text{minimum}}}{X_{\text{maximum}} - X_{\text{minimum}}} \tag{31.7}$$

Out of total 432 dataset, 324 datasets are used for training, and remaining 108 datasets are used for testing. In the present analysis, 5 inputs are used, viz. *FS*, D_f/B, $q_{d(\max)}/q_u$ (%), D_r (%), and *e*/*B*. For each input, two membership functions are used which make the number of rule equal to $2^5 = 32$. Grid partition method is used for the estimation of the number of rules. The Gaussian membership function is used in this study along with the hybrid training algorithm. The hybrid training algorithm is a combination of least square and back propagation techniques. The ANFIS structure is shown in Fig. 31.5. It is found that 200 epochs are sufficient for the present model to arrive at a constant accuracy point (Fig. 31.6).

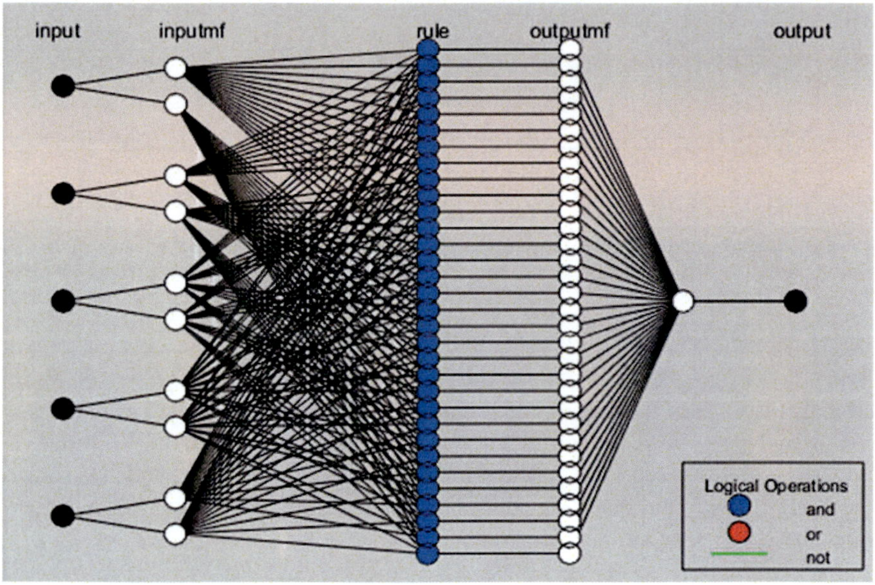

Fig. 31.5 Structure of the ANFIS model

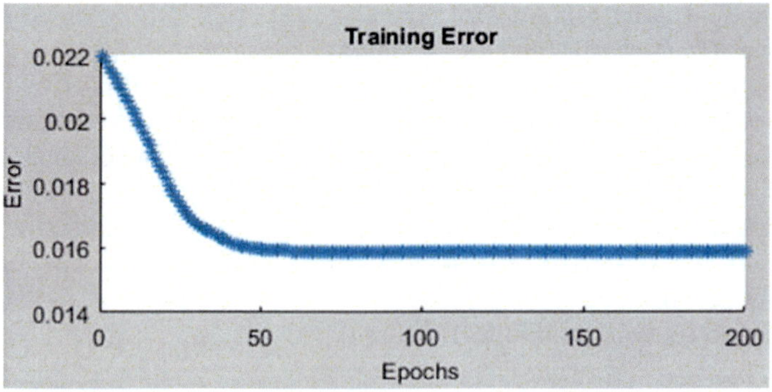

Fig. 31.6 Training state

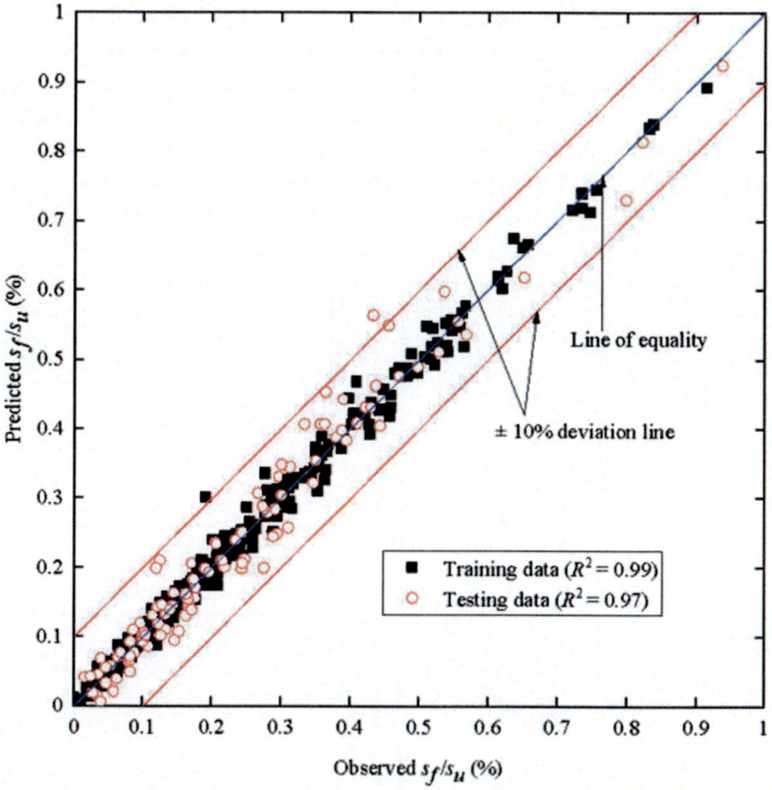

Fig. 31.7 Comparison of observed (BNWF model) and predicted (ANFIS) normalized outputs

Evaluation of Accuracy

The outcomes from the ANFIS model are compared with those from OpenSEES analysis and presented in Fig. 31.7. It is found that for training data, the coefficient of determination is 0.99, and for testing data, the coefficient of determination is 0.97. Hence, the developed model has not only good prediction ability but also has a very good generalization ability.

Parametric Analysis and Sensitivity Study

The variation of the membership functions for each of the input parameters is shown in Fig. 31.8. The membership function is an indication of degree of involvement of an input parameter. In Fig. 31.9, the surface plot is shown indicating the combined effect of two inputs on the response of footing. For this, any two parameters are varied from minimum to maximum value, and the corresponding changes in the output are observed. All other parameters other than the two in question are fixed at their median values. In Fig. 31.10, the influence of each individual parameter is shown, indicating the range of influence (Spider chart). It is found that *FS* (that determines the static load on foundation) is the most important parameter followed by; $q_{d(\max)}/q_u$ (%), D_r (%) D_f/B, and *e*/*B*. The settlement is inversely related to *FS* and directly related to $q_{d(\max)}/q_u$ (%). The other parameters influence the response nonlinearly.

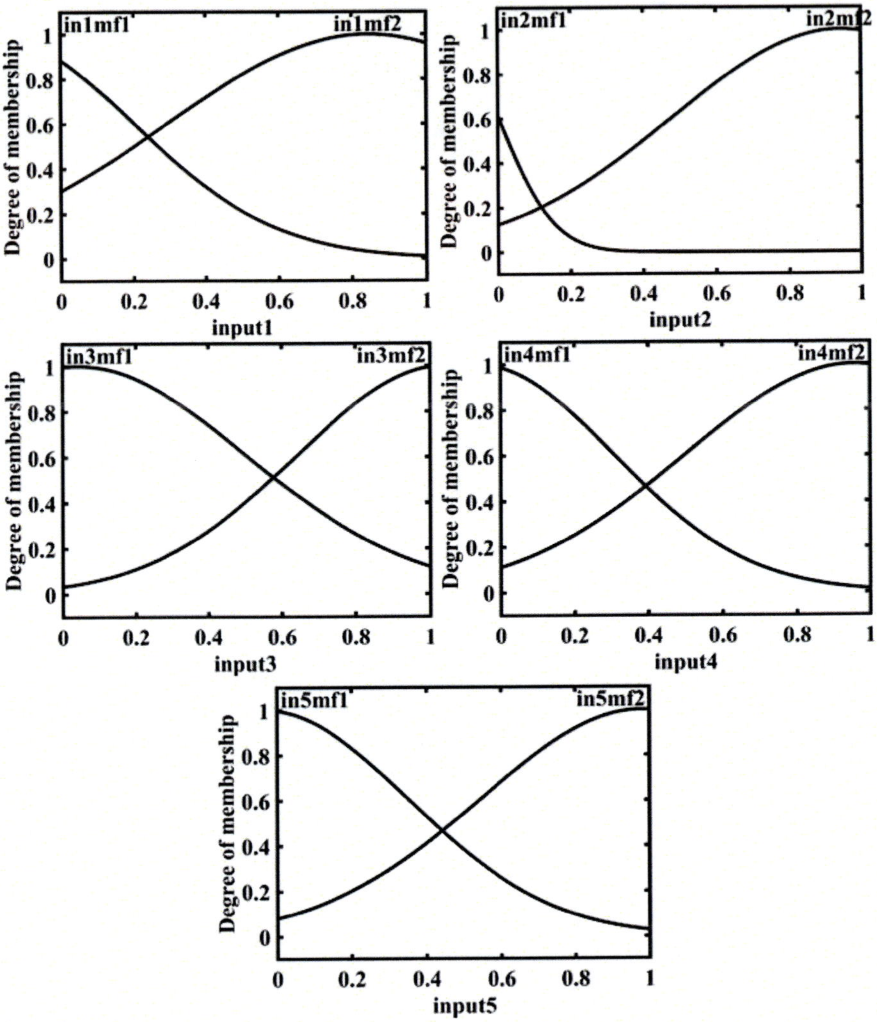

Fig. 31.8 Variations of the Gaussian membership functions (input 1 = *FS*, input 2 = D_f/B, input 3 = $q_{d(\max)}/q_u$ (%), input 4 = D_r (%) and input 5 = *e*/*B*)

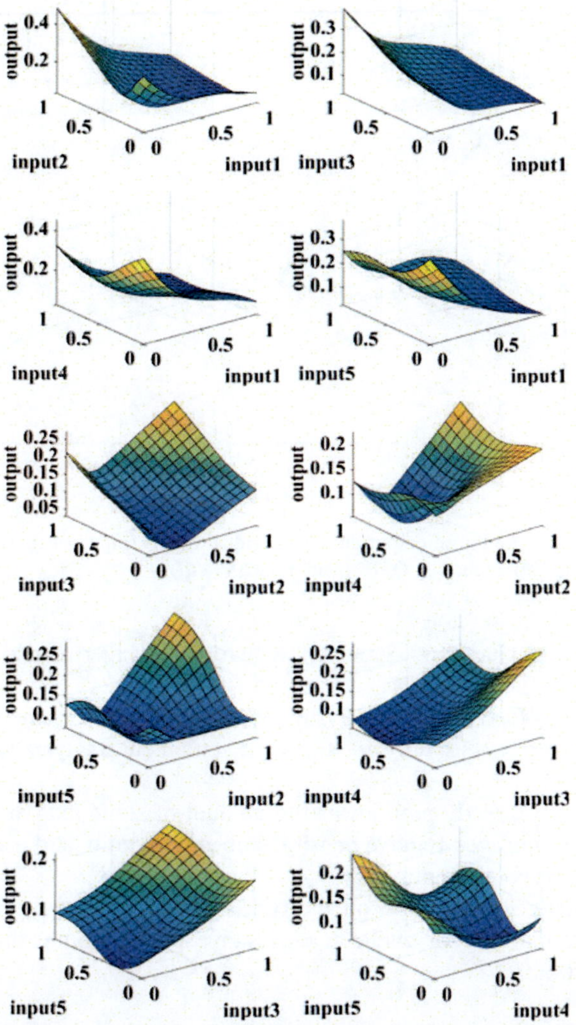

Fig. 31.9 Surface plot showing the combined influence of inputs (input 1 = *FS*, input 2 = D_f/B, input 3 = $q_{d(\max)}/q_u$ (%), input 4 = D_r (%) and input 5 = e/B)

Conclusions

A shallow foundation is modelled using FEM, and the effect of eccentrically applied allowable static load and one cycle of cyclic load is studied. The generated dataset for 432 number of conditions are analysed using ANFIS. From the numerical and statistical analysis, the following conclusions are drawn;

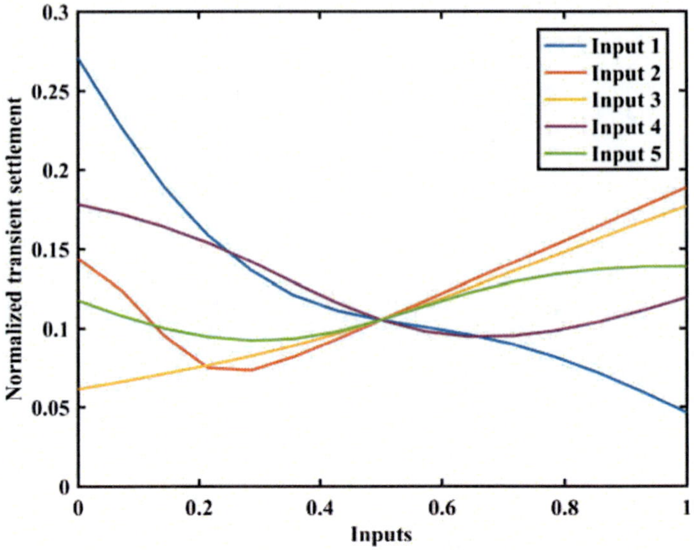

Fig. 31.10 Range of variation of output due to individual inputs (input 1 = *FS*, input 2 = D_f/B, input 3 = $q_{d(\max)}/q_u$ (%), input 4 = D_r (%) and input 5 = e/B)

- The BNWF model has good agreement with reported experimental and analytical results.
- Adaptive Neuro-Fuzzy Interface System with linear Gaussian membership function can be used successfully for transient settlement analysis under eccentric load.
- The s_f/s_u (%) is directly proportional to amount of static load and the intensity of pulse load. However, relative density, embedment ratio, and eccentricity ratio influence the response nonlinearly.
- Allowable static load applied on the foundation is the most important factor followed by; intensity of cyclic load, relative density, embedment ratio, and eccentricity ratio.

References

1. Asrari AA, Shahriar K, Ataeepour M (2015) The performance of ANFIS model for prediction of deformation modulus of rock mass. Arab J Geosci 8(1):357–365
2. Behera RN (2013) Behaviour of shallow strip foundation on granular soil under eccentrically inclined load. Doctoral dissertation, National Institute of Technology, Rourkela, India
3. Bransby MF (2001) Failure envelopes and plastic potentials for eccentrically loaded surface footings on undrained soil. Int J Numer Anal Meth Geomech 25(4):329–346
4. EPRI (1990) Manual on estimating soil properties for foundation design. Electric Power Research Institute, Palo Alto, California

5. Fedorovskii VG (2005) Bearing capacity of the granular bed of a strip foundation under an inclined eccentric load. Soil Mech Found Eng 42:111–119
6. Gajan S, Hutchinson TC, Kutter BL, Raychowdhury P, Ugalde JA, Stewart JP (2008) Numerical models for analysis and performance-based design of shallow foundations subjected to seismic loading. Pacific Earthquake Engineering Research Center, Berkeley
7. Gazetas G (1991) Formulas and charts for impedances of surface and embedded foundations. J Geotech Eng 117(9):1363–1381
8. Georgiadis M, Butterfield R (1988) Displacements of footings on sand under eccentric and inclined loads. Can Geotech J 25(2):199–212
9. Gottardi G, Butterfield R (1995) The displacement of a model rigid surface footing on dense sand under general planar loading. Soils Found 35(3):71–82
10. Harden C, Hutchinson T, Martin GR, Kutter BL (2005) Numerical modeling of the nonlinear cyclic response of shallow foundations. Rep No 2005/04 Pacific Earthquake Engineering Research Center (PEER), Berkeley, California
11. Kebria DY, Ghavami M, Javadi S, Goharimanesh M (2018) Combining an experimental study and ANFIS modeling to predict landfill leachate transport in underlying soil—a case study in north of Iran. Environ Monit Assess 190(1):26, 1–17
12. Lu L, Wang ZJ, Arai K (2014) Numerical and experimental analyses for bearing capacity of rigid strip footing subjected to eccentric load. J Central South Univ 21(10):3983–3992
13. Meyerhof GG (1963) some recent research on the bearing capacity of foundations. Can Geotech J 1(1):16–26
14. OpenSEES: Computer software, University of California, Berkeley
15. Patra CR, Behera RN, Sivakugan N, Das BM (2012) Ultimate bearing capacity of shallow strip foundation under eccentrically inclined load, Part I. Int J Geotech Eng 6(3):343–352
16. Saadat M, Bayat M (2019) Prediction of the unconfined compressive strength of stabilised soil by Adaptive Neuro Fuzzy Inference System (ANFIS) and Non-Linear Regression (NLR). Geomech Geoeng 1–12
17. Sahu R, Patra CR, Das BM, Sivakugan N (2016) Bearing capacity of shallow strip foundation on geogrid-reinforced sand subjected to inclined load. Int J Geotech Eng 10(2):183–189
18. Sargazi O, Hosseininia ES (2017) Bearing capacity of ring footings on cohesionless soil under eccentric load. Comput Geotech 92:169–178
19. Sasmal SK, Behera RN (2021) Transient response prediction of shallow foundation using ANFIS. In: Soil dynamics. Springer, Singapore, pp 389–397
20. Sethy BP, Patra CR, Sivakugan N, Das BM (2017) Application of ANN and ANFIS for predicting the ultimate bearing capacity of eccentrically loaded rectangular foundations. Int J Geosynthetics Ground Eng 3(4):35, 1–14
21. Trivedi R, Singh TN, Gupta N (2015) Prediction of blast-induced flyrock in opencast mines using ANN and ANFIS. Geotech Geol Eng 33(4):875–891
22. Yilmaz I, Kaynar O (2011) Multiple regression, ANN (RBF, MLP) and ANFIS models for prediction of swell potential of clayey soils. Exp Syst Appl 38(5):5958–5966

Chapter 32
1D, 2D, and 3D Seismic Response Analysis of Shallow and Deep Foundation Resting on Stratified Soil Deposit

Amit Kumar Ram, Anjali Verma, and Supriya Mohanty

Introduction

The earthquake is the key opponent of the loaded structures resting on the earth's crust. It has a remarkable effect in the northern and eastern regions of India whereas some parts of the west also come under its influence [1]. In the past two decades, India has experienced extremely damaging earthquakes in Gujarat (2001-Mw: 7.7), in Jammu and Kashmir (2005-Mw:7.6), and in Sikkim (2011-Mw:6.9). In the year 2020, all magnitude of earthquakes has been recorded and shows the higher frequency for the lower magnitude of earthquakes (Mw $\geq$ 6 (3), Mw: 5–6 (26), Mw: 4–5 (385), Mw: 3–4 (419), Mw: 2–3 (109)) [2]. The seismic wave (P and S wave) generally travels through the body of the earth and reaches the surface causing destruction of structures on the surface due to the amplification of the ground acceleration. The matching of frequency of the seismic wave with the natural frequency of the building is the reason behind amplification [3]. Earlier researchers were more interested in finding the way of forecasting earthquakes but no one come up with appreciable outcomes. Hence, the researchers started focusing on the earthquake-resistant design of structures [4]. In order to design an earthquake-resistant structure, it is crucial to have knowledge about site-specific ground response analysis. The safety of the new building can be ensured by designing the building as per the response spectra of that site which can be determined using ground response analysis. It also helps in the determination of the natural period of the site, amplification due to ground motion, and liquefaction

A. K. Ram (✉) · A. Verma · S. Mohanty
Department of Civil Engineering, IIT (BHU), Varanasi, India
e-mail: amitkumarram.rs.civ18@itbhu.ac.in

A. Verma
e-mail: anjaliverma.civ19@itbhu.ac.in

S. Mohanty
e-mail: supriya.civ@iitbhu.ac.in

K. Muthukkumaran et al. (eds.), *Soil Dynamics, Earthquake and Computational Geotechnical Engineering*, Lecture Notes in Civil Engineering 300,
https://doi.org/10.1007/978-981-19-6998-0_32

potential of the site prior to the construction [3]. The factors responsible for influencing the structural response subjected to seismic excitations are the geological condition of the topography, frequency of the seismic activity, and magnitude of the seismic event [5, 6]. The ground response analysis using linear, nonlinear, and equivalent linear approaches has been reported by a past study [7]. Several researchers had employed various computer programs for the response analysis such as SHAKE 2000 [8–10], DEEPSOIL [11, 12], QUAD-4 [13], FLUSH [14], and Cyclic 1D [15]. It was found that the responses of the nonlinear model are giving more realistic outcomes as compared to that of linear models [16]. The response study of the site can be performed in one (1D), two (2D), and three (3D) dimension ways depending on the required precision. In the 1D study, it is assumed that the ground boundaries are at infinity in all respective directions. The 2D and 3D study required higher time for the analysis and can solve all problems, which cannot be solved using 1D such as stratified soil profile, anisotropy, and irregular soil boundary. Basically, 2D analysis involves problems related to plain strain whereas 3D is useful boundary/motion changes in three dimensions [17]. Practically, there is a limited extent of the presence of homogeneous soil profiles, and the majority of existing ground contains stratification or heterogeneity. Das and Chakraborty [4] did a comparative study between homogeneous and heterogeneous soil profiles and found that for the high magnitude of input motion, the ground responses are significantly affected by the heterogeneity. Most of the studies focus on the 1D ground response analysis having a homogeneous soil profile and very limited studies have been performed considering the heterogeneity/stratifications in the soil profile. The objective of this study is to show the behaviour of ash deposit under seismic condition. So that its utilization can be extended to seismic-prone areas as well. Hence, in the present study, an attempt has been made to investigate the 1D, 2D, and 3D response of shallow and deep foundations resting on stratified soil-ash deposits under the action of different seismic excitations. The software or computer program used for 1D analysis was Cyclic1D, similarly, CyclicTP has been involved in the 2D analysis and OpenSeesPL has been used for the 3D analysis.

Materials

In the present study, a stratified soil-ash-foundation system has been considered for the ground response analysis. The soil domain having a width of 7.5 m and depth of 20 m subjected to two seismic excitations has been considered. For 2D analysis, a strip footing of 2 m width resting at a depth of 2 m whereas, for 3D analysis, a circular pile of 0.3 m diameter and length of 10 m has been employed. The foundation has a mass density of 2400 (kg/m^3), Young's modulus of 25,000,000 (kPa), and Poisson's ratio of 0.3. The basic input parameters needed for the analysis are listed in Table 32.1.

Table 32.1 Basic input parameters were considered for the present investigation

Layer No	Thickness (m)	Soil type	Shear wave velocity (m/s)	Mass density (kg/m^3)
1	5	Fly ash	116.62	1589
2	5	Medium, silt permeability	205	1900
3	5	Medium, sand permeability	205	1900
4	5	Medium-dense, sand permeability	225	2000

Seismic Input Motions

The Nepal and Uttarkashi earthquake input motion has been taken for the ground response analysis. The magnitude of Nepal and Uttarkashi earthquakes are 7.8 (in 2015) and 6.5 (1991), respectively. The peak ground acceleration of Nepal input motion is 0.164 g at 49.31 s whereas the peak ground acceleration of Uttarkashi input motion is 0.31 g at 30 s. The acceleration time history of both the considered input motion has been shown in Fig. 32.1.

Methodology and Convergence Study

All the 1D, 2D, and 3D analysis are based on the same liquefaction model which follows the structure of multi-yield-surface plasticity. In the clean medium to dense sand significant importance has been given in managing the magnitude of the cycle-by-cycle permanent shear strain accumulations. With this model, the dilation tendency and the cyclic mobility mechanism can be seen applying loading–unloading flow rules. The multi-yield surface of the material model has been shown in Fig. 32.2.

The only assumption made in Cyclic1D is that any response is primarily due to the component of shear waves that are polarized in the horizontal direction and propagate vertically upward from the underlying bedrock.

In order to determine the optimum number of elements in finite element mesh, there is a need for convergence study. For various numbers of elements, the numerical model was examined. For this, the number of elements has increased over time; the variation in the parameters to be examined for a specific location has remained relatively constant. The results are then considered to be converging, and the best number of elements for further analysis is chosen. The schematic convergence plot of all the analysis has been shown in Fig. 32.3.

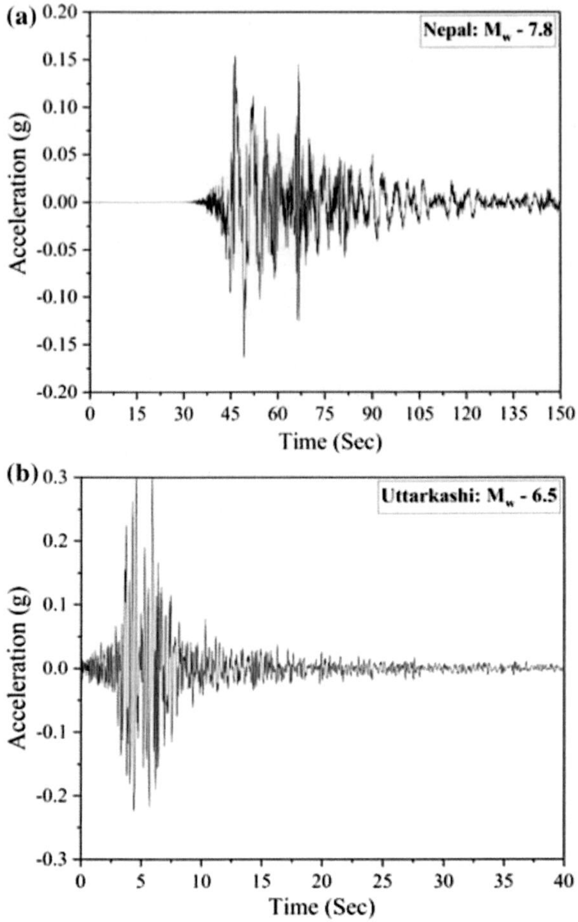

Fig. 32.1 Typical acceleration time history plot of **a** Nepal, and **b** Uttarkashi earthquake input motions

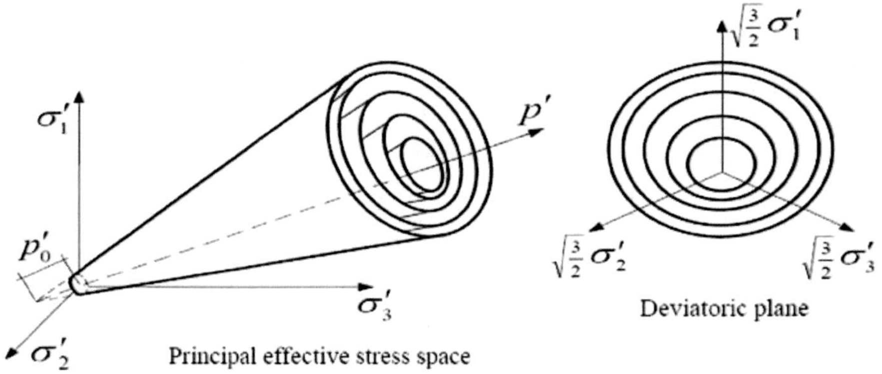

Fig. 32.2 Multi-yield surfaces in principal stress space and deviatoric plane [18]

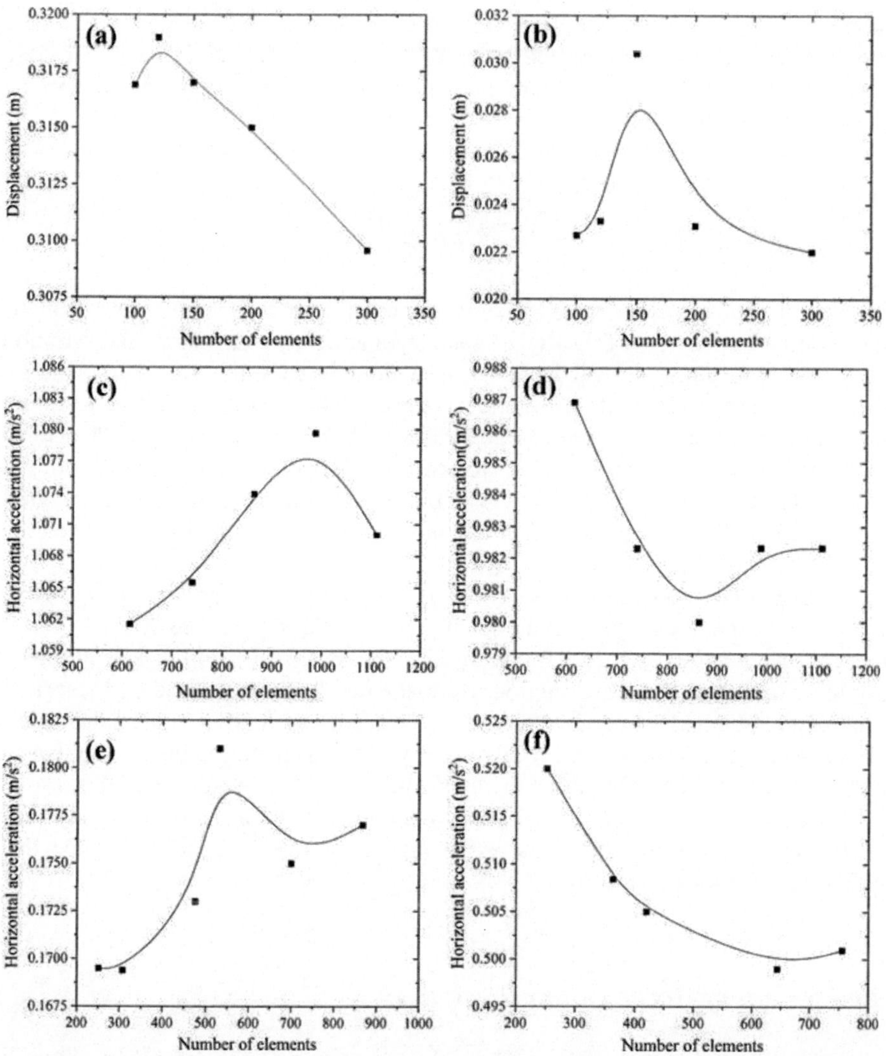

Fig. 32.3 Convergence plot of all the analysis subjected to Nepal (**a** 1D analysis), (**c** 2D analysis), (**e** 3D analysis), and Uttarkashi input motions (**b** 1D analysis), (**d** 2D analysis), (**f** 3D analysis)

Results and Discussion

The present study focuses on the ground response analysis of the soil-ash and soil-ash foundation system resting on stratified soil-ash deposits. The response analysis has been investigated by performing numerical analysis using finite element software Cyclic1D (for ground response analysis), CyclicTP (for shallow foundation), and OpenSeesPL (for deep foundation). Two past earthquakes, i.e. Nepal and Uttarkashi, have been adopted for the 1D, 2D, and 3D response analysis. The responses of all the three analyses have been presented in the form of acceleration, displacement, excess pore water pressure, and excess pore pressure ratio which has been explained in the subsequent sections.

Assessment of Displacement Response

The soil-ash domain has been divided into four equal parts consisting of fly ash on the top underlain by silt layer followed by sand layer and then dense sand. The variation of displacement along vertical depth has been plotted for Cyclic 1D (a, b), Cyclic TP (b, c), and OpenSeesPL (e, f) that is shown in Fig. 32.4. In 1D analysis, the maximum displacement has been observed at the interface of fly ash and silt layer. The Nepal earthquake was showing 0.316 m of displacement whereas the displacement of the Uttarkashi earthquake was observed to be 0.023 m. In the same way, in 2D analysis, the influence of loaded shallow foundation has been investigated and reported the displacement in either side of the footing.

The 1D and 2D analysis follows a more or less similar trend and found negligible change when moving away from the footing. In 3D analysis, the maximum displacement has been observed at 2.393 m away from the centre at the top surface of the fly ash layer for Uttarkashi input motion. Here, the displacement caused by Uttarkashi was found to be high as compared to that of Nepal motion. This may be due to the presence of higher peak ground acceleration of Uttarkashi motion than that of Nepal motion. The present investigation exhibits a higher value of response in the case of 1D and 2D analysis, which was also observed by Nautiyal et al. [19] and Reddy et al. [20]. The response study of 1D analysis always shows overestimated value than that of 2D and 3D analysis.

Assessment of Acceleration Responses

The plot of the variation of the acceleration response along with the depth under different earthquakes has been represented in Fig. 32.5 for Cyclic1D (a, b), CyclicTP (c, d), and OpenSeesPL (e, f) analysis. Here, in both 1D and 2D analysis the amplification of acceleration has been observed at or below the fly ash-silt interface. In Cyclic1D (1D analysis), the maximum value of acceleration for the Uttarkashi motion was recorded as 4.1336 m/s^2 at the depth of 15 m from the ground surface whereas for the Nepal input motion the maximum acceleration was found to be 3.395 m/s^2 at the depth of 5.1 m. In CyclicTP (2D analysis), the location of maximum accelerations for both Uttarkashi and Nepal motion was found to be the same as that in Cyclic1D. From the analysis of the domain using OpenSeesPL (3D analysis), it can be seen that the maximum acceleration for the Uttarkashi motion was recorded as 0.595 m/s^2 at the ground level at a distance of 2.393 m from the centre of the footing. Whereas, for

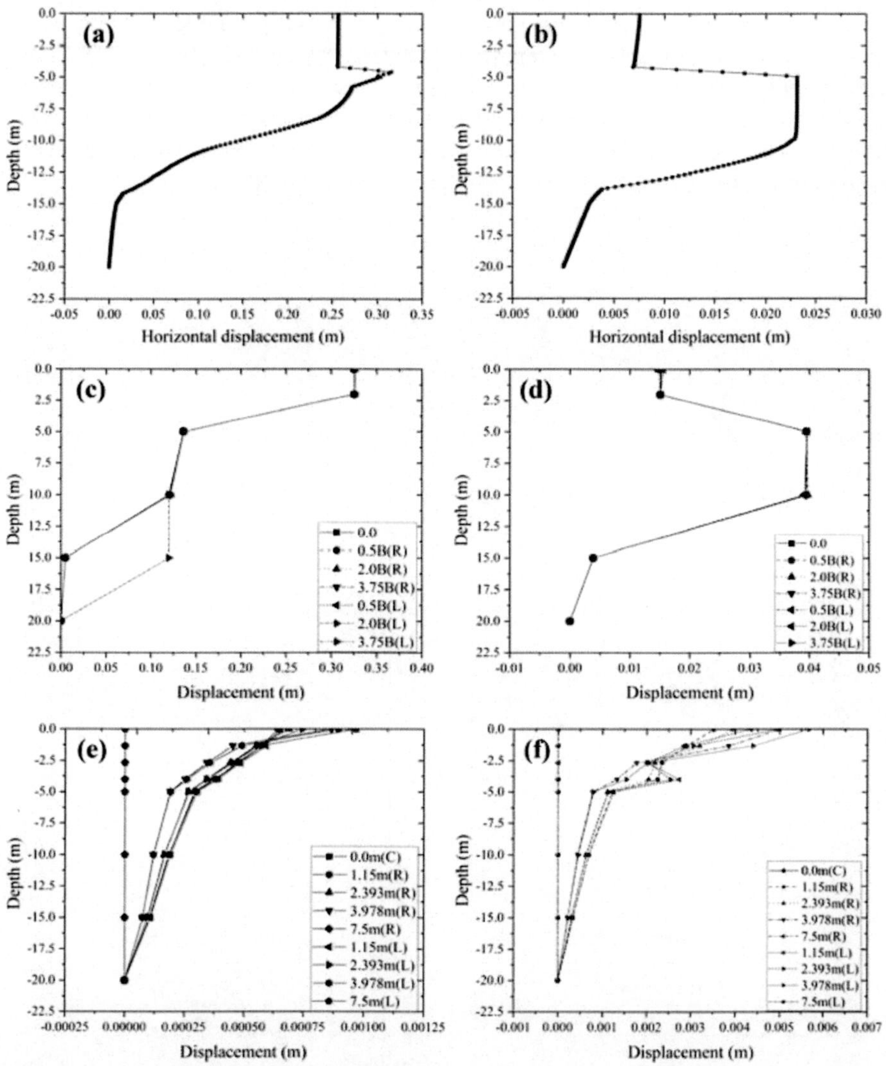

Fig. 32.4 Representation of displacement variation with depth considering all the analysis subjected to Nepal (**a** 1D analysis), (**c** 2D analysis), (**e** 3D analysis), and Uttarkashi input motions (**b** 1D analysis), (**d** 2D analysis), (**f** 3D analysis)

the Nepal input motion the maximum acceleration was found to be 0.188 m/s^2 at the depth of 0 m from the ground surface at the centre of the footing. After comparing the outcomes of 1D, 2D, and 3D analysis, it can be observed that the OpenSeesPL (3D analysis) is showing the reverse trend of acceleration with depth as compared with Cyclic1D and CyclicTP. After examining the seismic response of homogeneous and heterogeneous soil, Das and Chakraborty [4] found maximum acceleration below the ground surface for heterogeneous; however, homogeneous profile shows maximum acceleration at the top surface as reported by several researchers [16, 21–23]. In the present study, the amplification of acceleration was found to be high for a lower magnitude of the earthquake, and similar results have been reported by Ngo et al. [24]. The depletion in stiffness of soil because of nonlinear behaviour would be the reason for this result.

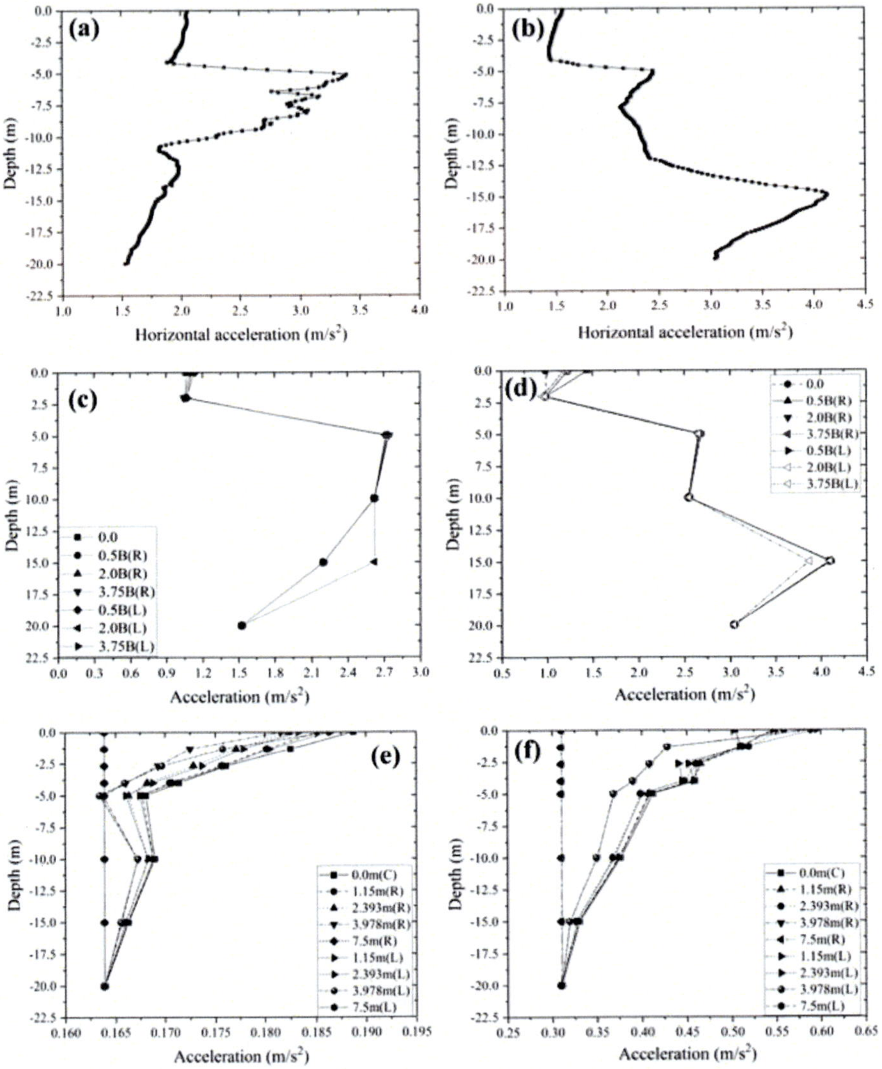

Fig. 32.5 Variation of acceleration with depth for all the analysis subjected to Nepal (**a** 1D analysis), (**c** 2D analysis), (**e** 3D analysis), and Uttarkashi input motions (**b** 1D analysis), (**d** 2D analysis), (**f** 3D analysis)

Assessment of Excess Pore Pressure Responses

The graphical representation of the excess pore water pressure with depth under two seismic excitations for all the analysis has been shown in Fig. 32.6. In 1D analysis, the maximum excess pore pressure value for the Uttarkashi motion was recorded as 156 kPa at the depth of 14.5 m from the ground surface whereas for the Nepal input motion the maximum excess pore pressure was found to be 158.32 kPa at the depth of 19.7 m below the ground surface. Moreover, in 2D analysis the maximum excess pore water pressure for the Uttarkashi motion was recorded as 195.305 kPa at the depth of 5 m from the ground surface at the centre of the footing. Whereas, for the Nepal input motion the maximum excess pore water pressure was found to be 186 kPa at the depth of 5 m below the ground surface at a distance of 1 m away from the centre of the footing. Similarly, in 3D analysis, the location of the maximum excess pore pressure was found to be the same as that of 2D analysis. The maximum pore pressure is usually found when the groundwater table is at ground level, which increases its probability of failure due to liquefaction phenomena [25]. Yoshimi and Tokimatsu [26] observed a higher value of pore pressure away from the centre of the footing which can be seen in the present study as well.

Assessment of Excess Pore Water Pressure Ratio

The change in excess pore water pressure ratio with depth subjected to Nepal and Uttarkashi earthquake in all the analysis has been presented in Fig. 32.7. From the plot of excess pore pressure ratio versus depth, it can be noticed that for all 1D, 2D, and 3D analysis the depth of maximum excess pore pressure ratio is the same that is 5 m. In 1D analysis, the maximum excess pore pressure ratio for the Uttarkashi motion was recorded as 0.998 at the depth of 5 m from the ground whereas for the Nepal input motion the maximum excess pore pressure ratio was found to be 1.005 at the depth of 5 m below the ground surface.

Similarly, in 2D analysis the maximum excess pore pressure ratio for the Uttarkashi motion was recorded as 2.607 at the depth of 5 m from the ground surface at the centre of the footing. Whereas, for the Nepal input motion the maximum excess pore pressure ratio was found to be 2.49 at the depth of 5 m below the ground surface at a distance of 1 m away from the footing centre. The excess pore pressure ratio acts as an indicator of the liquefaction probability of soil means if it is less than 1 then it can be considered that soil is not liquefiable for any particular input motion. However, in the present case, the ratio is greater than 1 for 2D and 3D analysis, which depicts a high potential of liquefaction for these input motions. Chiaradonna et al. [27] have noticed a similar depth of maximum value of excess pore pressure ratio as like present study.

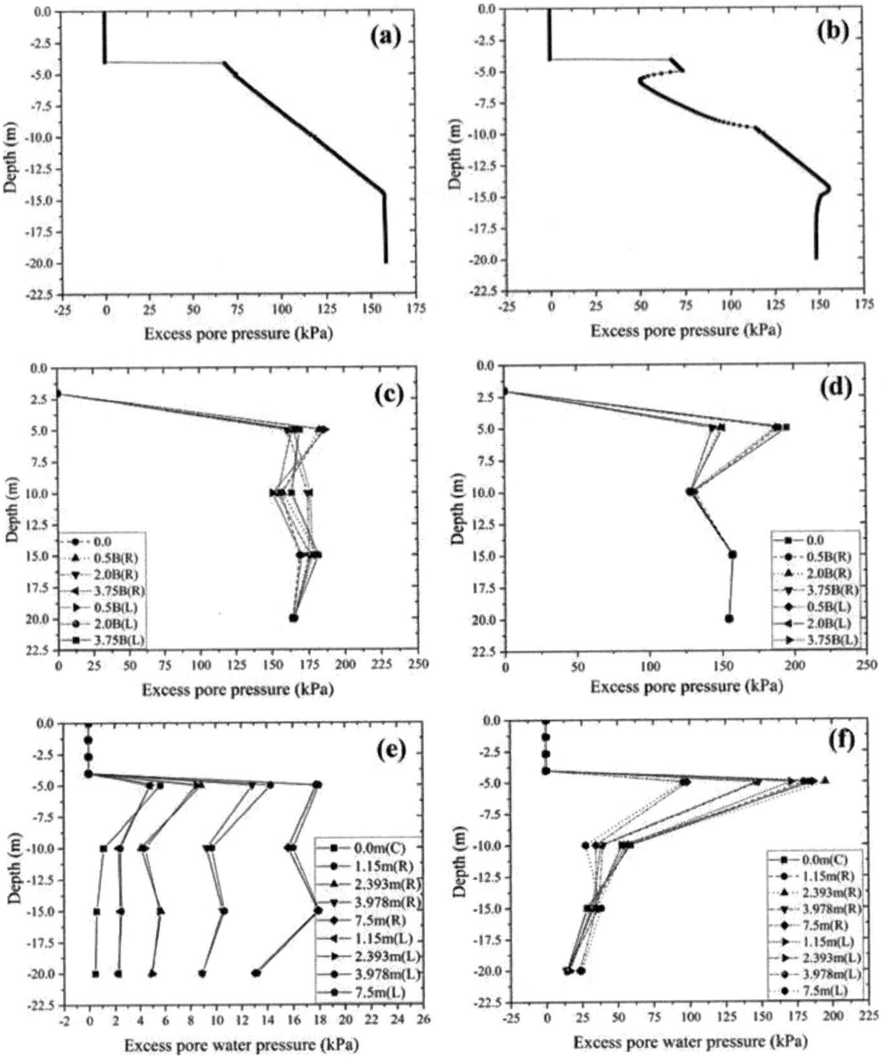

Fig. 32.6 Variation of excess pore water pressure with depth for all the analysis subjected to Nepal (**a** 1D analysis), (**c** 2D analysis), (**e** 3D analysis), and Uttarkashi input motions (**b** 1D analysis), (**d** 2D analysis), (**f** 3D analysis)

Conclusions

The present study has been conducted to compare the responses generated from the 1D, 2D, and 3D seismic analysis of the shallow and deep foundation resting on layered soil-ash deposit. After analysing the soil domain numerically, the following conclusion can be drawn.

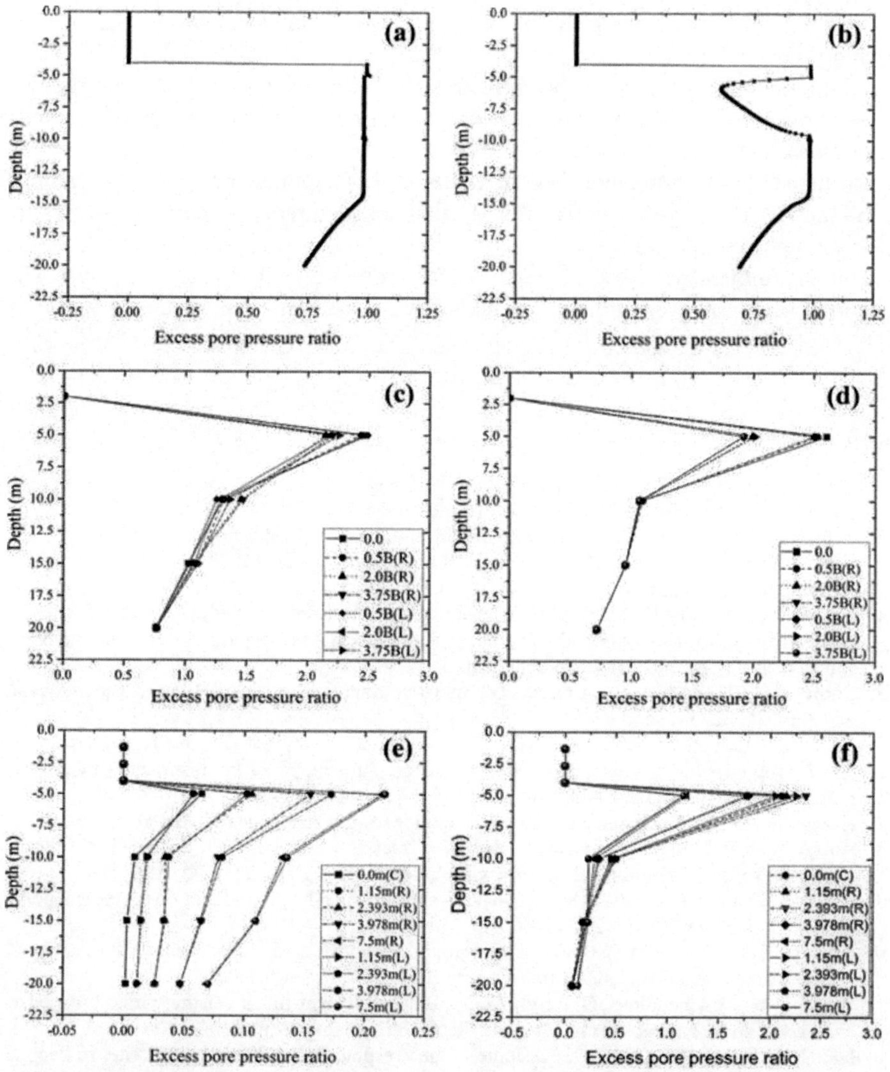

Fig. 32.7 Variation of excess pore water pressure ratio with depth for all the analysis subjected to Nepal (**a** 1D analysis), (**c** 2D analysis), (**e** 3D analysis), and Uttarkashi input motions (**b** 1D analysis), (**d** 2D analysis), (**f** 3D analysis)

- In 1D and 2D analysis, the maximum displacement has been observed at the interface of fly ash and silt layer. However, for 3D analysis the maximum displacement was observed at the ground surface.
- After comparing the outcomes of the 1D, 2D, and 3D analysis, it can be observed that the OpenSeesPL (3D analysis) is showing a reverse trend of acceleration

with depth as compared with that of Cyclic1D (1D analysis) and CyclicTP (2D analysis).

- A similar trend of excess pore pressure response for 2D and 3D analysis has been observed and found different peak locations of excess pore pressure when compared with 1D analysis.
- From the plot of excess pore pressure ratio versus depth, it can be noticed that for all the 1D, 2D, and 3D analysis the depth of maximum excess pore pressure ratio is the same that is at 5 m.
- It is recommended to use 3D analysis for ground response studies because 1D and 2D analysis overestimates the magnitude of response parameters.

References

1. Rautela P (2020) Unit-4 disaster profile of India. IGNOU
2. Volcanodiscovery.com/earthquakes/India/archive/2020.html. Last accessed 31/07/2021
3. Govinda Raju L, Ramana GV, Hanumantha Rao C, Sitharam TG (2004) Site-specific ground response analysis. Curr Sci 1354–1362
4. Das A, Chakrabortty P (2017) One dimensional seismic response analysis of heterogeneous layered soil. In: 15th international conference of international association for computer methods and advances in geomechanics, Wuhan, China
5. Boore DM (1972) A note on the effect of simple topography on seismic SH waves. Bull Seismol Soc Am 62(1):275–284
6. Nath SK, Thingbaijam KKS, Adhikari MD, Nayak A, Devaraj N, Ghosh SK, Mahajan AK (2013) Topographic gradient based site characterization in India complemented by strong ground-motion spectral attributes. Soil Dyn Earthq Eng 55:233–246
7. Kramer SL (1996) Geotechnical earthquake engineering. Prentice Hall, New York
8. Anbazhagan P, Thingbaijam KKS, Nath SK, Kumar JN, Sitharam TG (2010) Multi-criteria seismic hazard evaluation for Bangalore city, India. J Asian Earth Sci 38(5):186–198
9. Shukla J, Choudhury D (2012) Seismic hazard and site-specific ground motion for typical ports of Gujarat. Nat Hazards 60(2):541–565
10. Kumar SS, Krishna AM (2013) Seismic ground response analysis of some typical sites of Guwahati city. Int J Geotech Earthquake Eng (IJGEE) 4(1):83–101
11. Phanikanth VS, Choudhury D, Reddy GR (2011) Equivalent-linear seismic ground response analysis of some typical sites in Mumbai. Geotech Geol Eng 29(6):1109–1126
12. Naik N, Choudhury D (2013) Site specific ground response analysis for typical sites in Panjim city, Goa. In: Proceedings of Indian geotechnical conference, Roorkee, India
13. Idriss IM (1973) QUAD-4: a computer program for evaluating the seismic response of soil structures by variable damping finite element procedures, July
14. Lysmer J, Udaka T, Tsai CF, Seed HB (1975) FLUSH—a computer program for approximate 3-D analysis of soil-structure interaction problems. Report No. EERC 75-30, Earthquake Engineering Research Center, University of California, Berkeley, California, p 83
15. Reddy MRK, Mohanty S, Shaik R (2021) Comparative study of 1D, 2D and 3D ground response analysis of pond ash from Odisha under different earthquake motions. In: Geohazards. Springer, Singapore, pp 523–538
16. Puri N, Jain A, Mohanty P, Bhattacharya S (2018) Earthquake response analysis of sites in state of Haryana using DEEPSOIL software. Procedia Comput Sci 125:357–366
17. Kramer SL, Paulsen SB (2004) Practical use of geotechnical site response models. In: Proceedings of international workshop on uncertainties in nonlinear soil properties and their impact on modeling dynamic soil response. University of California, Berkeley, p 10

18. Elagamal A, Yang Z, Lu J (2015) User's manual (Cyclic 1D)—seismic ground response. University of California, San Diego
19. Nautiyal P, Raj D, Bharathi M, Dubey R (2019) Ground response analysis: comparison of 1D, 2D and 3D approach. In: Indian geotechnical conference, SVNIT, Surat
20. Reddy MRK, Mohanty S, Shaik R (2021) Comparative study of 1D, 2D and 3D ground response analysis of pond ash from Odisha under different earthquake motions. In: Geohazards. Springer, Singapore, pp 523–538
21. Anbazhagan P, Sitharam TG (2009) Estimation of ground response parameters and comparison with field measurements. Indian Geotech J 39(3):245–270
22. Erdik M (1987) Site response analysis. In: Strong ground motion seismology. Springer, Dordrecht, pp 479–534
23. Kumar SS, Dey A, Krishna AM (2018) Importance of site-specific dynamic soil properties for seismic ground response studies: ground response analysis. Int J Geotech Earthquake Eng (IJGEE) 9(1):78–98
24. Ngo VL, Lee C, Kim JM (2021) Effects of stratification on soil–foundation–structure interaction: centrifugal observation and numerical simulation. Appl Sci 11(2):623, 1–19
25. Kumar R, Mohanty S, Chethan K (2019) 3D seismic response analysis of shallow foundation resting on sandy soil. Int J Geotech Earthquake Eng (IJGEE) 10(1):61–76
26. Yoshimi Y, Tokimatsu K (1977) Settlement of buildings on saturated sand during earthquakes. Soils Found 17(1):23–38
27. Chiaradonna A, Tropeano G, d'Onofrio A, Silvestri F (2016) A simplified method for pore pressure buildup prediction: from laboratory cyclic tests to the 1D soil response analysis in effective stress conditions. Procedia Eng 158:302–307

Chapter 33
Study of Shallow Soil Deposit in East Coast of India by SPT, MASW, and Crosshole Tests

Ayush Kumar, Anbazhagan Panjamani, Ravinesh Kumar, and K. R. Lenin

Introduction

Earthquakes cause various hazards such as landslides, liquefaction, ground shaking, flooding, and tsunamis. Among these, liquefaction has been known to cause the failure of even well-designed and well-built structures as it is entirely a property of the subsurface soil. Liquefaction is defined as the transformation of granular material from solid to a liquefied state because of increased pore water pressure and reduced effective stress under seismic or cyclic loads.

In-situ seismic testing methods are commonly used for subsurface seismic velocity profiling which assists in the determination of low-strain stiffness of geological materials. With time, there has been an increase in the need for critical infrastructure; hence, the need for such tests is also increasing for efficient engineering design. With time and experience, in-situ test results have been used to estimate soil resistance against seismic hazards such as liquefaction. In-situ tests help determine the soil properties in their natural, undisturbed state. These tests are easy to perform at the site and reduce the dependency on laboratory tests which are often time-consuming, expensive, and require specific monitoring of the specimen properties [24].

The objective of this survey was to obtain the distribution of soil materials and S-wave velocities (V_S) of the shallow subsurface for the computation of elastic engineering properties of the subsurface strata. Standard penetration test and seismic tests of Crosshole Seismic Test (CHT) and MASW test were performed. Hammer energy correction was assigned to the equipment based on energy measurements from past studies on the same and similar equipment [3]. The velocity profiles obtained from

A. Kumar · A. Panjamani (✉) · R. Kumar
Indian Institute of Science, Bangalore, Karnataka 560021, India
e-mail: anbazhagan@iisc.ac.in

K. R. Lenin
Secon Pvt. Ltd, Bangalore, Karnataka, India

K. Muthukkumaran et al. (eds.), *Soil Dynamics, Earthquake and Computational Geotechnical Engineering*, Lecture Notes in Civil Engineering 300,
https://doi.org/10.1007/978-981-19-6998-0_33

MASW were coherent with the borehole records showing the gradual increase in the velocity values from gravelly silty sand at the top to the weathered limestone zone at the bottom of boreholes. Velocity profiles obtained from CHT showed higher velocity values than those obtained from MASW. This difference could have been due to the highly localized and confined nature of Crosshole testing between the boreholes and dispersed, and broader (global) coverage in the MASW test, where velocities measured are averaged out over the travel path length. N-values and V_S values thus obtained were used to estimate factor of safety against liquefaction (FS_L) for subsurface silty sand layers submerged under shallow groundwater.

Site Location and Description

The investigation was carried out in the coastal city of Tuticorin, located along the Eastern coast of Tamil Nadu, India. The testing site is located about 1.5 km west of seashore near Southern Petrochemical Industries Corporation (SPIC) Township consisting of level to gently sloping ground. The project area is covered with the fluvial, fluvio-marine, Aeolian, and unconsolidated marine sediments of the quaternary age, which are underlain by medium to coarse-grained tertiary sandstone and claystone. The groundwater table is present at 1.65 m depth below ground level. The site is in seismic zone II as per the seismic zoning map of India in IS: 1893 (Part I)-2016, which is based on past intensity and not based on possible seismic hazard parameters in the region [2].

Investigation Methods

This study involved the use of multiple exploration methods like SPT, CHT, and MASW. There have been limited studies in the Indian region with multiple approaches to determine subsurface properties. Since each method determines V_S differently and has its own interpretation, using a single method for important analysis such as liquefaction evaluation may not give sufficient estimation accuracy. These different site characterization tools have complementary roles in most situations and are most effectively used in combinations [9]. A brief description of the tests used and the profiles generated is presented in the following subsections.

Standard Penetration Test

The standard penetration test (SPT) is an in-situ dynamic penetration test designed to provide information on the geotechnical engineering properties of soil. This test is the

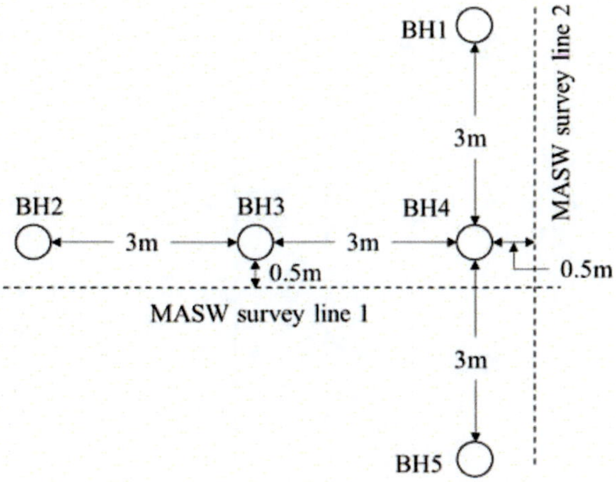

Fig. 33.1 Layout of boreholes for crosshole tests and location of MASW survey lines

most commonly used subsurface exploration method due to its workability and cost-effectiveness. The test provides samples for identification purposes and measures penetration resistance, which can be used for geotechnical design purposes.

SPT is of great use in cases where it is difficult to obtain undisturbed samples for testing, such as gravelly, sandy, silty, sandy clay, or weak rock formations. It is often used to approximate the in-situ density and angle of shearing resistance of cohesion-less soils and the strength of cohesive soils. Several other dynamic and static properties of subsoil layers are well correlated with SPT results [1, 4, 5]. The SPT is most widely used for empirical determination of a sand layer's susceptibility to soil liquefaction in India and also other places.

Hammer energy measurement is an integral part of the SPT procedure. Most of the empirical relations of N-value with static or dynamic soil properties are defined for a specific energy value [10]. If standard or any random energy value is assumed instead of the in-situ measurement, it will lead to erroneous calculations. For the same equipment and operator, the hammer energy ratio (measured energy/theoretical energy) (ER) values were measured previously in the range of 30–40%. The same range is assigned to the equipment, and analysis for liquefaction estimation has been carried out considering three ER values 30, 35, and 40%. In the current study, five boreholes were drilled in a T-shaped geometry (Fig. 33.1). SPT was performed in BH4. The depth profile of N-values with depth is presented in Fig. 33.2a. The N-values at refusal were estimated by extrapolation.

Crosshole Test

Crosshole test involves the determination of the velocity of horizontally travelling P-wave and S-wave in subsurface soil and rock strata between two or more boreholes. This method provides inputs for static/dynamic analyses as a means for computing

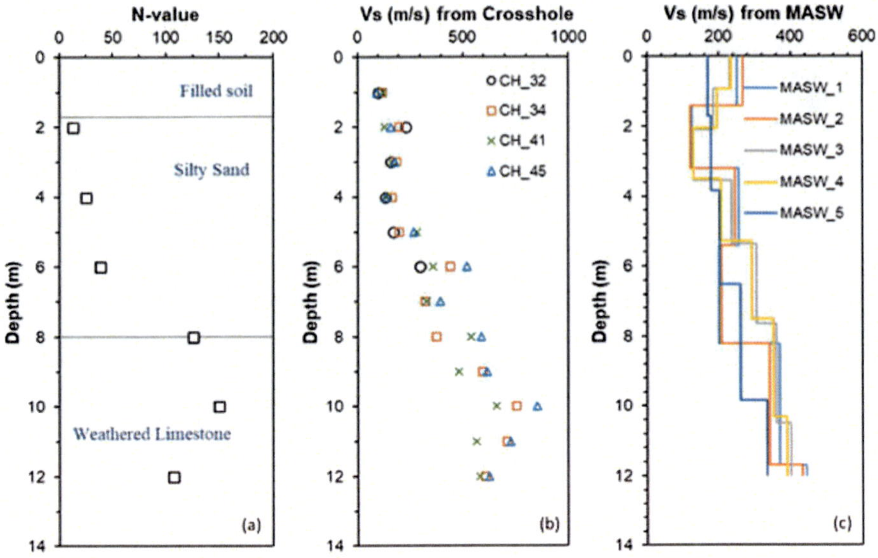

Fig. 33.2 **a** SPT N-values and V_S profiles at the site from **b** crosshole and **c** MASW tests

V_S, V_P, shear modulus, Young's modulus, and Poisson's ratio, or simply for the determination of anomalies that might exist between boreholes [22]. Typical test applications involve bridge/dam foundation analysis, in-situ materials testing, soil and rock mechanics, earthquake engineering, and liquefaction analysis.

Crosshole survey involves lowering a seismic source in one borehole and a 3 (or more) component geophone borehole receiver in the adjacent borehole(s) at the same depth, and impulses from the source are recorded in the receiver. The velocity of seismic waves is calculated by dividing distances between the boreholes by travel times of the recorded waveforms as per ASTM D4428-2000 and IS 13372-1992 part 2.

In the current study, CHT was conducted in four sets, denoted by CH_{32}, CH_{34}, $CH_{41,}$ and CH_{45}, where CH_{ij} means the source is lowered in BHi and the receiver in BHj. Figure 33.3 shows the field setup for the Crosshole test. BIS-SH sparker source and BGK5 borehole receiver (Getomographie GmBH) and Geode seismograph (Geometrics, Inc.) are used for data acquisition. The V_S profiles obtained from CHT are given in Fig. 33.2b. Although the tests were conducted till 20 m depth, the profile is shown only up to 12 m because of the scope of analysis in this study.

MASW Test

The MASW survey is a seismic method used for evaluating the low-strain properties of the subsurface. It utilizes the dispersive nature of the surface waves to determine

Fig. 33.3 Crosshole test setup in the field

the elastic properties of the half-space inverting the recorded field data [14, 15]. The MASW test has become a widely practised technique due to its use of multiple channels for data acquisition of surface waves. MASW technique offers high resolution and high signal-to-noise ratio. The Rayleigh waves offer the mode with the highest energy content among different waves produced during impact [23] and hence is used for the test. Post data acquisition and conventional signal processing techniques could be used for removing the ambient noise from the data and increase signal strength.

MASW system consisting of a 24-channel Geode seismograph with 24 numbers of vertical geophones of 4.5 Hz frequency has been used to carry out field experiments. MASW tests have been carried out with geophone intervals of 1 m. The source has been placed at 6 and 12 m from the nearest geophone to avoid the near-field and far-field effect [15]. The seismic waves are created by hitting a 17.6 lbs sledgehammer on a 300 mm × 300 mm metal plate with five stacks.

MASW tests were conducted near all five boreholes. Twenty-four geophones were used with 1 m receiver spacing and 10 m source distance from the nearest geophone. V_S profiles upto top 12 m of subsurface obtained from MASW test are presented in Fig. 33.2c. The seismic record for a typical MASW test is shown in Fig. 33.4. The phase velocity (V_φ)- frequency dispersion image and V_S profile from MASW for BH3, is shown in Fig. 33.5.

Evaluation of Liquefaction

In general, it is tedious to re-establish an in-situ stress state in the laboratory, and often soil specimens are too disturbed to yield relevant results. Sampling for undisturbed samples in granular soils requires specialized techniques, which are costly. Hence, field tests have become the state of practice for routine liquefaction investigations. Several methods have been developed to evaluate the safety of soil against liquefaction during a seismic event. Several field tests have been found to be helpful in evaluating liquefaction resistance, mainly SPT and V_S-based methods.

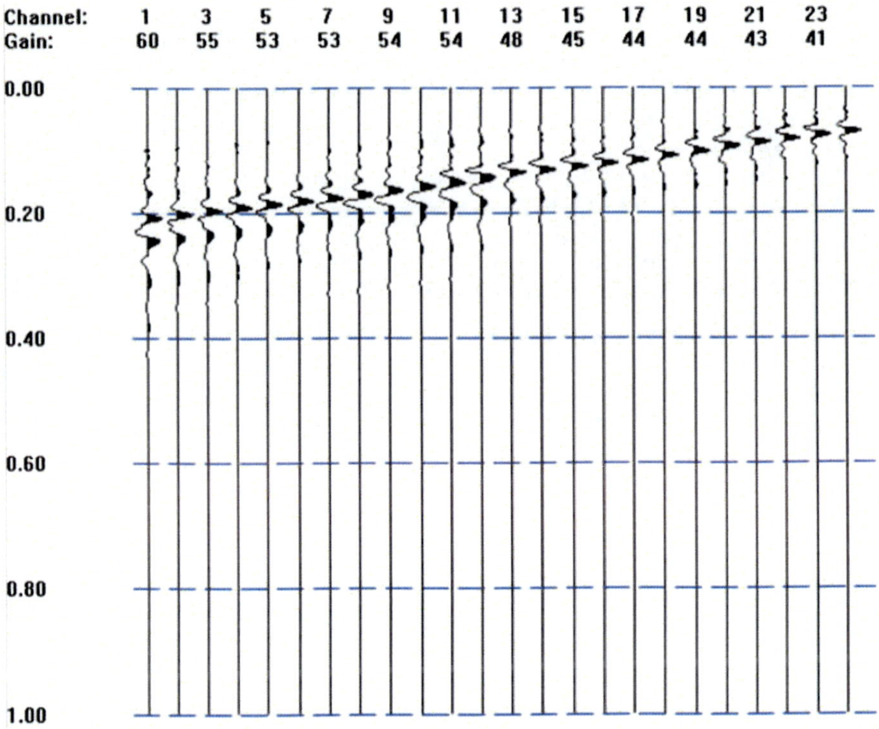

Fig. 33.4 Seismic record obtained from a typical MASW test at the site

Over the years, the 'simplified procedure' has evolved as the primary method for evaluating liquefaction resistance of soils. It was first developed by Seed and Idriss [18] as a stress-based approach and later improved by several researchers [6, 9, 16, 24] to enhance the applicability over various regions and improve the accuracy of soil behaviour prediction. For estimation of liquefaction resistance, two variables need to be calculated (1) seismic demand on the soil layer, also known as cyclic stress ratio (CSR) and (2) the capacity of the soil to resist liquefaction, known as cyclic resistance ratio (CRR). Evaluation of CSR is simple and needs the information of peak ground acceleration (0), total and effective stresses and depth of soil layers. The equation to calculate CSR [18] is given by

$$\mathrm{CSR} = \left(\frac{\tau_{\mathrm{av}}}{\sigma'_{v0}}\right) = 0.65\frac{a_{\max}}{g}\frac{\sigma_{v0}}{\sigma'_{v0}}r_d \tag{33.1}$$

where $a_{\max}$ = PHA at the ground surface due to earthquake, g = acceleration due to gravity, σ_{v0} = total vertical stress, σ'_{v0} = effective vertical stress, τ_{av} = effective shear stress (calculated for the effective number of loading cycles leading to liquefaction),

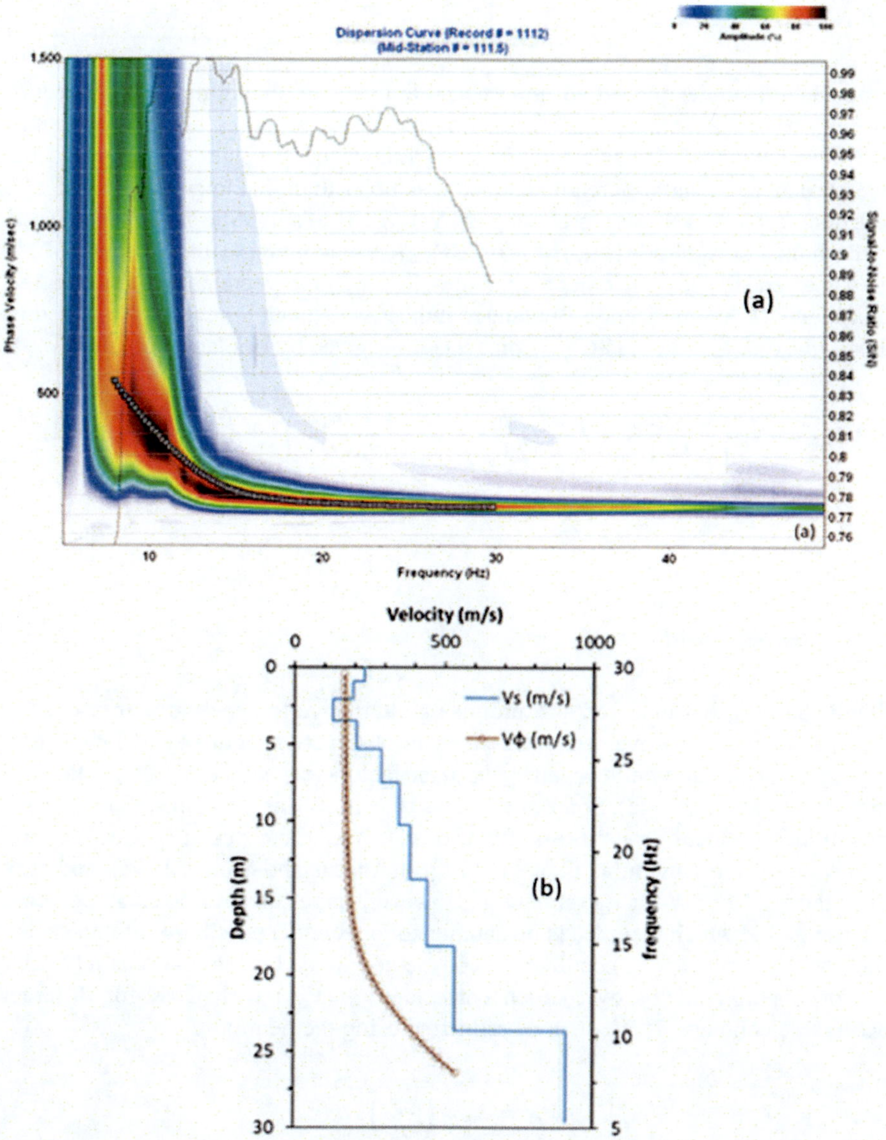

Fig. 33.5 **a** Dispersion image with dispersion curve and **b** 1D V_S profile with dispersion curve

r_d = stress reduction factor accounting for flexibility of soil with depth, estimated from Idriss [11].

For evaluation of CRR, two popular approaches are based on SPT and V_S.

SPT-Based Method

Liquefaction studies played an important role in standardizing the SPT procedure by improving SPT corrections such as overburden correction, energy correction, and fines content (FC) correction. In this method, the N-values are corrected for overburden, rod length, borehole diameter, sampler liner, hammer energy, and FC. The final corrected value is represented as $(N_1)_{60\,cs}$. The estimation of CRR is adjusted to the moment magnitude (M_W) of 7.5 ($CRR_{7.5}$). A magnitude scaling factor (MSF) is used to extend the usage to other magnitudes. Recent improvements in the calculation procedure are summarized in Boulanger and Idriss [9], and the same is followed in this study. Calculation of $CRR_{7.5}$ from $(N_1)_{60cs}$ is given by Boulanger and Idriss [9]:

$$CRR_{7.5} = \exp\left\{\frac{(N_1)_{60\,cs}}{14.1} + \left(\frac{(N_1)_{60\,cs}}{126}\right)^2 - \left(\frac{(N_1)_{60\,cs}}{23.6}\right)^3 + \left(\frac{(N_1)_{60\,cs}}{25.4}\right)^4 - 2.8\right\} \quad (33.2)$$

V_S-based Method

Small strain V_S has emerged as a promising alternative to penetration methods for liquefaction resistance evaluation. Based on the observation, both V_S and liquefaction resistance are similarly influenced by the same factors (e.g. void ratio, stress state and history, and geologic age). Some advantages of using V_S as a parameter are (1) easy in-situ measurement for hard-to-sample soils such as gravels and (2) direct relation to low-strain shear modulus G_{max} [7, 17, 20] (Andrus and Stokoe 1998), and easy procedure of MASW for V_S estimation. However, some concerns do exist regarding the use of V_S, which include (1) no sample collection, (2) possibility of undetected thin low V_S strata, and (3) confinement of method to the low-strain domain [7, 21].

This method relates overburden corrected V_S (V_{S1}) with CRR for different acceptable contents [7]. V_{S1} can be estimated using the relation

$$V_{s1} = V_s * \left(\frac{P_a}{\sigma_v'}\right)^{0.25} \quad (33.3)$$

Andrus and Stokoe [6] provided an upper limit V_{S1}* to V_{S1} for cyclic liquefaction occurrence considering different case studies in China, Japan, Taiwan, and USA. V_{S1}* can be calculated for different fine content, and CRR can be estimated if $V_{S1} \leq V_{S1}$*. $CRR_{7.5}$ from V_{S1} can be calculated using the relation Andrus and Stokoe [7]

$$CRR = \left\{a\left(\frac{V_{s1}}{100}\right)^2 + b\left(\frac{1}{V_{s1}^* - V_{s1}} - \frac{1}{V_{s1}^*}\right)\right\}$$

$$a = 0.0022, b = 2.8 \quad (33.4)$$

where

$$\begin{aligned} V_{s1}^{*} &= 215\,\text{m/s, for sands with FC} \leq 5\% \\ V_{s1}^{*} &= 215 - 0.5 * (\text{FC} - 5)\text{m/s, for sands with } 5\% < \text{FC} < 35\% \\ V_{s1}^{*} &= 200\,\text{m/s, for sands and silts with FC} \geq 35\% \end{aligned} \quad (33.5)$$

Results and Discussion

The Factor of Safety Against Liquefaction

Using SPT N-value data

Table 33.1 gives a sample calculation for correction of measured N-values at 35% ER and estimation of $(N_1)_{60cs}$. Sampler correction and borehole diameter correction are considered as unity as per standard recommendations [10]. Hammer energy corrections were taken from previously available measurements on the same equipment and operator. ER was found to be in the range of 30–40%, which concurs with previous measurements in India by Selvam et al. (2020), Anbazhagan and Sagar [3], and Anbazhagan et al. [4]. Hence, three different energy corrections for 30, 35, and 40% ER were applied, and changes in FS_L values were studied. It is to be noted that the Indian code for earthquake resistance design IS 1893 (Part 1) [13] suggests single standard correction values for different types of equipment. This could lead to incorrect estimation of FS_L as the exact hammer energy delivered remains unknown. Thus, $(N_1)_{60cs}$ remains ambiguous in such cases.

Based on FC data obtained from particle size classification, the procedure suggested by Boulanger and Idriss [9] has been utilized. The lowest FS_L was estimated at 3 m depth, where FS_L is calculated as 1.1 at 30% ER, and 1.21 at 35% ER, which can be considered potentially liquefiable. If the ER value, which was unmeasured but estimated using previous studies, could have been lower, this layer would show a drop in liquefaction resistance. Figure 33.6 shows FS_L values from SPT-based method for different ER values considered. For comparison with the Indian code provisions, ER = 80% for Donut hammer with trip/auto and ER = 45% for Donut hammer with rope and pulley [13] are also considered in the analysis. 45% ER value gives very high CRR and FS_L at and beyond 7 m (CRR, FS_L approaching infinity). 80% ER leads to very high CRR and FS_L beyond 3 m (CRR = 13.99 and FS_L = 76.31 at 4 m). Hence, those values are not plotted in Fig. 33.6. In general overview, the site seems safe when analysed using SPT-based method as per IS1893 [13] ER assumption. To check how the factor of safety against liquefaction varies when V_S based method is used, further analysis is presented in the next section.

Table 33.1 N-value correction for ER 35%

Depth (m)	N-value	ER %	Energy correction	Rod length correction	$(N_1)_{60}$	FC %	$(N_1)_{60cs}$
1	13	35	0.58	0.75	9.66	24	14.65
2	13	35	0.58	0.75	9.66	32	15.10
3	13	35	0.58	0.75	9.27	32	14.71
4	25	35	0.58	0.85	17.53	26	22.67
5	25	35	0.58	0.85	16.58	26	21.72
6	38	35	0.58	0.95	26.13	8	26.49
7	38	35	0.58	0.95	24.16	8	24.53
8	125	35	0.58	0.95	71.15	25	76.22
10	150	35	0.58	1	87.70	40	93.27
12	107	35	0.58	1	61.47	54	67.08

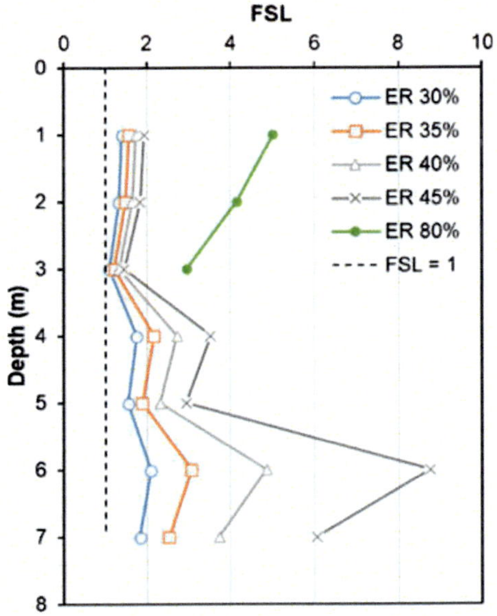

Fig. 33.6 Factor of safety against liquefaction estimated using N-values for ER values of 30, 35, 40, and 45%

Using V_S profiles

V_S from CHT

V_S profiles from CHT are shown in Fig. 33.2b. The liquefaction analysis results are summarized in Table 33.2. V_S profiles from all the CHT are considered separately, and then the average and minimum V_S values at all depths are considered. It can be observed that based on CHT data, some layers do show a prominent liquefiable tendency contrary to what was observed from the SPT data. Only the layers 1–2 m and 3–4 m have FS_L values close to 1 or ≤ 1, which can be a concern and need further investigation. For the layers with $V_{S1} > V_{S1}{}^*$, CRR could not be calculated as these

Table 33.2 Summary of FS_L calculated from CHT data

Depth (m)	CH_32	CH_34	CH_41	CH_45	Average	Minimum
	FS_L	FS_L	FS_L	FS_L	FS_L	FS_L
0–1	**1.14**	2.27	3.59	**0.92**	1.51	**0.92**
1–2	$V_{S1} > V_{S1}*$	$V_{S1} > V_{S1}*$	1.74	$V_{S1} > V_{S1}*$	$V_{S1} > V_{S1}*$	1.74
2–3	$V_{S1} > V_{S1}*$	$V_{S1} > V_{S1}*$	$V_{S1} > V_{S1}*$	$V_{S1} > V_{S1}*$	$V_{S1} > V_{S1}*$	$V_{S1} > V_{S1}*$
3–4	**0.95**	10.5	**1.11**	**1.06**	1.29	**0.95**
4–5	3.89	$V_{S1} > V_{S1}*$	$V_{S1} > V_{S1}*$	$V_{S1} > V_{S1}*$	$V_{S1} > V_{S1}*$	3.89
5–6	$V_{S1} > V_{S1}*$	$V_{S1} > V_{S1}*$	$V_{S1} > V_{S1}*$	$V_{S1} > V_{S1}*$	$V_{S1} > V_{S1}*$	$V_{S1} > V_{S1}*$

Table 33.3 Summary of FS_L calculated from MASW data

Depth (m)	MASW_1	MASW_2	MASW_3	MASW_4	MASW_5	Average	Minimum
	FS_L	FS_L	FS_L	FS_L	FS_L	FS_L	FS_L
0–1	$V_{S1} > V_{S1}*$	$V_{S1} > V_{S1}*$	$V_{S1} > V_{S1}*$	$V_{S1} > V_{S1}*$	$V_{S1} > V_{S1}*$	$V_{S1} > V_{S1}*$	$V_{S1} > V_{S1}*$
1–2	1.37	**1.21**	$V_{S1} > V_{S1}*$	$V_{S1} > V_{S1}*$	$V_{S1} > V_{S1}*$	$V_{S1} > V_{S1}*$	1.21
2–3	**0.94**	**0.83**	**1.06**	**1.04**	$V_{S1} > V_{S1}*$	1.32	**0.83**
3–4	$V_{S1} > V_{S1}*$	$V_{S1} > V_{S1}*$	**0.82**	$V_{S1} > V_{S1}*$	$V_{S1} > V_{S1}*$	$V_{S1} > V_{S1}*$	**0.82**
4–5	$V_{S1} > V_{S1}*$	$V_{S1} > V_{S1}*$	$V_{S1} > V_{S1}*$	$V_{S1} > V_{S1}*$	$V_{S1} > V_{S1}*$	$V_{S1} > V_{S1}*$	$V_{S1} > V_{S1}*$
5–6	$V_{S1} > V_{S1}*$	$V_{S1} > V_{S1}*$	$V_{S1} > V_{S1}*$	$V_{S1} > V_{S1}*$	$V_{S1} > V_{S1}*$	$V_{S1} > V_{S1}*$	$V_{S1} > V_{S1}*$

are not considered liquefiable [7]. FS_L values that can be critical are highlighted in bold.

V_S from MASW

V_S profiles from MASW are shown in Fig. 33.2c. The liquefaction analysis results are summarized in Table 33.3. It can be observed that based on MASW data, some layers do seem to be liquefaction susceptible, much like the CHT data. Layers 2–3 m and 3–4 m have FS_L values close to or ≤ 1. FS_L values that can be critical are highlighted in bold.

Differences in Results from Different Analyses

The difference in the results from analysis by different methods is quite visible. FS_L obtained from SPT shows higher values than the V_s*-based* method. To some extent, FS_L from the V_s-based methods does compare well with SPT for 3 m depth when ER is considered as actual without assumption as per IS1893 [13], resulting in a minimum FS_L. Moreover, for depths with V_{S1} exceeding $V_{S1}*$, comparison with SPT is not possible. Although the measured ER values help find the layer with the least FS_L, energy measurement should be an integral part of SPT and recommended standard values should not be used blindly, as evident from Fig. 33.6 and already discussed in the previous section. The FS_L obtained from two V_S-based methods also vary. The primary reason for variation between the two V_s-*based* calculations is the difference in the V_S profiles. The observed discrepancies between CHT and MASW could be attributed to the picking of the arrival time of the S-wave, as small changes in picked arrival times can lead to significant changes in the estimated V_S. Moreover, water in the borehole casing during the survey may generate tube waves that act as a secondary source of seismic waves inside the borehole. Tube waves are pressure pulses that propagate nearly unattenuated down (and up) the fluid column. These make picking of arrival times difficult.

Conclusion

This study highlights the importance of carrying out multiple geophysical tests to better interpret geotechnical site conditions, which is not possible to obtain using a single type of test. SPT, CHT, and MASW were performed at the same site to determine topographic layers and develop shallow subsurface V_S profile. The two seismic methods delivered different V_S profiles, with CHT giving higher velocities, although the nature of the profiles matched well with the soil profile from borelog, which showed a gradual increase in the stiffness values with depth till the weathered rock layer. N-values and V_S profiles are used to estimate liquefaction resistance using a simplified procedure. The SPT-based method resulted in higher FS_L values in the shallow depth than the V_s-based methods, which showed that the top and subsurface layers at the depth 2–4 m are liquefiable. These differences between different investigation methods are not well understood in the current state of the art and need to be further investigated. This study also highlighted the importance of hammer energy measurement during SPT, as for the same N-value, assuming high ER values (as per IS1893, 2016) would lead to higher FS_L, thus making the results unconservative. It also stresses the need to conduct multiple testing and confirm liquefaction potential to clear the ambiguity introduced because of uncertainty in ER values assumption.

Acknowledgements The authors thank the Dam Safety (Rehabilitation) Directorate, Central Water Commission for funding the project entitled 'Capacity Buildings in Dam Safety' under Dam Rehabilitation and Improvement Project. Authors also thank SERB, DST for funding project 'Development of correction factors for standard penetration test N-values in India through energy measurement and field experiments—Step towards a reliable Liquefaction Potential Assessment', Ref: SERB/F/198/2017-18 dated 11/05/2017.

References

1. Anbazhagan P, Anjali U, Moustafa SSR, Al-Arifi NSN (2016) Correlation of densities with shear wave velocities and SPT N values. J Geophys Eng 13:320–341. https://doi.org/10.1088/1742-2132/13/3/320
2. Anbazhagan P, Gajawada P, Moustafa SSR, Al-Arifi NSN, Parihar A (2014) Provisions for geotechnical aspects and soil classification in Indian seismic design code IS-1893. Disaster Adv 7(3):72–89
3. Anbazhagan P, Ingale SG (2021) Status quo of standard penetration test in India: a review of field practices and suggestions to incorporate in IS 2131. Indian Geotech J 51(2):421–434. https://doi.org/10.1007/s40098-020-00458-8
4. Anbazhagan P, Kumar A, Ingle SG, Jha SK, Lenin KR (2021) Shear modulus from SPT N-value with different energy values. Soil Dyn Earthq Eng 150:106925. https://doi.org/10.1016/j.soildyn.2021.106925
5. Anbazhagan P, Parihar A, Rashmi HN (2012) Review of correlations between SPT N and shear modulus: a new correlation applicable to any region. Soil Dyn Earthq Eng 36:52–69
6. Andrus RD, Stokoe KH II (1997) Liquefaction resistance based on shear wave velocity. In: Youd TL, Idriss IM (eds) NCEER workshop on evaluation of liquefaction resistance of soils, Salt Lake City, UT, Technical Report NCEER-97-0022. National Center for Earthquake Engineering Research, Buffalo, NY, pp 89–128
7. Andrus RD, Stokoe KH II (2000) Liquefaction resistance of soils from shear wave velocity. J Geotech Geoenviron Eng ASCE 126(11):1015–1025
8. ASTM D4428/D4428M (2014) Standard test methods for crosshole seismic testing. ASTM International, West Conshohocken, PA. www.astm.org
9. Boulanger RW, Idriss IM (2014) CPT and SPT based liquefaction triggering procedures. Report No. UCD/CGM-14/01, Center for Geotechnical Modeling, Department of Civil and Environmental Engineering, University of California, Davis, USA
10. Bowles JE (1996) Foundation analysis and design, 5th edn. The McGraw-Hill Companies Inc., New York
11. Idriss IM (1999) An update of the Seed-Idriss simplified procedure for evaluating liquefaction potential. In: Proceedings of TRB workshop on new approaches to liquefaction analysis, FHWA-RD-99-165, Transportation Research Board, Washington, DC
12. IS 13372-Part 2: Code of practice for seismic testing of rock mass, Part 2: Between the borehole, Bureau of Indian Standards, New Delhi (1992)
13. IS 1893-Part 1: Criteria for Earthquake Resistant Design of Structures, Bureau of Indian Standards, New Delhi, India (2016)
14. Park CB, Miller RD, Xia J (1999) Multichannel analysis of surface waves. Geophysics 64(3):800–808
15. Park CB, Miller RD, Xia J (2001) Offset and resolution of dispersion curve in multichannel analysis of surface waves (MASW). In: Proceedings of the SAGEEP 2001, Denver, Colorado, SSM4
16. Seed HB, Tokimatsu K, Harder L, Chung R (1985) Influence of SPT procedures in soil liquefaction resistance evaluations. J Geotech Eng 111:1425–1445. https://doi.org/10.1061/(ASCE)0733-9410(1985)111:12(1425)
17. Seed HB, Idriss IM, Arango I (1983) Evaluation of liquefaction potential using field performance data. J Geotech Eng ASCE 109(3):458–482
18. Seed HB, Idriss IM (1971) Simplified procedure for evaluating soil liquefaction potential. J Soil Mech Found Div ASCE 97(9):1249–1273
19. Sermalai S, Mukundan M, Alagirisamy S (2022) Standard penetration test (SPT) pitfalls and improvements. In: Satyanarayana Reddy CNV, Muthukkumaran K, Satyam N, Vaidya R (eds) Ground characterization and foundations. Lecture notes in civil engineering vol 167. Springer, Singapore, pp 363–375. https://doi.org/10.1007/978-981-16-3383-6_33

20. Stokoe KH II, Roesset JM, Bierschwale JG, Aouad M (1988) Liquefaction potential of sands from shear wave velocity. In: Proceedings of 9th world conference on earthquake engineering, Tokyo, Japan, vol III, pp 213–218
21. Verdugo R (2016) Experimental and conceptual evidence about the limitations of shear wave velocity to predict liquefaction. Soil Dyn Earthq Eng 91:160–174. https://doi.org/10.1016/j.soildyn.2016.09.046
22. Wightman WE, Jalinoos F, Sirles P, Hanna K (2003) Application of geophysical methods to highway related problems. Federal Highway Administration, Central Federal Lands Highway Division, Lakewood, CO, Publication No. FHWA-IF-04-021
23. Xia J, Miller RD, Park CB (1999) Estimation of near-surface shear-wave velocity by inversion of Rayleigh waves. Geophysics 64(3):691–700
24. Youd TL, Idriss IM, Andrus RD, Arango I, Castro G, Christian JT, Dobry R, Finn WDL, Harder LF Jr, Koester JP, Liao SSC, Marcuson WF III, Martin GR, Mitchell JK, Moriwaki Y, Power MS, Robertson PK, Seed RB, Stokoe KH II (2001) Liquefaction resistance of soils: summary report from the 1996 NCEER and 1998 NCEER/NSF workshops on evaluation of liquefaction resistance of soils. J Geotech Geoenviron Eng 127(10):817–833. https://doi.org/10.1061/(asce)1090-0241(2001)127:4(297)

Chapter 34
Evaluation of Site Effects Using HVSR Measurements Along Srinagar Metro Alignment, Jammu and Kashmir

Falak Zahoor, K. Seshagiri Rao, Fuzail Shoukat Wani, Syed Mehran Andrabi, and Neelima Satyam

Introduction

Microtremor HVSR (MHVSR) Method

Microtremors, also known as seismic noise, are the ground vibrations caused by sources like sea waves, wind, traffic, human activity, etc. [1]. Properties of the subsoil affect the microtremors or ambient noise measured at a site resulting in local site effects. These effects in turn cause amplification of waves when they travel from stiff bedrock to softer surficial layers. Nakamura's method [2] is most widely used for site response analysis using microtremor recordings [3, 4] and for seismic microzonation of cities [5, 6].

Peaks in HVSR curve correspond to the impedance contrast between two layers in the subsoil, and the H/V amplitude is a function of the magnitude of this impedance contrast. Hence, this property has been widely used to determine the depth of bedrock in several regions [7–9]. For this purpose, a joint fit between HVSR curves and dispersion curves from MASW survey is carried out using the information of shear wave velocity of the subsoil as a constraint for the analysis. Multiple frequency peaks are possible in HVSR curves suggesting the presence of more than one impedance contrasts at a site [10]. The lower frequency peak corresponds to impedance contrast

F. Zahoor (✉) · K. Seshagiri Rao
Indian Institute of Technology Delhi, Hauz Khas, Delhi 110016, India
e-mail: falak@nitsri.ac.in

F. Zahoor · F. S. Wani · S. M. Andrabi
National Institute of Technology Srinagar, Srinagar, Jammu and Kashmir 190006, India

N. Satyam
Indian Institute of Technology Indore, Simrol, Madhya Pradesh 452020, India

K. Muthukkumaran et al. (eds.), *Soil Dynamics, Earthquake and Computational Geotechnical Engineering*, Lecture Notes in Civil Engineering 300,
https://doi.org/10.1007/978-981-19-6998-0_34

at deeper depth, whereas higher frequency peak means a shallow depth strong impedance contrast.

It is common to correlate site amplification at high frequency with shear wave velocity of the soil up to 30 m depth (V_{s30}) [11]; however, recent studies contend that this correlation does not hold good and is poorly constrained [12–14]. It has been now understood that the complete shear wave velocity profile up to much deeper depths than 30 m and the impedance contrast between the layers in subsoil have a major role in site amplification [15, 16].

In practice, V_{S30} remains a widely used parameter for site classification despite these obvious limitations. This is mainly because of the simplicity of using a single parameter and its reliable determination from field testing. However, more refined site classification schemes have been recently introduced by several authors in terms of parameters other than V_{S30} [17–24]. The seismic codes around the world must weigh in the limitations of the use of V_{S30} for site classification and therefore incorporate changes to include better classification schemes. Keeping in view the limitations of using V_{s30} as a proxy, alternative site classification schemes need to be adopted.

Study Area

Srinagar is the summer capital of Jammu and Kashmir and is one of the most densely populated cities in the region. It is located in the basin of Kashmir valley, situated in the seismically active Himalayas. Jhelum river flows through the centre of the Srinagar city depositing its sediments in its floodplains in the form of silt, clay, and fine sand. These are young, unconsolidated sediments with high moisture content found mostly in the central part of the city on the banks of Jhelum. In addition to Jhelum, lakes like Dal and Anchar are present within the boundaries of the city. Isolated hills like Koh-i-Maran (Hari Parbat) and Sulaiman (Shankaracharya) hill outcrop within the city (Fig. 34.1). Towards the north, Karewa soils are exposed and uplifted forming a belt of land between Dal and Anchar [25]. This belt is known as Soura-Buchpora Karewa. Karewas are lacustrine deposits left behind due to the draining of the ancient lake that impounded the Kashmir basin [26].

Srinagar metro rail is set to be constructed to improve the transport facilities and connectivity within the city. Three corridors—Corridor I, Corridor II, and Corridor III—have been proposed along the Srinagar metro rail alignment spreading over a total length of ~50 km and connecting most parts within the city. In this study, single station microtremor horizontal-to-vertical spectral ratio (MHVSR) testing procedure [27] has been conducted at sites along the alignment of Srinagar metro (Fig. 34.1). Fundamental frequency (f_0) at each site has been determined which is an important soil dynamic parameter. Results from MHVSR have then been compared with the V_{S30} data obtained at the same sites in a previous study by our group [28] in order to understand if there is a correlation between V_{S30} and f_0, and whether V_{S30} can be used as a proxy for site classification in this region.

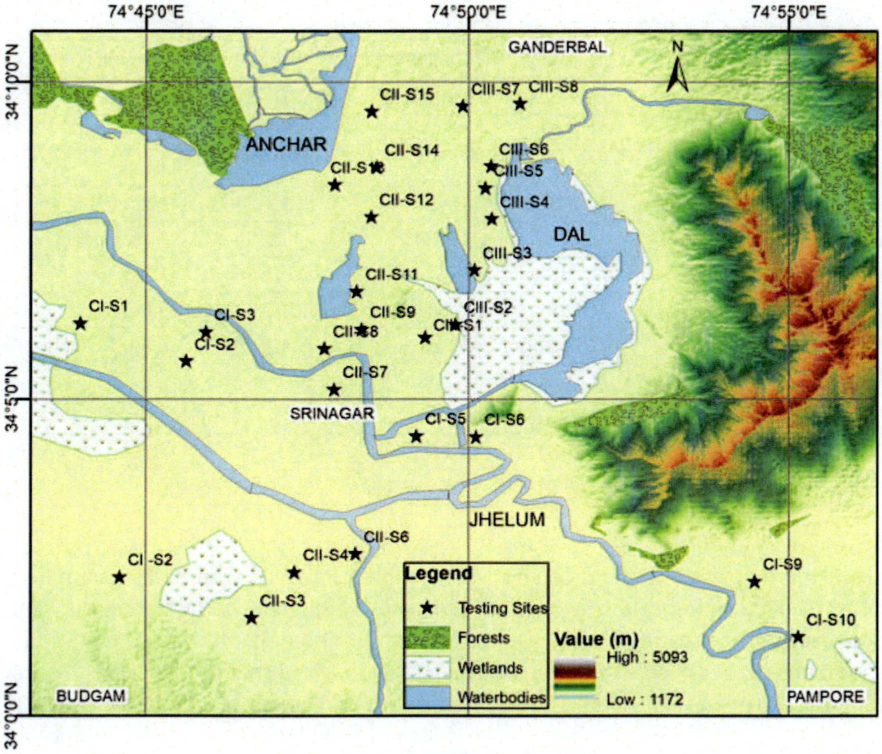

Fig. 34.1 Map of Srinagar city with superimposed topography (created using ArcGIS [29])

Data Acquisition and Analysis

Zahoor et al. [28] performed multichannel analysis of surface waves (MASW) testing at 33 sites along the three corridors (CI, CII, CIII) of the proposed Srinagar metro alignment. Using the resulting shear wave velocity profiles, they determined the V_{S30} parameter at each site and used it further to classify the sites according to NEHRP guidelines [30]. In the present study, we aimed at conducting single station microtremor testing at each of these 33 sites. However, due to certain limitations, recordings could be obtained only at 27 locations as presented in Table 34.1.

Data Acquisition and Field Testing

Ambient noise was recorded at all the 27 locations using a TROMINO® ENGY 3G portable equipment which works in the frequency range of 0.1–100 Hz. The instrument has spikes which facilitate its fixing in the ground (Fig. 34.2). The instrument

Table 34.1 Sites considered for microtremor testing along Srinagar metro alignment

Label	Location	Label	Location	Label	Location
CI-S1	HMT	CII-S2	Humhama	CIII-S1	Rainawari
CI-S2	Parimpora	CII-S3	Peerbagh	CIII-S2	Saidakadal
CI-S3	Qamarwari	CII-S4	Hyderpora	CIII-S3	Nigeen
CI-S5	Polo Ground	CII-S6	Barzulla	CIII-S4	Hazratbal
CI-S6	Sonwar	CII-S7	GMC	CIII-S5	Naseembagh
CI-S9	Sempora	CII-S8	Nawakadal	CIII-S6	Habak
CI-S10	Pampore	CII-S9	Razaikadal	CIII-S7	Zukura
		CII-S11	Zadibal	CIII-S8	Batpora
		CII-S12	Nowshera		
		CII-S13	Soura		
		CII-S14	90 ft Road		
		CII-S15	Osmanabad		

is then levelled with the help of an in-built bubble level at the top. The sampling rate for the recording of the seismic noise is set at 128 Hz. The instrument is kept undisturbed at the testing location for a duration of about 30–40 min to record the ambient noise. For each of the tests, care is taken to avoid recording at places or in times of large noise from nearby traffic, continuous movement of humans, and strong winds.

Data Processing and Analysis

Grilla software [31] was used to handle the acquired raw data and to process it in order to obtain the final HVSR curves. For the analysis of the recorded seismic noise, a frequency range of 0–20 Hz was chosen because this is the general range of waves generated during earthquakes. Nakamura [2] method was applied to the raw data in Grilla which uses the Fourier analysis to determine the amplitude spectrum for the two horizontal and one vertical recordings. A 20–40-s time window was selected for the calculation of Fourier spectra. A smoothing technique was applied based on 10% wide triangular window to get rid of spurious spikes in the HVSR curve. Nakamura method calculates the spectral ratio of the average of the two horizontal components to the vertical component (HVSR), removing the influence of source effect [32]. The H/V spectral ratio is calculated by making use of the following formula:

$$T_{H/V}(\omega) = \{[S_{\mathrm{NS}}(\omega) + S_{\mathrm{EW}}(\omega)]/2\}/S_V(\omega) \tag{34.1}$$

where $S_{\mathrm{NS}}(\omega)$ and $S_{\mathrm{EW}}(\omega)$ are the two perpendicular horizontal spectral components of the site spectra, and $S_V(\omega)$ is the vertical spectral component.

Fig. 34.2 Microtremor testing equipment (TROMINO®) setup in the field

Results and Discussion

HVSR Measurements

The HVSR curves were obtained from the analysis of microtremor recordings at each of the 27 sites in this study. A close inspection of the shapes and the *H*/*V* peaks in the HVSR images reveals a specific trend within the dataset. Sites clustered within an area having a particular geology and soil type tend to show similar shapes of HVSR curves (with a few exceptions). Moreover, within a particular cluster, the *H*/*V* peaks at various sites occur within the same range of fundamental frequency. The typical HVSR curves for each of these clusters are presented in Fig. 34.3. In the three corridors along the Srinagar metro alignment, the HVSR curves have been grouped into 6 clusters based on their shapes and number of peaks (Table 34.2), following the methodology of [33]. The table includes the range of fundamental frequencies (f_0) and *H*/*V* amplitudes (A_0) observed within each cluster. Fundamental frequency (f_0) is an important dynamic parameter of soils. In addition to this, V_{S30} values and corresponding NEHRP classifications of the sites have also been added for comparison taken from the work by Zahoor et al. [28]. These clusters have been found to be directly related to the geological formations and typical soils at the sites.

Cluster I (Jhelum Floodplains). Sites falling within this cluster lie in the floodplain of Jhelum which is a low-lying area in the central part of Srinagar city. These areas have deep deposits of young, unconsolidated, and soft riverine sediments mostly comprising fine sand, silt, and clay. Majority of the sites show HVSR curves having two major *H*/*V* peaks (Fig. 3a), one in the frequency range 0.21–0.29 Hz which is considered as the fundamental frequency (f_0) and a second larger peak (f_1) in the range 0.7–1.1 Hz. These sites have been grouped under Cluster Ia. f_1 is interpreted to be corresponding to the shallow interface between the soft, unconsolidated sediments and the stiffer sediments of Karewa below. On the other hand, f_0 is interpreted to be corresponding to the presence of a deep stiff layer underlying the top sediments. A few exceptions have been found in the floodplains and have been grouped under clusters Ib and Ic. Cluster Ib consists of sites lying in Jhelum floodplain which show three *H*/*V* peaks (Fig. 3b), while sites in Cluster Ic show only a single peak (Fig. 3c). These exceptions may signify localized changes in geology within the plains.

Cluster II (Soura-Buchpora Karewa). This cluster contains sites which lie on the exposed elevated areas of stiff Karewa soils in the northern part of Srinagar city known as Soura-Buchpora Karewa. The HVSR curve has a single major peak in the low frequency range 0.28–0.53 Hz (Fig. 3d). This *H*/*V* peak may correspond to the interface between the stiff Karewa sediments in this area and the underlying stiffer deposits.

Cluster III (Transition Zone near Hills). A few sites are located in the transition zone between the plains and the outcropping hills like Sulaiman (Shankaracharya hill) and Koh-i-Maran (Hariparbat). These sites show peculiar shape of HVSR curves (Fig. 3e) with a broad peak comprising two or more overlapped peaks, spreading over large frequency range.

Cluster IV (Pampore). This cluster includes sites at the southern end of the alignment near Pampore. These sites lie in Pampore Karewa located near Jhelum and show a single peak (Fig. 3f).

Comparison with V_{S30}

The results obtained from microtremor testing along the metro alignment in this study have been compared with V_{S30} values obtained at the same sites in a previous study by our group [28]. The comparison indicates that a direct correlation between V_{S30} and f_0 does not exist, as is clear from a close observation of Table 34.2. It was noted that certain sites with high V_{S30} classified as classes D and C using NEHRP guidelines showed low frequency peaks (Table 34.2). This has been repeatedly reported at a large number of sites in this study, suggesting that these cases are not merely isolated anomalies. For example, all sites in Soura-Buchpora Karewa grouped in Cluster II,

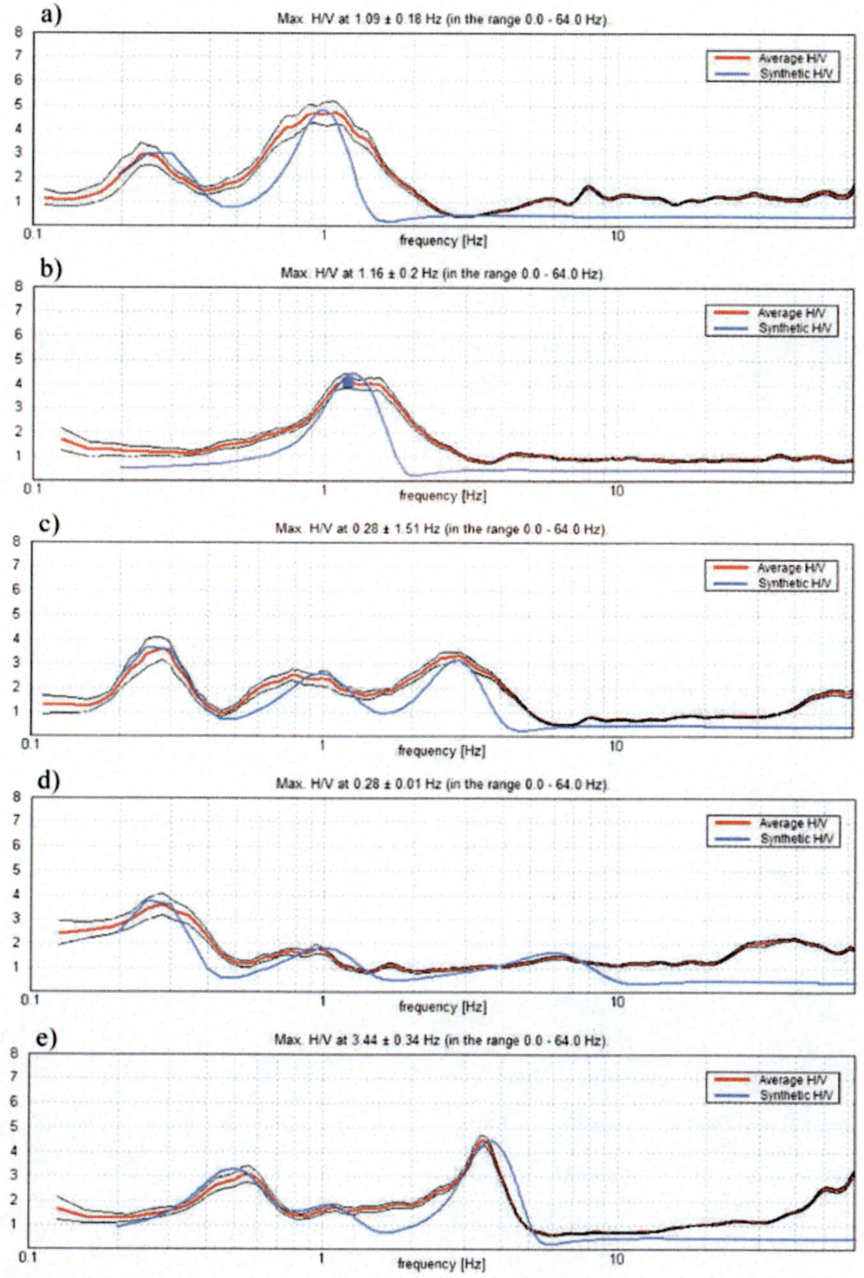

Fig. 34.3 Typical HVSR curves obtained in **a** Cluster Ia, **b** Cluster Ib, **c** Cluster Ic, **d** Cluster II, **e** Cluster III, and **f** Cluster IV

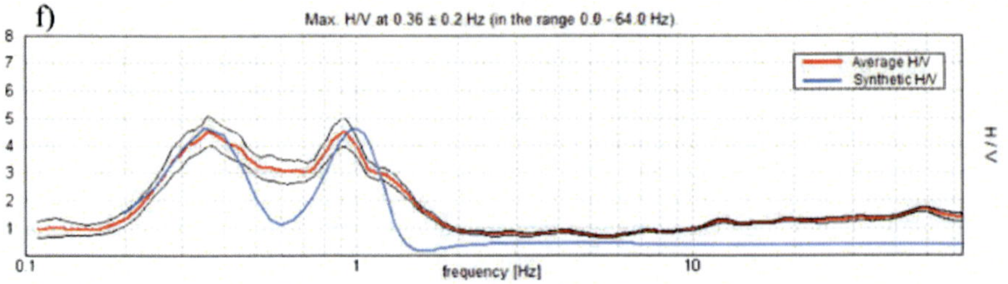

Fig. 34.3 (continued)

Table 34.2 Comparison of dynamic properties of soils in different clusters obtained from HVSR and MASW methods along the metro alignment

Cluster	Details	Sites	Range of V_{S30} (m/s)	Site class (NEHRP)	Range of f_0 (Hz)	Range of A_0 (Hz)
I (a)	Jhelum floodplains (double peak)	CI-S1, CI-S2, CI-S3, CII-S2, CII-S3, CII-S4, CII-S6	146–250	E, D	0.21–0.29	1.63–3.83
I (b)	Jhelum floodplains (triple peak)	CII-S7, CII-S8	175–184	E, D	0.28–0.31	1.94–2.67
I (c)	Jhelum floodplains (single peak)	CI-S5, CI-S6	140–234	E, D	0.47–1.16	3.79–4.16
II	Soura-Buchpora Karewa	CII-S12, CII-S13, CII-S14, CII-S15, CIII-S3, CIII-S4, CIII-S5, CIII-S6, CIII-S7, CIII-S8	351–429	D, C	0.28–0.53	3.14–4.99
III	Transition zone near Hills	CII-S9, CII-S11, CIII-S1, CIII-S2	196–446	D, C	0.58–0.99	2.59–3.09
IV	Pampore	CI-S9, CI-S10	138–256	E, D	0.36–0.65	2.97–3.69

despite showing stiff soils with V_{S30} as high as 351–429 m/s (classified as site classes D and C), have peak frequency as low as 0.28–0.53 Hz. The same is observed in the sites in the transition zone (Cluster III).

V_{S30} as Proxy for Site Amplification

The comparison of HVSR fundamental frequency (f_0) and V_{S30} at each site along the metro alignment clearly indicates that a significant correlation between these parameters does not exist. The results confirm the claims of Castellaro et al. [13], Park and Hashash [16], etc., who have stated that the use of V_{S30} as the sole parameter for site classification is inappropriate and insufficient. This means that stiff sites with high V_{S30} can show low fundamental frequency and vice versa. Keeping in view the limitations of using V_{s30} as a proxy, it is suggested that site classification schemes alternative to the existing ones in seismic codes need to be adopted.

Conclusions

Microtremor testing was conducted at 27 sites along the alignment of metro rail alignment proposed in Srinagar city of Jammu and Kashmir. Using Nakamura's method [2] of analysis of seismic noise, horizontal-to-vertical spectral ratio (HVSR) curves were obtained for each location. These images were then utilized to determine the fundamental frequency (f_0) of the subsoil and the H/V amplitude of the peak. The peaks in HVSR images correspond to the interface between the layers within the subsoil [7], and the amplitude of the peak is directly related to the magnitude of the impedance contrast. The HVSR curves obtained for all the sites were then grouped into 4 clusters (Clusters I, II, III, and IV) based on their shapes and number of peaks. It was observed that sites located within the same geological conditions and soils show similar shape of the HVSR curves. Further, a comparison of the fundamental frequency (f_0) and V_{S30} from a previous study [7] was carried out, in order to understand the correlation between the two parameters.

An irregularity was witnessed at sites having stiff soils (V_{S30} ~ 234–450 m/s) and falling in site classes C and D as per NEHRP guidelines. HVSR images attained at these locations showed peaks at low frequencies (0.21–1.16 Hz). Instead of being an isolated observation, this trend was consistently reflected at several sites. For example, all sites in the Soura-Buchpora Karewa like Batpora, Osmanabad, Habak, Hazratbal, Zukura, etc., and in the transition zone like Saidakadal, Rainawari, etc., presented an incoherency between the two parameters, whereby it was concluded that a correlation does not exist between the parameters (f_0 and V_{S30}), confirming the results of the likes of Castellaro et al. [13]. V_{S30} is a universally adopted parameter for seismic site classification at present, due to the ease of its determination through low-cost geophysical field tests and the simplicity of the site classifications based on this parameter. However, it has been proven that its use as a sole parameter for this purpose is questionable [15]. In fact, the complete shear wave velocity profile up to the bedrock must be considered for the correct determination of seismic amplification. As a result, advanced site classification schemes have been presented in several recent studies employing other parameters with or without combination with V_{S30} [18–24]. This study concludes that the use of V_{S30} alone for site classification is not sufficient, and hence, new classification schemes which take other parameters into consideration must be used to arrive at a better and more appropriate seismic site characterization for the region. This is important because this in turn affects the proper and safe design of buildings and other structures and a suitable seismic microzonation procedure.

References

1. Ibs-von Seht M, Wohlenberg J (1999) Microtremors measurements used to map thickness of soft soil sediments. Bull Seismol Soc Am 89:250–259
2. Nakamura Y (1989) A method for dynamic characteristics estimation of subsurface using microtremor on the ground surface. QR Railway Tech Res Inst 30(1):25–33
3. Field EH, Clement AC, Jacob KH, Aharonian V, Hough SC, Friberg PA, Babaian TO, Karapetian SS, Hovanessian SM, Abramian HA (1995) Earthquake siteresponse study in Giumri (Formerly Leninakan), Armenia using ambient noise observations. Bull Seismol Soc Am 85:349–353
4. Lachet C, Hatzfeld D, Bard P-Y, Theodulidis N, Papaioannou C, Savvaidis A (1996) Site effects and microzonation in the city of Thessaloniki (Greece) comparison of different approaches. Bull Seismol Soc Am 86:1692–1703
5. Rao S, Satyam N (2005) Site characterisation through microtremor studies for seismic microzonation of Delhi. In: Proceedings of 16th international conference on soil mechanics and geotechnical engineering. IOS Press, Osaka, Japan, pp 2703–2706
6. Rao KS, Rathod GW (2014) Seismic microzonation of Indian megacities: a case study of NCR Delhi. Indian Geotech J 44:132–148
7. Yamanaka H, Takemura M, Ishida H, Niwa M (1994) Characteristics of long-period microtremors and their applicability in exploration of deep sedimentary layers. Bull Seismol Soc Am 84(6):1831–1841
8. Giampiccolo E, Gresta S, Mucciarelli M, De Guidi G, Gallipoli MR (2001) Information on subsoil geo-logical structure in the city of Catania (Eastern Sicily) from microtremor measurements. Ann Geophys 44(1):1–11
9. Pappalardo G, Imposa S, Mineo S, Grassi S (2015) Evaluation of the stability of a rock cliff by means of geophysical and geomechanical surveys in a cultural heritage site (South-Eastern Sicily). Italian J Geosci 135(3):1–47
10. Macau A, Benjumea B, Gabàs A, Figueras S, Vila M (2015) The effect of shallow quaternary deposits on the shape of the H/V spectral ratio. Surv Geophys 36:185–208
11. Borcherdt RD (1994) Estimation of site-dependant response spectra for design earthquake (methodology and justification). Earthq Spectra 10:617–653
12. Lee VW, Trifunac MD, Todorovska MI, Novikova EI (1995) Empirical equations describing attenuation of the peaks of strong ground motion, in terms of magnitude, distance, path effects and site conditions, Los Angeles, California. Department of Civil Engineering, University of Southern California, Report CE 95-02
13. Castellaro S, Mulargia F, Rossi PL (2008) Vs30: proxy for seismic amplification? Seismol Res Lett 79(4):540–543

14. Lee VW, Trifunac MD (2010) Should average shear-wave velocity in the top 30 m of soil be used to describe seismic amplification? Soil Dyn Earthq Eng 30:1250–1258
15. Frankel AD, Carver DL, Williams RA (2002) Nonlinear and linear site response and basin effects in Seattle for the M 6.8 Nisqually, Washington, earthquake. Bull Seismol Soc Am 92:2090–2109
16. Park D, Hashash YMA (2004) Probabilistic seismic hazard analysis with nonlinear site effects in the Mississippi embayment. In: Proceedings of the 13th world conference on earthquake engineering, Vancouver, CD-Rom edition, Paper no. 1549
17. Steidl JH (2000) Site response in Southern California for probabilistic seismic hazard analysis. Bull Seismol Soc Am 90:149–169
18. Zhao JX, Irikura K, Zhang J, Fukushima Y, Somerville PG, Asano A, Ohno Y, Oouchi T, Takahashi T, Ogawa H (2006) An empirical site-classification method for strong-motion stations in Japan using H/V response spectral ratio. Bull Seismol Soc Am 96:914–925
19. Fukushima Y, Bonilla LF, Scotti O, Douglas J (2007) Site classification using horizontal-to-vertical response spectral ratios and its impact when deriving empirical ground-motion prediction equations. J Earthquake Eng 11:712–724
20. Ghasemi H, Zare M, Fukushima Y, Sinaeian F (2009) Applying empirical methods in site classification, using response spectral ratio (H/V): a case study on Iranian strong motion network (ISMN). Soil Dyn Earthq Eng 29:121–132
21. Luzi L, Puglia R, Pacor F, Gallipoli MR, Bindi D, Mucciarelli M (2011) Proposal for a soil classification based on parameters alternative or complementary to V_{S30}. Bull Earthq Eng 9:1877–1898
22. Di Alessandro C, Bonilla LF, Boore DM, Rovelli A, Scotti O (2012) Predominant-period site classification for response spectra prediction equations in Italy. Bull Seismol Soc Am 102:680–695
23. Pitilakis K, Riga E, Anastasiadis A (2013) New code site classification, amplification factors and normalized response spectra based on a worldwide ground-motion database. Bull Earthq Eng 11:925–966
24. Pitilakis K, Riga E, Anastasiadis A, Fotopoulou S, Karafagka S (2019) Towards the revision of EC8: proposal for an alternative site classification scheme and associated intensity dependent spectral amplification factors. Soil Dyn Earthq Eng 126:105173
25. Bhat MS (2017) In: Koul MN (ed) Geomorphological field guidebook on Kashmir Himalaya. Indian Institute of Geomorphologists, Allahabad, p 32
26. Bhatt DK (1975) On the quaternary geology of the Kashmir valley with special reference to stratigraphy and sedimentation, vol 24(1). Geological Survey of India, Miscellaneous Publications, pp 188–203
27. Molnar S, Cassidy JF, Castellaro S, Cornou C, Crow H, Hunter JA, Yong A (2018) Application of microtremor horizontal-to-vertical spectral ratio (MHVSR) analysis for site characterization: state of the art. Surv Geophys 39:1–19
28. Zahoor F, Rao KS, Malla SA, Tariq B, Bhat WA (2021) Seismic site characterization using MASW of sites along Srinagar metro rail alignment, Jammu and Kashmir. In: Patel S, Solanki CH, Reddy KR, Shukla SK (eds) Proceedings of the Indian geotechnical conference 2019, Lecture notes in civil engineering, vol 138. Springer, Singapore, pp 581–593
29. ESRI: ArcGIS Desktop: Release 10.5. Redlands, CA: Environmental Systems Research Institute (2016)
30. BSSC: NEHRP Recommended Provisions for seismic Regulations for New buildings and other Structures, Part 1: Provisions, FEMA 368, Federal Emergency Management Agency, Washington, D.C. (2003)
31. Micromed s.p.a. (2009). The short Tromino how to, p 30
32. Lermo J, Chávez-García FJ (1993) Site effect evaluation using spectral ratios with only one station. Bull Seismol Soc Am 83(5):1574–1594
33. Bonnefoy-Claudet S, Baize S, Bonilla LF, Berge-Thierry C, Pasten CR, Campos J, Volant P, Verdugo R (2009) Site effect valuation in the basin of Santiago de Chile using ambient noise measurements. Geophys J Int 176:925–937

Chapter 35
Assessment of Underground Structure–Soil Interaction Under Dynamic Conditions

K. S. Amith and S. Ganesh Kumar

Introduction

Over the past few years, construction of underground structures for parking and storage, telecommunication services, transportation, etc. has increased and played a prominent role in infrastructure development [1]. These underground structures found to be more efficient in its operation than super-structures constructed on ground due to the confinement effect offered by surrounding ground. Hence, it has been understood from general perception that, underground structures are not much vulnerable to earthquakes or other dynamic conditions. However, several earthquakes such as San Fernando (1971), Kobe Earthquake (1995) [2], Chi-Chi Earthquake (1999) [3], Wenchuan earthquake (2008) [4] caused significant deformation and damage to the underground structures. Similarly, the behaviour of tunnel mainly depends on type of soil surrounding the tunnel, type of tunnel lining, depth of tunnel embedment, and maximum ground acceleration induced due to earthquake. In assessing the dynamic performance of tunnel structures, the experimental studies on scaled down model tunnel through 1-g shaking table tests contributed greater insights on tunnel–soil interaction. The results concluded that, with the increase in acceleration response, dynamic earth pressure and settlement increased causing more deformation and displacement [5–8].

In India, several tunnel constructions have been planned by border road construction (BRO) out of which 60% of the tunnel construction planned in the eastern mountainous states of Arunachal Pradesh and Jammu and Kashmir which falls in

K. S. Amith (✉) · S. Ganesh Kumar
Academy of Scientific and Innovative Research (AcSIR), Ghaziabad 201002, India
e-mail: amith.cbri20j@acsir.res.in

S. Ganesh Kumar
e-mail: ganeshkumar@cbri.res.in

Geotechnical Engineering Division, CSIR-CBRI, Roorkee 247667, India

K. Muthukkumaran et al. (eds.), *Soil Dynamics, Earthquake and Computational Geotechnical Engineering*, Lecture Notes in Civil Engineering 300,
https://doi.org/10.1007/978-981-19-6998-0_35

high seismic zone. This motivated the authors to assess the dynamic performance of tunnel–soil interaction under dynamic events. In addition to this, the continuous occurrence of foreshock and after shock events associated during earthquake incidence is also simulated in this study in the form of repeated incremental acceleration loading conditions. The tunnel–soil interaction is evaluated in terms of acceleration response at different depths and developed dynamic soil pressure values. Also, the maximum tunnel displacement and strain development is estimated using digital image correlation. Based on observations, parameters influencing the performance of tunnel system under repeated incremental shaking conditions were identified and discussed.

Methodology

Shaking Table

The experimental model tests were conducted on shaking table facility at CSIR-CBRI. The shaking table has dimensions 2.0 × 2.0 m with maximum load capacity and actuator displacement of 3 T tones and ± 160 mm, respectively. The operating frequency and acceleration range with which shake table can work is 0.01 to 50 Hz and 0.001 to 1 g, respectively. For testing, test tank having dimensions 1.7 × 0.75 × 1 m was used and mounted over the shaking table. To minimize boundary effects, PU foam having thickness 50 mm was placed inside the tank perpendicular to shaking directions.

Soil Selected for the Study

In the current research work, locally available Solani sand was selected and used. The soil was characterized as per IS 2720. The sand was characterized as poorly graded sand as per IS 2720: Part III and IV. [9–11] The maximum and minimum density of sand was found to be 16.1 and 13.9 kN/m^3, respectively. The other soil properties are listed in Table 35.1.

Experimental Setup and Testing Conditions

For experimental testing, ground density 60% and having 600 mm thickness was prepared using dry pluviation technique. The quantity of soil required to achieve the selected height was estimated and prepared layer by layer to achieve maximum uniformity. After placing first layer having thickness 600 mm, the scaled down model

Table 35.1 Properties of soil

S No.	Soil characteristics	Value
1	Specific gravity	2.67
2	Uniformity coefficient (C_u)	2.6
3	Coefficient of curvature (C_C)	1.14
4	Void ratio corresponding to 60% relative density	0.75
5	Soil type	Poorly graded

has been placed. After placement of scaled down tunnel model, the sand raining was continued up to the targeted height.

A scaled down model tunnel having dimension 270 × 270 mm with 730 mm length and having a thickness of 30 mm was prepared with Gypsum material. For casting mix, ratio of 1:0.7 was used with the targeted strength of 0.4 MPa which is in accordance with published literature [12]. The model was prepared with a scale ratio of 1:10 corresponding to prototype conditions.

The objective of the study is to evaluate the effect of continuous shaking events on tunnel system; hence, the testing conditions were selected accordingly. For this, sinusoidal loading having 0.3 and 0.4-g intensity with 5 Hz frequency having 40 s duration was selected and applied. The selected intensity was similar to high to very high earthquake loading in actual conditions. Similarly, the loading was applied to the prepared ground sequentially one after the other. Through series of sensors and instrumentation, acceleration and soil pressure were monitored. Similarly, the maximum deformation and strain development within the tunnel model under repeated shaking events also monitored.

Instrumentation

To investigate the tunnel–soil interaction under repeated incremental loading events, 3 accelerometers and 2 earth pressure cells were placed vertically as shown in Fig. 35.1. To measure the acceleration response of the tunnel, 2 accelerometers were placed at 360 mm interval at top surface of the tunnel. For assessing displacement, 2-dimensional DIC was used.

Digital Image Correlation

Digital image correlation is pixel-based image correlation technique which allows assessing the displacement and strain over a virtual grid defined on a digital image. The main advantage of DIC is that simplicity of needed devices, and it is a non-contact full-field strain and displacement measurement technique. In the current research, the

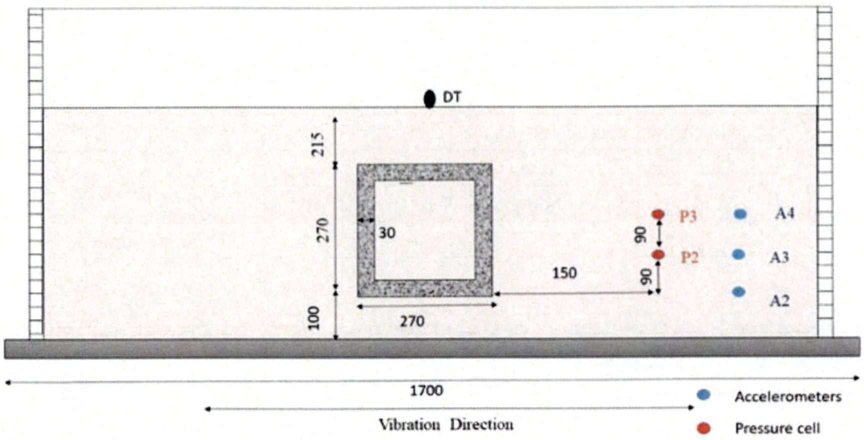

Fig. 35.1 Sensor arrangement adopted for experimental study on tunnels

DIC system involved mainly a digital camera, two LED lights and VIC 2D software meant exclusively to measure strain and deformation using correlation technique. The measurement of strain or deformation using DIC involved (a) preparation of tunnel surface by spraying a white paint above which speckles are marked using a marker. (b) Tunnel deformation due to input motion are captured through series of images using a camera by considering the first image as reference image. The remaining series of images captured are correlated with respect to the reference image. Then, the images are calibrated and analysed by specifying the area of interest. Displacement and strains are computed at each point by tracking selected subset through employing VIC 2D software. The test arrangement is shown in Fig. 35.2.

Results and Discussions

Acceleration Response of Soil and Tunnel

To investigate the behaviour of tunnel–soil interaction during repeated shaking events, the acceleration response was studied by installing 5 numbers of accelerometers. 2 accelerometers were placed at front end (A0) and rear end (A1) of the tunnel model with 360 mm c/c distance. The other three accelerometers were placed at 100 mm (A2), 190 mm (A3), and 280 mm (A4) from bottom of the tank, respectively. Figure 35.1 shows the arrangement of accelerometers. The observed acceleration response from A0 and A1 found to have uniform characteristics. However, this response found to be higher when compared to the acceleration response from A4 which is placed at 280 mm depth. Also, the observed acceleration response from A4, A3, and A2 located within the ground showed varying acceleration response

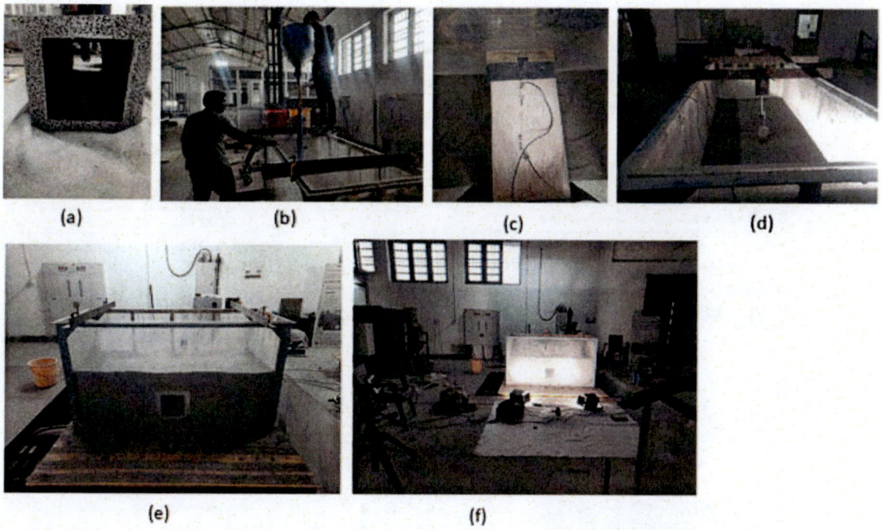

Fig. 35.2 **a** Tunnel speckling for DIC measurement. **b** Sample preparation using dry pluviation technique. **c** Instrumentation on tunnel top. **d** Installation of displacement transducer. **e** Tunnel-embedded ground. **f** 2-DIC instrumentation

suggesting the interaction effect of embedded tunnel model. The accelerometer (A3) located adjacent to the mid-depth of tunnel model showed higher acceleration response compared to shallow depth and deeper depth acceleration response. The obtained higher response highlighted the interaction between tunnel–soil system. Similarly, the ground was then subjected to subsequent shaking, i.e. 0.4 g and acceleration response was monitored. Similar trend was observed as seen during 0.3-g shaking. With the increment in intensity of acceleration shaking, the acceleration response also increases. Though soil compaction was observed after initial acceleration loading, the improvement in density does not contributed significantly in minimizing the acceleration response. This highlights the influence of repeated shaking in underground systems. Similarly, the obtained values from Fourier spectrum also highlighted the variation in peak amplitude response during repeated loading events and its influence in affecting the dynamic characteristics of tunnel system (Figs. 35.3 and 35.4).

Effect of Soil Displacement

To measure the soil displacement during repeated shaking testing, displacement transducers were placed over the surface of the ground. Figure 35.5 shows the obtained soil displacement test results during 0.3 and 0.4-g loading. It can be seen that, with the increase in acceleration loading, soil displacement also increases. However, the

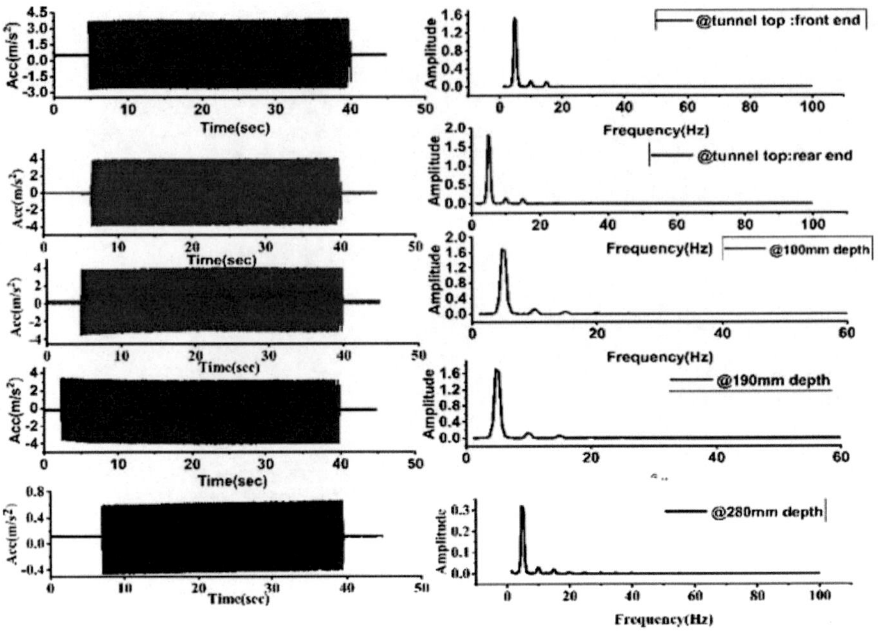

Fig. 35.3 Plot of acceleration time history and corresponding Fourier spectra due to 0.3-g input motion

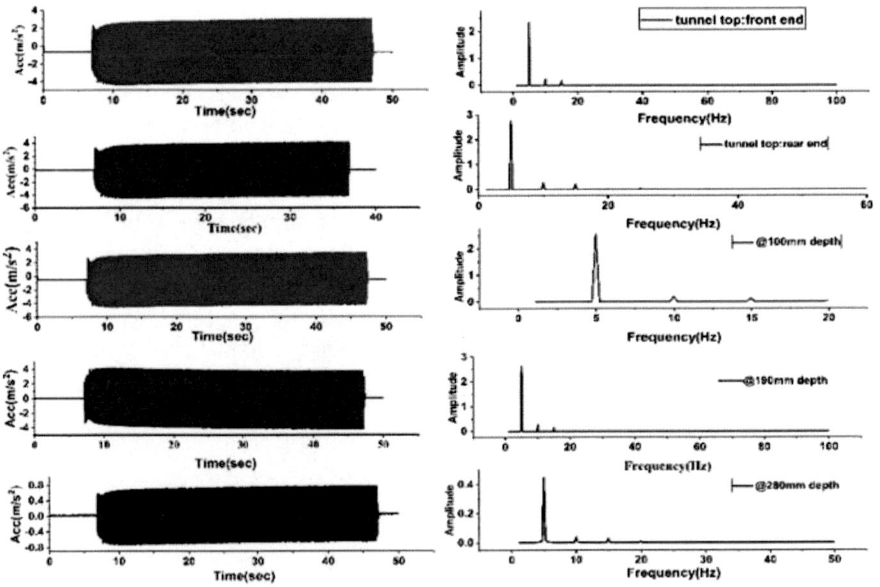

Fig. 35.4 Acceleration time history and Fourier spectra corresponding to 0.4-g input motion

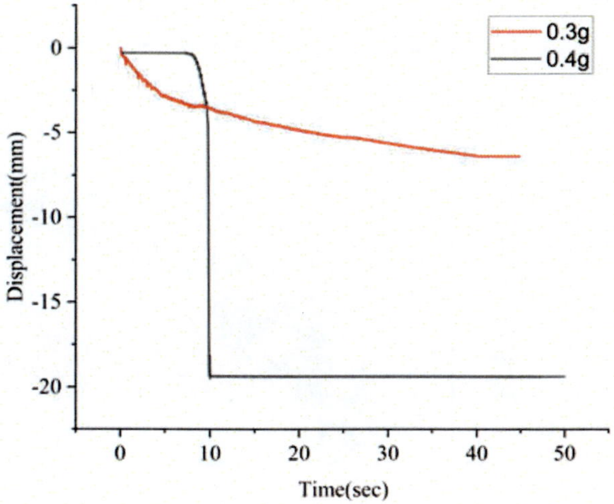

Fig. 35.5 Displacement of soil due to 0.3 and 0.4-g input motion

soil displacement does not contribute in improving the confinement characteristics of the tunnel model which was evident from the acceleration response as discussed in previous section. At 0.3-g loading, maximum displacement of 6.8 mm was observed whereas at 0.4-g shaking, 19.38-mm displacement was observed. About 1.85 times higher increment in soil displacement was observed when tested under repeated shaking events. The obtained soil displacement verified the occurrence of soil densification in the prepared ground. The stresses induced due to this displacement resulting in the development of strains and displacement of the embedded tunnel model. This was discussed in the subsequent section.

Dynamic Soil Pressure

The generated dynamic soil pressure within the ground and tunnel system was monitored using soil pressure transducers. Due to operational error, the data received from soil pressure transducer got corrupted during 0.3-g shaking conditions. However, the data obtained during 0.4-g shaking condition is shown in Fig. 35.6. It can be seen that the generated soil pressure found maximum at 190 mm depth. Due to variation in soil stratification at shallow depth during repeated shaking, the response obtained at 280 mm (P3) depth is lower than the response obtained from transducer (P2) placed at deeper depth. Around 60% variation in soil pressure values was found from top to bottom, i.e. P2 to P3. Due to this, the confinement characteristics of the ground-tunnel system reduced at shallow depth reduces. However, the transducer placed at 190 mm depth showed higher earth pressure values which validates the improvement in soil densification. This induces strain development at the bottom of the tunnel model during repeated loading. Thus, the obtained results elucidate that with the increase in acceleration loading disturbs soil at shallow depth continuously. Also,

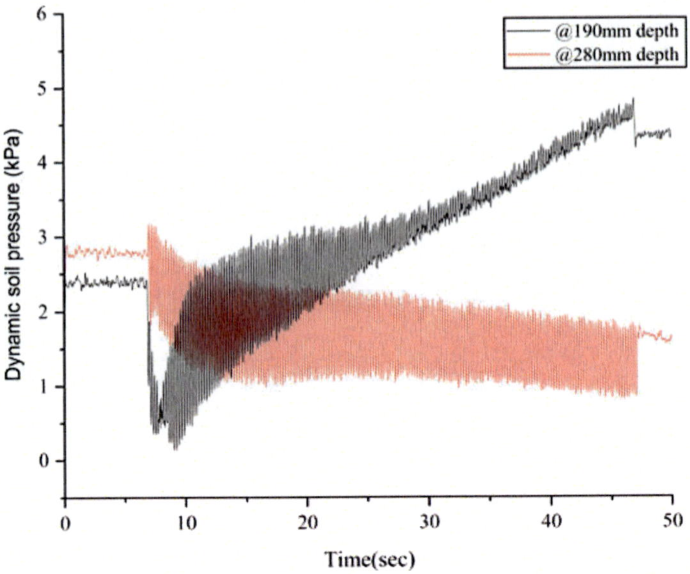

Fig. 35.6 Soil pressure variation at different depths due to 0.4-g input motion

this repeated loading induces soil densification at deeper depths which caused stress accumulation on the tunnel embedded in the ground. Due to this, tunnel undergoes more displacement and strain development during repeated shaking conditions.

Displacement and Strains

The displacement and strains developed inside the tunnel model were measured using 2-DIC technique. With the increase in acceleration during repeated loading, displacement of the tunnel also increases. At 0.3-g loading, maximum displacement of 6 mm was observed whereas at 0.4-g shaking, 8-mm displacement was obtained. Around 33% increment in displacement was observed when tunnel tested under repeated shaking conditions.

For measuring the strains at different locations of tunnel, four corners of the tunnel (P0, P1, P2, P3) and midpoint of two side walls (P4 and P5) were selected. Results from digital image correlation showed that, maximum peak strains were developed at left-corner bottom (P2) and right-corner bottom (P3) of the tunnel model. Comparatively, bottom corner showed maximum strain development due to continuous soil deposition and its associated soil densification. As seen from dynamic soil pressure values, maximum values were seen at the deeper depth compared to shallow depth. Due to this, more stresses induced at the bottom portion of the tunnel model which results in strain accumulation. However, due to instability of soil grains at

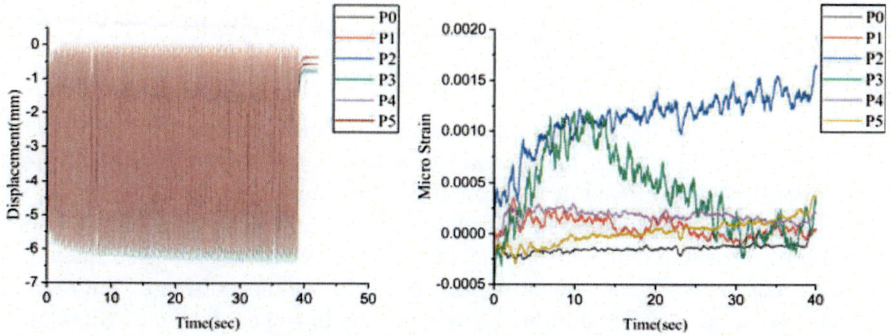

Fig. 35.7 Displacement of soil bed along with tunnel during 0.3-g input motion

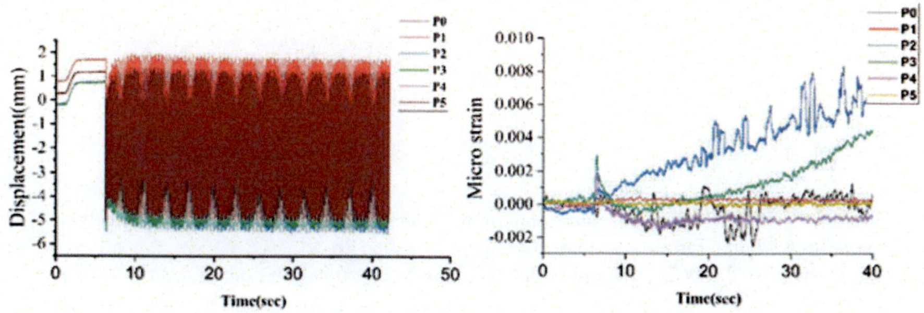

Fig. 35.8 Displacement of soil bed along with tunnel during 0.4-g input motion

shallow depth during repeated loading, minimum stresses transferred to the tunnel top portion. This can be verified from the tunnel model on obtained strain values at the top portion of the tunnel model. Around 2.5% variation in strain values were observed when compared bottom portion of the tunnel model with top portion. It is to be noted that, irrespective of the depth, strain development increases with the increase in acceleration loading during repeated shaking conditions. As evident from acceleration response, soil displacement and dynamic soil pressure results; confinement characteristics reduce during repeated shaking events at shallow depth. However, the occurrence of soil densification during repeated loading induces stresses to the embedded tunnel system. Due to this, continuous strain development occurs within the tunnel model and affects the stability of the tunnel system (Figs. 35.7 and 35.8).

Conclusions

In this study, repeated shaking table tests were conducted on the ground embedded with tunnel model subjected to 0.3 and 0.4-g sinusoidal loading to investigate the seismic soil tunnel interaction. The analyses were carried out in terms of acceleration response, dynamic soil pressure, and deformation of tunnel. Based on the observations, following conclusions were made.

(1) Under repeated incremental acceleration loading conditions, acceleration response increases with the increase in acceleration intensity irrespective of the density improvement in the ground achieved during repeated loading.
(2) Tunnel displacement increases with the increase in acceleration loading. Around 33% increment in tunnel displacement was observed during repeated shaking.
(3) Maximum strain development was observed in bottom portion of the tunnel model compared to top portion due to variation in soil confinement. Similarly, the strain development also increases when tested under repeated shaking.
(4) Dynamic design of tunnel system should incorporate the effect of repeated dynamic loading conditions such that stability of the tunnel system can be ensured for longer period.

References

1. Tsinidis G, et al (2020) Seismic behaviour of tunnels: from experiments to analysis. In: Tunneling and underground space technology
2. Che A, Iwatate T (2002) Shaking table test and numerical simulation of seismic response of subway structures. Struct Mater
3. Wang WL, Wang TT, Su JJ, Lin CH, Seng CR, Huang TH (2001) Assessment of damage in mountain tunnels due to the Taiwan Chi-Chi Earthquake. In: Tunneling and underground space technology
4. Yu H, Chen J, Bobet A, Yuan Y (2016) Damage observation and assessment of the Longxi tunnel during the Wenchuan earthquake. In: Tunneling and underground space technology
5. Ding X, Zhang Y, Wu Q, Chen Z, Wang C (2020) Shaking table tests on the seismic responses of underground structures in coral sand. In: Tunneling and underground space technology
6. Ding X, Zhang Y, Wu Q, Chen Z, Wang C (2020) Shaking table tests on the seismic responses of underground structures in coral sand. In: Tunnelling and Underground space technology
7. Ding X, Feng L, Wang C, Chen Z, Han L (2020) Shaking table tests of the seismic response of a utility tunnel with a joint connection. Soil Dyn Earthq Eng
8. Zhang W, et al (2020) Study on seismic behaviors of a double box utility tunnel with joint connections using shaking table model tests. Soil Dyn Earthq Eng
9. Bureau of Indian Standards (Part 14) (1983) Methods of test for soils: determination of density index (relative density) of cohesionless soils. IS 2720
10. Bureau of Indian Standards (Part 4) (1983) Methods of test for soils: grain size Distributions. IS 2720
11. Bureau of Indian Standards (Part 3) (1937) Methods of test for soils Determination of specific gravity. J AOAC Int 20(3):535–542. https://doi.org/10.1093/jaoac/20.3.535
12. Xu H, Li T, Xia L, Zhao JX, Wang D (2016) Shaking table tests on seismic measures of a model mountain tunnel. Tunn Undergr Sp Technol 60(June):197–209. https://doi.org/10.1016/j.tust.2016.09.004

Chapter 36
Significance of Impedance Ratio in Case of Infilled-Pile Barrier for Vibration Isolation

Jahnavi Thentu and Sanjit Biswas

Introduction

Vibrations are resulted from massive natural disasters like earthquake to day-to-day activities like construction works, vehicle movements machine vibrations, etc. The necessity of curbing these vibrations comes as soil under laying and structures adjoining these vibratory sources get disturbed and may lead to distortion.

The passive method of isolation trenches is extensively used earlier. Although in shallow depths the performance of trenches is satisfactory but in soft soils and soils with high water table trenches are not quite suitable. Woods [12] experimented with the alternate solution of putting piles into use for vibration isolation. Subsequently numerous researchers have worked on various parameters of piles and their effect on isolation. According to Avilés-Sesma [1], Kattis [5, 6], Tsai [11], Lu [7] and Xu [13] as the length of pile increases the overall effectiveness of vibration increases. Pu [9] remarks that as radius of pile increases width of the attenuation zone increases. Avilés and Sánchez-Sesma [1], Kattis [5], Ding [2], Lu [7], Xu [13] altogether collectively mentioned that as spacing decreases the magnitude of isolation gets enhanced. It is an obvious fact that as number of piles increase the isolation effectiveness increase and it is evident by works of Ding [2], Huang [4], Xu [13] and Pu [9].

However, the focus is laid less on impedance ratio and its effect on different isolation materials in the literature till now. Hence, the sole intention of this present study is to understand the impact of impedance ratio on vibration isolation using the parameters like amplitude reduction factor and attenuation zone by numerical analysis using finite element method.

J. Thentu (✉) · S. Biswas
Department of Civil Engineering, National Institute of Technology, Warangal 506004, India
e-mail: jahnavithentu@gmail.com

S. Biswas
e-mail: sbiswas@nitw.ac.in

K. Muthukkumaran et al. (eds.), *Soil Dynamics, Earthquake and Computational Geotechnical Engineering*, Lecture Notes in Civil Engineering 300,
https://doi.org/10.1007/978-981-19-6998-0_36

Table 36.1 Properties of the materials

S No.	Material	Unit weight (kN/m^3)	E (Mpa)	Υ	c (kN/m^2)	ϕ	Impedance ratio
1	Clay	19	20	0.35	30	10^0	1
2	Geofoam	0.15	2.4	0.1	–	–	0.034
3	Tyre chips	8.63	3.21	0.3	–	–	0.27
4	Fly ash	5	25	0.35	–	–	0.57

Numerical Modelling

The present numerical study is done using the finite element software Plaxis 3D [8]. Clay is considered in the main surrounding soil medium and model dimensions are considered as 40 × 20 × 20 m. For isolation, total 7 barrier piles are used in a single row. The diameter of piles is taken as 1 m and spacing between the piles is taken as 1.5 m. Isolation materials considered are Fly ash, Geofoam and Tyre chips and the properties are taken from Pu [9], Elsayed [3] and Siringi [10], respectively. Impedance ratio of the materials with the clay is also obtained using the following equation.

$$\text{Impedance ratio (I.R)} = \frac{\zeta_p v_p}{\zeta_s v_s} \tag{36.1}$$

where ζ_p, v_p are the density and velocity of pile material, respectively, and ζ_s, v_s are the density and velocity soil medium, respectively. The material properties and impedance ratio for different materials with clay is listed in Table 36.1.

Absorbent boundaries are taken at all sides of the model so that the reflected waves will no longer disturb the soil. Fixed supports are used for the bottom boundary and roller supports are used for the side boundaries and top surface is free in all directions. For meshing, 10-noded tetrahedral solid elements are used. The prepared finite element model is shown in the Fig. 36.1.

Results and Discussion

Validation

The procedure of the model preparation is validated using the results presented in Pu [9]. The study considered 10 Fly ash barrier piles embedded in soil consisting of six layers. Loading includes a point source of 1kN with the frequency of 41 Hz.

From the analysis, the surface displacements are taken at all the locations in an axis passing through the loading point and results are compared with displacements reported in Pu [9]. The comparison curve is present in Fig. 36.2. It is clearly observed

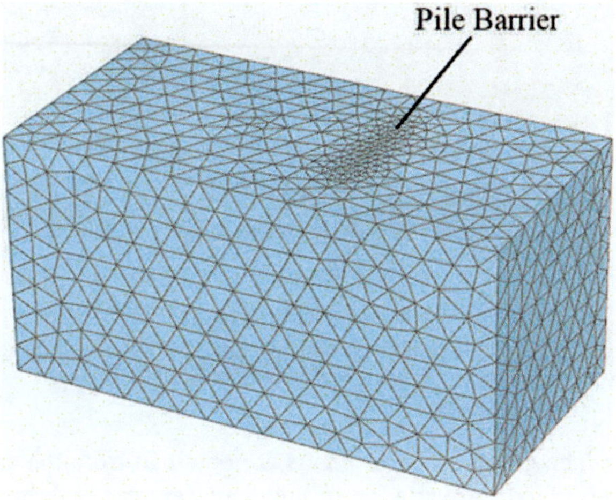

Fig. 36.1 The finite element model created using Plaxis 3D

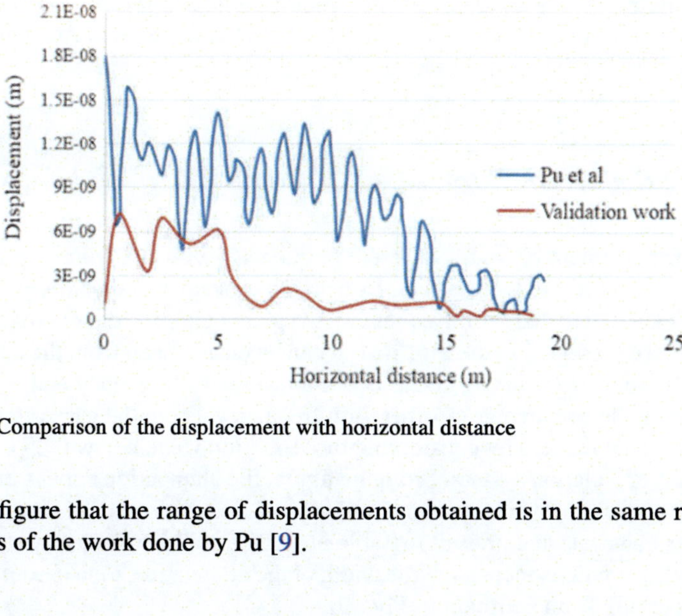

Fig. 36.2 Comparison of the displacement with horizontal distance

from the figure that the range of displacements obtained is in the same range with the results of the work done by Pu [9].

Variation of Displacement with Horizontal Distance

To simulate the reciprocating machine vibration loading, a sinusoidal point load of 1kN is taken with varying range of frequency starting from 10 to 60 Hz. The loading is applied at a distance of 10 m from the pile barrier. The ground displacements are measured at a point 5 m away from the pile barrier line in the opposite side of the loading. Results are obtained for all the pile barrier material and without pile barrier.

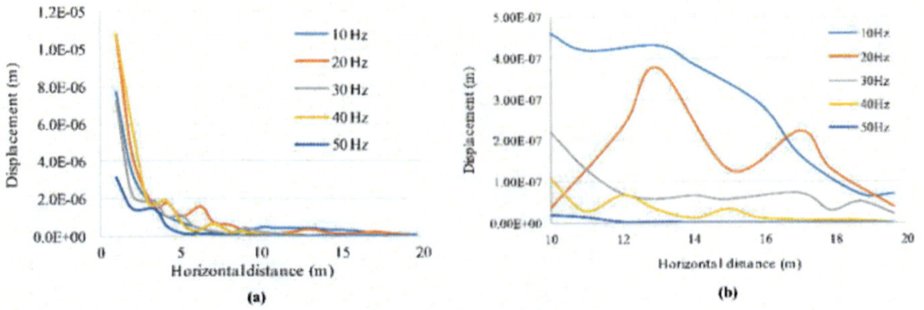

Fig. 36.3 Variation of displacements along the horizontal distance form geofoam piles in clay. **a** 0–20 m horizontal distance. **b** 10–20 m horizontal distance

The ground displacements are obtained from the analysis for different frequencies. A typical variation of the displacements with horizontal distance is presented in Fig. 36.3 for geofoam barrier piles. It is found from the figure that there is a constant decrease in displacement as the distance increases. It can be also seen that the displacements are continuously decreasing with the increase in frequencies due to damping effect of the barrier system.

Amplitude Reduction Curves

Based the displacement curve for different frequencies with and without pile barrier system, the amplitude reduction factor of barrier system as compared to without barrier piles are calculated with frequency. A typical curve for the geofoam barrier system piles are presented in the Fig. 36.4. It can be clearly seen from the curves that the amplitude reduction curve reduces continuously with frequency and reaches its minimum value before increase further with frequency. From the curves the attenuation zones can be marked by considering the reduction value below 0.25 for a righteous method of isolation system. From the figure, the attenuation zone is marked as 23–26 Hz for geofoam barrier system. Similarly, the attenuation zone is also obtained for the other materials and present in Table 36.2. The Table 36.2 clearly shows that as the impedance ratio is increasing the width of the attenuation zone is continuously increasing starting from geofoam to Fly ash.

Conclusions

In this study, it is aimed to study the effectiveness of the pile barrier system as an isolation system. Another aspect of the study is to recommend the best suitable

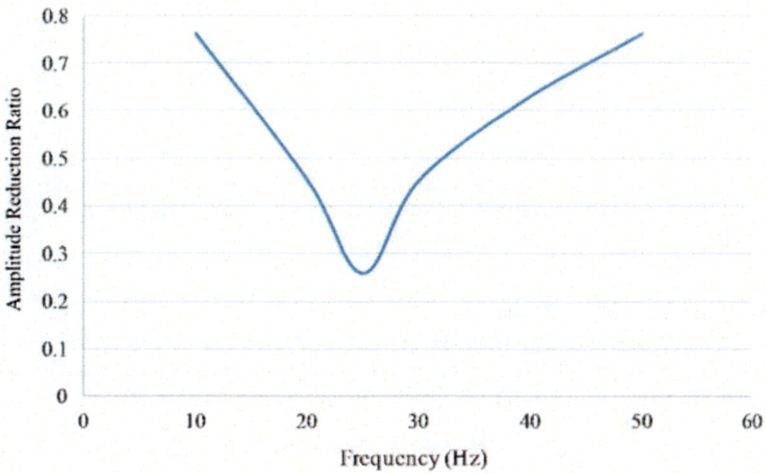

Fig. 36.4 Variation of amplitude reduction ratio with frequency for geofoam

Table 36.2 Attenuation zones for all materials

Material	Attenuation zones	Impedance ratio
Geofoam	23–26 Hz	0.034
Tyre chips	18–23 Hz	0.27
Fly ash	18–23 Hz and 35–43 Hz	0.57

isolation material for the infilled-pile barrier system in clay. Based on the study, the following conclusions are made.

- The pile barrier system is one of most effective isolation methods for both surface as well as deeper waves.
- As the impedance ratio is increased, a wider attenuation zone is created.
- Fly ash is best suited isolation material for the clay because the range of attenuation zone is wider in this case.

References

1. Avilés J, Sánchez-Sesma FJ (1983) Piles as barriers for elastic waves. J Geotech Eng 109:1133–1146
2. Ding G, Xu G (2008) Screening of plane S waves by an array of rigid piles in poroelastic soil. J Zhejiang Univer Sci A: Appl Phys Eng 9:589–599. https://doi.org/10.1631/jzus.A071494
3. Elsayed SM, Sorour TM, Fathy AM, Numerical analysis of heave reduction using EPS geofoam inclusions. IOSR J Mech Civ Eng (IOSR-JMCE) 17:41–49
4. Huang J, Shi Z (2013) Application of periodic theory to rows of piles for horizontal vibration attenuation. Int J Geomech 13:132–142

5. Kattis S, Polyzos D, Beskos D (1999) Modelling of pile wave barriers by effective trenches and their screening effectiveness. Soil Dyn Earthq Eng 18:1–10
6. Kattis S, Polyzos D, Beskos D (1999) Vibration isolation by a row of piles using a 3-D frequency domain BEM. Int J Numer Meth Eng 46:713–728
7. Lu JF, Xu B, Wang JH (2009) A numerical model for the isolation of moving-load induced vibrations by pile rows embedded in layered porous media. Int J Solids Struct 46:9771–9781
8. PLAXIS Version 2012 Scientific Manual, vol 1, pp 720–766. http://www.plaxis.nl/
9. Pu X, Shi Z, Xiang H (2017) Feasibility of ambient vibration screening by periodic geofoam-filled trenches. Soil Dyn Earthq Eng 104:228–235
10. Siringi GM, Abolmaali A, Aswath PB (2013) Properties of concrete with crumb rubber replacing fine aggregates (sand). Adv Civ Eng Mater 2(1):218–232. https://doi.org/10.1520/ACEM20120044.ISSN2165-3984
11. Tsai P, Feng Z, Jen T (2008) Three-dimensional analysis of the screening effectiveness of hollow pile barriers for foundation-induced vertical vibration. Comput Geotech 35:489–499
12. Woods RD, Barnett NE, Sagesser R (1974) Holography—a new tool for soil dynamics. J Geotech Eng 100:1231–1247
13. Xu B, Xu MQ (2014) Numerical analysis of vibration isolation using pile rows international journal of engineering mathematics. Against the vibration due to moving loads in a viscoelastic medium, Article ID 810525

Chapter 37
Seismic Vulnerability of Residential Buildings in Jammu City, Jammu and Kashmir

Abdullah Ansari, Falak Zahoor, K. Seshagiri Rao, Arvind K. Jain, and Tanzeel Ur Riyaz

Introduction

The seismic vulnerability is the inability to resist any earthquake event resulting into structural and substructural damages. Indian subcontinent has faced many near-field as well as far-field devastating earthquakes having magnitude ≥ 7.5, viz. 2004 Sumatra–Andaman tsunamigenic earthquake; 2005 Muzaffarabad earthquake; 2015 Gorkha earthquake and most recent 2015 Hindu Kush earthquake [1, 2]. The 1993 Latur and 2001 Bhuj earthquakes are the most catastrophic events that occurred in the stable continental region [3–5]. The most devastating earthquake events that occurred in and around the Jammu and Kashmir region are the 1552 Kashmir earthquake; 1885 Pattan earthquake; 2005 Muzaffarabad earthquake and the 2015 Hindu Kush earthquake.

Due to the 2005 Muzaffarabad earthquake, more than 780,000 structures were destroyed or damaged beyond repair, with many more rendered useless [6]. The death toll crossed 0.1 million and affected 0.5 million families. The liquefaction features in the form of sand boils, ground cracks, fissures in embankments were even observed in Simbal village in Jammu city, about 240 km from the epicentre of the earthquake [7, 8]. After this deadly earthquake, the government and administration started looking to study the seismic vulnerability of existing structures in Jammu and Kashmir. In this paper, an attempt has been made to predict the seismic vulnerability

A. Ansari (✉) · F. Zahoor · K. Seshagiri Rao · A. K. Jain
Department of Civil Engineering, Indian Institute of Technology Delhi, New Delhi 110016, India
e-mail: ansariaa@civil.iitd.ac.in; aamomin183@gmail.com

F. Zahoor
Department of Civil Engineering, National Institute of Technology Srinagar, Srinagar, Jammu and Kashmir 190006, India

T. U. Riyaz
Department of Civil Engineering, RIMT University, Punjab 147301, India

K. Muthukkumaran et al. (eds.), *Soil Dynamics, Earthquake and Computational Geotechnical Engineering*, Lecture Notes in Civil Engineering 300,
https://doi.org/10.1007/978-981-19-6998-0_37

of residential buildings in and around Jammu city based on data collected during the field survey.

Seismic Vulnerability of Structures

For better urbanization, human safety, public health, and structural safety are the three main parameters [5]. The regional seismic vulnerability assessment methodology is a critical tool for governments and decision-makers to use in allocating resources and mitigating the effects of earthquakes [9]. Existing structures may become seismically weak when seismic design code standards are regularly improved as technical knowledge advances [4, 10, 11]. One of the most essential metrics for assessing earthquake potential damages in urban environments is building vulnerability [12]. Underground structures constructed in Jammu and Kashmir like seismically active area are subjected to seismic forces [1] and can undergo a different form of deformation depending on the nature in which seismic waves propagate in soil medium [5, 13]. According to the seismic hazard zoning map of India [14], the Jammu city falls in the seismic zone IV, which is characterized by moderate risk with an expected peak ground acceleration of 0.24 g. The rapid construction of buildings in Jammu city forces detailed evaluation of the seismic vulnerability of residential buildings for controlling and handling property damages due to any upcoming earthquake event in this region.

Field Survey and Assessment Factors

Jammu is the winter capital of Jammu and Kashmir lying on the bank of river Tawi as shown in Fig. 37.1. It is the second-most populous city in Jammu and Kashmir having an area of 240 km^2. The Jammu is surrounded by the Himalayas which is one of the most seismically active regions in the world. Jammu is divided into 75 wards under the control of Jammu Municipal Corporation (JMC). An intensive field survey was conducted on 225 residential buildings covering 75 wards for collecting the data required for seismic vulnerability assessment. In this study, demographic framework and structural specifications are considered as major factors [15] to assess the seismic vulnerability of residential buildings in and around Jammu city.

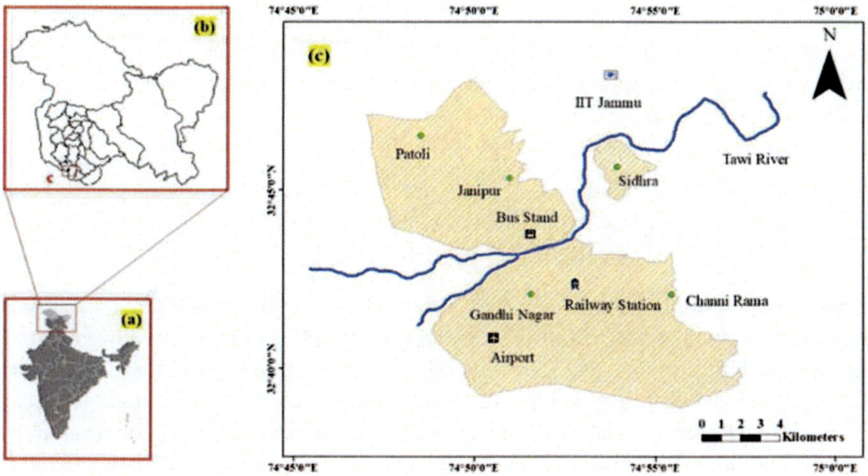

Fig. 37.1 Showing the location of the study area (Jammu, Jammu, and Kashmir) considered for field survey

Demographic Framework

The population and building density are considered under the demographic framework [16]. Based on building density data, vulnerability divided into five categories (a) Very High (building density > 3000), (b) High (building density 2300–3000), (c) Moderate (building density 1900–2200), (d) Low (building density 1000–1800), and (e) Very Low (building density < 1000). A higher value of building density in any area suggests higher seismic vulnerability and more chances of structural damages. Depending on the population ratio, regional vulnerability is fixed under three broad classes (a) More Populous (population ratio > 0.7), (b) Moderate Populous (population ratio 0.3–0.7), and (c) Less Populous (population ratio < 0.3).

Structural Specifications

The data relating to building height, building age, and materials used for construction purposes is scanned under the roof of structural specifications [17, 18]. The seismic vulnerability of surveyed residential buildings can be defined using three classes (a) High (storey 3 +), (b) Moderate (storey 2–3), and (c) Low (storey 1) corresponding to building height data. Considering damages observed during the 2005 Muzaffarabad earthquake, all buildings constructed after the 2005 earthquake are considered less vulnerable. This helped to categorize the vulnerability into (a) Very High (building age before 1980), (b) High (building age 1981–1990), (c) Moderate (building age 1991–2000), (d) Low (building age 2001–2005), and (e) Very Low (building age

2006–Present). According to the materials used for construction purposes, seismic vulnerability of residential buildings can be grouped into three classes (a) High (building material brick and wood), (b) Moderate (building material brick and iron), and (c) Low (building material concrete).

Result and Discussion

The data for 225 buildings collected from 75 wards during the field survey was used to evaluate the seismic vulnerability of residential buildings located in Jammu city. According to building density, 15 wards covering 20% of the total area fall under the 'Very Low Vulnerability' category; 19 wards covering 25.33% of the total area fall under the 'Low Vulnerability' category; 12 wards cover 16% of the total area fall under 'Medium Vulnerability' category; 17 wards covering 22.67% of the total area fall under 'High Vulnerability' category, and 12 wards covering 16% of the total area fall under 'Very High Vulnerability' category as shown in Fig. 37.2. About 8%, 54.67%, and 37.33% of wards fall under the group, low vulnerability, medium vulnerability, and high vulnerability, respectively, as per population ratio.

Based on building age, 29 wards covering 38.67% of the total area fall under the 'Very Low Vulnerability' category; 3 wards covering 4% of the total area fall under the 'Low Vulnerability' category; 5 wards covering 6.67% of the total area fall under 'Medium Vulnerability' category; 8 wards covering 10.67% of the total area fall under 'High Vulnerability' category, and 30 wards covering 40% of the total area fall under 'Very High Vulnerability' category. About 25.33, 34.67, and 40% of wards fall under the class, low vulnerability, medium vulnerability, and high vulnerability, respectively, as per building height. According to the quality of construction material,

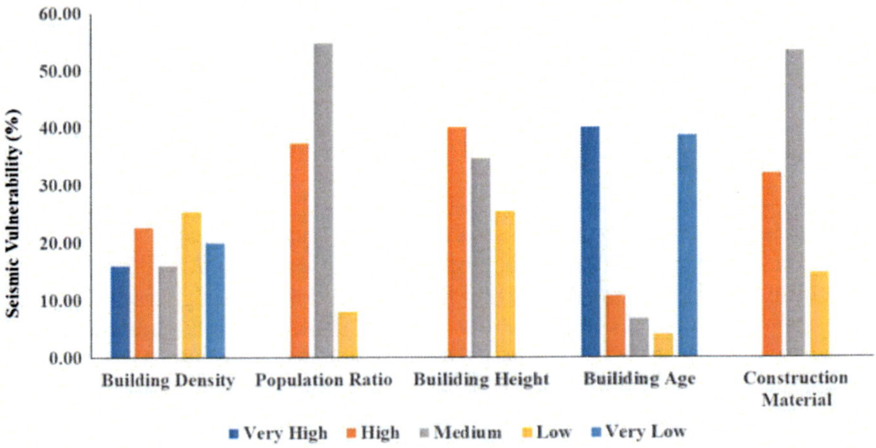

Fig. 37.2 Seismic vulnerability of residential buildings in Jammu based on field survey data

out of 75, 11 wards fall under low vulnerability, 40 wards under medium vulnerability, and 24 wards under the high vulnerability category. For building density, the maximum part of the Jammu city falls under the 'Medium Vulnerability' category as shown in the following Fig. 37.3a. The population ratio for outer areas of Jammu city is less which leads to falling under the 'Low Vulnerability' category presented in Fig. 37.3b. To assess the seismic vulnerability based on structural specifications, three important factors, building height, building age, and material for construction taken under consideration. Due to urbanization, the construction of new buildings with multi-storey has become a trend. In Jammu, most of the buildings have 3 + storey which leads the maximum part of the city to fall under the 'High Vulnerability' zone as presented in Fig. 37.4a. As shown in Fig. 37.4b, old structures are mostly located in the central part of the city. Areas in NE and SW direction fall under the 'Very Low Vulnerability' class. In the south portion of the city, new structures were constructed after the 2005 Muzaffarabad earthquake. Areas, where maximum structures construed after the 2005 earthquake, can resist any future earthquake and falls under the low to very low vulnerability zone. The northern part of the city where brick and wood-based houses were observed during the field survey falls under the 'High Vulnerability' category.

For assessing the seismic vulnerability, a logic tree framework was used to avoid any conflict among the selected parameters [3]. The weight factor of 0.5 is given to demographic framework and structural specifications [19]. Figure 37.5 illustrates that by considering the impact of all paraments, 14, 17, 31, 21, and 17% parts of the Jammu city fall under very high, high, medium, low, and very low vulnerability zone, respectively. The northern part of the city which lies on the right bank of the Tawi River is more vulnerable compared to the southern part as shown in Fig. 37.6.

Conclusion

In this study, an attempt has been made to assess the seismic vulnerability of residential buildings in Jammu city. An intensive field survey was conducted, and data was collected from 75 wards related to demographic framework and structural specifications. Separate seismic vulnerability maps were prepared based on building density, population ratio, building height, building age, and construction material. All these five factors together play a significant role to decide the seismic vulnerability of any structural element. To study the combined effect of all these five parameters, the logic tree framework was adopted with an equal weightage of 0.5 for both demographic framework and structural specifications. One-third part of the Jammu city falls under high to very have vulnerability class. The northern part of the city which lies on the right bank of the Tawi River is more vulnerable compared to the southern part. In a way to avoid any catastrophic structural damage due to any future either far-field or near-field earthquake, an urgent step should be taken to keep them safe for the people who live in them. Through the process of retrofitting, the vulnerability of these residential buildings can be reduced. The results presented in this study can

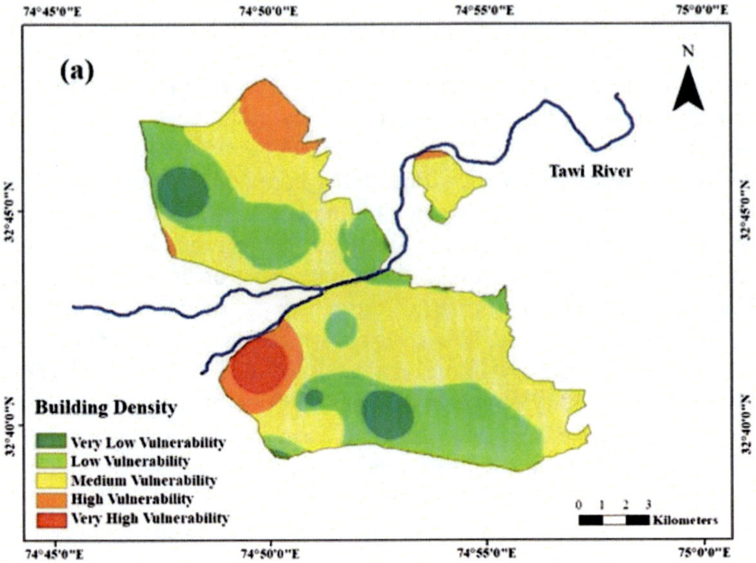

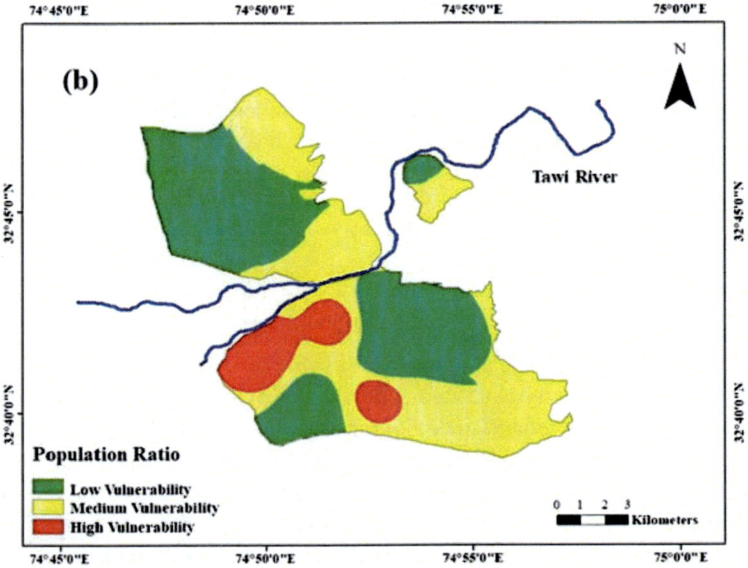

Fig. 37.3 Seismic vulnerability of residential buildings in Jammu based on demographic framework considering **a** building density and **b** population ratio

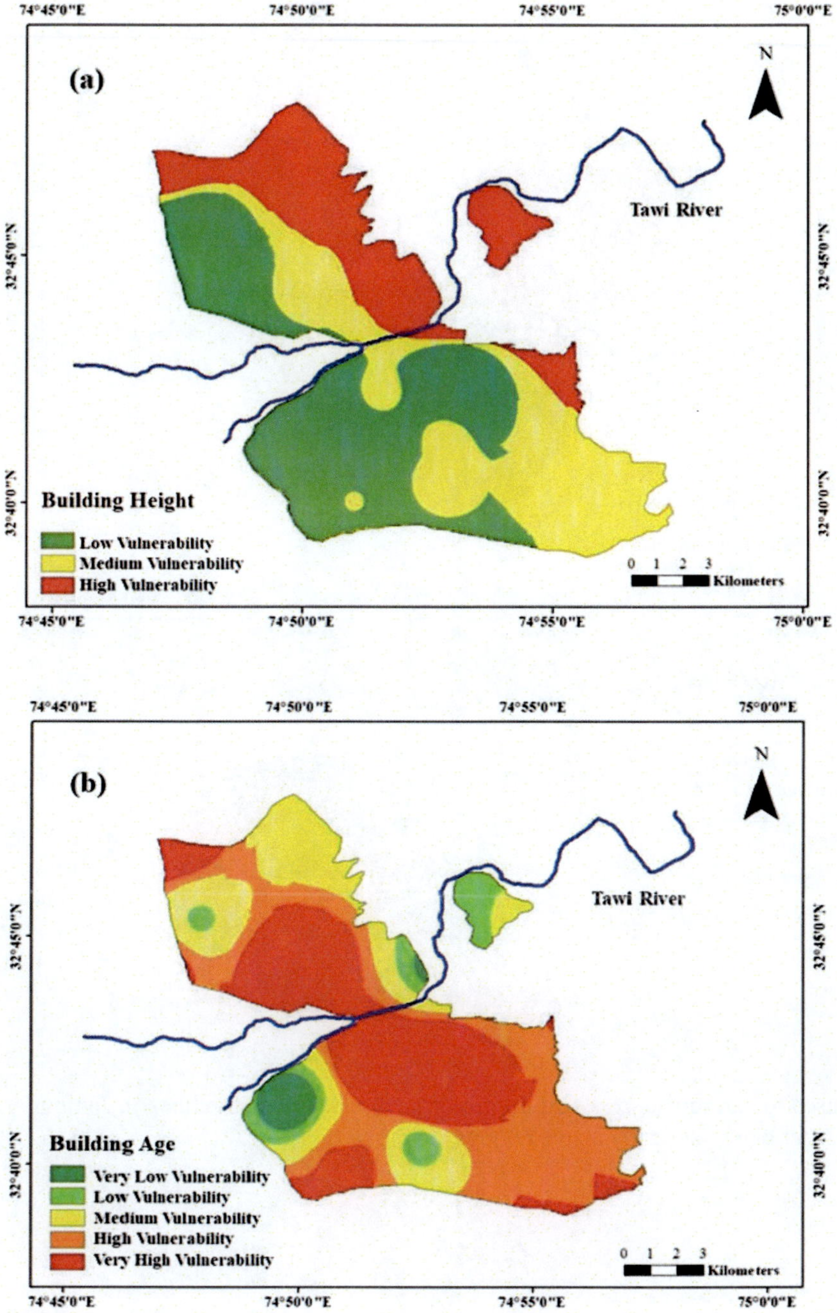

Fig. 37.4 Seismic vulnerability of residential buildings in Jammu based on structural specifications considering **a** building height, **b** building age, and **c** construction material

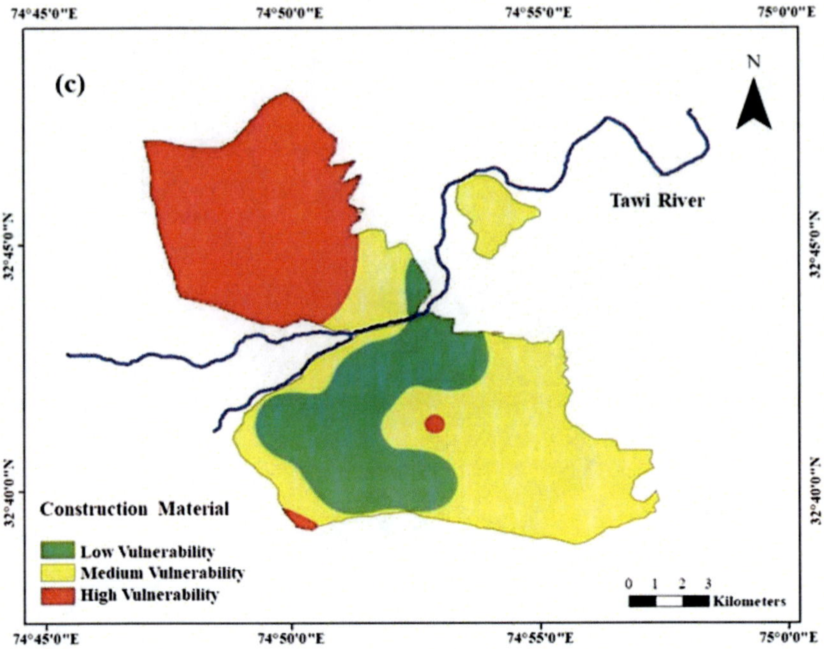

Fig. 37.4 (continued)

Fig. 37.5 Percentage distribution of Jammu city in different vulnerability classes

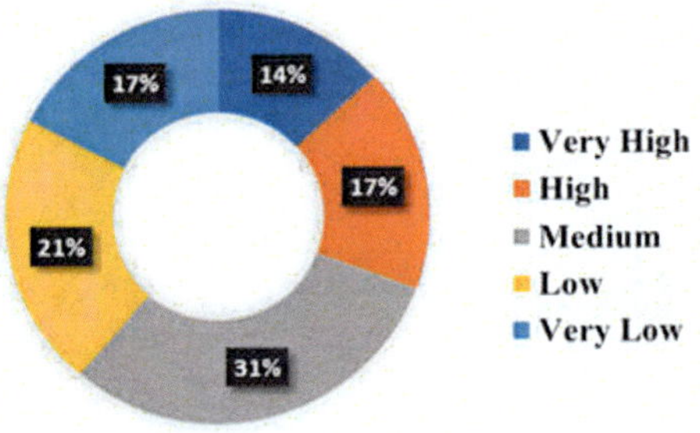

be used to earthquake-resistant design of new residential structures in Jammu and Kashmir like seismically active regions.

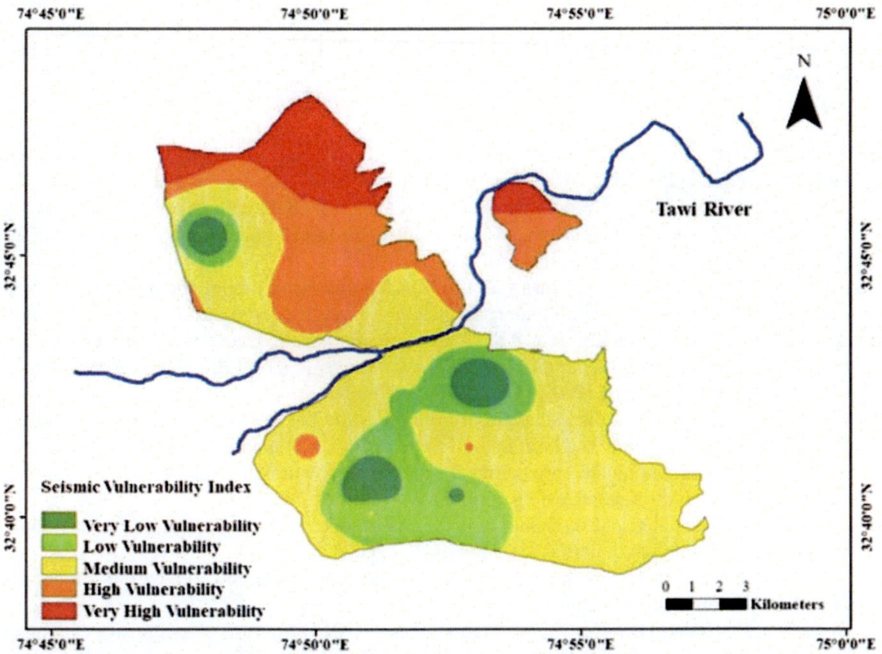

Fig. 37.6 Seismic vulnerability assessment result for residential buildings in Jammu considering demographic framework and structural specifications

References

1. Ansari A, Rao KS, Jain AK (2022) Damage assessment of tunnels in seismic prone zone during earthquakes: a part of hazard evaluation. In: Earthquakes and Structures (pp 161–169). Springer, Singapore. https://doi.org/10.1007/978-981-16-5673-6_13
2. Ansari A, Rao KS, Jain AK (2022) Liquefaction hazard assessment in a seismically active region of Himalayas using geotechnical and geophysical investigations: a case study of Jammu region, Jammu and Kashmir. Bull Eng Geol Environ 81(349):1–19. https://doi.org/10.1007/s10064-022-02852-3
3. Alizadeh M, Hashim M, Alizadeh E, Shahabi H, Karami MR, Pour AB, Pradahan B, Zabihi S (2018) Multi-criteria decision making (MCDM) model for seismic vulnerability assessment (SVA) of urban residential buildings. ISPRS Int J Geo Inf 7(11):444
4. Ansari A, Rao KS, Jain AK (2021) Seismic Hazard and risk assessment in maharashtra: a critical review. In: Sitharam TG, Kolathayar S, Sharma ML (eds) Seismic hazards and risk, lecture notes in civil engineering, vol 116. Springer, Singapore, pp 33–45. https://doi.org/10.1007/978-981-15-9976-7_4
5. Ansari A, Daigavane PB (2021) Analysis and modelling of slope failures in municipal solid waste dumps and landfills: a review. Nat Environ Pollut Technol 20(2):825–831. https://doi.org/10.46488/NEPT.2021.v20i02.045
6. Maqsood ST, Schwarz J (2008) Analysis of building damage during the 8 October 2005 earthquake in Pakistan. Seismol Res Lett 79(2):163–177
7. Ansari A, Rao KS, Jain AK (2022) Seismic analysis of shallow tunnels in soil medium. In: Stability of Slopes and Underground Excavations (pp 343–352). Springer, Singapore. https://doi.org/10.1007/978-981-16-5601-9_29

8. Malik JN, Sahoo AK, Shah AA, Rawat A, Chaturvedi A (2007) Farthest recorded liquefaction around Jammu caused by 8 October, 2005 Muzaffarabad earthquake of Mw=7.6. J. Geo Soc India 69:39–41
9. Tesfamariam S, Saatcioglu M (2008) Risk-informed seismic evaluation of reinforced concrete buildings. Earthq Spectra 24(3):95–821
10. Alam N, Alam MS, Tesfamariam S (2012) Buildings' seismic vulnerability assessment methods: a comparative study. Nat Hazards 62(2):405–424
11. Singh NN, Deviprasad BS, Krishna PH, Kumar GK (2015) Probabilistic seismic hazard analysis for warangal considering single seismogenic zoning. In: Indian geotechnical conference 2020 (IGC-2020). Visakhapatnam, Indias
12. Ansari A, Rao KS, Jain AK, Ansari A (2022) Deep learning model for predicting tunnel damages and track serviceability under seismic environment. Model Earth Syst Environ
13. Ansari A, Zahoor F, Rao KS, Jain AK (2022) Seismic hazard assessment studies based on deterministic and probabilistic approaches for the Jammu region, NW Himalayas. Arab J Geosci 15(11):1-26. https://doi.org/10.1007/s12517-022-10330-z
14. BIS 1893 (2002) Part 1: Indian standard criteria for earthquake resistant design of structures, part—1, general provisions and buildings. Bureau of Indian Standards, New Delhi, India
15. NEHRP (1994) Recommended provisions for the development of seismic regulations for new buildings. In: Building seismic safety council, Washington, DC
16. ATC-21 (1988) Rapid visual screening of buildings for potential seismic hazards: a handbook. In: Applied technology council. Redwood city, CA, USA
17. Census of India (2011) Series 2: district census handbook Jammu, directorate of census operations. Jammu and Kashmir
18. Singh NN, Deviprasad BS, Kumar GK, Hari Krishna P (2015) Analysis of earthquake catalogue for seismic hazard analysis of Warangal city. Discovery 41:136–142
19. FEMA (1992) NEHRP handbook for the seismic evaluation of existing buildings. Building Seismic Safety Council, Washington, DC

Chapter 38
Effect of Earthquake on the Footing Placed on the Different Location of the Slope Surface

Shantanu Saraswat and Manendra Singh

Introduction

For the construction in Himalayan areas, which comprise about 15% of the total area of India [1], which is also under the seismic zone IV and V [2], it is necessary to check the behaviour of footings on slopes during the earthquake. There are two criteria for the failure of footings: shear failure criteria and settlement failure criteria. Large research has been done regarding shear failure to calculate the bearing capacity of footings. Terzaghi [3], Meyerhof [4], Hansen [5], Sreenivasulu [6], Reddy [7], Desai [8], Waheed [9], Ibrahim [10] calculated the bearing capacity on plain ground using different methodologies. Some researchers like Hansen [5], Keskin [11], Wasem Azzam [12], Acharyya [13] also calculated the bearing capacity of footings when they are placed on the surface of slope or near the slope.

After the 1990's area of interest shifted towards the effect of the earthquake as per the requirements, hence researches have been done by Richards [14], Soubra [15], Subba Rao [16], Tung [17], and many others to calculate the seismic bearing capacity of footings on plain ground. Some researchers like Xiaming et al. [18], Azzam [19], Alzabeebee [20], Maheswari [21] checked the effect of earthquake considering settlement criteria. In these researches, displacement of the footing was observed during earthquake loading by varying different parameters. Sarma and Chen [22], Castelli and Motta [23], Kumar [24] have studied the seismic response of footings on the slope, in most of the cases, the footing was placed on the crest of the slope at various distances from the slope surface. Chakraborty [25] and Raj

S. Saraswat
Department of Earthquake Engineering, Indian Institute of Technology Roorkee, Roorkee 247667, India

M. Singh (✉)
Department of Civil Engineering, National Institute of Technology, Hamirpur, HP 177005, India
e-mail: manendra@nith.ac.in

K. Muthukkumaran et al. (eds.), *Soil Dynamics, Earthquake and Computational Geotechnical Engineering*, Lecture Notes in Civil Engineering 300,
https://doi.org/10.1007/978-981-19-6998-0_38

Table 38.1 Soil parameters [27]

Soil parameters	Unit	Values
Unit weight (γ)	kN/m^3	20.4
Modulus of elasticity (*E*)	kN/m^2	132,000
Poison's ratio (υ)	– —	0.3
Peak cohesion (c)	kN/m^2	21.04
Residual cohesion (c')	kN/m^2	17.26
Peak friction angle (φ)	Degree	33.2
Residual friction angle	Degree	29

[26] placed the footing on the surface of the slope and calculated the seismic bearing capacity.

In this present research, the earthquake effect has been examined when footings were placed at different locations of the slope surface.

Problem Description

A Typical section of the slope in the Himalayan region located along the National Highway-58 between Rishikesh and Devprayag (Uttarakhand) has been taken for the analysis. Properties of the soil have been taken from the study of Siddique and Pradhan [27]. There are eight risky debris slopes available at this site, but the weakest soil data is chosen for current research. Properties of soil have been presented in Table 38.1.

The damping ratio of the soil has been taken as 10%. The study has been carried out on a 55 m high hill. Figure 38.1 shows the geometry of the slope with different locations of footing. Slope angle (β) has been varied from 20 to 35°. Six footing positions have been chosen on the slope's surface. These footings were designated as *A*, *B*, *C*, *D*, *E* and *F*, which are placed below the crest at 1, 5, 10, 20, 30, and 40 m, respectively. Each footing is placed at least 1 m below the nearest surface of the slope. The width of the footing has been taken as 1 m.

This site of the slope is situated about 80 km away from Uttarkashi, where the earthquake was stuck in 1991. Hence acceleration versus time history data of this earthquake, as given in Fig. 38.2, has been used for the analysis in the present study. The peak ground acceleration value of the Uttarkashi earthquake was 2.37 m/s^2.

Methodology and Modelling

Numerical analysis has been done using the finite element method in PLAXIS 2D V20 software [28]. Mohr–Coulomb model has been used to define soil material

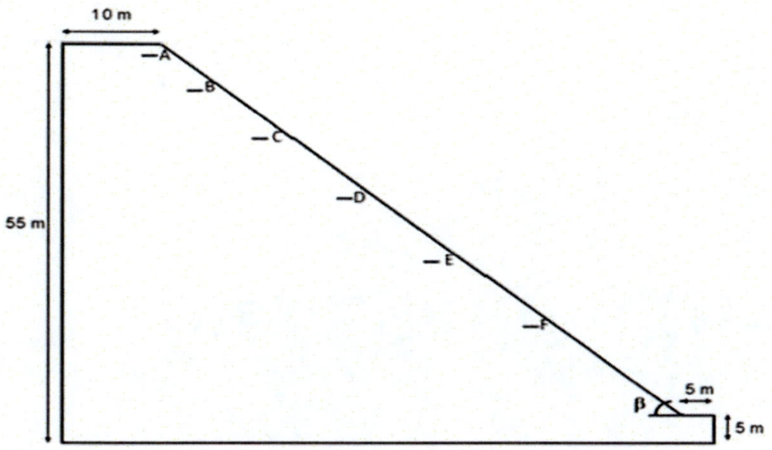

Fig. 38.1 Geometric model of the slope with footings at different locations

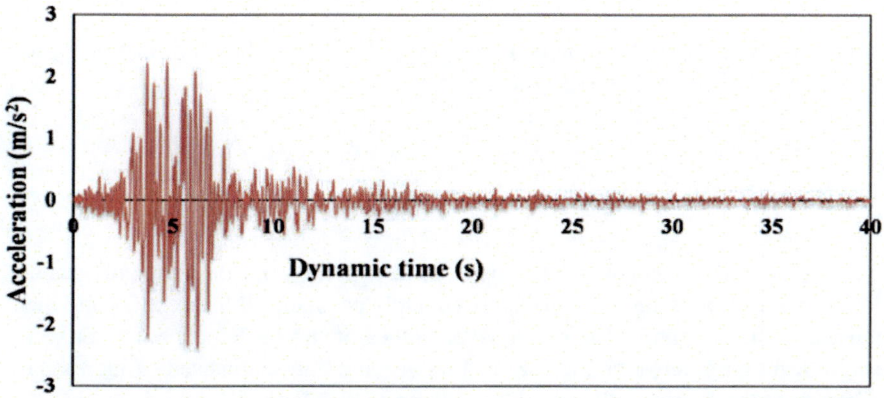

Fig. 38.2 Acceleration versus time history of the horizontal component of the 1991 Uttarkashi earthquake

properties. As the soil surface is not horizontal hence initial stresses in soil mass were generated by gravity loading. Footings were simply made with line segments to make them free from self-weight and made fixed in the horizontal direction so that they behave as rigid in this direction. These footings were loaded with a safe load of 300 kN/m^2. To avail, the in-situ ground condition vertical boundaries were selected as horizontally fixed. For dynamic analysis, viscous boundary [29] was adopted so that seismic waves should not reflect back into the soil body. The bottom boundary was fixed in both directions, and the top and slope surfaces were free in both directions. A prescribed displacement of 1 m has been applied at the bottom of soil mass in the horizontal direction, and earthquake time history data was used as a multiplier to this prescribed displacement. Automatic meshing generation has been used with

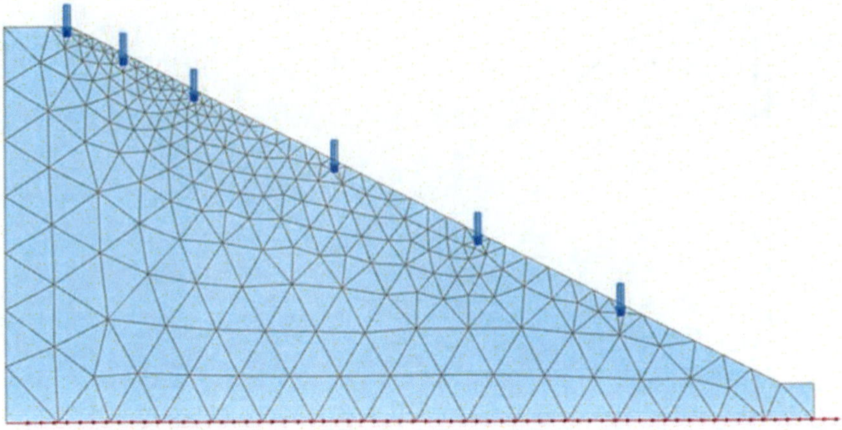

Fig. 38.3 Finite element mesh of the slope-foundation system

medium meshing with the 15-noded triangular element as shown in Fig. 38.3, for optimization to discretize the soil mass.

Results and Discussion

When the slope is subjected to earthquake, as time passes, the amplitude of acceleration increases till 4 s and remains approximately constant till the next 6 s and then decreases until the end of the earthquake. Values of the settlement were found to be significant only between 2 and 12 s. This acceleration has caused an increase in settlement of the footings placed on the surface of the slope.

Figure 38.4 shows the settlement of footings on 20° slope during earthquake loading. From Fig. 38.4, it can be stated that the settlement of each footing seems not to be considerable up to a time of 3.5 s, but after this time, a sudden increase in settlement of the footings was observed until a certain time, and thereafter the settlement becomes constant. When the effect was observed with the position of the footing, then it can be seen that settlement of the footing at positions *A* and *B* were very large (about 200 mm), whereas at positions below *D*, there was little effect of earthquake and settlement was observed below 10 mm.

Figure 38.5 depicts the effect of the earthquake on footings placed on the 25° slope. It can be observed that the behaviour of footings at each location was very similar to the footings on the 20° slope, but the values of settlement at locations *A*, *B*, *C*, and *D* have increased, but below *D*, settlement has not much increased.

The settlement of footings on 30° slope is shown in Fig. 38.6. It can be observed from Fig. 38.6 that settlement at positions *A* and *B* goes on increasing till the end of the

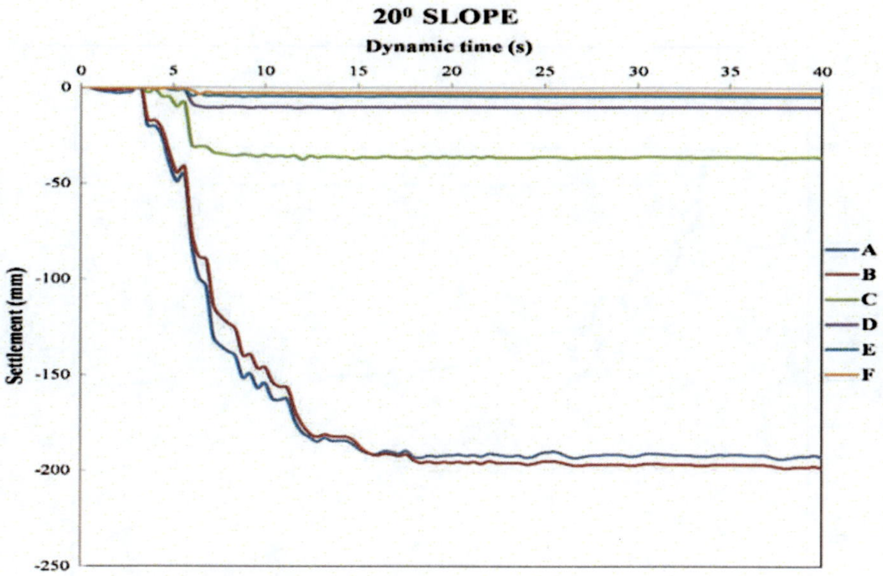

Fig. 38.4 Settlement of footings on 20° slope

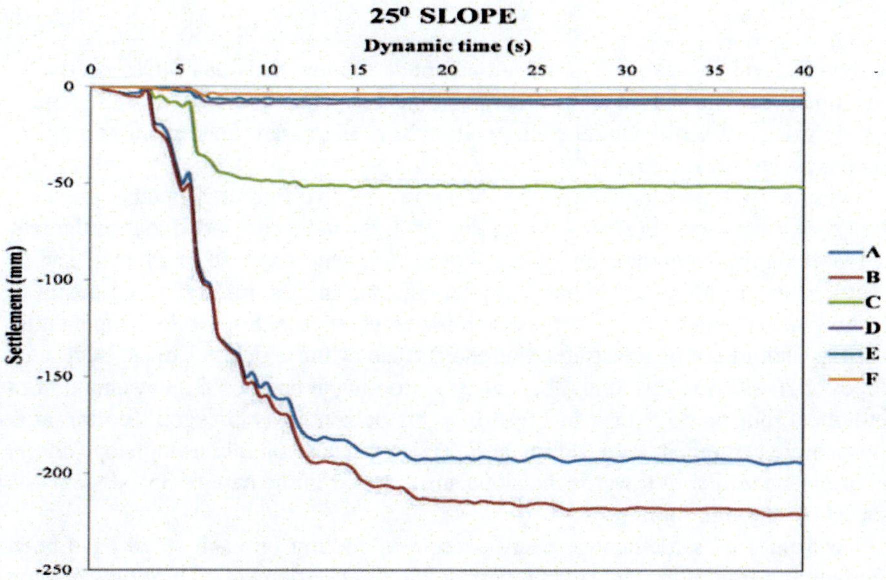

Fig. 38.5 Settlement of footings on 25° slope

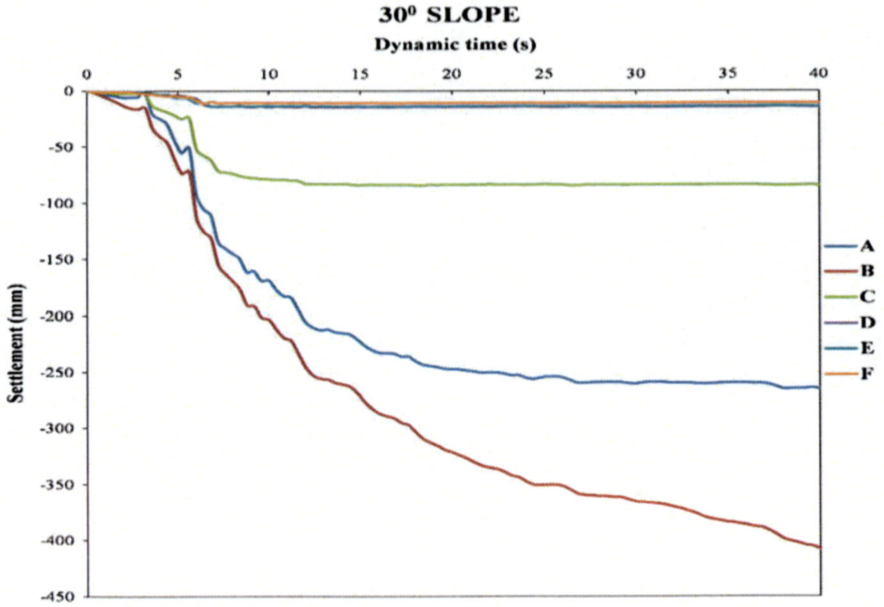

Fig. 38.6 Settlement of footings on 30° slope

earthquake, and very large value of settlement of 265 mm and 408 mm, respectively, were obtained. The pattern at the other positions is same as at the 20° and 25° slopes, but the values of settlement at each position have increased. Settlement at point *D* has reached up to 15 mm.

The effect of the earthquake on the 35° slope is much different from other previous slopes, as observed in Fig. 38.7. From Fig. 38.7, it can be concluded that settlement increases significantly for each footing. Position *F*, which is the least affected, shows a settlement of 130 mm, which is very significant, and the most affected position, *A*, shows a settlement of 760 mm, which is the worst case. So, for 35° slope angle, footings should not be constructed on the surface of those debris slopes. In case of slopes of 20–30° in Figs. 38.4, 38.5 and 38.6, footing at position *B* shows maximum settlement, but on 35° slope in Fig. 38.7, this behaviour is changed, footing at *A* shows more settlement than footing at *B*. This is because on increasing slope angle, failure plane will shift towards the slope surface, and in the case of 35° slope it has passed close to the footing *A*.

The values of settlement at each location of footing on each slope have been compared in Fig. 38.8. Figure 38.8 depicts the final settlement of footings on each slope due to earthquake loading. The settlement increases drastically with an increase in slope angle. This effect is worse on a slope of 35°, and even at position *F*, which is least affected, shows a very large settlement. Therefore, this slope is not safe for construction purposes. For slopes of 20° and 25°, the settlement at positions *A* and *B* is roughly equal, or it can be stated that the footing at *B* has experienced a slightly

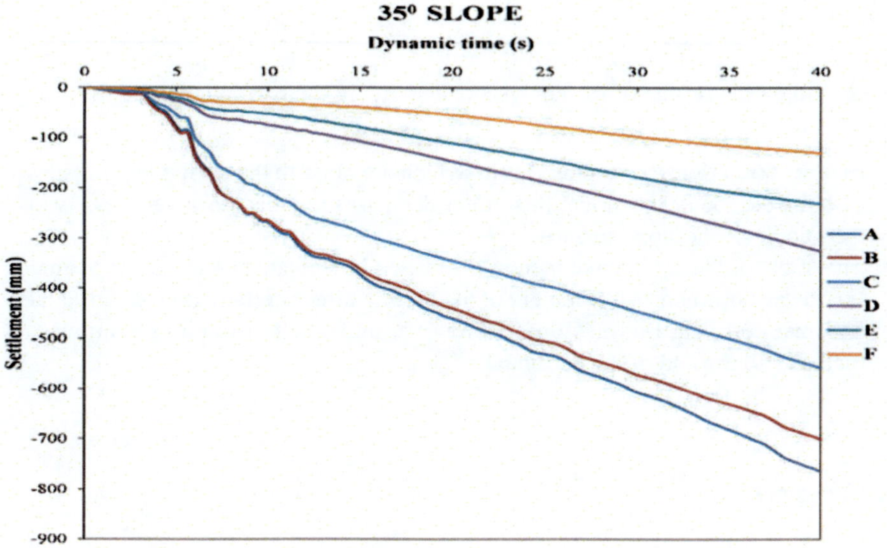

Fig. 38.7 Settlement of footings on 35° slope

higher settlement than the footing at *A*, but on steeper slopes, position *A* is more vulnerable to the earthquake. Because these two positions are so close to the crest of the slope, construction of footings should be avoided in these areas.

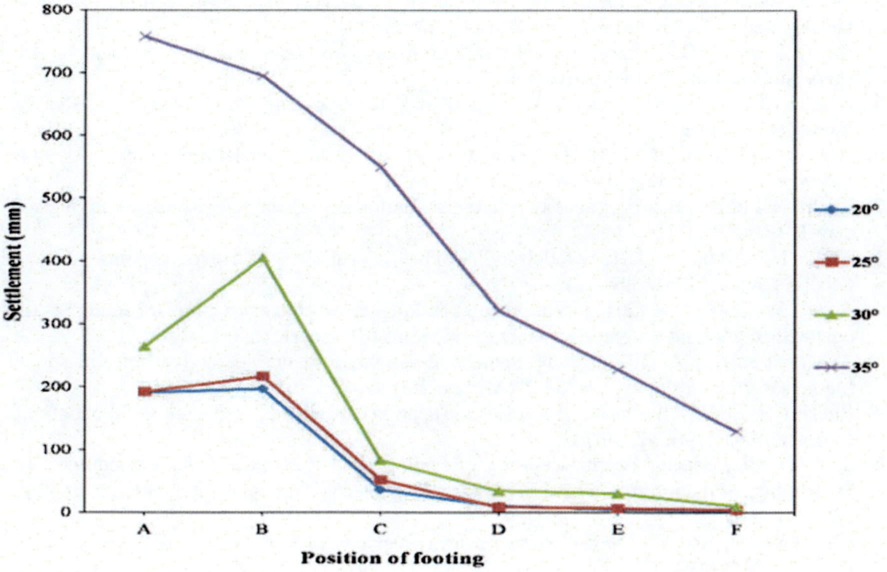

Fig. 38.8 Comparison of final settlement of footings on each slope

Conclusion

After finding the results of the analysis and studying their variations, the following conclusions can be made:

- Earthquakes affect the footings more, which are close to the crest. Ongoing down towards the toe of the slope, this effect decreases and becomes very less below the middle of the slope surface.
- The effect of the earthquake increases drastically with an increase in slope angle.
- For lower slope angles, settlement of the footing after a certain time of earthquake becomes constant. But on slopes above 30°, settlement of footings continues to increase till the end of the earthquake.

References

1. Indian Himalayan Region—ENVIS centre on himalayan ecology, GBPIHED, http://www.Gbpihedenvis.nic.in. Last accessed 04 Jun 2021
2. IS 1893-Part 1 (2002) Indian standard criteria for earthquake resistant design of structures (Fifth Revision)- General Provisions and Buildings. Bureau of Indian Standards, New Delhi
3. Terzaghi K (1943) Theoretical soil mechanics. Wiley, New York
4. Meyerhof F (1963) Discussion of "Bearing Capacity of Foundations". J Soil Mech Found Div 89(1):319–320
5. Hansen BJ (1970) A revised and extended formula for bearing capacity. Geotechnical Institute Bulletin, vol 28. Copenhagen
6. Sreenivasulu V, Ranganatham B (1971) Bearing capacity of anisotropic non-homogeneous medium under Φ=0 condition. Soils Found 11(2):17–27
7. Reddy, Rao K (2016) Bearing capacity of strip footing on anisotropic and nonhomogeneous clays. Soils Found 21(1):99–110
8. Desai VGM, Desai AVM (2017) Analysis of strip footing using PLAXIS 2D. Int J Res Sci Innov 4(10):33–38
9. Waheed MQ, Asmail NM (2018) Parametric study of shallow foundation bearing capacity in clayey soil. Int J Civ Eng Technol 9(10):1223–1330
10. Ibrahim KI (2016) Bearing capacity of circular footing resting on granular soil overlying soft clay. HBRC J 12(1):71–77
11. Keskin MS, Laman M (2013) Model studies of bearing capacity of strip footing on sand slope. KSCE J Civ Eng 17(4):699–711
12. Azzam W, El Wakil A (2016) Experimental and numerical studies of circular footing resting on confined granular subgrade adjacent to slope. Int J Geomech 16(1)
13. Acharyya R, Dey A (2017) Finite element investigation of the bearing capacity of square footings resting on sloping ground. INAE Lett 2(3):97–105
14. Richards R, Elms D, Budhu M (1993) Seismic bearing capacity and settlements of foundations. J Geotech Eng 119(4):662–674
15. Soubra (1997) Seismic bearing capacity of shallow strip footings in seismic conditions. In: Proceedings of the institution of civil engineers—geotechnical engineering, vol 125, no 4, pp 230–241
16. Choudhury D, Rao KS (2005) Seismic bearing capacity of shallow strip footings. Geotech Geol Eng 23(4):403–418
17. Tung S, Mukherjee S, Roy N, Mukherjee S (2021) Seismic bearing capacity of shallow footing in layered soil. Lecture Notes Civ Eng 487–499

18. Yuan Y, Meng S, Sun R (2003) A semi-empirical formula for evaluating residual strain of soils under earthquake loading. Earthq Eng Eng Vib 2(2):307–313
19. Azzam W (2015) Finite element analysis of skirted foundation adjacent to sand slope under earthquake loading. HBRC J 11(2):231–239
20. Alzabeebee S (2020) Seismic settlement of a strip foundation resting on a dry sand. Nat Hazards 103(2):2395–2425
21. Maheshwari P, Yadav NK, Sahoo J (2020) Seismic performance of footings on stone columns treated dry sand beds. KSCE J Civ Eng 24(11):3230–3236
22. Sarma S, Chen Y (1995) Seismic bearing capacity of shallow strip footings near sloping ground. In: 5th SECED conference on European seismic design practice. Chester, England, pp 505–512
23. Castelli, Motta E (2009) Bearing capacity of strip footings near slopes. Geotech Geol Eng 28(2):187–198
24. Kumar J, Ghosh P (2006) Seismic bearing capacity for embedded footings on sloping ground. Géotechnique 56(2):133–140
25. Chakraborty, Kumar J (2015) Seismic bearing capacity of shallow embedded foundations on a sloping ground surface. Int J Geomech 15(1):04014035
26. Raj D, Singh Y, Shukla S (1943) Seismic bearing capacity of strip foundation embedded in c—ϕ soil slope. Int J Geomech 18(7):04018076
27. Siddque T, Pradhan S (2018) Stability and sensitivity analysis of Himalayan Road cut debris slopes: an investigation along NH-58, India. Nat Hazards 93(2):577–600
28. PLAXIS 20V. User's manual of PLAXIS, A.A. Balkema Publishers
29. Lysmer J, Kuhlemeyer R (1969) Finite dynamic model for infinite media. J Eng Mech Div 95(4):859–877

Chapter 39
Site-Specific GRA to Quantify Ground Motion Amplification for Bettiah Site: A Case Study

Suryakant Kumar Sisodiya, Prabhakar Kumar, and Shiv Shankar Kumar

Introduction

Site-specific ground response analysis (GRA) is one of the major concerns to evaluate the ground motion amplification for seismic hazard assessment or seismic microzonation of any area. Site-specific GRA provides the ground motion amplification at any required depth, which is further utilized to develop the design response spectra for the design of earthquake resistance structures [1]. Therefore, considering the present scenario of earthquakes such as the Bhuj earthquake (2001), the Sikkim earthquake (2011), and the Nepal earthquake (2015), the seismic GRA can be performed at any construction site, however, is of utmost importance in an earthquake-prone area. During the Nepal earthquake (2015), many structures in Bihar were severely damaged though the epicentre was very far from the centre of Bihar. As per IS:1893 [2], Bihar comes under the seismic zone III and zone IV, therefore, site-specific GRA becomes one of the significant aspects to foresee the potential consequences for amplification or amplification of the strong ground motions at the ground surface before the design of the structure. Seismic GRA can be analysed using one-dimensional (1D) linear, 1D equivalent linear or nonlinear approach with the assumptions that the soil layers and materials below the soil layers are horizontal and extend to infinity in the lateral direction. The equivalent linear (EL) approach is based on total stress conditions, whereas nonlinear soil behaviour is approximated by equivalent linear response of soil, whereas the nonlinear approach is based on effective stress conditions wherein

S. K. Sisodiya · P. Kumar (✉) · S. S. Kumar
Department of Civil Engineering, National Institute of Technology, Patna 800005, India
e-mail: prabhakark.phd20.ce@nitp.ac.in

S. K. Sisodiya
e-mail: suryakants.pg19.ce@nitp.ac.in

S. S. Kumar
e-mail: k.shiv.ce@nitp.ac.in

K. Muthukkumaran et al. (eds.), *Soil Dynamics, Earthquake and Computational Geotechnical Engineering*, Lecture Notes in Civil Engineering 300,
https://doi.org/10.1007/978-981-19-6998-0_39

nonlinear soil behaviour is accurately modelled to predict the GRA with water pressure generation model as well as the coupling of pore pressure generation response and soil behaviour. GRA studies carried out by several researchers, mentioned in Table 39.1, also indicate that the ground response analysis is one of the fundamental requirements for the design of earthquake-resistant structures [1–12]. Kumar et al. [13] have performed the seismic vulnerability of Kathmandu City (Nepal) based on the Nepal 2015 earthquake. The range of peak ground acceleration and peak spectral acceleration was found to be in the order of 0.21–0.42 g and 0.74–1.50 g, respectively. Based on the liquefaction potential index, it was concluded that 6 sites out of 10 sites are at high risk of liquefaction. Kumar et al. [14] have performed one-dimensional equivalent linear GRA for Guwahati city and recommended that GRA of any region considering its regional dynamic soil properties is the best option to obtain more realistic outcomes. During the literature study, it was found that seismic GRA for the Bihar region was scanty, and therefore, GRA was carried out to observe the site-amplification at the construction site of Bettiah, Bihar.

Methodology

Seismic GRA can be performed either by 1D linear or 1D equivalent linear or 1D nonlinear analysis to estimate the seismic response of stratified soil profile at any specific site in terms of ground acceleration profile, response spectrum, and stress–strain histories. Every mythology has its advantage and disadvantages. In the linear approach, the shear modulus (G) and damping ratio (ξ) are assumed to be constant throughout a particular layer. The equivalent linear (EL) approach approximates the actual nonlinear variation of G and ξ of soil with shear strain, whereas in nonlinear (NL), the realistic behaviour of soil, i.e. nonlinear response of soil is incorporated accurately. In the present study, one-dimensional nonlinear GRA has been performed at the selected location, i.e. Bettiah site in Bihar using the DEEPSOIL software [15] with 18 numbers of borehole data. In this approach, the nonlinear inelastic stress–strain behaviour of soil is considered to evaluate the equations of motion in small time steps which have been referred to obtain the appropriate soil properties to be used in that time steps.

Nonlinear Time Domain Analysis

In contrast with comparable direct recurrence area investigation, the time–space nonlinear strategy is better in anticipating solid ground movement and it additionally permits significant pore water strain to be incorporated. The precision of forecasts which relies upon the constitutive model depiction of soil conduct and for powerful vertical pressure investigation, exactness of expectation additionally relies upon the precision of pore water pressure age reaction model just as the coupling of pore water

Table 39.1 Summary of the literature review

Authors	Location	Software used	Methodology	Remarks
Akhila et al. [3]	Kolkata	DEEPSOIL	1D equivalent linear	PGA of the city range was reported in the range of 0.1–0.34 g
Bashir et al. [4]	Agartala	DEEPSOIL	1D nonlinear	To confirm the sanctity of exact-based evaluation of liquefaction potential. A critical spatial variety of ground reaction risk over the entire region is expected
Puri et al. [5]	Harayana	DEEPSOIL	1D equivalent linear	Cyclic reaction accounted for utilizing standard shear modulus corruption and damping curves. PGA (g) values for selected 77 sites vary from 0.073 to 0.396 g
Gupta et al. [6]	Harayana	DEEPSOIL	1D equivalent linear	The analysis is done using spectrum compatible acceleration time histories and the amplification factor varies from 1.25 to 1.84
Jishnu et al. [7]	Kanpur	SHAKE2000 and OPENSEES	1D nonlinear and 2D nonlinear	Examined the conduct of a more profound soil layer under solid ground movement as far as the overabundance pore pressure improvement, liquefaction potential, and post-liquefaction settlement

(continued)

Table 39.1 (continued)

Authors	Location	Software used	Methodology	Remarks
Kumar et al. [8]	Guwahati	DEEPSOIL	1D equivalent linear	Seven different recorded earthquake motions were adopted of M_w range 6.1–8.1. Fourier amplification at the surface varies from 4.14 to 8.99 for a frequency of 1.75–3.13 Hz
Ranjan et al. [9]	Dehradun	SHAKE2000	MASW	Spectral acceleration varies from 0.06 to 0.37 g for the frequency range of 1–10 Hz. Different zones of shear wave velocity and spectral acceleration have been classified
Kumar and Krishna [10]	Guwahati	DEEPSOIL	1D equivalent linear	Fourier amplification of ground motion at the surface is in the range of 4.14–8.99 for a frequency band of 1.75–3.13 Hz for the range of input motion PGA 0.1–0.34 g
Singhai et al. [11]	Guwahati	DEEPSOIL	1D nonlinear and 1D equivalent linear	The generation of excess PWP indicates a reduction in stiffness and shear strength of soil resulting in changes in the soil behaviour. PGA values of NL approach with excess PWP dissipation are approx. 35–95% lesser than the EL approach

(continued)

Table 39.1 (continued)

Authors	Location	Software used	Methodology	Remarks
Desai et al. [12]	Mumbai	SHAKE2000	1D equivalent linear	Seismic amplification varies from 2.53 to 4.14 for a frequency band of 1.75–3.5 Hz

pressing factor and conduct of the soil. In nonlinear response methodology, to represent the nonlinear behaviour of soil, the following expressions (dynamic equilibrium and equations of motion) are required to be solved discrete-time augmentations in the time domain.

$$[M]\{\ddot{u}\} + [C]\{\dot{u}\} + [K]\{u\} = -[M]\{I\}\ddot{u}_g \tag{39.1}$$

where $[M]$ represents the mass matrix, $[C]$ represents the damping matrix, $[K]$ represents the stiffness matrix, $\{\ddot{u}\}$ represents the relative acceleration, $\{\dot{u}\}$ represents the relative velocity, $\{u\}$ represents the relative displacements, $\{I\}$ represents the unit vector, and $\ddot{u}_g$ represents the acceleration of the base of the soil column.

Excess Pore Water Pressure Generation and Dissipation Model

Excess pore water pressure generation and dissipation exist at the same time during ground shaking. Due to strong ground shaking, excess pore pressure induces in the soil which creates excessive settlement, loss of strength in the soil, and other damages due to a decrease in effective stress and shear strength of the soil. Excess pore pressure dissipation can remarkably affect the magnitude of excess pore pressure achieved during ground shaking which depends on the permeability of soil and drainage conditions. The loss of excess pore pressure model is assumed to occur only in the vertical direction and is based on Terzaghi's 1D consolidation theory.

$$\frac{\partial u}{\partial t} = C_v \frac{\partial^2 u}{\partial z^2} \tag{39.2}$$

where C_v represents the coefficient of consolidation in the vertical direction. In the present study, the pore water pressure model proposed by Matasovic and Dobry (1993) for sand and proposed by Matasovic and Vecetric (1995) for clay was used to determine a curve fitting procedure in the absence of laboratory results. These curve fitting parameters have been considered as per Hashash et al. [15].

Pressure-Dependent Hyperbolic Model

Pressure-dependent modified hyperbolic model has been used to define the stress–strain relationship developed in the soil due to the cyclic loading–unloading phenomenon. The modified hyperbolic model, proposed by Matasovic (1993), is based on the strain-independent hyperbolic model proposed by Konder and Zelasko (1963). This modified hyperbolic model is also called modified Matasovic-Konder-Zelasko (MKZ) and expressed as

$$\tau = \frac{G_o \gamma}{1 + \beta\left(\frac{\gamma}{\gamma_r}\right)^s} \tag{39.3}$$

where G_o represents the initial shear modulus, τ represents the shear strength, γ represents the shear strain, β, s, and γ, are the model parameters. β and s are utilized to adjust the shape of the backbone curve. Further, the shear stress and confining pressure are coupled to make soil profile pressure dependent as follows [15]

$$\gamma_r = \text{Reference strain}\left(\frac{\sigma'_v}{\text{Reference stress}}\right)^b \tag{39.4}$$

where σ'_v represents the effective vertical stress, reference stress is the vertical effective stress, γ_r is reference stress, and b represents the curve fitting parameter of the backbone curve.

Modulus Reduction and Damping Ratio Curves

Due to the unavailability of site-specific curves, representing dynamic soil properties, the modulus reduction and damping ratio curves existing in the DEEPSOIL were utilized in the analysis. In the present investigation, to characterize the dynamic feature of sands and clays, the standard curves for modulus reduction and damping ratio proposed by Seed and Idriss [16] and Vucetic and Dobry [17] have been considered from the database of DEEPSOIL [15]. The plasticity index (PI) values were chosen in the range of 20–23% for clay layers.

Shear Wave Velocity Profile for Soils

Shear wave velocity (V_s) is one of the significant input parameters in seismic GRA studies. In the present study, the correlations between V_s with SPT-N values, listed in Table 39.2, were used to evaluate V_s profiles. Due to uncertainty in the selection of V_s profiles, the average value of the correlations, listed in Table 39.2, have been used.

Table 39.2 Correlations for estimation of average V_s

S No.	Author(s) name	Correlations	Soil type
1	Imai and Tonouchi [18]	$V_s = 97\ N^{0.314}$	Applicable for all soils
2	Iyisan [19]	$V_s = 51.5\ N^{0.516}$	
3	Kiku et al. [20]	$V_s = 68.3\ N^{0.292}$	
4	Hasencebi and Ulusay [21]	$V_s = 90\ N^{0.309}$	
5	Maheshwari et al. [22]	$V_s = 95.64\ N^{0.301}$	
6	Hanumantharao [23]	$V_s = 82.6\ N^{0.43}$	
7	Dikmen [24]	$Vs = 58\ N_{0.39}$	

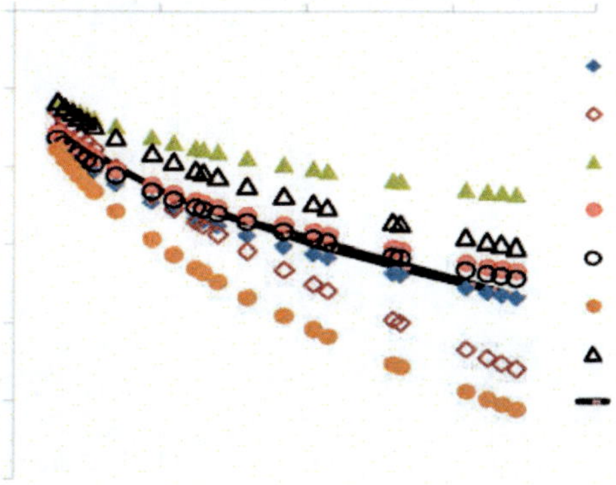

Fig. 39.1 Variation of shear wave velocity with SPT-N values for BH-PN-2

Further, the average value of V_s profile, shown in Fig. 39.1, was used at BH-PN-2. Similarly, the average correlations between V_s with SPT-N were evaluated for all 18 boreholes and utilized for seismic GRA.

Study Area

Bettiah is situated near the Indo-Nepal border and 225 km from the northwest of Patna, Bihar (India). Bettiah is an administrative head office of the West Champaran, Bihar, which lies between 26°48′05"N latitude and 84°30′10"E longitude. The district comes under the seismic zone IV, as per IS-1893 [2], and therefore has a moderate to high damage risk from earthquakes. To characterize the soil properties, soil exploration data was obtained from 18 boreholes (namely, BH-AL, BH- PL-1, BH-PL-3, BH- PL-5, BH- PL-7, BH-AN, BH- PN-2, BH- PN-4, BH- PN-6, BH-AM, BH-PM-1, BH- PM-3, BH- PM-5, BH-AB, BH- PB-0, BH-PB-2, BH-PB-4, BH- PB-6) between Bettiah -Kumar Bagh Station or Road Ch-29.00 km of NH-727 and Ch-0.00 of SH-105 in the Samastipur Division (Fig. 39.2). It was observed from all of the boreholes, having a terminating depth of 40 m, that the site consists of

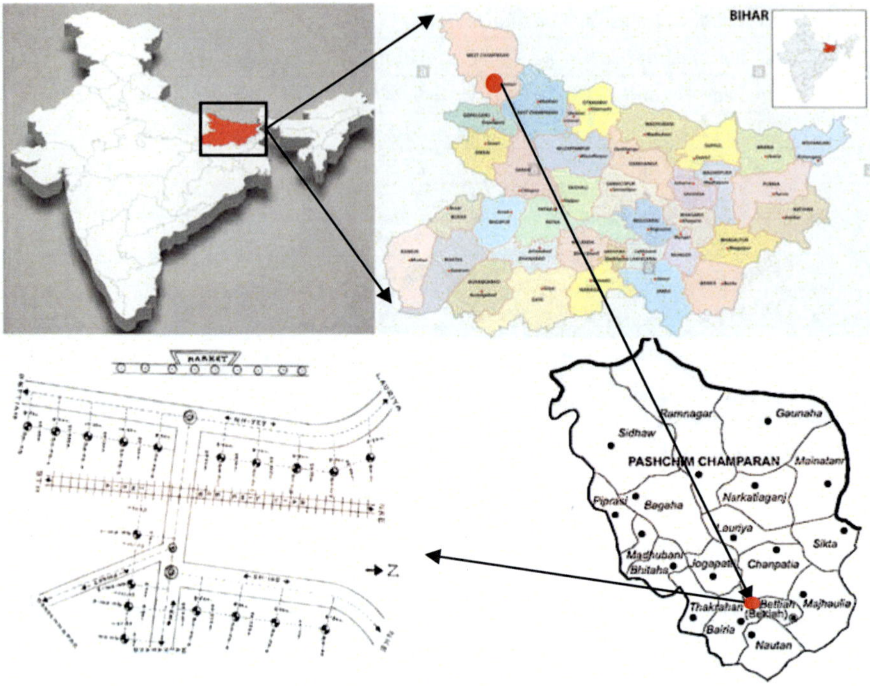

Fig. 39.2 Location of Bettiah Site and all 18 boreholes

clayey, silty, and sandy soil with a remarkably high-water table between 0.8 and 1.6 m from the ground surface. Also, the bulk density at all 18 boreholes varies from 18–19 kN/m^3 and the plastic limit for silty and clayey soil was in the range of 20–23%.

Input Strong Ground Motion

In this study, three acceleration time histories obtained from the 2011-Sikkim earthquake (M_w 6.9) of PGA 0.16 g (recorded motion), 0.286 g (scaled motion), and 0.43 g (scaled motion) are chosen as input motion to represent low, moderate high, and very high seismic hazards. The acceleration time histories of all three input motions are presented in Fig. 39.3.

Fig. 39.3 Acceleration time histories of input motion

Results and Discussions

One-dimensional nonlinear seismic GRA was carried out for all 18 boreholes at the site, Bettiah (Bihar), with the consideration of excess pore water pressure generation and dissipation using the DEEPSOIL computer programme. Utilizing three acceleration time histories, the results were obtained in terms of the variations in acceleration, maximum shear strain, shear stress ratio, and pore water pressure ratio along with depth for all boreholes, which is presented and discussed in the following sections.

Significance of Input Motion on GRA

To analyse the effect of earthquakes of different energy levels (reflected by different PGA), one borehole BH-PN-2 has been selected among 18 boreholes. At borehole BH-PN-2, a clayey layer was found at the depth of 35 m below the ground surface. Figure 39.4a–d presents the variations in acceleration, shear strain, shear stress ratio, and pore water pressure ratio along with depth at BH-PN-2 using acceleration time histories of PGA = 0.16, 0.286, and 0.43 g. From Fig. 39.4a, it can be seen that at a specific depth between 35.4 and 37.5 m, variation of PGA varies in the range of 0.231–0.545 g which shows the chance of wave amplification. Figure 39.4b indicates higher shear strain at the depth of 35 m, which might be due to the relatively softer soil. It has also been observed that the value of strain subjected to PGA = 0.286 g is higher than that obtained from the input motion of PGA = 0.16 and 0.43 g may be due to the high energy content of input motion. The shear stress ratio is one of the important indexes to predict the liquefaction susceptibility behaviour of stratified soil. The stress ratio greater than 1 at any depth reflects the shear stress employed at the soil is greater than the effective vertical stress causing liquefaction.

Figure 39.4c represents the distribution of shear stress ratio along with depth by using input motion of PGA = 0.16, 0.286, and 0.43 g. It has been observed that the distribution of shear stress ratio within the stratified soil layers is less than 1, which shows there is no chance of the occurrence of liquefaction. If the PWP ratio at any depth of stratified layered soil is less than 1, then no liquefaction occurred at any depth from the ground surface. However, if the excess PWP ratio becomes 1 then liquefaction is inevitable. Figure 39.4d presents the variations of PWP ratio along with depth using input motions of PGA = 0.16, 0.286, and 0.43 g for BH-PN-2.

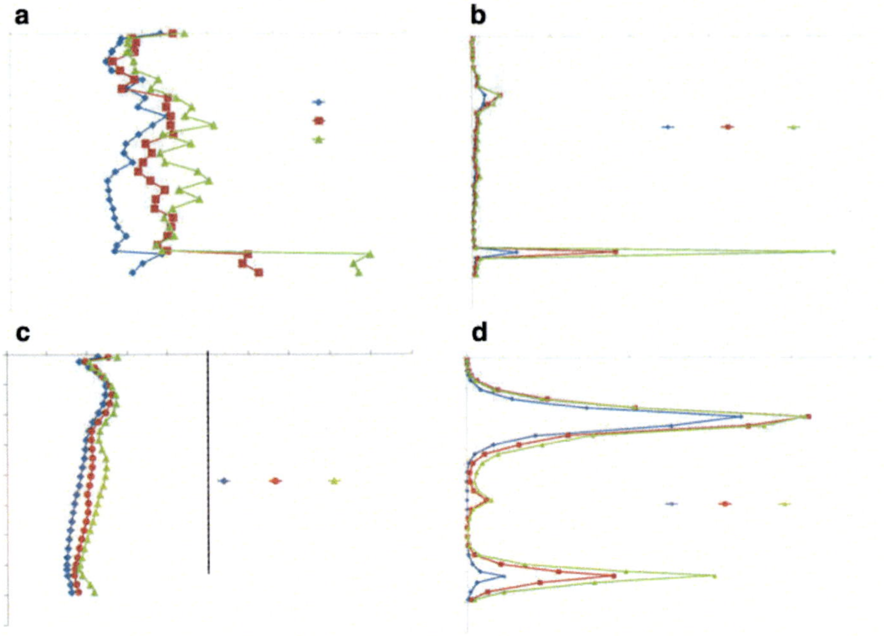

Fig. 39.4 Effects of variation **a** maximum acceleration, **b** max. strain, **c** max. stress ratio, and **d** max PWP ratio with depth using 0.16, 0.286, and 0.43 g input motion at BH-PN-2

From Fig. 39.4d, the PWP ratio is found to be 0.67 at the depth of 11.25 m using 0.16 g PGA motion, which shows there is no chance of occurrence of liquefaction. However, with input motions of PGA 0.286 and 0.43 g, the PWP ratio was nearly 0.83 and 0.84, respectively, at depth 9.75 m below the ground surface which might hint the initiation of the liquefaction stage.

Significance of Subsoil Profiles on GRA

GRA considering a single borehole might give some idea about the amplification or de-amplification of seismic waves but cannot predict the response for the larger area. Therefore, the seismic-resistant design based on the response of a single borehole will not be judicious since the soil, as well as soil conditions, varies from place to place. Figure 39.5a–d presents the variations in acceleration, shear strain, shear stress ratio, and pore water pressure ratio along with depth for all boreholes using acceleration time histories of PGA = 0.286 g. The variations in acceleration at the ground surface were found to be in the range of 0.18–0.42 g, in Fig. 39.5a. Therefore, it can be stated that the amplification factor (AF-ratio of obtained maximum acceleration at a different depth to the maximum acceleration of input motion) varies from 0.708

to 1.7 for a PGA of 0.286 g, which shows that the soils of a specific site of Bettiah are capable of amplifying during earthquake motions. It has also been observed that acceleration at the ground surface for borehole BH-PB-4 is higher than the other one. Figure 39.5b shows the accumulation of high shear strain at 14.25 m at BH-PL-5 and 35.7 m at BH-PN-2 which indicates more energy loss during seismic loading. The development of high shear strain may cause permanent deformation in the ground. Therefore, it is recommended to evaluate the strain-dependent dynamic soil behaviour before the construction.

Figure 39.5c represents the distribution of maximum shear stress ratio along with depth at all 18 boreholes using input motion of PGA = 0.286 g. The maximum shear stress ratio at the ground surface was found to be in the range of 0.18 to 0.42. The increasing values of the stress ratio also hint at the increase in the tendency of soil liquefaction or cyclic mobility. From Fig. 39.5d, considering input motion of PGA = 0.286 g, it can be seen that the maximum increase in PWP ratio is found 0.84 in BH-PM-5 and 0.939 in BH-AN at depth 8.25 m, 0.93 in BH-PN-4 at depth 11.25 m, 0.919 in BH-PN-6 at depth 17.25 m from the ground surface and 0.89 in BH-PM-5 at depth 15.75 m. These values of the PWP ratio indicate the initiation of

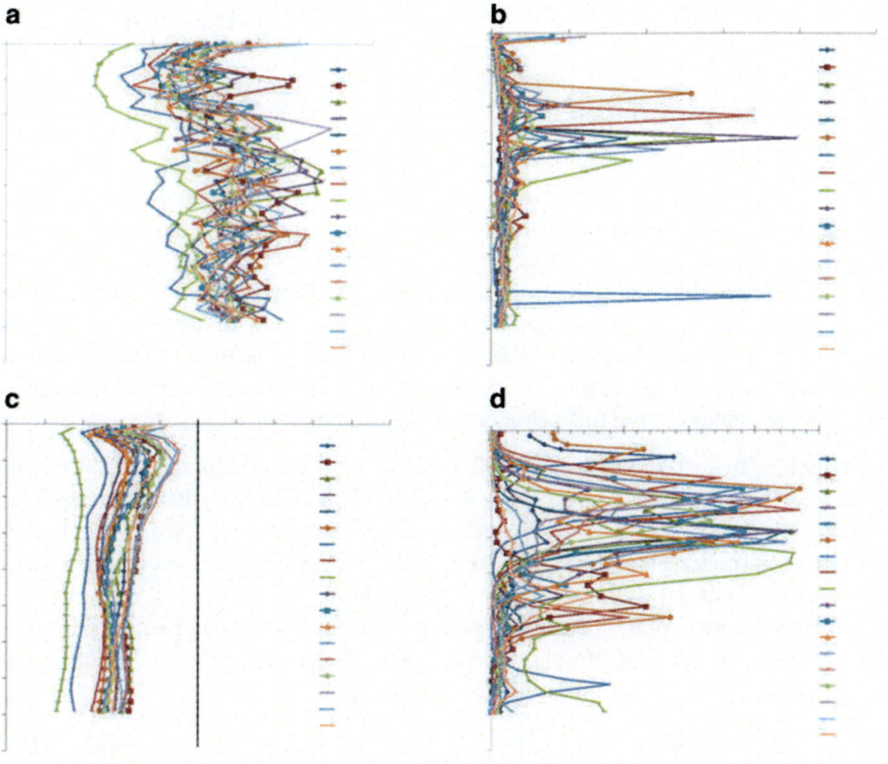

Fig. 39.5 Effects of variation **a** maximum acceleration, **b** max strain, **c** max stress ratio, and **d** max PWP ratio with depth using 0.286 g input motion at all 18 boreholes

Table 39.3 Amplification of PGA

Input PGA (g)	0.16	0.286	0.43
AF of all boreholes	1.178–2.452	0.708–1.7	0.528–1.194

soil liquefaction at most of the borehole when analysed with higher input motion. Moreover, the conditions of soils (such as loose or stiff) at any depth or any location are more responsible for the higher or lower amplification, higher or lower shear strain, higher or lower stress ratio, and higher or lower PWP ratio. Hence, for an accurate assessment of ground acceleration variations, stress ratio variations, strain variations, or liquefaction potential, a detailed study needs to be performed.

Nonlinear GRA has also been carried out for 18 boreholes with the input motion of PGA = 0.16 and 0.43 g. However, the results of the same have not been presented here due to brevity. During analysis, it was observed that AF varies from 1.178 to 2.452 for PGA of 0.16 g, from 0.708 to 1.7 for PGA of 0.286 g, and from 0.528 to 1.194 for PGA of 0.43 g (Table 39.3). These AF values indicate that the amplification of seismic wave is higher corresponding to the lower value of input motion and vice versa. Thus, it can be stated that there is ample chance of amplification of seismic waves at the Bettiah site. Therefore, it can be recommended that the GRA needs to perform before the construction of any soil-supported structures in the Bettiah region.

Conclusions

One-dimensional nonlinear seismic GRA was performed, with the consideration of excess pore water pressure generation and dissipation, using the DEEPSOIL computer programme for 18 boreholes located in the Bettiah district (Bihar). Three acceleration time histories obtained from the 2011-Sikkim earthquake of $M_w = 6.9$, of PGA 0.16 g (recorded motion), 0.286 g (scaled motion), and 0.43 g (scaled motion) have been chosen to represent low, moderate high, and very high seismic hazards. Based on the analysis, the following conclusions have been drawn.

- AF varies from 1.178 to 2.452 for PGA of 0.16 g, from 0.708 to 1.7 for PGA of 0.286 g, and from 0.528 to 1.194 for PGA of 0.43 g. Therefore, it can be stated that there is ample chance of amplification of seismic waves at the Bettiah site; hence, it is recommended that the GRA is required to perform before the construction of any soil-supported structures in the Bettiah region.
- The development of high shear strains (up to 5%) may cause permanent deformation in the ground. Therefore, it is recommended to evaluate the strain-dependent dynamic soil behaviour before the construction.
- The maximum shear stress ratio at the ground surface was found to be in the range of 0.18 to 0.42, which indicates that the tendency of cyclic mobility or soil liquefaction might increase with increasing stress ratio.

- The maximum increase in PWP ratio was 0.84 in BH-PM-5 and 0.939 in BH-AN at depth 8.25 m, 0.93 in BH-PN-4 at depth 11.25 m, 0.919 in BH-PN-6 at depth 17.25 m from the ground surface, and 0.89 in BH-PM-5 at depth 15.75 m, which indicate the initiation of soil liquefaction at most of the borehole when analysed with higher input motion.

References

1. Kramer SL (1996) Geotechnical earthquake engineering. Pearson Education India
2. IS: 1893 (Part I) (2016) Indian standard criteria for earthquake resistant design of structures. Bureau of Indian Standards, New Delhi
3. Akhila M, Ghosh C, Satyam DN (2012) Detailed ground response analysis at park hotel in Kolkata city. India, vol 15. WEEE, Lisboa
4. Bashir et al (2017) Assessment of liquefaction potential using empirical equations and 1D ground response analysis for Agartala city. IGC-2017, GeoNEst, IIT Guwahati, India
5. Puri N, Jain A (2017) Deterministic seismic hazard analysis for the State of Haryana, India. Indian Geotech J 46(2):164–174
6. Gupta R, Puri N, Jain A (2017) Equivalent linear ground response analysis of typical sites in state of Haryana using spectrum compatible ground motions. In: 6th Indian young geotechnical engineers conference. NIT Trichy, India
7. Jishnu RB, Naik SP, Patra NR, Malik JN (2013) Ground response analysis of Kanpur soil along Indo-Gangetic plains. Soil Dyn Earthq Eng 51:47–57
8. Kumar SS, DeyA, Krishna AM (2013) Equivalent linear and nonlinear ground response analysis of two typical sites at Guwahati city. In: Proceeding Indian geotechnical conference (IGC-2013), Kakinada, India
9. Ranjan R (2005) Seismic response analysis of Dehradun city. Thesis submitted to The International Institute for Geo-Information and Earth- Observation, India
10. Kumar SS, Krishna AM (2013) Seismic ground response analysis of some typical sites of Guwahati City. Int J Geotech Earthq Eng (IJGEE) 4(1):83–101
11. Singhai A, Kumar SS, Dey A (2016) Site-specific 1-D nonlinear effective stress GRA with pore water pressure dissipation. In: 6th international conference on recent advances in geotechnical earthquake engineering and soil dynamics, New Delhi, pp 1–11
12. Desai SS, Choudhury D (2015) Site-specific seismic ground response study for nuclear power plants and ports in Mumbai. Nat. Hazards Rev 16(4):1–13
13. Kumar SS, Acharya P, Dammala PK, Krishna AM (2020) Characterization of ground response and liquefaction for Kathmandu City based on 2015 earthquake using total stress and effective stress approach. Int J Geotech Earthq Eng (IJGEE) 11(2):1–25
14. Kumar SS, Dey A, Krishna AM (2018) Importance of site-specific dynamic soil properties for seismic ground response studies: ground response analysis. Int J Geotech Earthq Eng (IJGEE) 9(1):78–98
15. Hashash Y, Park D, Tsai CC, Philips C, Groholski DR (2016) DEEPSOIL-1D wave propagation analysis program for geotechnical site response analysis of deep soil deposits. In: Version v7, tutorial and user manual. University of Illinois, Urbana—Campaign
16. Seed HB, Idriss IM (1970) Soil moduli and damping factors for dynamic response analysis. In: Report No. EERC 70-10. University of California, Berkeley
17. Vucetic M, Dobry R (1991) Effect of soil plasticity on cyclic response. J Geotech Eng Div ASCE 111(1):89–107
18. Imai T, Tonouchi K (1982) Correlation of N-value with S-wave velocity and shear modulus. In: Proceeding 2nd European symposium of penetration testing (Amsterdam), vol 14, pp 57–72

19. Iyisan R (1996) Correlations between shear wave velocity and in-situ penetration test results. Tech J Chamber Civil Eng Turkey 7:1187–1199
20. Kiku H, Yoshida N, Yasuda S, Irisawa T, Nakazawa H, Shimizu Y, Ansal A, Erkan A (2001) In-situ penetration tests and soil profiling in Adapazari, Turkey. In: Proceeding ICSMGE/TC4
21. Hasancebi N, Ulusay R (2006) Empirical correlations between shear wave velocity and penetration resistance for ground shaking assessments. Bull Eng Geol Environ 66:203–213
22. Maheswari UR, Boominathan A, Dodagoudar GR (2008) Development of empirical correlation between shear wave velocity and standard penetration resistance in soils of Chennai City. In: Proceedings of the 14th world conference on earthquake engineering. Beijing, China
23. Hanumantharao C, Ramana GV (2008) Dynamic soil properties for microzonation of Delhi, India. J Earth Syst Sci 117:719–730
24. Dikmen U (2009) Statistical correlations of shear wave velocity and penetration resistance for soils. J Geophys Eng 6:61–72

Chapter 40
"Liquefaction Potential Study of Subsurface Soil in Southern Jorhat, Assam"

Kapil Saikia, Sahidul Haque, Krishan Borah, and Rituparna Goswami

Introduction

Jorhat, an administrative district, located in North-eastern India, in the state of Assam with a population of 1,092,256 as per the 2011 census, covers an area of about 2852 km^2. Jorhat lies in the central part of the Brahmaputra valley, which is further divided into three zones, viz. Central, East, and West. Jorhat is located at 26.75° N 94.22° E, within an average elevation of 116 m (Fig. 40.1).

Seismicity

Jorhat lies in the North-eastern part of India, where three major plates interact along two convergent boundaries; the Himalayas and the Indo-Burma range, which meet at the Assam syntax. As per the plate tectonics, Assam is in the easternmost projection of the Indian plate where the plate is thrusting underneath the Eurasian plate, creating a subduction zone and the Himalayas. Hence, the Indian Standard Code of practice (IS 1893 (Part-1) 2016) identified North-East India, including Assam, as a highly seismic zone by placing it in the highest seismic zonal level, i.e., Zone V. The state has experienced two major earthquakes in the year Assam earthquake 1897 and Assam-Tibet earthquake 1950 of intensities 8.0 and 8.6 on the Richter scale, respectively, causing a large-scale damage and ground liquefaction.

K. Saikia (✉) · S. Haque · K. Borah · R. Goswami
Department of Civil Engineering, Jorhat Engineering College, Jorhat 785007, India
e-mail: kapilsaikia17@gmail.com

K. Muthukkumaran et al. (eds.), *Soil Dynamics, Earthquake and Computational Geotechnical Engineering*, Lecture Notes in Civil Engineering 300,
https://doi.org/10.1007/978-981-19-6998-0_40

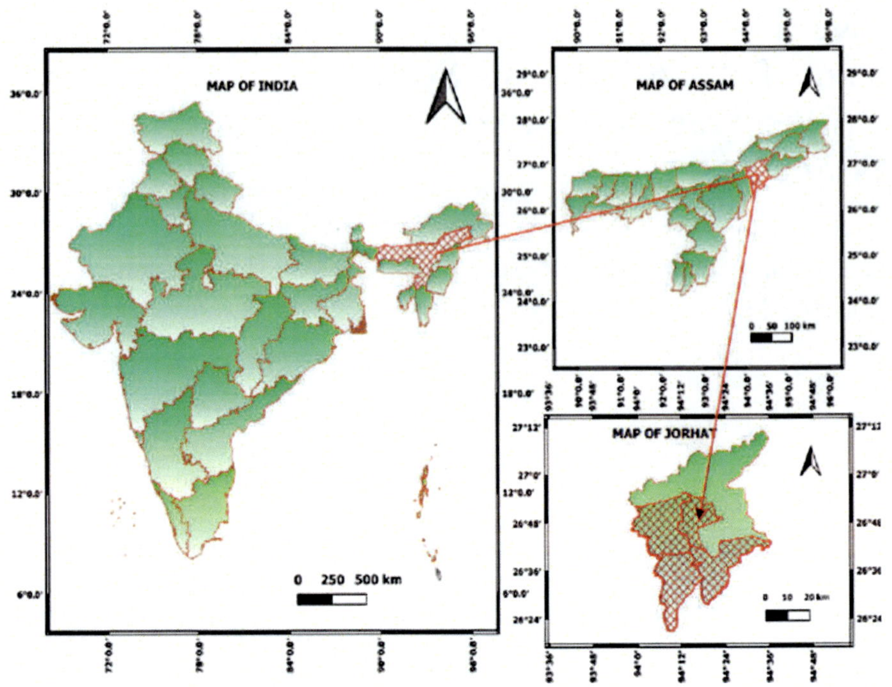

Fig. 40.1 Area of study (Constructed using QGIS for the present study)

Existing literature

Iwasaki et al. [1]

Iwasaki et al. proposed a simplified method based on both a liquefaction resistance factor, FL and a liquefaction potential factor, PL for the evaluation of soil liquefaction potential at a given depth of a site, and the factor PL indicates the one at a site. To prove the effectiveness of the described simplified method, liquefaction potential was investigated by calculating the factor at 64 liquefied sites and 23 non-liquefied sites during the Nobi Earthquake (1891), the Tonankai earthquake (1944), the Fukui earthquake (1948), the Niigata earthquake (1964), the Tokachi-oki earthquake (1968) and the Miyagi-ken-oki earthquake (1978), and shaking table tests. They concluded that the values of FL are less than 1 for liquefied layers and larger than 1 for non-liquefied layers.

Seed et al. [2]

Seed et al. presented field data for the sites which were known to have liquefied or not liquefied during earthquakes in the USA, Japan, China, Guatemala, Argentina, and other countries to establish a criterion for evaluating the liquefaction potential of sands. The results of their study were then extended to other magnitude earthquakes using a combination of laboratory and field data. Available information on the liquefaction resistance of silty sands was also reviewed and a simple procedure for considering the influence of silt content was proposed. They presented a method for using the field data to evaluate the possible magnitude of pore water pressure generation in sands and silty sand which remain stable during earthquake shaking. Finally, the applicability of other in situ field tests, such as the static cone penetrometer, shear wave velocity, and electrical measurements for evaluating the liquefaction resistance of soils was examined.

Seed et al. [3]

Seed et al. tried to clarify the meaning of the values of standard penetration values used in correlation of fields observation of soil liquefaction with N measured in SPT tests. The field data were reinterpreted and plotted in terms of a newly recommended standard, (N1)60, determined in SPT tests where the driving energy is 60% of theoretical free fall energy. Energies associated with different methods of performing SPT tests in different countries and can readily be used to convert any measured N-value to the standard (N1)60 values and with different fines content were proposed. Finally, suggestions are made concerning the significance of the term 'Liquefaction' as it is often used in conjunction with field evidence of this phenomenon.

Juang et al. [4]

Juang et al. presented an assessment of existing and new probabilistic methods for liquefaction potential evaluation. They emphasised on the comparison of probabilities of liquefaction calculated with two different approaches, logistic regression, and Bayesian mapping. In their study, the simplified procedure for soil liquefaction evaluation, including the Seed-Idriss, Robertson-Wride, and Andrus-Stokoe methods, based on the standard penetration, cone penetration, and shear velocity measurement, respectively, were used as the basis for developing Bayesian mapping functions. The study showed that the Bayesian mapping approach is preferred over the logistic regression approach for estimating the site-specific probability of liquefaction, although both methods give comparable probabilities. The paper also compares the three simplified methods in the context of probability of liquefaction and argued for the use of probability-based procedures for evaluating liquefaction potential.

Sushama et al. [5]

Sushama et al. evaluated the liquefaction potential for Ramgarh Taal in Gorakhpur, which lies in the area of high seismic probability (Seismic Zone IV), using the Indian Standard method. In their study, the liquefaction potential of Ramgarh Taal was evaluated using IS 1893 (Part-1): 2016. Further comparison was done between liquefaction potential obtained using the IS method with the simplified method. It was found that there was a significant variation in the factor of safety calculated from the IS code method and simplified method.

Liquefaction

Soil liquefaction is a phenomenon whereby a saturated or partially saturated soil substantially loses strength and stiffness in response to applied stress, usually earthquake shaking or other sudden change in stress conditions, causing it to behave like a liquid.

A more precise definition as given by Sladen et al. [6] states that "Liquefaction is a phenomenon wherein a mass of soil loses a large percentage of its shear resistance, when subjected to monotonic, cyclic, or shocking loading, and flows in a manner resembling a liquid until the shear stresses acting on the mass are as low as the reduced shear resistance".

Location of Different Bore Log Data

For the purpose of calculation of the liquefaction potential, different subsoil investigation data was used, which were collected by a private firm in Jorhat district during past years. A description of all the 37 bore holes is given in Table 40.1, including the ground water table depth along with the latitudes and longitudes.

Evaluation of Liquefaction Potential

The simplified procedure for evaluation of liquefaction potential, which is included in the Annexure F IS 1893(Part 1):2016 [7], is used for the purpose of evaluation of liquefaction potential.

Evaluation of Liquefaction Potential, Taking into Account the Assam-Tibet Earthquake 1950 (Mw = 8.6)

Bore-log (BL): ONGC, Cinamara, Jorhat: The sub-soil investigation data, which includes fines content, unit weight and SPT blow Count-N for the sub-soil, collected for this location are given in Table 40.2.

The liquefaction potential calculations for the location of ONGC, Cinamara, Jorhat is given in Table 40.3 with respect to depth by considering Assam-Tibet earthquake 1950 (Mw = 8.6). Similarly, the liquefaction potential for all other 36 boreholes is calculated by following the same procedure. And all the evaluated factor of safety with respect to depth by considering a Mw = 8.6 are listed in Table 40.4 in reference to the bore holes designated in Table 40.1.

Table 40.1 Description of the bore holes

Borehole No.	Locations	Groundwater table (M)	Latitude	Longitude
BH1	ONGC Chinamara	1.00	26° 42′ 25.53	94° 14′ 52.22
BH2	TJFS	1.00	26° 27′ 2.52"	94° 9′ 14.18
BH3	Dholi, Titabor	1.00	26° 36′ 31.20"	94° 14′ 13.79
BH4	Titabor Stadium	1.30	26° 35′ 49.92"	94° 12′ 21.87"
BH5	Titabor ITI	1.50	26° 44′ 57.96"	94° 13′ 21.88"
BH6	Swimming pool at Bapuji Stadium	1.00	26° 39′ 23.69"	94° 19′ 9.94"
BH7	Bongal Pukhuri	1.00	26° 44′ 42.20"	94° 12′ 48.18"
BH8	Doha Bora Chowk	1.00	26° 45′ 5.23"	94° 12′ 54.29"
BH9	Gar Ali	1.00	26° 45′ 22.76"	94° 13′ 9.26"
BH10	Punjab Engineering	1.80	26° 44′ 42.05	94° 13′ 24.10
BH11	Jorhat Law College	0.80	26° 44′ 58.60	94° 12′ 43.95
BH12	Lmtc Garmur	1.00	26° 44′ 43.85	94° 14′ 38.61
BH13	Dulia GAON beyond JEC	2.00	26° 44′ 50.71"	94° 14′ 54.88"
BH14	Girls' Hostel JEC	2.00	26° 44′ 48.50	94° 15′ 4.84
BH15	Mariani seleghat	1.00	26° 39′ 11.63	94° 18′ 32.55
BH16	In front of JEC	1.00	26° 44′ 49.52	94° 14′ 53.31
BH17	Project Jail	1.00	26° 43′ 31.56	94° 12′ 30.82
BH18	Thana Road	1.00	26° 45′ 19.53	94° 12′ 49.42
BH19	Raja Maidam	1.00	26° 44′ 47.66	94° 12′ 9.54
BH20	Women University	1.00	26° 43′ 7.13	94° 11′ 7.06
BH21	Das & co.	1.00	26° 42′ 45.33	94° 10′ 28.63
BH22	Jail Road	1.00	26° 44′ 47.29"	94° 12′ 5.73
BH23	Ckb college	1.00	26° 45′ 9.46	94° 12′ 25.42
BH24	Circuit House	0.50	26° 45′ 10.14	94° 12′ 28.02
BH25	Rani Hotel	1.50	26° 45′ 33.98	94° 12′ 43.11
BH26	Chinamara	1.00	26° 41′ 39.52	94° 14′ 4.29
BH27	Club Road	1.50	26° 44′ 12.98	94° 13′ 6.50
BH28	IOC Chinamara	0.80	26° 43′ 21.71	94° 13′ 47.71
BH29	AG Hospital	1.00	26° 44′ 47.30	94° 12′ 5.71
BH30	Judge quarter	1.00	26° 45′ 31.57	94° 12′ 20.48
BH31	Ramdhenu path	1.80	26° 44′ 23.69"	94° 12′ 46.38
BH32	Beparichuk	0.50	26° 33′ 38.51"	94° 10′ 4.85
BH33	Vetenary near Titabor	1.80	26° 27′ 34.55"	94° 9′ 42.04
BH34	Soikota, Gandhinagar	0.40	26° 39′ 5.66"	94° 19′ 41.73

(continued)

Table 40.1 (continued)

Borehole No.	Locations	Groundwater table (M)	Latitude	Longitude
BH35	PHE Jorhat	1.00	26° 45′ 17.26"	94° 12′ 42.35
BH36	Nigamananda	0.50	26° 44′ 58.69"	94° 13′ 42.75
BH37	Nagajanka	0.40	26° 36′ 36.24"	94° 20′ 18.72

Table 40.2 Bore-log

Depth in metre	Observed penetration resistance	Effective unit wt. of soil, γ (In t/m^3) (Bulk density above W.T. and submerged density below W.T.)	Effective overburden pressure, (σ) (In t/m^2)	Corrected penetration resistance
1.0–1.3	2	2.15	2.15	2
2.0–2.3	5	1.15	3.30	5
3.0–3.3	7		4.45	7
4.0–4.3	13		5.60	15.3
5.0–5.3	5		6.75	5
6.0–6.3	3		7.90	3
7.0–7.3	4		9.05	4
8.0–8.3	3		10.20	3
9.0–9.3	4		11.35	4
10.0–10.3	4		12.5	4

Similar calculations are done for all the other earthquake magnitudes, viz. Assam earthquake 1897 (Mw = 8.0), Assam earthquake 1947 (Mw = 7.3), Dhubri earthquake 1930 (Mw = 7.1), and two arbitrary earthquake magnitude Mw = 6.5 and 6, the evaluated liquefaction potential of the bore holes is presented in the form of maps using QGIS to evaluate the variation of liquefaction potential of various locations within the district of Jorhat in the four revenue circles with respect to depth for different earthquakes.

Results and Discussion

Discussion of Result with Respect to Depth for Different Earthquake Magnitude

From the Figs. 40.2, 40.3, 40.4, 40.5, 40.6 and 40.7, it is seen that as the depth increases the susceptibility towards liquefaction decreases as the overburden pressure on the subsoil layer increases.

Table 40.3 Liquefaction potential for different depths determined using SPT-N blow counts

Depth	Effective overburden pressure	Total overburden pressure	Stress reduction factor	CSR	N_{60}	$(N_1)_{60}$	Fines content	$(N_1)_{60CS}$	CRR	FOS	Status
1	2.150	2.150	0.992	0.232	2.000	3.400	100	4.580	0.507	0.236	Liquefiable
2	3.300	4.300	0.985	0.300	5.000	8.500	92	10.700	0.464	0.141	Liquefiable
3	4.450	6.450	0.977	0.331	7.000	10.595	92	13.213	0.464	0.104	Liquefiable
4	5.600	8.600	0.969	0.348	13.000	17.539	71	21.547	0.348	0.062	Liquefiable
5	6.750	10.750	0.962	0.358	5.000	6.144	99	7.873	0.502	0.074	Liquefiable
6	7.900	12.900	0.954	0.365	3.000	3.408	100	4.589	0.507	0.064	Liquefiable
7	9.050	15.050	0.946	0.368	4.000	4.245	100	5.594	0.507	0.056	Liquefiable
8	10.200	17.200	0.939	0.370	3.000	2.999	100	4.099	0.507	0.050	Liquefiable
9	11.350	19.350	0.931	0.371	4.000	3.791	100	5.049	0.507	0.045	Liquefiable
10	12.500	21.500	0.907	0.365	4.000	3.612	100	4.835	0.507	0.041	Liquefiable

Table 40.4 Factor of safety with respect depth for various places, pertaining to the Assam-Tibet earthquake 1950 (Mw = 8.6)

Bore hole No.	Factor of safety with respect to depth									
	1M	2M	3M	4M	5M	6M	7M	8M	9M	10M
BH1	0.209	0.280	0.303	0.476	0.186	0.133	0.146	0.125	0.137	0.137
BH2	0.257	0.269	0.271	0.175	0.159	0.128	0.142	0.281	0.388	0.670
BH3	0.257	0.271	0.167	0.233	0.184	0.149	0.142	0.215	0.408	0.348
BH4	N/A	0.393	0.412	0.384	0.216	0.131	0.164	0.218	0.467	0.515
BH5	0.266	0.315	0.291	0.217	0.170	0.181	0.171	0.182	0.213	0.228
BH6	0.543	1.132	1.008	0.949	0.925	0.568	0.520	0.565	0.477	0.599
BH7	0.419	0.276	0.233	0.371	0.408	0.182	0.199	0.697	0.811	0.160
BH8	0.257	0.235	0.233	0.328	0.331	0.304	0.201	0.170	0.208	0.227
BH9	0.308	0.778	1.017	0.943	0.184	0.889	0.867	0.214	0.245	0.280
BH10	0.308	0.336	1.233	1.122	0.652	1.011	0.9760.	0.956	0.940	0.940
BH11	0.230	0.250	0.192	0.227	0.180	0.858	0.179	0.191	0.841	0.856
BH12	0.419	0.235	0.233	0.204	0.162	0.152	0.126	0.140	0.152	0.120
BH13	N/A	0.203	0.165	0.171	0.204	0.189	0.159	0.152	0.147	0.162
BH14	0.257	0.365	0.202	0.367	0.234	0.266	0.324	0.276	0.359	0.448
BH15	0.308	0.276	0.268	0.126	0.138	0.173	0.912	0.905	0.902	0.917
BH16	N/A	N/A	0.959	0.717	0.772	0.716	3.996	3.996	4.009	4.097
BH17	0.308	0.361	0.589	0.148	0.917	0.893	0.880	0.871	0.873	0.894
BH18	0.308	0.237	0.168	0.177	0.186	0.174	0.146	0.124	0.153	0.168
BH19	0.147	0.314	1.016	0.951	0.911	0.884	0.869	0.155	0.133	0.132
BH20	0.217	0.319	0.305	0.233	0.211	0.174	0.127	0.109	0.153	0.186
BH21	0.147	0.236	0.168	0.204	0.237	0.949	0.206	0.158	0.171	0.222
BH22	0.362	0.240	0.200	0.152	0.141	0.134	0.113	0.126	0.109	0.138
BH23	0.257	0.279	0.479	0.599	0.733	0.863	0.390	0.177	0.153	0.186
BH24	0.108	0.163	0.248	0.217	0.173	0.231	0.288	0.227	0.179	0.142
BH25	N/A	N/A	N/A	0.158	0.145	0.157	0.311	0.109	0.138	0.137
BH26	0.308	0.366	0.233	0.176	0.186	0.114	0.145	0.123	0.152	0.135
BH27	0.258	0.315	0.291	0.217	0.170	0.181	0.171	0.182	0.213	0.228
BH28	0.126	0.171	0.286	0.140	0.129	0.103	0.116	0.096	0.109	0.124
BH29	0.257	0.236	0.200	0.150	0.091	0.946	0.929	0.216	0.170	0.151
BH30	0.147	0.199	0.233	0.972	0.932	0.898	0.876	0.868	0.864	0.529
BH31	N/A	0.336	0.877	0.861	0.652	1.984	0.618	0.116	0.125	0.111
BH32	N/A	0.210	0.136	0.131	0.144	0.114	0.128	0.123	0.136	0.118
BH33	N/A	0.336	0.877	0.861	0.652	1.984	0.618	0.116	0.125	0.111
BH34	N/A	0.193	0.181	0.139	0.131	0.255	0.286	0.293	0.397	0.505

(continued)

Table 40.4 (continued)

Bore hole No.	Factor of safety with respect to depth									
	1M	2M	3M	4M	5M	6M	7M	8M	9M	10M
BH35	N/A	0.198	0.200	0.176	0.322	0.268	0.285	0.214	0.229	0.191
BH36	N/A	0.153	0.173	0.161	0.309	0.256	0.195	0.147	0.160	0.124
BH37	N/A	0.193	0.181	0.138	0.131	0.255	0.286	0.293	0.397	0.505

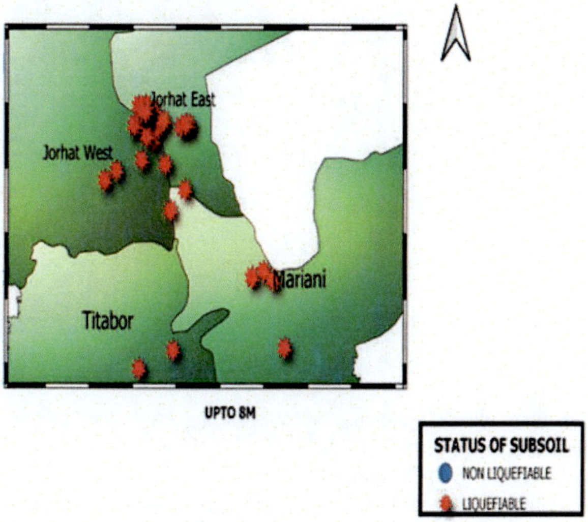

Fig. 40.2 Places liquefiable up to a depth of 8 m for earthquake magnitude 8.6 (constructed using QGIS for the present study)

Conclusion

This study was conducted to evaluate the liquefaction potential of 37 different sites in the Jorhat district, lying in the banks of the river Brahmaputra, for six different earthquake magnitudes, viz. Assam earthquake 1897 (Mw = 8.0), Dhubri earthquake 1930 (Mw = 7.1), Assam earthquake 1947 (Mw = 7.3), Assam-Tibet earthquake 1950 (Mw = 8.6), and two arbitrary earthquakes of magnitude 6 and 6.5. In this study, the results are presented with respect to different depths of soil deposits for a given earthquake.

The following conclusions can be drawn from our analysis are

1. The factor of safety against liquefaction for most of the sites in Southern Jorhat lies below 1 for earthquake of magnitudes 8.6, 8.1, 7.3, 7.1, 6.5, 6.0, indicating to the susceptibility of soil to liquefaction in this region.
2. From the Figs. 40.2, 40.3, 40.4, 40.5, 40.6 and 40.7, it is seen that as the depth increases the susceptibility towards liquefaction decreases as the overburden pressure on the subsoil layer increases.

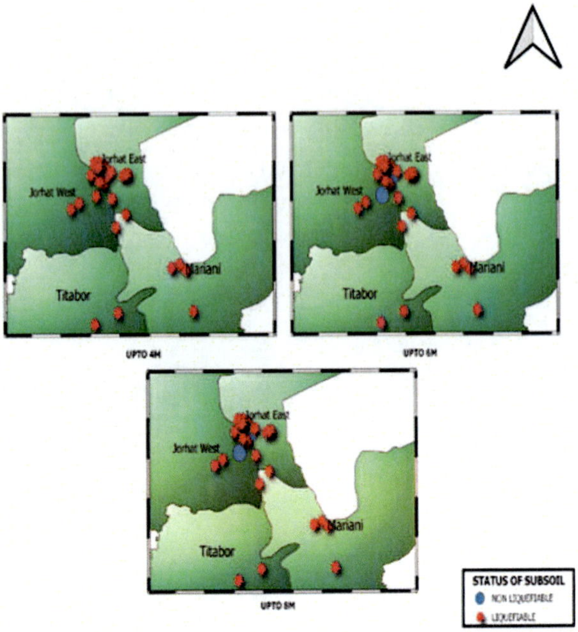

Fig. 40.3 Places liquefiable for earthquake magnitude 8.0 with respect to depth up to an explored depth of 8 m. (constructed using QGIS for the present study)

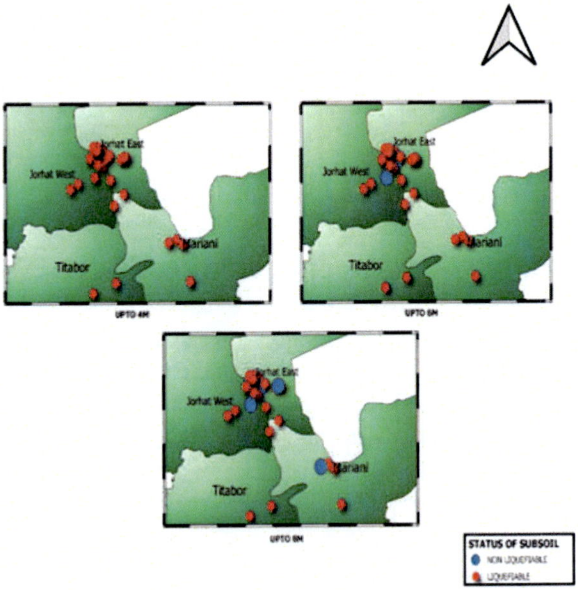

Fig. 40.4 Places liquefiable for earthquake magnitude 7.3 with respect to depth up to an explored depth of 8 m. (constructed using QGIS for the present study)

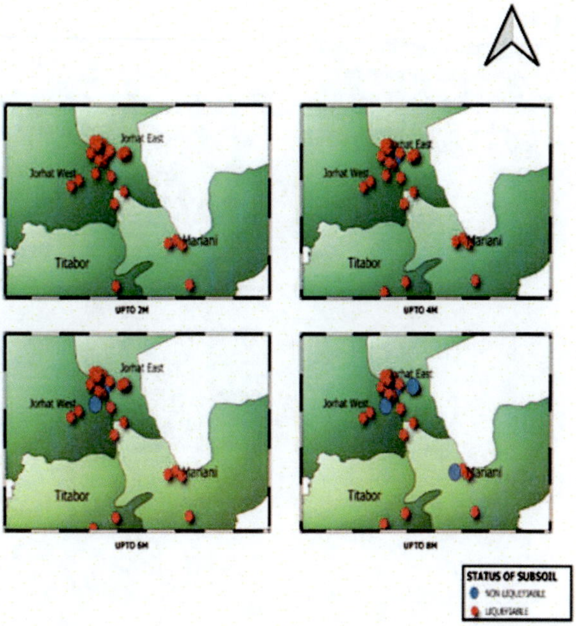

Fig. 40.5 Places liquefiable for earthquake magnitude 7.1 with respect to depth up to an explored depth of 8 m. (constructed using QGIS for the present study)

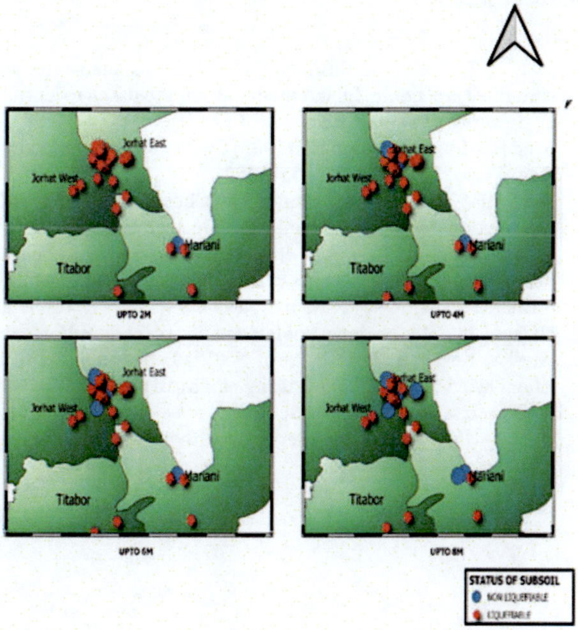

Fig. 40.6 Places liquefiable for earthquake magnitude 6.5 with respect to depth up to an explored depth of 8 m. (constructed using QGIS for the present study)

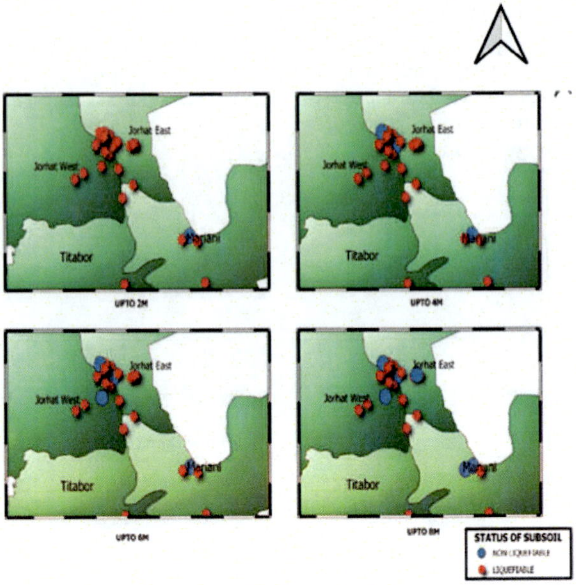

Fig. 40.7 Places liquefiable for earthquake magnitude 6.0 with respect to depth up to an explored depth of 8 m. (Constructed using QGIS for the present study)

References

1. Iwasaki T, Tokida K, Tatsuoka F (1981) Soil liquefaction potential evaluation with use of the simplified procedure. In: International conference on recent advances in geotechnical earthquake engineering and soil dynamics, vol 12
2. Seed HB, Idriss IM, Arango I (1983) Evaluation of liquefaction potential using field performance data. J Geotech Eng 109(3):458–482
3. Seed HB, Tokimatsu K, Harder LF, Chung RM (1985) Influence of SPT procedure in soil liquefaction resistance evaluations. J Geotech Eng 111(12):1425–1445
4. Juang CH, Jiang T, Andrus RD (2002) Assessing probability based method for liquefaction potential evaluation. J Geotech Geo Environ Eng 128(7):580–589
5. Sushama SNM, Muley P, Ram S, Jawaid SMA (2019) Liquefaction potential evaluation using IS code method. In: National conference on recent development in civil engineering infrastructure
6. Ates A, Keskin I, Totic E, Yesil B (2014) Investigation of soil liquefaction potential around Efteni lake in Duzce Turkey: using empirical relationship between shear wave velocity and SPT blow count (N). Adv Mater Sci Eng
7. IS 1893 (Part-1) (2016), Criteria for earthquake resistant design of structures: Bureau of Indian Standards

Chapter 41
Determination of Seismic Bearing Capacity of Soil Using Experimentally Obtained Failure Surface

Annoy Kumar Ghosh and Ashim Kanti Dey

Introduction

The capacity of soil to sustain an applied load is termed as the bearing capacity. Another way to express the bearing capacity of soil is to obtain the maximum contact pressure over the area of a foundation prior to shear failure of the soil. However, during or after the seismic event, the bearing capacity's value changes dramatically, and now, this new value is known as the seismic bearing capacity of the soil. Estimating the seismic bearing capacity is a challenging task as it plays a crucial part in foundation design.

In soil mechanics, idealizations are generally considered to achieve solutions that can be easily implemented, and the solutions for seismic bearing capacity are also based on some idealizations. The assumptions leading to the idealization of soil turn the seismic bearing capacity solutions impractical or practical for application purposes. Many researchers have developed and used numerous theories to obtain the solution for the seismic bearing capacity [1–4, 6–8, 10]. Additionally, in most of the available studies [1–6, 8, 10], failure surfaces were predefined or assumed. This idealization leads to some error in the value of the seismic bearing capacity of the soil.

In this study, a shallow foundation idealizing a particular geographical location has been considered, and a small-scale experimental study has been carried out in the laboratory. Under a specific type and value of dynamic loading, the failure surface has been demarcated experimentally. The dynamic loading has been applied using a 1D shaker. The failure surface has been demarcated by using coloured sand. The failure

A. K. Ghosh (✉) · A. K. Dey
Department of Civil Engineering, National Institute of Technology, Silchar, India
e-mail: akg.nitd@gmail.com

A. K. Dey
e-mail: akdey@civil.nits.ac.in

K. Muthukkumaran et al. (eds.), *Soil Dynamics, Earthquake and Computational Geotechnical Engineering*, Lecture Notes in Civil Engineering 300,
https://doi.org/10.1007/978-981-19-6998-0_41

Table 41.1 Properties of soil obtained from experiments

Parameters	Values
Uniform coefficient (C_u)	2.43
Curvature coefficient (C_c)	0.98
Specific gravity	2.67
Bulk unit weight for test (kN/m^3)	16.54
Internal friction angle	30°
Unified system classification	SP

surface has been used to obtain the expressions for N_c, N_q, and N_γ with application of the limit equilibrium method. The obtained values have been tabulated and compared with the past studies.

In this experimental study, a specific type of soil, i.e. sandy soil, has been used. The reason for choosing sandy soil is that failure surface due to shaking is prominently visible for this type of soil. However, the value and methodology of determining seismic bearing capacity depend on the local soil type, site of study from nearby fault, type of foundation, seismic vulnerability, pore pressure, the magnitude of the earthquake, and topographical features. Additionally, the non-homogenous nature of soil makes it difficult for obtaining a universal solution.

Experimental Components

Properties of Soil

Standard experiments have been conducted to obtain the properties of the soil used in the present study. The properties are given in Table 41.1.

Colouring of Sand

The sand is coloured using food colour available in the market. Allura Red AC (Magenta Red) is used. As the name indicates, the Allura Red AC is red in colour. It has several market names out of which the most popular name is FD&C Red 40. E number of this food dye is E129. The usual supply of the dye is as red sodium salt but other forms like calcium and potassium salts are also available. All the salts are water soluble. The sand is dried in sunlight in order to remove extra water from the sand. The coloured sand is used to demarcate the sand undergoing deformation due to shaking.

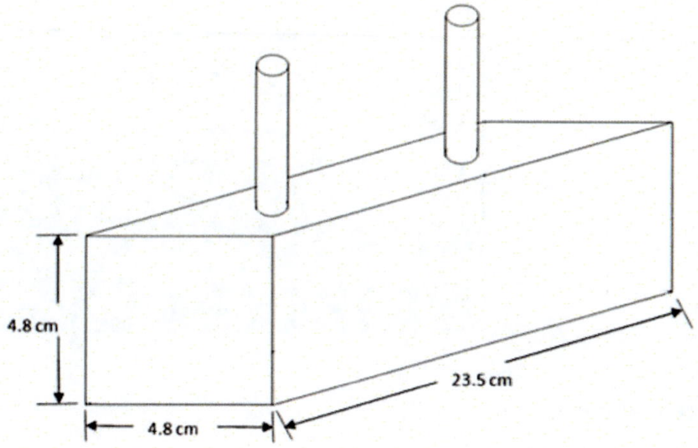

Fig. 41.1 Sketch of strip footing used in the experiment

Acrylic Container with Graph

A container is prepared for a small-scale laboratory experiment. The container is of cuboidal shape and has a square cross section. The dimension of the container is 240 × 240 × 135 mm height. The container is firmly fixed to the shaker through bolts.

The centimetre graph is printed on OHP sheets. Overhead projector (OHP) sheets, also called transparent sheets, are a unique range of sheets manufactured using a superior grade of transparent, lightweight, and non-shaded material.

This transparent sheet is glued over the acrylic box (transparent) to perform the layering of sand and coloured sand uniformly and to derive the failure surface's geometry easily.

Model Foundation

A strip foundation is made using the iron. The two rods have been provided in order to add extra load if needed. The foundation size is shown in Fig. 41.1.

1D Shaker

A horizontal 1D shaker has been used for this experiment. A frequency of 4 Hz and amplitude of 2 mm has been used for excitation purposes.

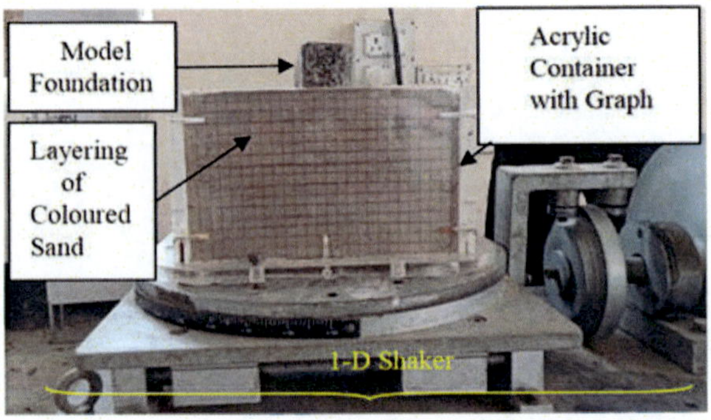

Fig. 41.2 Acrylic container filled with layered sand fixed on shake table

Experimental Setup

The sand and coloured sand have been arranged in such a way shown in the Fig. 41.1. The OHP sheet printed with a graph is attached to the outer surface of the acrylic container to obtain the failure surface's geometry after shaking. It has also helped to maintain a uniform thickness of sand and coloured sand.

The experimental setup shown in Fig. 41.2 has been used. The dynamic load has been applied through a one-dimensional shaker to a model strip footing placed on dry sand. The shaker uses a fly wheel of non-uniform thickness which transmits a horizontal shaking motion to the base plate. The amplitude of horizontal shaking depends upon the variation of the thickness of the fly wheel. The peak amplitude of vibration for the present study is made as 2 mm. The frequency of motion (f_d) can be varied through a frequency controller, and the operating frequency has been varied between 0 and 4 Hz. The model container fitted to the base plate of the shaker has a plan area of 240 × 240 mm and height 135 mm. Using the experimental setup as stated, it was possible to determine the dynamic bearing capacity.

Analysis of Experimental Result

A series of experiments have been performed. In the initial series of experiments, it has been found that the layering of coloured sand need not be performed up to the bottom end, as the loading and shaking have a negligible effect at the bottom end.

The best-achieved deformation among a series of experiments has been shown (Figs. 41.3 and 41.4).

The movement of the coloured sand particles is observed. A triangular wedge is also visible. Furthermore, following the trajectory of the sand particles, the failure plane has been drawn, as shown in Fig. 41.5.

Fig. 41.3 Deformation of sand layers after shaking, along with graph

Fig. 41.4 Deformation of sand layers after shaking, graph sheet has been removed

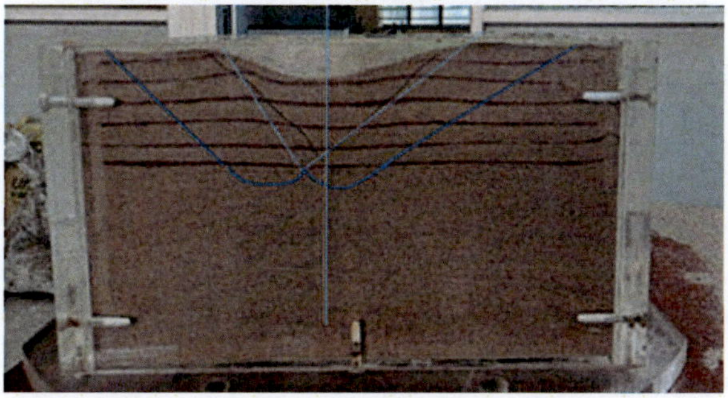

Fig. 41.5 Geometry of failure plane achieved after shaking

Method of Analysis

In the present study, pseudo-static approach under limit equilibrium method has been used to determine the seismic bearing capacity. The soil is assumed to be homogenous and isotropic subjected to a uniform surcharge load. Mohr–Coulomb failure criterion is adopted considering the soil to be rigid and perfectly plastic. It is further assumed that the soil parameters like unit weight, γ, cohesion, c, and angle of internal friction, ϕ, are not affected by the dynamic load. A uniform horizontal seismic acceleration with coefficient, k_h, is assumed to act over the entire domain. A non-symmetrical elastic wedge is formed under the footing due to the dynamic load. It is assumed that the passive resistance has been fully mobilized along the failure surfaces on both sides of the footing.

Determination of k_h

k_h is the coefficient of horizontal seismic acceleration.
f = Peak frequency of 1D shaker = 4 Hz.
A = Peak amplitude at peak frequency = 2 mm.
$u(t)$ = displacement at time t.
$\omega = 2\pi f = 8\pi$.
$g = 9810$ mm/s^2

$$u(t) = A \sin 8\pi t = 2 \sin 8\pi t$$

$$\dot{u} = \frac{du}{dt} = 16\pi \cos 8\pi t$$

$$\ddot{u} = \frac{d^2u}{dt^2} = (16\pi)(8\pi) \sin 8\pi t$$

$$\ddot{u}_{max} = 1262 \text{ mm/s}^2$$

$$k_h = \frac{\ddot{u}_{max}}{g} = 0.1286$$

α_1 and α_2 are obtained from geometry of series of experiment. One of the calculations from Fig. 41.6 is shown as follow:
$\alpha_1 = -\tan^{-1}\frac{6}{8.5} = -35.22°$; $\alpha_2 = \tan^{-1}\frac{6}{3.8} = 57.65°$.
Considering Fig. 41.7, the notations are as follows:

$P_{p\gamma d_1}$ = Unit weight component of the total seismic resistance.

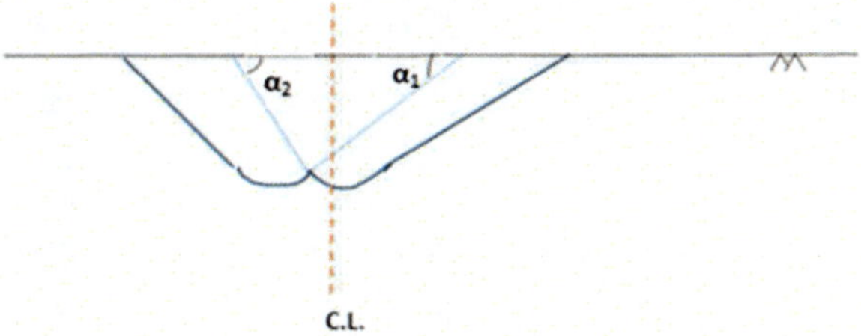

Fig. 41.6 Sketch of failure surface obtained from the experiment

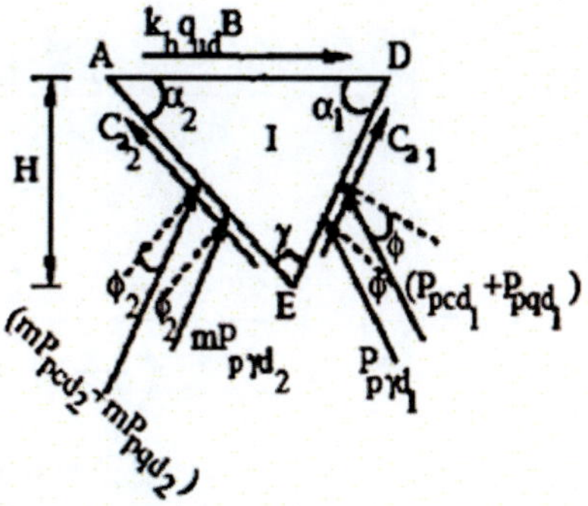

Fig. 41.7 Forces considered in the analysis

P_{pqd_1} = Surcharge component of the total seismic resistance.
P_{pcd_1} = Cohesive component of the total seismic resistance.

$m = \frac{\tan \emptyset_2}{\tan \emptyset}$; m = mobilization factor
$\emptyset$ = friction angle of sand
$\emptyset$ = mobilized friction angle

} It is assumed that $m = 1$

$k_{p\gamma d_1}$ = The coefficient of seismic passive earth pressure for unit weight acting normally on the faces DE.
$k_{p\gamma d_2}$ = The coefficient of seismic passive earth pressure for unit weight acting normally on the faces AE.
k_{pqd_1} = The coefficient of seismic passive earth pressure for surcharge loading acting normal to the face DE.
k_{pqd_2} = The coefficient of seismic passive earth pressure for surcharge loading acting normal to the face AE.
k_{pcd_1} = The coefficient of seismic passive earth pressure for cohesion component acting normal to the face DE.
k_{pcd_2} = The coefficient of seismic passive earth pressure for cohesion component acting normal to the face AE.
N_{cd} = Cohesion component of the seismic bearing capacity factors.
N_{qd} = Surcharge component of the seismic bearing capacity factors.
$N_{\gamma d}$ = Unit weight component of the seismic bearing capacity factors.

The equation of seismic bearing coefficients is obtained from [5]. The coefficients are as follows:

Table 41.2 Values of $k_{p\gamma d}$, k_{pqd}, and k_{pcd} corresponding to values of k_h and k_γ

k_h	k_v	k_{pcd}	k_{pqd}	$k_{p\gamma d}$
0.0	0.0	1.103	1.779	1.803
0.5	0.0	1.107	1.006	1.573
0.1286	0.0	1.104	1.5802	1.7438

$$N_{cd} = \frac{1}{k_h}\left[\frac{\frac{k_{pcd_1}}{\cos\emptyset}\sin(\alpha_1-\emptyset)-\frac{mk_{pcd_2}}{\cos\emptyset_2}\sin(\alpha_2-\emptyset_2)}{\frac{1}{\tan\alpha_1}+\frac{1}{\tan\alpha_2}}+\frac{\sin\alpha_1\tan\emptyset_2\cos\alpha_2}{\sin(\alpha_1+\alpha_2)\tan\emptyset}-\frac{\sin\alpha_2\cos\alpha_1}{\sin(\alpha_1+\alpha_2)}\right] \quad (41.1)$$

$$N_{\gamma d} = \frac{1}{k_h}\left[\frac{\frac{k_{p\gamma d_1}}{\cos\emptyset}\sin(\alpha_1-\emptyset)-\frac{mk_{p\gamma d_2}}{\cos\emptyset_2}\sin(\alpha_2-\emptyset_2)}{\left(\frac{1}{\tan\alpha_1}+\frac{1}{\tan\alpha_2}\right)^2}\right]-\frac{1}{\frac{1}{\tan\alpha_1}+\frac{1}{\tan\alpha_2}} \quad (41.2)$$

$$N_{qd} = \frac{1}{k_h}\left[\frac{\frac{k_{pqd_1}}{\cos\emptyset}\sin(\alpha_1-\emptyset)-\frac{mk_{pqd_2}}{\cos\emptyset_2}\sin(\alpha_2-\emptyset_2)}{\frac{1}{\tan\alpha_1}+\frac{1}{\tan\alpha_2}}\right] \quad (41.3)$$

The value of $k_{p\gamma d1}$, $k_{p\gamma d2}$, k_{pqd1}, k_{pqd2}, k_{pcd1}, and k_{pcd2} is obtained from [4]. The exact value was not available for corresponding value of k_h. That is why the value of the above coefficients has been interpolated from Table 41.2 available in [4].

Result and Discussion

A series of experiment have been performed among which three have been considered for analysis purpose. The geometry of the triangular wedge has been considered for calculating the dynamic bearing capacity factors. Notations are as per Fig. 41.7. Using the angles shown in Fig. 41.6 and using the formulas of Eqs. (41.1), (41.2), and (41.3), the values of dynamic bearing capacity factors have been calculated and tabulated in Table 41.3. Negative sign for α_1 signifies that it is opposite to the direction of α_2. Clockwise direction is considered to be positive.

Minimum of N_{cd}, N_{qd}, and $N_{\gamma d}$ (Set 2) have been considered as final result.

For $\varnothing = 30°$, the value of N_{cd}, N_{qd}, and $N_{\gamma d}$ has been obtained by above Eq. (41.1), (41.2), and (41.3) and have compared with the past literatures, as given in Table 41.4.

$$q_{u_{static}} = cN_c + qN_q + \frac{1}{2}B\gamma N_\gamma \quad (41.4)$$

Table 41.3 Experimental results

S. No.	Experiment	α_1	α_2	N_c	N_q	N_γ
1	Set 1	−35.22	57.65	37.43	27.24	30.22
2	Set 2	−33.46	55.34	36.54	25.01	26.39
3	Set 3	−36.98	59.96	38.25	29.41	34.22
4	Set 4	−34.16	56.49	36.60	25.76	27.54
5	Set 5	−35.92	59.34	36.94	27.76	30.77

Table 41.4 Comparison of experimental results with past literatures

Literature	N_{cd}	N_{qd}	$N_{\gamma d}$
Present study	36.54	25.01	20.39
Prakash and Saran [7]	36.6	22.4	19.4
Vesic [10]	30.13	18.4	22.4
Saran and Agarwal [8]	37.2	22.5	29.4
Soubra [9]	30.25	18.46	21.81
Choudhury and Rao [5]	29.86	18.35	20.03

$$q_{u_{\text{dynamic}}} = cN_{\text{cd}} + qN_{\text{qd}} + \frac{1}{2}B\gamma N_{\gamma d} \tag{41.5}$$

$$q_{u_{\text{static}}} = 8.89\,\text{kN/m}^2$$

$$q_{u_{\text{dynamic}}} = 8.08\,\text{kN/m}^2$$

$$\%\ \text{difference} = \frac{q_{u_{\text{static}}} - q_{u_{\text{dynamic}}}}{q_{u_{\text{dynamic}}}} \times 100 = 10\% \tag{41.6}$$

Conclusion

- For the same angle of internal friction, the dynamic bearing capacity is less than the static bearing capacity. The value of dynamic bearing capacity for angle of friction = 30° is equal to the static bearing capacity corresponding to angle of internal friction = 29°. As per the calculation shown above, 10% difference is observed.
- Good conformity of the test results has been obtained with Saran et al. [8]. The reasonably nearer value is primarily because [8] had used the limit equilibrium method and preassumed the failure plane's exact geometry as achieved in the experimental study.

- The value of seismic bearing capacity depends on the shape and size of the failure plane.
- But as per the past literatures, the nature of the failure plane depends upon the local soil type, location of the site of study from nearby fault, type of foundation, seismic vulnerability, pore pressure, the magnitude of the earthquake, and topographical features. So, in order to obtain a holistic view, effect of all the factor and their relative interactions should be considered,

References

1. Budhu M, Al-Karni A (1993) Seismic bearing capacity of soils. Géotechnique 43(1):181–187
2. Castelli F, Motta E (2010) Bearing capacity of strip footings near slopes. Geotech Geol Eng 28(2):187–198
3. Qin C, Chian SC (2018) Seismic bearing capacity of non-uniform soil slopes using discretization based kinematic analysis considering Rayleigh waves. Soil Dyn Earthq Eng 109:23–32
4. Choudhury D, Subba Rao KS (2002) Seismic passive earth resistance for negative wall friction. Canadian Geotech J 39(5):971–981
5. Choudhury D, Subba Rao KS (2005) Seismic bearing capacity of shallow strip footings. Geotech Geol Eng 23(4):403–418
6. Ghosh P (2008) Upper bound solutions of bearing capacity of strip footing by pseudo-dynamic approach. Acta Geotech 3(2):115–123
7. Prakash S, Saran S (1971) Bearing capacity of eccentrically loaded footings. J Soil Mech Found Eng Div Am Soc Civ Eng 97(1):95–117
8. Saran S, Agarwal RK (1991) Bearing capacity of eccentrically obliquely loaded footing. J Geotech Eng Div Am Soc Civ Eng 117(11):1669–1690
9. Soubra AH (1999) Upper-bound solutions for bearing capacity of foundations. J Geotech Geo-environ Eng 125(1):59–68
10. Vesic AS (1973) Analysis of ultimate loads of shallow foundations. J Soil Mech Found Div 99(sm1):43–73

Chapter 42
Impact of Low-Frequency Ground Motions on Local Site Effect of Guwahati

R. Arunsiva and Abhishek Kumar

Introduction

Earthquake causes sudden release of energy into Earth's crust which travels as seismic wave fronts along the Earth's surface. Such seismic waves generated during earthquakes undergo change in their original characteristics while travelling through various subsoil medium. The extent and intensity of damage caused by an EQ might vary from epicentral region to remote locations, depending on subsoil features, according to the reports from throughout the world. The local site effect is a phenomenon that has significant impact on ground-motion parameters such as frequency parameters, amplitude content, and the duration of input ground motion while travelling to the ground surface from the bedrock level [1, 2, 3]. Across the world, there are several examples of ground-motion modification, caused by local soil conditions. These modified ground motion results in triggering induced effects such as liquefaction and landslides [4, 5]. According to Chavez-Garcia and Bard [6], the 1985 Michoacan EQ of magnitude 4.8 caused catastrophic damage to numerous sites located around 360 km distant from the epicentre. As a result, amplification scale of up and around to 50 were observed between bedrock-level and the ground-level movements in the frequency bandwidth of 0.25–0.7 Hz. Similarly, during the 1989 Loma Prieta earthquake (Mw-6.9), local soil response led to massive damage in the San Francisco-Oakland area, which is roughly 80 km away from the epicentre. In another incidence during 1999 Chamoli EQ (M-6.8) caused significant damage far from the epicentre. Although the earthquake's epicentre was between the greater and lesser Himalayas, it wreaked havoc on buildings in Delhi and Dehradun regions,

R. Arunsiva (✉) · A. Kumar
Indian Institute of Technology, Guwahati, India
e-mail: arunedu95@gmail.com

A. Kumar
e-mail: abhiak@iitg.ac.in

K. Muthukkumaran et al. (eds.), *Soil Dynamics, Earthquake and Computational Geotechnical Engineering*, Lecture Notes in Civil Engineering 300,
https://doi.org/10.1007/978-981-19-6998-0_42

which were more than 200 km away from the epicentre. 2001 Bhuj EQ (M-7.7), which shook Gujarat's western region significantly, has caused massive structural and ground collapses as well as liquefaction in the towns of Umedpur and Lodai, which are roughly 50 km from the epicentre. Sandstorms were also produced near Chobari, some 100 km away to the epicentre per Rajendran et al. [7]. Large-scale amplification of the soil deposits was also recorded in Ahmedabad about 350 km from the epicentre, leading to significant damage to higher buildings [8].

Above stated examples clearly indicate the importance of local site effect in amplifying the earthquake generated ground motions. It addition, it also raises the importance of studying local site amplification by ground response analysis. It should be highlighted here that earthquake hazards can be reduced if the probable damages from an earthquake can be forecasted, by means of ground response analysis. A site-specific ground response analysis can be done in three different numerical ways. Among the available numerical analysis, equivalent linear analysis is computationally easier and also provides approximate result of that of nonlinear analysis. In equivalent linear analysis, the initially considered dynamic properties of soil are modified up until the degree of strain generated in the soil is compatible with shear strain in successive steps. In one-dimensional ground response analysis, the ground level and any other subsoil strata are assumed to be levelled, extending endlessly in all lateral directions. Transfer functions are used in equivalent linear method to relate required parameters to available parameters. The thickness, rigorousness, and hampering properties of subsoil strata affect the nature of transfer function.

Study Area

The present study area of Guwahati is one of the significant cities in north-eastern part of India. Guwahati city is situated on the south-east edge of Assam's Kamrup district, on the bank of the river Brahmaputra. The city is extending in about 600 km^2 area with its centre at 26°08′ N and 91°40′ E. According to the 2011 census, Guwahati has a population of about 1 million. Guwahati is north-east's largest business centre. It is located in seismic zone V as per IS 1893 (Part 1): 2016. Seismic hazard of Guwahati can be explained by the fact that it is located near to the Shillong Plateau (SP). While studying the seismic potential of the SP, Baro and Kumar [9] concluded that there exists 4 active seismic sources exist within 500 k radial distance around the Shillong Plateau, which are the sources of numerous earthquakes of earthquakes were located. These seismic sources are;

1. Shillong Plateau–Assam Valley Thrust or Zone
2. Indo-Burma Ranges Thrust or Zone
3. The Bengal Basin Thrust or Zone and
4. Eastern Himalaya Thrust or Zone

Owing to the existence of tectonically active blocks on all edges, the Guwahati city is susceptible to higher-magnitude earthquakes inside the range of 100 km. These include the Himalayan plate which is at the northern boundary and north-eastern boundary, as well the Naga-Disang fault or thrust zone which lies at the eastern boundary, and the active Shillong plateau lies at south [10, 11]. The North Assam Earthquake of 1950, which was triggered near Po Chu fault in Arunachal Pradesh, resulted in multiple landslides and liquefaction sideways the Brahmaputra River [12]. Even the flow of the River Brahmaputra was obstructed during the earthquake by numerous descending rocks rolled-out from the nearby hill. This earthquake also caused liquefaction and ground subsidence at a number of locations in Guwahati. Furthermore, during the 1897 Assam earthquake, ejections of substructure strata up to a height of 1 m above the existing ground surface were observed to the eastern parts of Dhubri location in Assam, thus confirming the function of local soil at a regional scale as per Kumar and Mondal [13]. Similar damages were also reported by numerous earthquakes in Guwahati, highlighting the importance of local soil effect understanding.

Input Soil Properties

As per Geological Survey of India (GSI), geology in the majority of Guwahati locales is newer (Holocene age), with Pleistocene deposits. Along with the newly formed layers, the northern and southern areas of Guwahati have Archean and Proterozoic geological deposits. This clearly shows that with the except a few areas in the north and south, majority of Guwahati is made up of relatively younger geological layers. The underground contour map for Guwahati city which was developed by DST [14] highlighted the subsurface features beneath the city. The map shows a surge in subsoil thickness from western part to eastern part, attaining an extreme value of 300 m. The thickness of the overburden varies from Brahmaputra River's bank to the slopes of nearby mountain ranges. DST [14] also mentioned that seven paleoseismic channels, that were once streams of the Brahmaputra River, have dumped alternative films of sediment. Overall, the subsoil information suggests that the majority of the subsurface in Guwahati is clay with widely varied plasticity, followed by the presence of silty sand strata. Seismic site classification study carried out by Kumar et al. [15] shows that subsoil at Guwahati belongs to seismic site class (SSC) *D* and SSC *E*. For this reason, only site class *D* and site class *E* are taken into account subsequently in the current analysis. The soils were grouped under different site class based on their SPT-N values. Refer Table 42.1 for site classification.

Table 42.1 Site classification (after IBC 2009)

Range of N	Site class
$N < 15$	*E*
$15 < N < 50$	*D*
$N > 50$	*C*

Under the Guwahati Metro Rail scheme, Guwahati Metropolitan Development Authority (GMDA) excavated boreholes. The subsoil information collected for those projects data such as soil type, N-SPT value at depths till 30 m below ground surface, water table level, among other things. A typical borelog is given in Table 42.4. All boreholes were drilled to a diameter of 150 mm in accordance with IS: 1892 [16], and standard penetration measurement was recorded at every 1.5 m in accordance with the IS: 2131. [17]. According to IS: 2132 [18], disturbed samples and undisturbed samples also obtained from boring and split spoon sampler, respectively, at various depth below ground level. The index properties are calculated by laboratory tests at soil lab with the collected distressed soil samples. All the testings were done as stated in IS: 1498 [19]. In addition to these 175 borehole, some additional boreholes were collected by Kumar et al. [15]. This way, Kumar et al. [15] employed a total of 244 borelogs to investigate subsurface lithology in Guwahati. Out of these 244 boreholes, 45 boreholes are chosen randomly for the present study. These selected boreholes are from important locations, where various infrastructure is yet to be developed in the near future. Of 45 borelogs, 20 borehole belongs to site class *D*, and 25 boreholes belong to site class *E* (Refer Tables 42.2 and 42.3).

In order to perform the equivalent linear ground response of soil, it is essential to provide dynamic properties of soils involved. Dynamic soil properties include the

Table 42.2 Selected borehole location of site class *D* for analysis

Borehole no	Location
1	Zoo Road Tiniali (Near Hangout Restaurant)
2	PWD Office, Fancy Bazar
3	KKH Hostel, Cotton College, Panbazar
4	GMC Water Supply, Panbazar
5	Silpukhuri (Campus of Shrinkhala Bora)
6	Hostel of Cotton Collegiate H.S School
7	A.S.T.C Office, Rupnagar
8	Six Mile, Jayanagar
9	Japarigog
10	Hatigaon, Near High School
11	Rukminigaon
12	Ganeshguri
13	Public Health Office, Hengrabari
14	Kahilipara (Campus of Deepak Chakravarti)
15	Dakhin Beltola High School
16	Swadeshi Academy, Barbari, V.I.P
17	Bagharbari, Panjabari
18	Bagharbari, Panjabari
19	Opposite to TeteliaBodi LP School
20	HiteswarSaikia College, Sixmile, Panjabari

Table 42.3 Selected borehole location of site class *E* for analyses

Borehole no	Location
1	Near Ulubari Overbridge, Vivekananda Path
2	Assam Textile Institute, Ambari
3	Near State Zoo, Zaparigog Road (Campus of Dipankar Saikia)
4	Gandhibasti
5	Zoo Narengi Road, Hem Baruah path tiniali, House No-2
6	Jonali (Campus of Mrinal Talukdar)
7	Bhetapara
8	Kahilipara, Dakhin Chuba
9	Ghoramara
10	Hatigaon, Bhetapara
11	Rupnagar
12	Navajyoti Nagar, Panjabari
13	Six Mile, V.I.P. Road
14	Beltola Buniyadi Vidyalaya
15	Dalibari, Satgaon
16	Panchadeep Jatiya Vidyalaya, PatharQuarry
17	Satgaon
18	Narengi High School, Narengi
19	Noonmati LP School
20	Boragaon, Shankardev
21	Raju Hydraulic Automobiles, Near Boragaon
22	Sasal garbage dumping ground
23	Complex of Niranjan Deka (Near Balaji)
24	Himanshu Printers and Publications
25	Saraswati Bidya Niketan, Mathgaria

shear strain-dependent damping ratio β and the modulus reduction ($G/G_{\max}$) curves. The amplification of the input ground motions is directly related to these dynamic soil properties. These strain-dependent dynamic soil properties curve can be drafted by conducting dynamic soil tests at the laboratory using cyclic triaxial test, resonant column test, shear test (simple shear or torsional shear), etc. [20, 21]. In case such properties are not available for site-specific soil due to inability of necessary testing facilities, some of the available standard strain-dependent dynamic property curves can be used from the literature. Those standard curves are developed after numerous test results and can be incorporated into ground response analysis. Selection of the curves is based on the type of soil such as sand or clay and with their plasticity index value, and over consolidation ratio values in case of clayey soil. For the present work, referring to the work by Kumar et al. [15], shear strain-dependent dynamic

Table 42.4 Typical borelog located in Zoo Road Tinali provided by GMDA

Location: Zoo Road Tiniali (Near Hangout Restaurant)					
Latitude: 26.17378° N; Longitude: 91.77418° E					GWT—0.20 m
Depth (m)	Bulk density (KN/m^3)	Total overburden pressure, σ_o (KN/m^2)	Effective overburden pressure, σ_o' (KN/m^2)	N-value	IS soil classification
1.5	17.836	26.754	14.014	7	CI
3	18.424	54.39	26.95	5	CI
4.5	18.424	82.026	39.886	12	CI
6	18.424	109.662	52.822	15	CI
7.5	18.424	137.298	65.758	20	CI
9	18.522	165.081	78.841	23	CI
10.5	18.3554	192.614	91.674	20	SP
12	18.3554	220.147	104.507	16	SP
13.5	18.0908	247.283	116.943	13	CI
15	18.0908	274.42	129.38	11	CI
16.5	18.228	301.762	142.022	15	CI
18	18.13	328.957	154.517	11	CI
19.5	18.424	356.593	167.453	27	CI
21	21.56	388.933	185.093	50	CI
22.5	21.56	421.273	202.733	50	CI
24	21.952	454.201	220.961	50	CI
25.5	22.05	487.276	239.336	50	SP
27	17.738	513.883	251.243	26	SP
28.5	17.836	540.637	263.297	28	CI
30	17.836	567.391	275.351	28	CI

GWT—Ground Water Table

soil properties curve for sandy soil are taken from Seed and Idriss [22], and for clay soil, it is taken from Sun et al. [23].

Selection of Input Motion

Bedrock motion is a critical element in any site response study. Seismograph installation in India began in the mid of 1980’s though a developing nation with seismic occurrence in the range of low to extremely high. Since then, there has been no major EQ in the region. In the absence of region-specific seismogram, standard earthquake motions and past historical earthquake motion like the El-Centro earthquake of 1940,

Mexico earthquake of 1985, Loma Prieta earthquake of 1989, Northridge earthquake of 1994, Kobe earthquake of 1995, and Chi-Chi earthquake of 1999 have been in use around the world, including in India as stated by Kumar et al. [2, 24]. Carrying out ground response analysis from the available input motion parameters of any other locations might not be acceptable in case of non-availability of regional-specific ground-motion data. Furthermore, the results of such a site response study will be restricted in terms of measuring induced impacts like liquefaction and landslides, as well as providing an accurate estimate of damage scenarios during future EQ. Furthermore, for site response analysis, the creation of artificial ground movements based on past earthquakes is widely accepted [25], which is termed as synthetic ground motions. For the present study, the synthetic ground motions are developed using EXSIM method which is a stochastic extended finite-fault ground-motion simulation algorithm. EXSIM is a FORTRAN-based open-source stochastic extended finite-source simulation method that creates earthquake ground-motion time series [26, 27, 28]. To account for finite-fault effects in ground-motion modelling (such as faulting geometry, distributed rupture, and rupture inhomogeneity), EXSIM splits an earthquake's fault surface into a grid of subsources, each of which may be handled as a point source. The point-source stochastic model created by Boore [29, 30] and popularized by the stochastic-method SIMulation (SMSIM) computer code [30, 31] is used to model time series from the subsources. From developing synthetic ground motions for Guwahati city, motions whose predominant frequency is lesser than 5 Hz and those motions whose magnitude and hypocentral distances in accordance with the past earthquake occurred in and around the Guwahati city are selected for the present study. In total, 30 such motions are sorted for the equivalent linear ground response analysis (refer to Table 42.5). Motion number mentioned in the later discussion is with reference to Table 42.5.

Analyses

As discussed earlier, the equivalent linear analysis method is used to analyse the impact of low-frequency earthquake motions. An elaborated procedure on the method can be referred in Kramer [32] and is not presented here. Considering the length of the paper, a brief analysis procedure is discussed here. In general, equivalent linear response analysis approach is performed by assuming values for the strain-dependent shear modulus (G) and shear strain-dependent damping ratio (β) corresponding to low strain values γ and then using them in finding the shear strain time history of the soil. Using the computed strain time history, the effective strain γ_{eff} is calculated based on the below correlation.

$$\gamma_{\text{eff}}^{i} = R_{\gamma}\gamma_{\max_j}^{i} \tag{42.1}$$

Kramer [33] in which i denotes the number of iterations

j denotes the layer number

Table 42.5 Synthetic ground motions (SGM)

Motion no.	Hypocentral distance (km)	Predominant frequency (Hz)	PHA (cm/s^2)
1	514.02	2.19	0.0003
2	512.41	3.91	0.0001
3	512.41	3.91	0.0001
4	485.93	1.86	0.0003
5	516.98	2.49	0.0007
6	488.14	2.18	0.0006
7	516.97	2.33	0.0006
8	519.83	2.10	0.0033
9	521.6	0.28	0.0006
10	521.59	0.28	0.0005
11	491.71	0.61	0.0011
12	491.65	0.98	0.0008
13	525.88	2.38	0.0027
14	525.87	0.74	0.0016
15	496.77	0.21	0.0061
16	491.65	2.51	0.0010
17	460.66	0.27	0.0082
18	536.48	0.46	0.0071
19	528.6	1.63	0.0050
20	460.82	0.34	0.0084
21	503.27	0.47	0.0094
22	545.14	0.06	0.0025
23	545.11	0.10	0.0022
24	544.98	0.16	0.0031
25	467.75	0.08	0.0048
26	548.32	0.13	0.0121
27	513.46	0.09	0.0109
28	512.78	0.18	0.0109
29	469.93	0.13	0.0228
30	469.93	0.13	0.0182

R_γ denotes the effective strain to maximum strain ratio as per Schnabel et al. [34] it is considered as 0.65 for analysis.

From the calculated γ_{eff}, a new value is assigned for the dynamic soil properties (G and β) from the backbone reduction curve and the damping curve. The above process is repeated till the difference between the calculated damping ratio and shear modulus in two consecutive iterations are less than 1%. To do the analysis, a recently

Table 42.6 Input parameters for analysis

Layer depth (m)	1.5	1.5	1.5	1.5	1.5	1.5	1.5	1.5
Unit weight (kN/sq. m)	18.13	18.424	21.56	21.56	21.952	22.05	17.738	17.836
Shear wave velocity (m/s)	205.90	359.30	526.47	526.47	526.47	526.47	350.99	367.49
Code	6	4	4	4	4	6	6	6

developed MATLAB code by Kumar and Mondal [13] intended for finding the site-specific response of soil using the numerical approach, equivalent linear site response analysis is used. The input data for the MATLAB code is listed in Table 42.6.

The maximum layer depth is restricted to 1.5 m, and relaxation is given for layer having ground water table. In addition, code value is given based on soil type, and based on this code, the damping curve and modulus reduction curve are chosen for the analyses. The code values are given based on Kumar and Mondal [13]. The correlation between SPT-N value and shear wave velocity (Vs) established for Lucknow area by Anbazhagan et al. [35] is used in this analysis. Initially, 1% damping ratio is assigned to each soil layer. Based on the outcomes of the analyses (Refer to Table 42.7), it is observed that on an average, the subsoils in Guwahati city which belongs to site class *D* and site class E have the capacity to amplify the peak acceleration of input motion by 3.2 times and 3.13 times, respectively, whereas the maximum amplification ratio of soil is observed as much as 4.97 times for site class *D* soil and 5.38 times for site class *E* soil (Refer to Table 42.7a and b). It is also observed that the motions for which the soil exhibits maximum response have the predominant frequency in the range of 0.21–2.5 Hz in case of site class *D* soils. Similarly, in case of site class *E* soil, this range is even low (0.08–2.5 Hz) independent of hypocentral distance. As an interesting fact, it is noted that when the predominant frequency of input motion being the lowest in both site class *D* and site class E analysis, the amplification ratio is still being in the highest range of around 3 times (Refer Table 42.7a and b). This clearly shows that even low-frequency earthquake motions can cause significant damage to the super structures whose foundation lies in such type of soils.

It is observed based on the peak acceleration value that the PHA value of input motion is amplified from a minimum of 1.6 to maximum of 4 times when reaching the ground surface (Refer Table 42.8). This amplification is due to local site effect. Typical variation of PGA with respect to depth is shown in Figs. 42.1 and 42.2.

Conclusion

As numerous seismic sources exist in and around the Guwahati region, present work attempted to understand the effect of low-frequency ground motions on local soil in terms of ground-motion amplification. Since the predominant frequency of selected ground motions is in close range with the natural frequency of subsoil medium, input

Table 42.7 **a** Maximum amplification ratio of site class *D* soils for different input motions, **b** Maximum amplification ratio of site class *E* soils for different input motions

Motion no.	Hypocentral distance (km)	Predominant frequency	Maximum amplification ratio
(a)			
1	514.02	2.19	3.93
4	485.93	1.86	4.16
6	488.14	2.18	3.55
8	519.83	2.10	3.61
14	525.87	0.74	2.84
15	496.77	0.21	2.82
16	491.65	2.51	2.57
17	460.66	0.27	4.97
(b)			
1	514.02	2.19	4.22
4	485.93	1.86	3.13
5	516.98	2.49	3.23
6	488.14	2.18	4.31
7	516.97	2.33	3.71
8	519.83	2.10	2.49
9	521.60	0.28	2.86
14	525.87	0.74	5.38
15	496.77	0.21	3.30
17	460.66	0.27	2.65
25	467.75	0.08	3.89
27	513.46	0.09	2.60

ground motion showed manifold increase in the amplitude [1]. The output of this study emphasizes that the low-frequency ground motions in concurrence with the local site effect. This will have more significant effect on building design and other induced effects. Usually, earthquake hazards like liquefaction and landslides can be triggered by this kind of low-frequency input ground motion by local site effect as stated in [4, 5]. Hence, it is brought to notice that while construction of a structure the site should be analysed not only for higher-frequency ground motions but must also include low-frequency input motions which may be strong enough to trigger the soil failure.

Table 42.8 Peak acceleration at ground surface of borehole 1 for 30 synthetic ground motions

PGA at ground surface (g)	Input PHA (g)	Amplification ratio
0.0014	0.000279	3.96
0.0003	$6.97 \times e^{-05}$	3.72
0.0004	$9.1 \times e^{-05}$	3.68
0.0014	0.000302	3.80
0.0026	0.000675	2.89
0.0021	0.00059	2.61
0.0020	0.000552	2.70
0.0157	0.003344	3.68
0.0019	0.000554	2.49
0.0022	0.000529	3.14
0.0035	0.001132	2.06
0.0032	0.000763	3.17
0.0085	0.002659	2.19
0.0074	0.001645	3.50
0.0310	0.006107	4.08
0.0046	0.001035	3.41
0.0371	0.008216	3.51
0.0330	0.007115	3.63
0.0132	0.005022	1.63
0.0387	0.008369	3.62
0.0321	0.009356	2.43
0.0080	0.002512	2.18
0.0074	0.002241	2.30
0.0121	0.00309	2.92
0.0192	0.004782	3.02
0.0397	0.012097	2.29
0.0427	0.010869	2.93
0.0519	0.010908	3.76
0.0981	0.02279	3.30
0.0806	0.018226	3.42

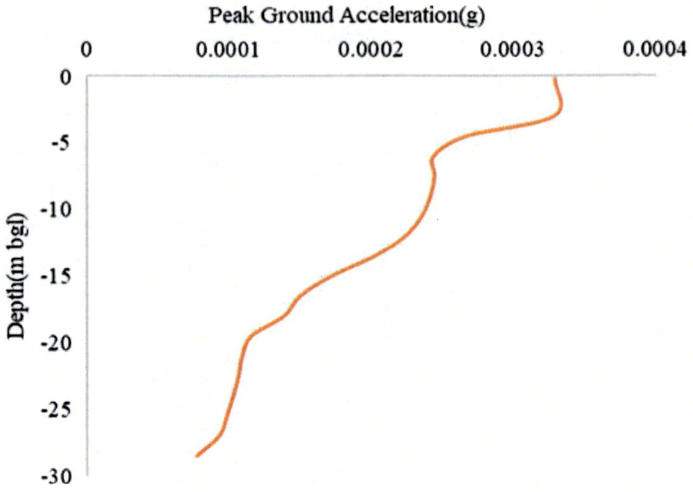

Fig. 42.1 PGA variation with depth borehole 1 of site class D for motion 2

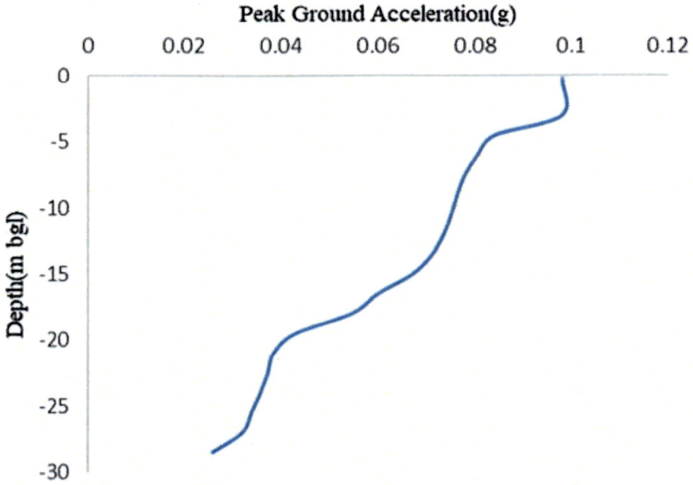

Fig. 42.2 PGA variation with depth borehole 1 of site class D for motion 29. Refer Tables 42.2 and 42.3 for borehole location of corresponding number

References

1. Kumar A, Harinarayan NH, Baro O (2015) High amplification factor for low amplitude ground motion: assessment for Delhi. Disaster Adv 8(12):1–11
2. Kumar A, Baro O, Harinarayan N (2016) Obtaining the surface PGA from site response analyses based on globally recorded ground motions and matching with the codal values. Nat Hazards 81(1):543–572
3. Anbazhgan P, Thingbaijam KKS, Nath SK, Kumar JN, Sitharam TG (2010) Multi-criteria

seismic hazard evaluation for Bangalore city, India. J Asian Earth Sci 38(5):186–198
4. Pitilakis K (2004) Site effects, recent advances in earthquake geotechnical engineering and microzonation. Geotech Geol Earthq 1:139–197
5. Topal T, Doyuran V, Karahanoglu N, Toprak V, Suzen ML, Yesilnacar E (2003) Microzonation for earthquake hazards: Yenisehir settlement, Bursa Turkey. Eng Geol 70(1):93–108
6. Chávez-García Francisco J, Bard PY (1994) Site effects in Mexico City eight years after the September 1985 Michoacan earthquakes. Soil Dyn Earthq Eng 13(4):229–247
7. Rajendran K, Rajendran CP, Thakur M, Tuttle MP (2001) The 2001 Kutch (Bhuj) earthquake: coseismic surface features and their significance. Curr Sci 80(11):1397–1405
8. Narayan JP, Sharma ML (2004) Effects of local geology on damage severity during Bhuj, India earthquake. In: Proceeding of the 13th World conference on earthquake engineering, Vancouver, paper no 2042
9. Baro O, Kumar A (2017) Seismic source characterization for the Shillong Plateau in Northeast India. J Seismol 21(5):1229–1249
10. Evans P (1964) The tectonic framework of Assam. J Geol Soc India 5:80–96
11. Gupta ID (2006) Delineation of probable seismic sources in India and the neighbourhood by a comprehensive analysis of seismotectonic characteristics of the region. Soil Dynam Earthq Eng 76:766–790
12. Reddy DV, Kumar NP, D, et al (2009) The great1950 Assam earthquake revisited: field evidences of liquefaction and search for paleoseismic events. Tectono physics 474:463–472. https://doi.org/10.1016/j.tecto.2009.04.024
13. Kumar A, Mondal JK (2017) Newly developed MATLAB based code for equivalent linear site response analysis. Geotech Geol Eng 35:2303–2325
14. DST Nath SK (ed) (2007) Seismic microzonation atlas of Guwahati Region. Department of Science and Technology, Government of India, New Delhi
15. Kumar A, Harinarayan NH, Verma V et al (2018) Seismic site classification and empirical correlation between standard penetration test N value and shear wave velocity for Guwahati based on thorough subsoil investigation data. Pure Appl Geophys 25:50. https://doi.org/10.1007/s00024018-1858-1
16. IS 1892 (1974) Indian standard code of practice for subsurface investigation for foundations. Bureau of Indian Standards, New Delhi
17. IS 2131 (1981) Indian Standard, method for standard penetration test for soils, vol 1st revision. Bureau of Indian Standards, New Delhi
18. IS 2132 (1986) Indian Standard code of practice for thin-walled tube sampling of soils, vol 2nd revision. Bureau of Indian Standards, New Delhi
19. IS 1498 (1970) Indian Standard Classification and identification of soils for general engineering purposes. vol 1st revision. Bureau of Indian Standards, New Delhi
20. Dorourdian M, Vucetic M (1995) A direct simple shear device for measuring small-strain behavior. Geotech Test J 18(1):69–85
21. Stewart JP, Liu AH, Choi Y, Baturay MB (2001) Amplification factors for spectral acceleration in active regions. Pacific Earthquake Engineering Research Centre, PEER report 2001/10
22. Seed HB, Idriss IM (1970) Soil moduli and damping factors for dynamic response analysis. Report no. EERC 70-10, University of California, Berkeley
23. Sun JI, Golesorkhi R, Seed HB (1988) Dynamic moduli and damping ratios for cohesive soils. Report no. EERC 88-15, University of California, Berkeley
24. Phanikanth VS, Choudhury D, Reddy GR (2011) Equivalentlinear seismic ground response analysis of some typical sites in Mumbai. Geotech Geol Eng 29:1109–1126
25. Kennedy R, Short S, Merz K, Tokarz F, Idriss I, Power M, Sadigh K (1984) Engineering characterization of ground motion-Task I: effects of characteristics of free field motion on structural response. U.S. Nuclear Regulatory Commission, Washington
26. Motazedian D, Atkinson GM (2005) Stochastic finite-fault modeling based on a dynamic corner frequency. Bull Seismol Soc America 95:995–1010. http://dx.doi.org/10.1785/0120030207
27. Assatourians K, Atkinson G (2012) EXSIM12: A stochastic finite- fault computer program in FORTRAN. http://www.seismotoolbox.ca (last accessed April 2019)

28. Boore DM (2009) Comparing stochastic point-source and finite-source ground-motion simulations: SMSIM and EXSIM. Bull Seismol Soc Am 99:3202–3216
29. Boore D (1983) Stochastic simulation of high-frequency ground motions based on seismological models of the radiated spectra. Bull Seismol Soc Am 73:1865–1894
30. Boore DM (2003) Simulation of ground motion using the stochastic method. Pure Appl Geophys 160:635–675
31. Boore DM (2005) SMSIM—Fortran Programs for Simulating Ground Motions from Earthquakes: Version 2.3—A Revision of OFR 96–80– A. http://www.daveboore.com/software_online.html (last accessed April 2019)
32. Kramer SL (1996) Geotechnical Earthquake Engineering, Pearson Education
33. Kramer SL (2011) Geotechnical earthquake engineering. Pearson Education, Upper Saddle River
34. Schnabel PB (1973) Effect of local geology and distance from source on earthquake ground motion. Ph. D. Thesis, University of California, Berkeley, California
35. Anbazhagan P, Kumar A, Sitharam TG (2013) Seismic site classification and empirical correlation between standard penetration test N value and shear wave velocity for deep soil sites in Indo-Gangetic Basin. Pure appl Geophys 170(3):299–318
36. Kumar A, Mondal JK (2017) Impact of higher frequency content of input motion upon equivalent linear site response analysis for the study area of Delhi. Geotech Geol Eng 35:959–981
37. Kumar A, Harinarayan NH, Baro O (2017) Nonlinear soil response to ground motions during different earthquakes in Nepal, to arrive at surface response spectra. Nat Hazards 87:13–33
38. Kumar A, Suman H, Olympa B (2018) Approximation of equivalent linear ground response analysis at low strain by strain dependent linear ground response analysis for typical site at Delhi, India. In: Geotechnical earthquake engineering and soil dynamics V GSP 291
39. Chang C-Y, Mok CM, Tang HT (1996) Inference of dynamic shear modulus from Lotung downhole data
40. Anbazhagan P (2009) Liquefaction hazard mapping of Bangalore, South India. Disaster Adv
41. Muhammad K (2012) Case history-based analysis of liquefaction in sloping ground
42. Ganapathy GP, Rajawat AS (2012) Evaluation of liquefaction potential hazard of Chennai city, India: using geological and geomorphological characteristics. Nat Hazards
43. Rama Subba Rao GV, Usha Sai B (2018) Seismic liquefaction analysis of capital region of Andhra Pradesh state, India. Civil Eng J 2(13)
44. Ishihara K (1993) Liquefaction and flow failure during earthquakes. Geotechnique 43(3):351–415
45. GSI (2000) Seismotectonic atlas of India and its environs. Geological Survey of India
46. GMDA (2006) New revised building byelaws for Guwahati Metropolitan Area 2006. Guwahati Metropolitan Development Authority, 80
47. DST (2008) Report on seismic microzonation of Guwahati region.
48. Baro O, Kumar A (2015) A review on the tectonic setting and seismic activity of the Shillong Plateau in the light of past studies. Disaster Adv 8(7):34–45
49. Baro O, Kumar A, Ismail-Zadeh A (2018) Seismic hazard assessment of the Shillong Plateau India. Geom Nat Hazards Risk 9(1):841-861
50. Kumar A (2013) Seismic microzonation of Lucknow based on region specific GMPE's and geotechnical field studies. Ph.D. Thesis, Indian Institute of Science, Bangalore
51. Kumar A, Baro O (2016) In-direct estimation of local soil response in the light of past as well as recent earthquakes in the Shillong Plateau. In: Proceeding of the Indian geotechnical conference IIT Madras, Chennai, India
52. Kumar A, Mittal H, Sachdeva R (2012) Indian strong motion instrumentation network. Seismol Res Lett 83(1):59–66
53. Kumar A, Baro O, Narayan LM (2014) Estimation of surface PGA and determination of target value for no liquefaction at Guwahati city. In: Proceedings of geo-innovations, Indian Institute of Science, Bangalore, India

54. Mahajan AK, Virdi KS (2001) Macro seismic field generated by 29 March, 1999 Chamoli earthquake and its seismotectonic. J Asian Earth Sci 19(4):507–516
55. Mondal JK, Kumar A (2016) Impact of higher frequency content of input motion upon equivalent linear site response analysis for the study area of Delhi. Geotech Geolog Eng. https://doi.org/10.1007/s10706-016-0153-0
56. NDMA (2010) Development of probabilistic seismic hazard map of India. Technical report by National Disaster Management Authority, Government of India
57. NDMA (2011) Geotechnical/Geophysical investigation for Seismic microzonation studies of urban centres in India. Technical report by National Disaster Management Authority, Government of India
58. Nihon (2011) Liquefaction induced damages caused by the M 9.01 Japan mega Earthquake on March 11, 2011, Tokyo Metropolitan University, Hisataka Tano, Nihon University, Koriyama Japan, with cooperation of save Earth co. and Waseda University
59. Oldham T (1882) A catalogue of Indian earthquakes from the earliest time to the end of A.D. 1869, by the late Thomas Oldham edited by R. D. Oldham, Mem. Geol. Surv., India, 19, 163–215. Mem Geol Surv India 19:1–88
60. Oldham RD (1899) Report on the great earthquake of 12 June 1897. Mem Geol Surv India 29:1–379
61. Raghukanth STG, Dash SK (2010) Evaluation of seismic soil liquefaction at Guwahati city. Environ Earth Sci 61:355–368
62. Raghukanth STG, Sreelatha S, Dash SK (2008) Ground motion estimation at Guwahati city for an Mw 8.1 earthquake in the Shillong plateau. Tectonophysics 448:98–114
63. RaghuKanth STG, Dixit J, Dash SK (2009) Estimation of site amplification factors for Guwahati city. In: Proceedings of Indian geotechnical conference 2009, Guntur, India

Chapter 43
Effect of Footing Size on the Dynamic Behaviour of Cohesionless Soil-Foundation System

Gobinda Das and Priyanka Ghosh

Introduction

The vibrations induced by machine foundations affect the smooth working condition of highly sensitive equipment installed in the vicinity and disturb the nearby structures. Therefore, machine foundations need to be appropriately designed to cut down unwanted vibrations to certain permissible limits. Many researchers performed theoretical and experimental studies to understand the behaviour of machine foundations [1–5]. The dynamic response of machine foundations are affected by various aspects such as footing size and shape, soil properties and profile, embedment depth of foundations, frequency, and type of loading [6, 7]. Baidya and Rathi [8] studied the behaviour of dynamically loaded model footings placed on a sand layer of finite thickness. Kumar and Reddy [9] performed experimental studies to observe the behaviour of the spring-mounted base of different sizes. Ramesh and Prathap Kumar [10] conducted vertical vibration tests on a finite sand layer using model footings of different masses and sizes. However, limited studies are available on large-scale field tests under the vertical vibration on cohesionless soil considering different footing sizes. The present study attempts to understand the impact of footing size on the behaviour of foundations subjected to machine vibration, resting on cohesionless soil by conducting large-scale field vibration tests on rigid concrete footings of two different sizes.

G. Das · P. Ghosh (✉)
Indian Institute of Technology, Kanpur, India
e-mail: priyog@iitk.ac.in

K. Muthukkumaran et al. (eds.), *Soil Dynamics, Earthquake and Computational Geotechnical Engineering*, Lecture Notes in Civil Engineering 300,
https://doi.org/10.1007/978-981-19-6998-0_43

Experimental Details

Setup of Experiment

In the current work, a block resonance test setup is designed as per the guidelines prescribed in IS 5249 [11], which majorly consists of a rigid RCC block foundation, mechanical oscillator, DC-operated motor, and speed control system, tachometer, accelerometer, and digital vibration metre. Two RCC square footings, (a) footing 1 (F1): 550 mm × 550 mm × 200 mm and (b) footing 2 (F2): 650 mm × 650 mm × 200 mm, made of M20 grade concrete, are used as rigid block foundations. The footings have suitable arrangements for the easy mounting of the vibrating oscillator and clamping the square-shaped mild steel plates of size 450 mm × 450 mm × 16 mm to achieve the desired static load per unit area under the footing. The general layout of the block resonance experimental setup is presented in Fig. 43.1.

A mechanical oscillator capable of generating sinusoidal harmonic vertical vibrations is used to simulate a high-speed rotary machine. The dynamic load generated by the oscillator is a function of the angular frequency and the eccentricity angle of the unbalanced eccentric masses. It can be computed using the following expression.

$$F_0 = m_e e\omega^2 = 0.119\omega^2 \sin\left(\frac{\theta}{2}\right) \tag{43.1}$$

In the above expression, F_0 stands for the dynamic load, m_e represents the unbalanced eccentric mass, e is the radius of rotation. However, ω and θ are the angular frequency and eccentricity angle, respectively. The mechanical oscillator can be driven at different operating frequencies by a DC-operated motor connected to it

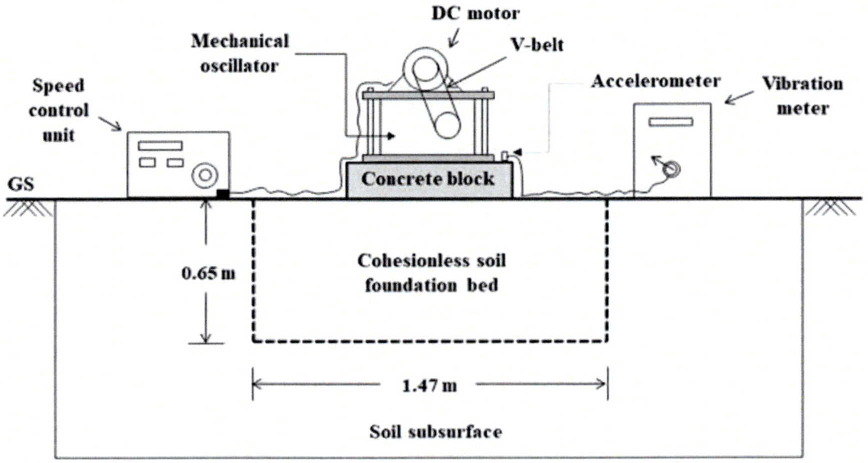

Fig. 43.1 General layout of block resonance experimental setup

using a V-type belt. The rotational speed of the motor is varied by a speed control system and is measured using a non-contact type digital laser tachometer. The acceleration response of the foundation is measured using a low 'g', high output MEMS accelerometer, where g is the acceleration due to gravity. The output is displayed using a digital vibration metre connected with the accelerometer. The outputs are recorded in the form of the peak amplitudes of acceleration, velocity, and displacement.

Test Pit

The test pit was prepared near the site of Geotechnical Field Laboratory in the premises of IIT Kanpur UP, India. The soil at the location primarily consists of clayey silt with a considerably low percentage of sand [12]. During the tests, in-situ density and moisture content were found to be 17.3 kN/m^3 and 4.5%, respectively, up to a depth of 1.5 m. Sankaran et al. [13] investigated the effect of pit size. They found that the dynamic behaviour remains unaffected as the sides of the excavated pit are equal or greater than 2.25 times the corresponding sides of the footing. Therefore, the pit dimensions were kept as 1.47 m $\times$ 1.47 m $\times$ 0.65 m, satisfying the above criteria for both the footings used in this study.

Test Materials and Preparation of Foundation Bed

The foundation bed consists of uniformly graded dry river sand taken from the bed of the Yamuna River at Kalpi (UP). The sand was sieved (keeping particle size in between 1 and 0.425 mm) to match the grade—II specifications of IS 650 [14]. The sand of such grade is the most efficient material for conventional rainfall method to produce uniform and repeatable sand beds [15]. Figure 43.2 presents the grain size distribution curve of sand taken for preparing the foundation bed.

The foundation bed was filled by pouring the sand from a height of 40 cm using an aluminium hand hopper to attain a relative density of up to 93%. The placing dry density and the peak friction angle of the sand were 15.3 kN/m^3 and 41°, respectively. The fixed volume method was adopted to check the relative density. In this method, samples were collected using small containers having a diameter and height of 5 cm. The containers were placed at various locations in the foundation bed while pouring the sand. The uniformity of sand density was ensured using a dynamic penetrometer. Once the test pit was filled, dynamic penetration tests were conducted at various places in the sand bed, and the corresponding penetration values were monitored. A proper uniformity in the soil properties was ensured before performing each series of experiments. The aluminium hopper and the dynamic penetrometer are similar to those used by Srivastava et al. [16]. A fresh new bed was prepared before placing each footing.

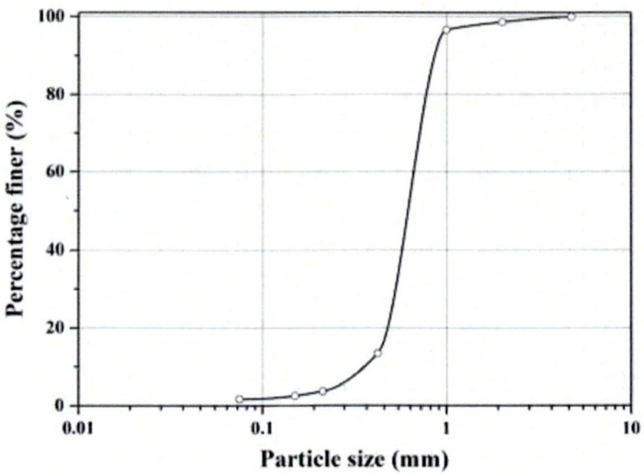

Fig. 43.2 Gradation curve for the sand used for foundation bed preparation

Testing Procedures and Experimental Program

Two model block foundations of different sizes are used in this investigation. After preparing the foundation bed, the footings were placed centrally over the surface without disturbing the foundation bed. A base plate of mild steel was fixed tightly on top of the footing for the easy mounting of the oscillator. Further, the oscillator, cover plate, and DC motor were fixed such that the entire block vibration assembly acted as a single unit. A static load per unit area of 10.75 kN/m^2 was maintained under the footings (F1 and F2). The eccentricity angle (θ) was varied to required values to simulate different dynamic force levels (F_0). For each footing, three eccentricity angles ($\theta = 16°$, 24°, and 32°) were used, and the magnitudes of the corresponding eccentric force settings ($m_e e$) are 0.017, 0.025, and 0.033 N-sec^2, respectively. The vertical acceleration response of the footing was measured using the accelerometer placed on the surface of the footing. The rotational frequency (rpm) of the motor was measured using the digital tachometer, whereas the corresponding displacement amplitude was recorded by the vibration metre. For an accurate measurement of the resonance, the responses were recorded at 50–100 rpm frequency intervals, and stable readings were noted at each frequency. A total of eight tests (listed in Table 43.1) were conducted in this study, out of which, two tests were performed to verify the repeatability of tests.

Results and Discussions

The dynamic response of both F1 and F2 in the form of displacement amplitude versus frequency response at different eccentric force settings is presented in Fig. 43.3a and b, respectively. It is worth mentioning that as the eccentric force setting increases, the resonant displacement amplitude at resonance increases, whereas the

Table 43.1 Details of the experimental program

Test series	Footing number	Eccentricity angle (θ)	Number of tests
A	F1	16°, 24°, and 32°	$3 + 1^*$
B	F2	16°, 24°, and 32°	$3 + 1^*$
Total number of tests			8

* Represents the repeatability tests

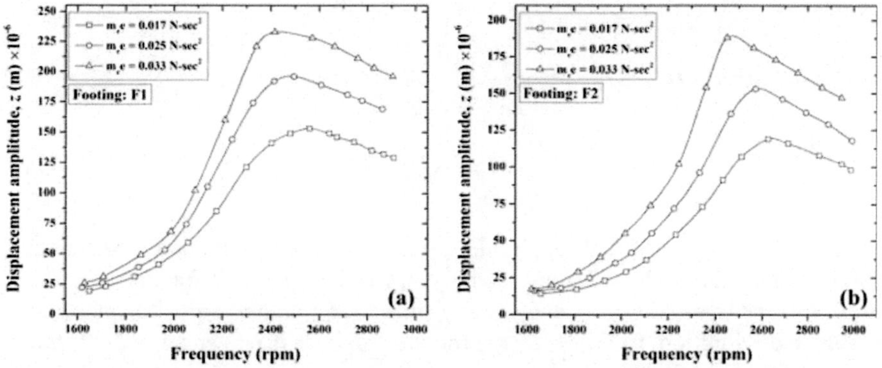

Fig. 43.3 Displacement amplitude versus frequency response for footings **a** F1 and **b** F2

resonant frequency decreases, irrespective of the size of the footings. In contrast, for a given eccentric force setting the resonant frequency increases, and the resonant displacement amplitude decreases with an increase in the footing size.

The essential dynamic parameters such as equivalent dynamic stiffness (k), damping ratio (D), natural frequency of the soil (f_n), frequency ratio (f_r), and magnification factor (M) can be determined using the following expressions proposed by Richart et al. [6].

$$\frac{z_r}{\left(\frac{m_e e}{m}\right)} = \frac{1}{2D\sqrt{1-D^2}} \tag{43.2}$$

$$f_{mr} = \frac{1}{2\pi\sqrt{1-2D^2}}\sqrt{\frac{k}{m}} \tag{43.3}$$

$$f_r = \frac{f}{f_n} \tag{43.4}$$

$$f_n = \frac{1}{2\pi}\sqrt{\frac{k}{m}} \tag{43.5}$$

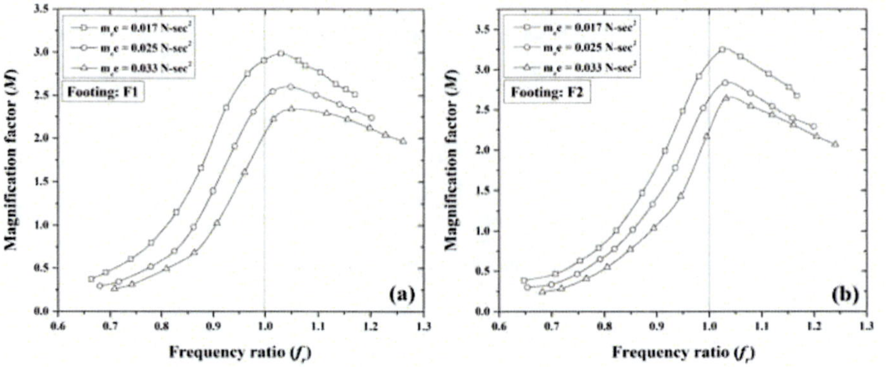

Fig. 43.4 Variation of M with f_r for footings **a** F1 and **b** F2

$$M = \frac{z}{\left(\frac{m_e e}{m}\right)} \tag{43.6}$$

where z_r is the displacement amplitude at resonance, m is the combined mass of footing and mechanical oscillator assembly, f_{mr} is the operating frequency at resonance, f is the operating frequency at any time, and z is the vertical displacement amplitude of vibration. In Fig. 43.4, the footing response is presented in the form of M plotted against f_r.

$$(f_r)_{resonance} = \frac{1}{\sqrt{1 - 2D^2}} \tag{43.7}$$

The peak values of the M can be observed beyond $f_r = 1.0$. It is worth mentioning that a similar response is also observed in reported literature [6, 17]. It may be attributed to the following facts: (a) for a rotating mass type steady–state vibration system, the maximum magnification can be observed at $f_r = (f_r)_{resonance}$ as given in Eq. 43.7, and (b) the magnitude of $(f_r)_{resonance}$ is greater than unity for any value of D lesser than $1/\sqrt{2}$ [17]. Hence, the peak values of the magnification factor occur slightly to the right of the resonance. As the eccentric force setting increases from 0.017 N-sec^2 to 0.033 N-sec^2, the maximum magnification factor varies from 3.0 to 2.3, and from 3.2 to 2.6 for footings F1 and F2, respectively.

Figure 43.5 shows the variation of f_{mr} and f_n of the soil with the eccentric force setting ($m_e e$). It can be observed that both resonant and natural frequencies decrease with increasing eccentric force setting, irrespective of the footing size. However, as the footing size varies, the natural frequency follows a trend similar to that of resonant frequency. Keeping the static load intensity constant, the increase in the footing size from F1 to F2 results in an average increase of 71 and 87 rpm in the resonant and natural frequency, respectively, for all three eccentric force settings.

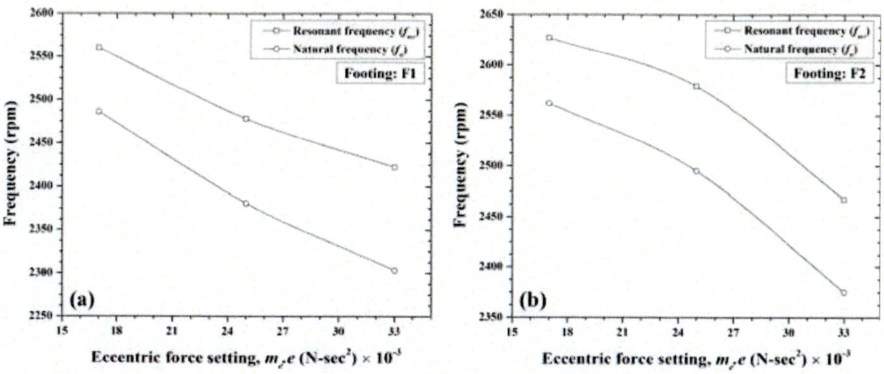

Fig. 43.5 Variation of resonant frequency (f_{mr}) and natural frequency (f_n) of soil with eccentric force setting ($m_e e$) for footings **a** F1 and **b** F2

At different eccentric force settings ($m_e e$), the total damping ratio (D) and equivalent dynamic stiffness (k) are obtained using Eqs. (43.2) and (43.3), respectively. The variation of D and k with eccentric force setting is shown in Fig. 43.6a and b, respectively. The damping ratio varies from 16.9% to 21.9% and 15.6% to 19.1% for F1 and F2, respectively. Further, a comparison is made between the present results and the results presented by Kumar and Reddy [9]. The experimental conditions considered in the study of Kumar and Reddy [9] were different from those of the present study. However, the loading type (rotating mass type vibration) adopted in both cases is similar. Both the studies show a good agreement in the trends. From Fig. 43.6, it is noted that the damping ratio increases irrespective of the footing size, and the equivalent dynamic stiffness decreases as the eccentric force setting increases. This is in line with the findings of existing studies [5, 8, 9]. However, for a given eccentric force setting, the damping ratio tends to decrease, and the soil stiffness increases when the size of the footing increases from F1 to F2. Similar trends were reported by Swain and Ghosh [5].

Summary and Conclusions

This paper emphasizes the influence of footing size on the dynamic response of the machine foundation system resting on cohesionless soil. Two sets of large-scale block resonance tests are performed at three different eccentric force settings for two different footing sizes. The foundation behaviour is presented in the form of displacement amplitude and frequency at resonance, the natural frequency of soil, the damping ratio, and the equivalent dynamic stiffness of soil. The results indicate that with an increase in the footing size, the resonant and natural frequency along with the equivalent dynamic stiffness of the soil increase, whereas the resonant displacement amplitude and the damping ratio decrease. However, an opposite trend

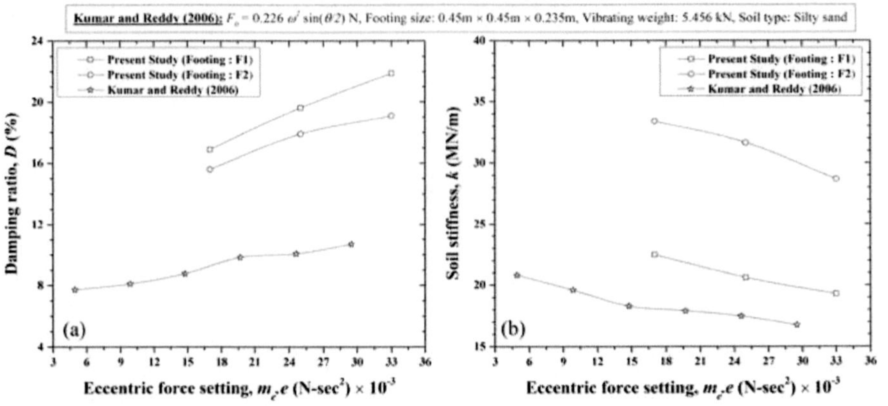

Fig. 43.6 Present results compared with Kumar and Reddy [9] for **a** damping ratio and **b** dynamic stiffness of soil

can be noticed irrespective of the footing size when the variation of eccentric force setting is considered. Therefore, it can be inferred that the footing size remarkably affects the dynamic behaviour of machine foundation systems resting on cohesionless soil and should be taken into consideration for the design process.

References

1. Reissner E (1936) Stationary, axially symmetrical vibrations of a homogeneous elastic half-space, excited by a shaking mass. Eng Arch 7(6):381–396
2. Lysmer JFER, Richart FE Jr (1966) Dynamic response of footings to vertical loading. J Soil Mech Found Div 92(1):65–91
3. Erden SM (1974) Influence of shape and embedment on dynamic foundation response. Ph. D. thesis, University of Massachusetts, Amherst
4. Bhoumik K (1989) Coupled vibration of footings on elastic half-space. M. Tech. thesis, Indian Institute of Technology, Kanpur, India
5. Swain A, Ghosh P (2015) Experimental study on dynamic interference effect of two closely spaced machine foundations. Can Geotech J 53(2):196–209
6. Richart FE, Hall JR, Woods RD (1970) Vibrations of soils and foundations. Prentice-Hall, Englewood Cliffs, NJ
7. Gazetas G (1991) Formulas and charts for impedances of surface and embedded foundations. J Geotech Eng 117(9):1363–1381
8. Baidya DK, Rathi A (2004) Dynamic response of footings resting on a sand layer of finite thickness. J Geotech Geoenviron Eng 130(6):651–655
9. Kumar J, Reddy CO (2006) Dynamic response of footing and machine with spring mounting base. Geotech Geol Eng 24:15–27
10. Ramesh HN, Prathap Kumar MT (2013) Stiffness of finite sand stratum under vertical vibrations. Proc Inst Civ Eng—Geotech Eng 166(3):299–309
11. IS 5249 (1992) Method of test for determination of in-situ dynamic properties of soils. Bureau of Indian Standards, New Delhi, India
12. Jishnu RB, Naik SP, Patra NR, Malik JN (2013) Ground response analysis of Kanpur soil along Indo-Gangetic Plains. Soil Dyn Earthq Eng 51:47–57

13. Sankaran KS, Krishnaswamy NR, Nair PB (1980) Effect of test pit size on vibrations of footings. Can Geotech J 17(2):292–295
14. IS 650 (1991) Standard sand for testing cement specification. Bureau of Indian Standards, New Delhi, India
15. Dave TN, Dasaka SM (2012) Assessment of portable traveling pluviator to prepare reconstituted sand specimens. Geomech Eng 4(2):79–90
16. Srivastava S, Srinivasan V, Ghosh P (2016) Effective utilization of dynamic penetrometer in determining the soil resistance of the reconstituted sand bed. Japan Geotech Soc Spec Publ 2(7):331–334
17. Kameswara Rao NSV (1998) Vibration analysis and foundation dynamics. Wheeler, New Delhi

Chapter 44
Plasticity-Based Liquefaction Prediction Using Support Vector Machine and Adaptive Neuro-Fuzzy Inference System

Sufyan Ghani and Sunita Kumari

Introduction

Liquefaction is an earthquake-induced seismic hazard which has always been considered as one of the major disastrous failures. During liquefaction, strength of soil significantly decreases which affects the capacity of a soil to withstand any geotechnical structure. Due to threat and risk, this phenomenon has been extensively studied by many researchers over decades. Simplified approaches for determining liquefaction potential of sandy soils were provided by Seed and Idriss. A simplified approach based on the Standard Penetration Test (SPT) was utilized in a number of projects to determine liquefaction potential. Previously, the term liquefaction was solely used to describe sandy soil deposits; however, measurements conducted during multiple earthquakes revealed liquefaction in soil with fine content ranging from medium to low plasticity. Non-plastic or plastic materials, such as silt or clay, tend to make a significant influence in the cyclic strength. (Wang [24] observed liquefaction of silty sand during the Haicheng and Tangshan earthquakes in 1975 and 1976, and suggested the Chinese Criteria. Multiple episodes of failure that led to high damage to structures in silty and clayey soils during the several earthquakes in between 1994 and 1999, questioned the effectiveness and accuracy of Chinese Criteria. Andrews and Martin analysed empirical methodologies and proposed the Modified Chinese Criteria, a new assessment index that renamed the old Chinese Criteria to conform

S. Ghani (✉)
Department of Civil Engineering, Sharda University, Greater Noida, Uttar Pradesh 201310, India
e-mail: sufyan.ghani@sharda.ac.in

S. Kumari
Department of Civil Engineering, National Institute of Technology Patna, Patna, Bihar 800005, India
e-mail: sunitafce@nitp.ac.in

K. Muthukkumaran et al. (eds.), *Soil Dynamics, Earthquake and Computational Geotechnical Engineering*, Lecture Notes in Civil Engineering 300,
https://doi.org/10.1007/978-981-19-6998-0_44

to US standards. It is worth noting that the effect of soil clay percentages on liquefaction susceptibility is unreliable and perplexing. This prompted the researchers to look into the impact of soil plasticity on liquefaction reaction. According to several findings in the literature, liquefaction resistance reduces as PI rises. Furthermore, various researchers believe that plasticity and cyclic strength of soil are interdependent, leading to the usage of fine content (FC), liquid limit (LL), and plasticity index (PI) as fundamental criteria for evaluating liquefaction susceptibility of soil deposits. Evaluation of liquefaction potential is done either theoretically or empirically; both of which have their own set of limitations [1–6, 9, 13, 15, 16, 20, 21]. As a result, relying on the findings of experimental tests in the field may produce an inaccurate prognosis. These empirical approaches take a long time to complete. Analytical methods, such as the finite element approach, are not ideal for complicated issues like liquefaction where the parameters required are high and professional knowledge to get an accurate result [17]. As a result, in the field of geotechnical engineering, computational approaches based on artificial intelligence (AI) are being adopted since they decrease the aforementioned obstacles and give an easy technique for analysing any complex problems such as liquefaction. Researchers widely accept AI-based models to predict the liquefaction potential of soil deposits [10, 14, 17–19, 22, 26]. Despite efforts by academics to identify liquefiable soil types using a variety of computer methodologies, no qualitative or quantitative assessment has been proposed that may be utilized as a tool for identifying soil liquefaction susceptibility. In comparison with theoretical and empirical methodologies, a sophisticated categorization model is required that delivers more exact, realistic, qualitative, and quantitative assessments for soil liquefaction. In the light of this, the current study proposes adaptive neuro-fuzzy inference system (ANNFIS) and support vector machine (SVM)-based classification models for classifying liquefaction susceptibility of soil deposits using basic geotechnical and seismic factors identified in the literature as input variables.

Theoretical Detail of Soft Computing Models

In many fields of research and engineering, computational models have shown to be invaluable. It is a branch of study that enables computers to learn without the assistance of humans. The techniques for soft computing models applied in this study are detailed in the next section. The theoretical elements of the semi-empirical approach employed are described, and further methodological approach for soft computing models is explained, which includes the ANFIS and SVM.

Empirical Approach Based on Idriss and Boulanger [8] Method

Liquefaction estimation for soil deposit using *CSR* that illustrates the earthquake loading on the soil has been presented as:

$$(\mathrm{CSR})_{7.5} = 0.65\left(\frac{\sigma_{vo} a_{\max}}{\sigma'_{vo}}\right)\frac{r_{\mathrm{d}}}{\mathrm{MSF}}\frac{1}{K_{\sigma}} \tag{44.1}$$

In which, σ_{vo} represents total effective overburden pressure and σ'_{vo} presents effective vertical overburden stress at a depth *z*; $a_{\max}$ represents peak ground acceleration in horizontal direction; r_{d} presents stress reduction factor; MSF represents the magnitude scaling factor; and K_{σ} represents the correction factor for effective overburden.

Mathematical expression to evaluate CRR and corresponding FOS generated from the corrected blow count $(N_1)_{60}$ are:

$$\mathrm{CRR} = \exp\left[\frac{(N_1)_{60\mathrm{cs}}}{14.1} + \left\{\frac{(N_1)_{60\mathrm{cs}}}{126}\right\}^2 - \left\{\frac{(N_1)_{60\mathrm{cs}}}{23.6}\right\}^3 + \left\{\frac{(N_1)_{60\mathrm{cs}}}{25.4}\right\}^4 - 2.8\right] \tag{44.2}$$

In which, term $(N_1)_{60\mathrm{cs}}$ depends on $(N_1)_{60}$ and given as:

$$(N_1)_{60\mathrm{cs}} = (N_1)_{60} + \Delta(N_1)_{60} \tag{44.3}$$

$$\Delta(N_1)_{60} = \exp\left(1.63 + \frac{9.7}{\mathrm{FC}+0.1} - \left(\frac{15.7}{\mathrm{FC}+0.1}\right)^2\right) \tag{44.4}$$

$$\mathrm{FOS} = \frac{\mathrm{CRR}}{\mathrm{CSR}} \tag{44.5}$$

Adaptive Neuro-Fuzzy Inference System (ANFIS)

ANIFS models are a hybrid learning technique that uses if-then rules and reasoning processes to model qualitative components of human knowledge without relying on accurate quantitative assessments. It combines the learning capability of neural network and knowledge representation capability of fuzzy logic. This approach has a strong generalization capability in comparison with ANN and other machine learning approaches. ANFIS model comprises an input layer that comprises input nodes, a hidden layers with hidden nodes, and in the end an output layers with output nodes.

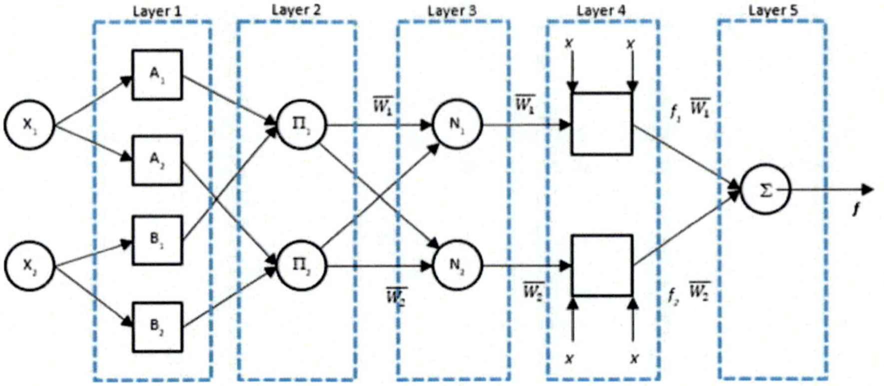

Fig. 44.1 A typical representation of ANFIS Structure

The following is a classic rule set with two fuzzy if–then rules:

$$\text{Rule 1}: \text{if } a \text{ is } A_1 \text{ and } b \text{ is } B_1, \text{ then } f_1 = p_1 a + q_1 b + r_1 \tag{44.6}$$

$$\text{Rule 2}: \text{if } a \text{ is } A_2 \text{ and } b \text{ is } B_2, \text{ then } f_2 = p_2 a + q_2 b + r_2 \tag{44.7}$$

where a and b are the two crisp inputs, and A_1, A_2, B_1, and B_2 are MF related with inputs x and y related with the node function. The parameters p_1, q_1, r_1 and p_2, q_2, r_2 are associated with output functions f_1 and f_2, respectively. A representative ANFIS assembly is presented in Fig. 44.1. More details can be found in many publications like [10–12, 22, 26].

Support Vector Machine (SVM)

SVM is a type of classification system that categorizes data grounded on its features and was instigated from statistical learning theory. The best line is selected by computing the maximum margin from equidistant support vectors. The line that divides the data perfectly and keeps the margin between the support vectors is technically called hyperplane. Presently, SVM is one of the best robust machine learning algorithms, which gained its popularity due to its good performance and attractive features. Khan and Coulibaly investigated the support vector machine's (SVM) capability in long-term lake water level prediction. Wang et al. [23] developed a support vector machine (SVM)-based approach for automatically identifying cracking types. In comparison with a neural network technique, SVM-based methodologies aimed at classification and prediction have been employed in various other civil engineering fields and have been found to work effectively [18, 25, 27]. Figure 44.2 illustrates the support vectors, hyperplane used to classify data in conventional SVM model.

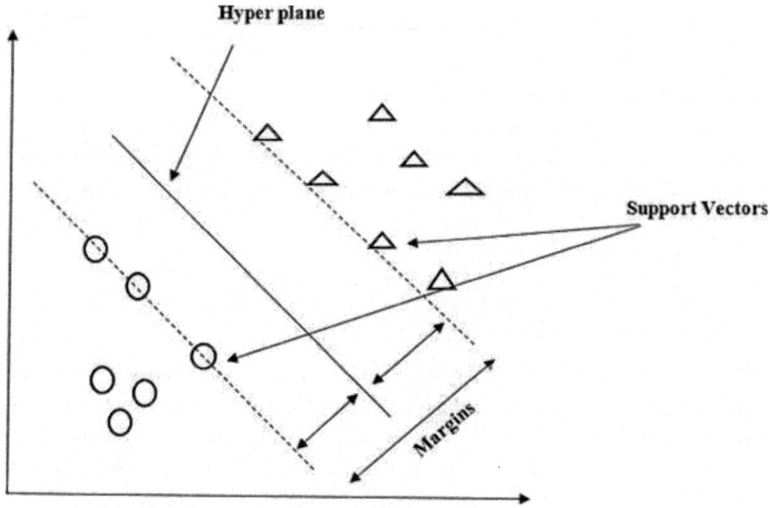

Fig. 44.2 Hyperplane-based classification in SVM model

Study Area and Data Collection

The state of Bihar in India is a developing state with significant amount of construction works being carried out across the state and has dense population. The state's history demonstrates its vulnerability to high-intensity earthquakes and liquefactions, the biggest of which occurred in 1934 with a magnitude of 8.4 and resulted in widespread destruction and a massive loss of human life. The state is located on the Himalayan tectonic plates borderline, near the Bihar–Nepal border, which comprises seismic zones III, IV, and V. Liquefaction was also detected in the Bihar district of Madhubani during the Nepal earthquakes of 2015. Since its cities are located near the banks of various rivers, the state also serves as an excellent example of alluvial soil deposits with a high percentage of fine particles. The state's annual floods result in massive depositions of plastic and non-plastic silts, increasing the possibility for liquefaction of soil deposits. As a result, it is critical to assess the region's liquefaction potential using the most up-to-date framework so that effective mitigation strategies can be strengthened to protect structures. In the light of the foregoing, the data acquired from investigation sites in the Bihar (India) districts of Darbhanga (Longitude 26.1542° N and Latitude 85.8918° E) and Muzaffarpur (Longitude 26.1197° N and Latitude 85.3910° E) are used to study the liquefaction behaviour of soil in this paper. Both regions are in seismic zone V, which is the most vulnerable to earthquakes. A total of 90 data samples were gathered from both sites, followed by laboratorial testing to determine their geotechnical parameters and analyse their liquefaction behaviour. Finally, the 90 data samples were used to create and test soft computing models.

Data Handling and Statistical Parameters

Normalization of datasets is regarded one of the most important processes to perform before applying any computational analysis. The purpose of normalization is to eliminate the dimensionality effect of the variables. As a result, before creating any model, the input and output variables were standardized between 0 and 1 using the formula:

$$x_{\text{NORMALISED}} = \left(\frac{x - x_{\min}}{x_{\max} - x_{\min}}\right) \tag{44.8}$$

For any given parameter (x), the maximum and minimum values are represented by $x_{\max}$ and $x_{\min}$, respectively. The 'min–max' normalization technique is the name for this method. The dataset is further divided into training and testing subsets after it has been normalized. After the development of the models, five statistical indicators are evaluated to measure the performance of the established computational models, and their expressions are:

$$\text{RMSE} = \sqrt{\frac{1}{n}\sum_{i=1}^{n}(\text{LS}_{\text{A}} - \text{LS}_{\text{P}})} \tag{44.9}$$

$$R^2 = \frac{\sum_{i=1}^{n}(\text{LS}_{\text{A}} - \text{LS}_{\text{MEAN}})^2 - \sum_{i=1}^{n}(\text{LS}_{\text{A}} - \text{LS}_{\text{P}})^2}{\sum_{i=1}^{n}(\text{LS}_{\text{A}} - \text{LS}_{\text{MEAN}})^2} \tag{44.10}$$

$$\text{VAF}(\%) = \left(1 - \frac{\text{var}(\text{LS}_{\text{A}} - \text{LS}_{\text{P}})^2}{\text{var}(\text{LS}_{\text{A}})}\right) \times 100 \tag{44.11}$$

$$\text{PI} = \text{adj}.R^2 + 0.01\text{VAF} - \text{RMSE} \tag{44.12}$$

$$\text{MAE} = \frac{1}{n}\sum_{i=1}^{n}|(\text{LS}_{\text{A}} - \text{LS}_{\text{P}})| \tag{44.13}$$

where LS_{A} and LS_{P} are the actual and predicted liquefaction susceptibility, respectively, and n is the total number of data samples. The mean of the values is denoted by LS_{MEAN}. Higher R^2, VAF, and PI values show a better predictive power of the developed model, while lower RMSE and MAE values suggest error-free behaviour of the developed model.

Results and Discussions

After the division of datasets into training and testing parts, the soil data is first analysed using [8] method. The CSR and CRR corresponding to the soil layers were evaluated using Eq. (44.1) and Eq. (44.6), and further FOS was determined with the help of Eq. (44.9). Soil deposits with FOS $\geq$ 1 are termed as non-liquefiable soil whereas soil layers with FOS < 1 are liquefiable. The FOS obtained for different soil layers of the proposed site are a form of continuous data, and in order to train a classification model, these continuous datasets were classified as 'class 0' and 'class 1'. Soil that liquefied is assigned 'class 0', and soil layers that did not liquefy are defined as 'class 1'.

Figure 44.3 represents the relation between PI values and liquefaction behaviour of soil for empirical method. Observation drawn from the figure clearly highlights that for higher plasticity values soil tends to fall on the safer zone. Similar trend was seen when liquefaction behaviour was represented along with SPT blow count values $(N_1)_{60}$ in Fig. 44.4.

Furthermore, two robust classification models were trained which use several soil parameters like PI, LL, SPT-N values $(N_1)_{60}$, and few seismic parameters like peak ground acceleration (PGA) and magnitude of earthquake (M_w). Figures 44.5 and 44.7 present the relation between liquefaction susceptibility and PI, when evaluated using ANFIS and SVM models, respectively. It is evident from the illustrations that PI value has a strong correlation with liquefaction response of soil. As plasticity of the soil increases, liquefaction potential tends to decrease and soil layers fall on the safer zone. Comparable results were observed when liquefaction behaviour was studied in relation to $(N_1)_{60}$ and have been illustrated in Figs. 44.6 and 44.8 for ANFIS and SVM models, respectively. The statistical performance parameters defined in

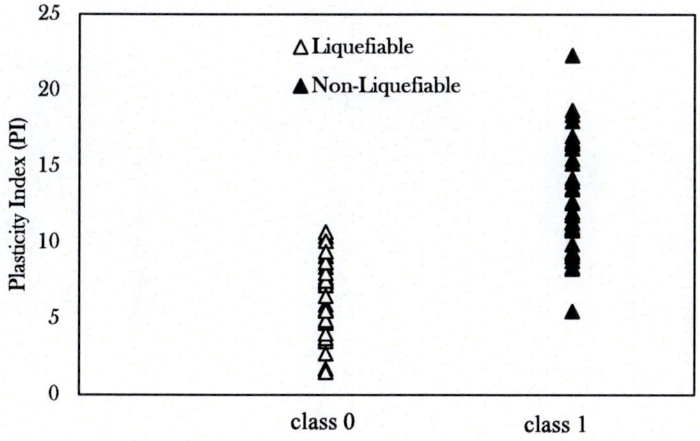

Fig. 44.3 Liquefaction correlation with PI evaluated using empirical approach

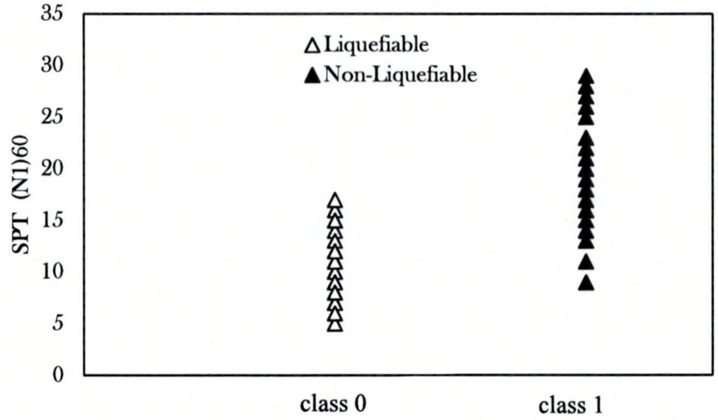

Fig. 44.4 Liquefaction correlation with $(N_1)_{60}$ evaluated using empirical approach

aforementioned sections are evaluated for both computational model and have been tabulated in Tables 44.1 and 44.2 for training and testing datasets, respectively.

Figure 44.9 illustrates graphical comparison among two developed classification models. It highlights the fact that SVM Model has better classification capability as compared to ANFIS model. When the proposed site was evaluated using Empirical method, 50% of the soil layers belonged to class 1, i.e. safe against liquefaction and the remaining 50% belonged to class 0, i.e. liquefiable. As empirical methods are time-consuming and require large datasets to predict soil's liquefaction response. Hence, ANFIS model and SVM model were developed, tested, and compared to [8] method. According to ANFIS model, 43% of the soil layers tends to fall on the safer side, i.e. it belongs to class 1 and 57% of soil layers tends to fall in liquefiable zone, i.e. class 0. This observation clearly challenges the results drawn from

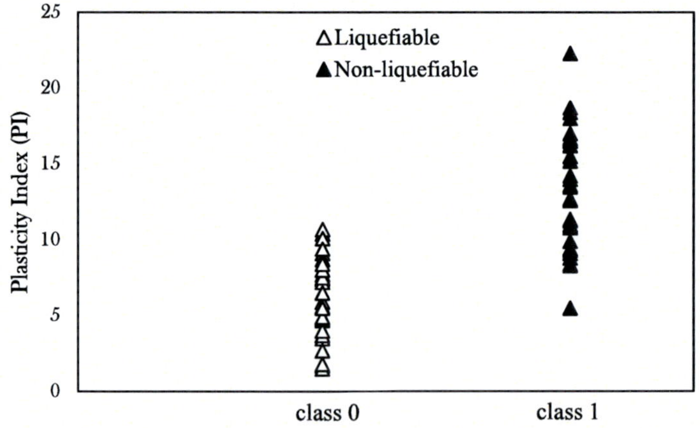

Fig. 44.5 Liquefaction correlation with PI evaluated using ANFIS model

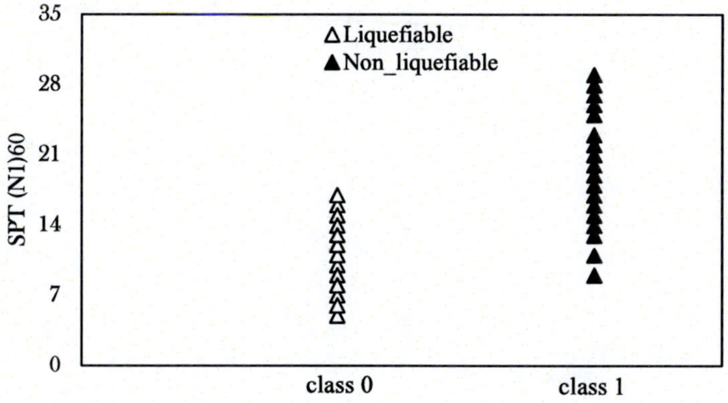

Fig. 44.6 Liquefaction correlation with $(N_1)_{60}$ evaluated using ANFIS model

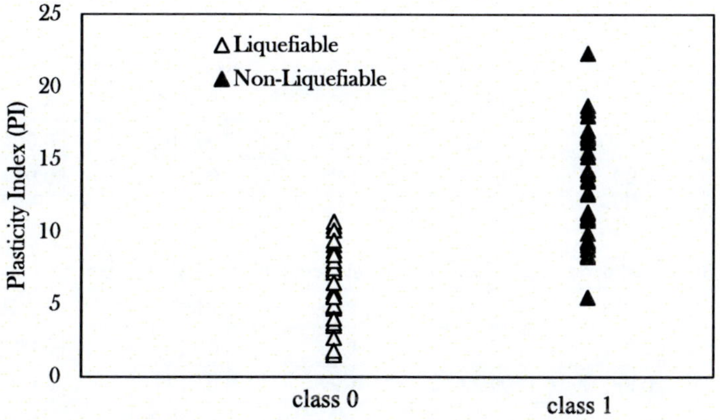

Fig. 44.7 Liquefaction correlation with PI evaluated using SVM model

empirical approach. Whereas, when liquefaction is evaluated using SVM model, no such inconsistency was observed. According to SVM model, 51% of the soil layers tends to fall in class 1 and 49% belongs to class 0, which is in similarity to empirical approach. Therefore, it may be concluded that SVM replicates empirical model in the most accurate form with high accuracy and least error as compared to ANFIS model.

The performance and efficacy of these models are evaluated in meticulous and comprehensive fashion. To comprehend the ability of developed models, five performance indices, namely R^2, VAF, PI, RMSE and MAE, were determined and presented in Tables 44.1 and 44.2 for training and testing datasets, respectively. The observation made suggests that SVM model is superior in comparison with the ANFIS model with $R^2 = 0.99$ and RMSE $= 0.002$ in the training stage and $R^2 = 0.99$ and RMSE $= 0.012$ in the testing phase. The present study also emphasizes on the use of *PI* for studying

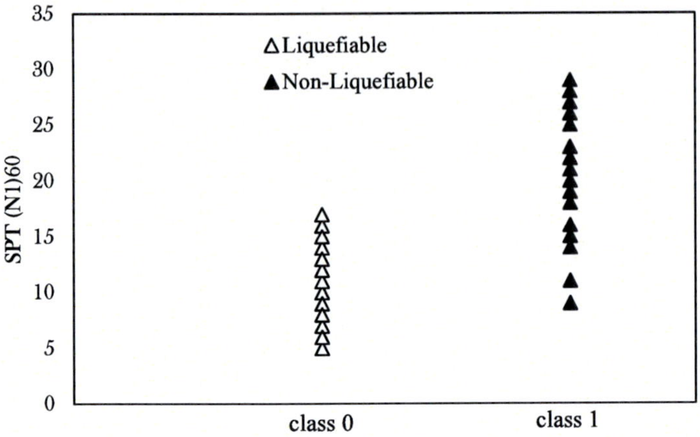

Fig. 44.8 Liquefaction correlation with $(N_1)_{60}$ evaluated using SVM model

Table 44.1 Statistical parameters of computational model during training phase

Models	R^2	VAF	PI	RMSE	MAE
ANFIS	0.97	98.26	1.86	0.096	0.06
SVM	0.99	99.94	1.97	0.002	0.01

Table 44.2 Statistical parameters of computational model during testing phase

Models	R^2	VAF	PI	RMSE	MAE
ANFIS	0.94	97.96	1.80	0.12	0.09
SVM	0.99	99.28	1.97	0.012	0.01

the liquefaction response of soil. According to the results of this study, as soil flexibility improves, its liquefaction potential diminishes, which is consistent with the findings of other studies. Many scholars have studied and investigated the relationship between PI and soil liquefaction potential. According to [7], using PI to analyse liquefaction is a reliable and acceptable method, and soils with greater plasticity indices are less prone to liquefy. Ghani and Kumari [4] also highlighted the significance of PI on liquefaction behaviour of soil. Similar trend was observed for $(N_1)_{60}$ values; with an increase in N value, liquefaction susceptibility tends to decrease. Therefore, it can be concluded that SVM model replicates the empirical approach with accuracy and overcomes their drawbacks while highlighting the significance of PI and $(N_1)_{60}$ on liquefaction susceptibility of soil.

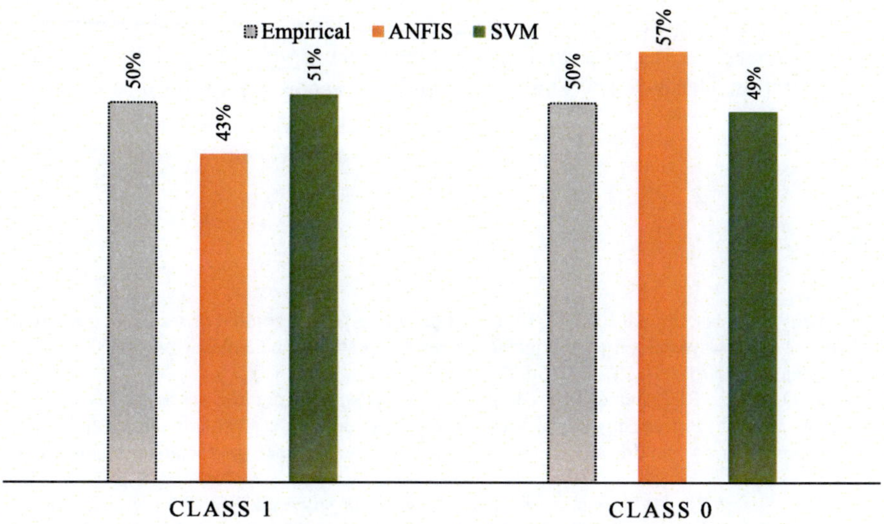

Fig. 44.9 Graphical illustration of all the classification models used in the study

Conclusion

To evaluate soil liquefaction response for seismically active region of Bihar, two computation-based classification models ANFIS and SVM were developed. The fundamental benefit of these models is that they have a high degree of generalization and a low computing cost. In the training and testing stages, the performance of the proposed models was shown to be in good accord with that of the empirical approach. These computational models overcome all of the empirical method's limitations and shortcomings. The impacts of PI, LL, $(N_1)_{60}$, PGA, and M_w were investigated in this study, and the following conclusions were drawn based on the results and findings:

1. The study highlights the fact that lengthy and complex calculations involved in empirical approach can be conveniently replicated by training computational models which can further categorize liquefiable and non-liquefiable soil deposit with accuracy and precision.
2. The SVM model provides the best prediction performance, according to the results of performance indices with $R^2 = 0.99$ and RMSE $= 0.002$ in the training stage and $R^2 = 0.99$ and RMSE $= 0.012$ in the testing phase.
3. This study also highlights a significant relation between liquefaction and PI of soil deposits, criteria which has been discussed by many researchers over past few years.
4. It was observed that as PI value increases liquefaction susceptibility of soil decreases. This dependency of liquefaction on soil's plasticity provides a new insight to researchers and geotechnical engineers about liquefaction response of plastic soils.

5. Overall, the suggested SVM model can be considered a novel classification tool for analysing the response of soil liquefaction during planning, designing, and construction phase of any engineering project, according to the results of the performance analysis.

References

1. Akbari Paydar N, Ahmadi MM (2016) Effect of fines type and content of sand on correlation between shear wave velocity and liquefaction resistance. Geotech Geol Eng 34(6):1857–1876. https://doi.org/10.1007/s10706-016-9995-8
2. Bray JD, Sancio RB, Riemer MF, Durgunoglu T (2004) Liquefaction susceptibility of fine-grained soils. In: Proceedings of the 11th international conference on soil dynamics and earthquake engineering and 3d international conference on earthquake geotechnical engineering, pp 655–662
3. Bray JD, Sancio RB (2006) Assessment of the liquefaction susceptibility of fine-grained soils. J Geotech Geoenviron Eng 132(9):1165–1177. https://doi.org/10.1061/(ASCE)1090-0241(2006)132:9(1165)
4. Ghani S, Kumari S (2021) Insight into the effect of fine content on liquefaction behavior of soil. Geotech Geol Eng 39(1):1–12. https://doi.org/10.1007/s10706-020-01491-3
5. Ghani S, Kumari S (2021b). Liquefaction susceptibility of high seismic region of Bihar considering fine content. In: Basics of computational geophysics, pp 105–120. https://doi.org/10.1016/b978-0-12-820513-6.00012-6
6. Ghani S, Kumari S, Choudhary AK, Jha JN (2021) Experimental and computational response of strip footing resting on prestressed geotextile-reinforced industrial waste. Innov Infrastruct Solutions 6(2):1–15. https://doi.org/10.1007/s41062-021-00468-2
7. Gratchev IB, Sassa K, Fukuoka H (2006) How reliable is the plasticity index for estimating the liquefaction potential of clayey sands? J Geotech Geoenviron Eng 132(1):124–127. https://doi.org/10.1061/(asce)1090-0241(2006)132:1(124)
8. Idriss IM, Boulanger RW (2006) Semi-empirical procedures for evaluating liquefaction potential during earthquakes. Soil Dyn Earthq Eng 26(2–4):115–130. https://doi.org/10.1016/j.soildyn.2004.11.023
9. Idriss IM, Boulanger RW (2012) Examination of SPT-based liquefaction triggering correlations. Earthq Spectra 28(3):989–1018. https://doi.org/10.1193/1.4000071
10. Kaya Z (2016) Predicting liquefaction-induced lateral spreading by using neural network and neuro-fuzzy techniques. Int J Geomech 16(4):1–14. https://doi.org/10.1061/(ASCE)GM.1943-5622.0000607
11. Kayadelen C (2011) Soil liquefaction modeling by genetic expression programming and neuro-fuzzy. Expert Syst Appl 38(4):4080–4087. https://doi.org/10.1016/j.eswa.2010.09.071
12. Kumar V, Venkatesh K, Tiwari RP (2014) A neurofuzzy technique to predict seismic liquefaction potential of soils. Neural Netw World 24(3):249–266. https://doi.org/10.14311/NNW.2014.24.015
13. Marto A, Tan CS, Makhtar AM, Ung SW, Lim MY (2015) Effect of plasticity on liquefaction susceptibility of sand-fines mixtures. Appl Mech Mater 773–774:1407–1411. https://doi.org/10.4028/www.scientific.net/amm.773-774.1407
14. Muduli PK, Das SK, Bhattacharya S (2014) CPT-based probabilistic evaluation of seismic soil liquefaction potential using multi-gene genetic programming. Georisk 8(1):14–28. https://doi.org/10.1080/17499518.2013.845720
15. Polito C (2001) Scholars' mine plasticity based liquefaction criteria plasticity based liquefaction
16. Prakash S, Sandoval JA (1992) Liquefaction of low plasticity silts. Soil Dyn Earthq Eng 11(7):373–379. https://doi.org/10.1016/0267-7261(92)90001-T

17. Sabbar AS, Chegenizadeh A, Nikraz H (2019) Prediction of liquefaction susceptibility of clean sandy soils using artificial intelligence techniques. Indian Geotech J 49(1):58–69. https://doi.org/10.1007/s40098-017-0288-9
18. Samui P, Sitharam TG (2011) Machine learning modelling for predicting soil liquefaction susceptibility. Nat Hazards Earth Syst Sci 11(1):1–9. https://doi.org/10.5194/nhess-11-1-2011
19. Samui P (2007) Seismic liquefaction potential assessment by using relevance vector machine. Earthq Eng Eng Vib 6(4):331–336. https://doi.org/10.1007/s11803-007-0766-7
20. Sandoval-Shannon JA (1989) Liquefaction and settlement characteristics of silt soils
21. Seed RB, Cetin KO, Moss RES, Kammerer AM, Wu J, Pestana JM, Riemer MF, Sancio RB, Bray JD, Kayen RE, Faris A (2003) Recent advances in soil liquefaction engineering : a unified and consistent framework. In: Report No. EERC 2003-06, Earthquake Engineering Research Center, pp 1–72. https://doi.org/10.1007/978-3-319-07118-3_19
22. Venkatesh K, Kumar V, Tiwari RP (2013) Appraisal of liquefaction potential using neural network and neuro fuzzy approach. Appl Artif Intell 27(8):700–720. https://doi.org/10.1080/08839514.2013.823326
23. Wang L, Wu C, Li Y, Liu H, Zhang W, Chen X (2019) Probabilistic risk assessment of unsaturated slope failure considering spatial variability of hydraulic parameters. KSCE J Civ Eng 23(12):5032–5040. https://doi.org/10.1007/s12205-019-0884-6
24. Wang (1979) Some findings in soil liquefaction. Water Conservancy and Hydroelectric Power Research Institute, Beijing
25. Xue X, Yang X, Li P (2017) Application of a probabilistic neural network for liquefaction assessment. Neural Netw World 27(6):557–567. https://doi.org/10.14311/NNW.2017.27.030
26. Xue X, Yang X (2013) Application of the adaptive neuro-fuzzy inference system for prediction of soil liquefaction. Nat Hazards 67(2):901–917. https://doi.org/10.1007/s11069-013-0615-0
27. Zhao HB, Ru ZL, Yin S (2007) Updated support vector machine for seismic liquefaction evaluation based on the penetration tests. Mar Georesour Geotechnol 25(3–4):209–220. https://doi.org/10.1080/10641190701702303

Chapter 45
Effect of Shaking Pattern on Reliquefaction Potential of Solani Sand

Gowtham Padmanabhan and B. K. Maheshwari

Introduction

Soil liquefaction during earthquakes resulted in ground/building settlement, foundation failures/damages and incurred excessive damages to existing structures and built environment (e.g. 1964 Niigata earthquake, Japan and 2001 Bhuj earthquake, among others). Initial time research works paid close attention on first liquefaction phenomenon and the mitigation measures, whereas reliquefaction works are not reported/limited. Occurrence of historical earthquakes such as 2010–2011 Canterbury earthquakes series and 2011 Tohoku, Japan have shown that a site liquefied once can be reliquefied due to the occurrence of successive earthquake events. This invited the attention of researchers to work on this complex and interesting research topic in geotechnical earthquake engineering. Reliquefaction works started gaining importance and many researchers started investigating the mechanism through the cyclic triaxial, 1 g shake table and centrifuge modelling experiments [1–5]. These studies mainly focused on identifying the reason behind the increased/reduced reliquefaction resistance to its initial liquefaction and concluded that the resistance was mainly influenced by anisotropy of sand, liquefaction history, preshaking and the presence of foreshocks/aftershocks. The extensive works carried out by the researchers [6–8] through experiments concluded the sand reliquefaction behaviour influenced by paleo seismic shaking, preshaking history and changes in fabric anisotropy. However, these laboratory evidences alone are not sufficient to examine the mechanism of reliquefaction and to analyse the several factors influencing the increased or decreased reliquefaction potential.

From observations of 2011 Japan earthquake, liquefaction susceptibility was found to be lesser for the deposits attributed to intense preshaking and/or geologic

G. Padmanabhan (✉) · B. K. Maheshwari
Department of Earthquake Engineering, IIT Roorkee, Roorkee, Uttarakhand, India
e-mail: gowtham@eq.iitr.ac.in

K. Muthukkumaran et al. (eds.), *Soil Dynamics, Earthquake and Computational Geotechnical Engineering*, Lecture Notes in Civil Engineering 300,
https://doi.org/10.1007/978-981-19-6998-0_45

age by the previous earthquakes [9]. Similar observations were reported by [10–12] through field observations and experimental evidences for different earthquakes. In addition, the researchers [13] examined the field observations and reported that the liquefaction susceptibility of field deposits found to be increased rather than decreased when subjected to subsequent earthquake events. Based on the above studies, it is concluded that the beneficial effects of preshaking and geologic age will completely be lost during the occurrence of full liquefaction and will results in reduced liquefaction resistance. From both the field and experimental evidences, the contrasting nature of reliquefaction phenomenon is observed by the researchers through field evidences, which substantiate the complex and interesting behaviour of reliquefaction phenomenon.

Motivated by the complex reliquefaction behaviour of field deposits, Padmanabhan and Maheshwari [14] presented the influence of preshaking on reliquefaction potential with the help of existing literatures. In the analysis, Fukuoka-ken Seiho-oki earthquake, 2005 and Kumamoto earthquake series, 2016 are considered and these two earthquakes possess contrasting earthquake and liquefaction history. The case study concluded that the effect of preshaking influenced by magnitude of foreshocks/aftershocks and liquefaction history and that has resulted in the increased reliquefaction resistance in Kumamoto region despite several large earthquake events.

Scope of the Paper

The significance/influence of different shaking pattern on reliquefaction potential of sand deposits investigated using 1 g shake table apparatus. Reliquefaction potential and liquefaction resistance was examined on the basis of generated EPWP subjected to sequential/successive shaking events. In addition, various factors influencing the reliquefaction, the present study considers shaking pattern and examined the influence through laboratory experiments. Earthquakes such as 2016 Kumamoto, Japan and 2011 Tohoku, Japan followed the increasing and decreasing shaking pattern respectively. This study helps to examine and predict the occurrence of reliquefaction and in the development of liquefaction resistance. The findings of the study can be correlated with the successive earthquakes of similar pattern and in the selection of appropriate ground improvement system.

Experimental Procedure

Experimental Apparatus

Experiments were conducted in an indigenously fabricated 1 g uniaxial shake table available in Soil dynamics lab, IIT Roorkee. The shaking table can operate at a 0–10 Hz frequency and an acceleration amplitude (0.05–1 g). To prepare the saturated sand bed, a water-tight container of dimension 1050 × 600 × 600 mm was used. The entire setup of the shaking table experiment was given in Fig. 45.1. Three number of pore pressure transducers was used in the experiments to examine the generated EPWP time history at different locations. The location of measuring sensors at three different locations are: 560, 400, and 240 mm from the top of the tank for the bottom (*B*), middle (*M*), and top (*T*) piezometers. Figure 45.2 shows schematic diagram of three pore pressure transducers used in the study.

The excess pore water pressures were recorded for each second till the complete dissipation takes place. The excess pore water pressure U_{excess} at that particular depth will be calculated by multiplying the saturated unit weight of the sample to the rise in pore water height.

Fig. 45.1 Shaking table apparatus

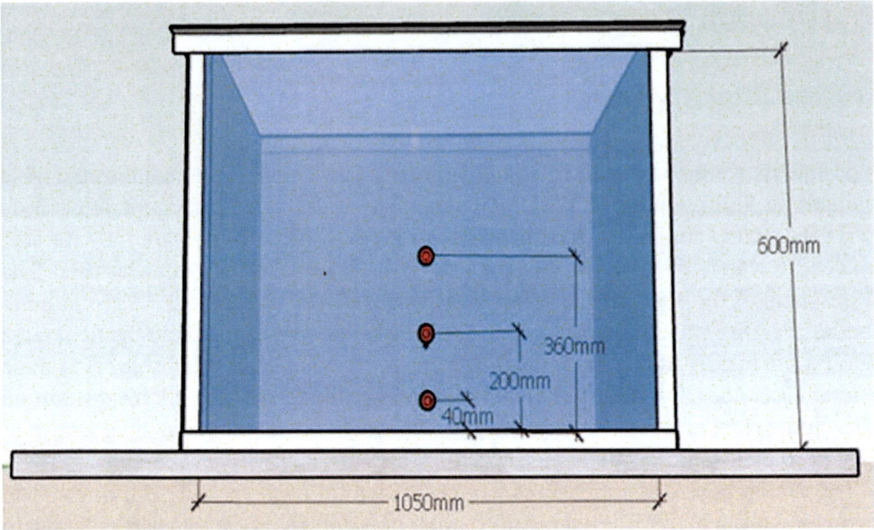

Fig. 45.2 Location of pore pressure transducers

Material Properties

Sand procured from the Solani riverbed (Roorkee) Uttarakhand was used in the sample preparation. The sand used in the experiments was poorly graded as per [15] and susceptible to reliquefaction under multiple shaking events. Index properties of the Solani sand was estimated as per the guidelines of Indian standards and are listed in the Table 45.1. Sample was prepared to a depth of 570 mm. Preparation of sand bed for liquefaction model studies are critical. Proper care has to be taken to achieve uniform sand deposition throughout the depth. The total quantity of sand essential to fill the tank was estimated using the procedure adopted by [16]. The same procedure was followed throughout the entire experimental programme. The calculated amount of sand and water was divided in three layers, so as to obtain uniform densification throughout the depth. The preparation of sand bed was carried out using air pluviation method to achieve maximum uniformity in sand density. A specially designed hopper arrangement was used to pour sand from a determined constant height to obtain relative equal sand density. Height at which the sand is poured is kept constant till the complete preparation of the sand bed, which helps to achieve the uniform sand bed. Earlier, laboratory experiment was conducted to determine the height of fall at which the dry sand need to be poured [17]. The height of fall was 10 cm to obtain the desired relative density of 25%.

Table 45.1 Soil properties

S. no.	Property/characteristic		Value
1	Type		Poorly graded sand
2	Specific gravity (G)		2.68
3	Coefficient of uniformity (C_u)		1.80
4	Curvature coefficient (C_c)		1.00
5	Particle size	D_{50}	0.25 mm
		D_{10}	0.15 mm
6	Max. void ratio (e_{max})		0.87
7	Min. void ratio (e_{min})		0.55
8	Density (D_r)		25%
9	Void ratio (e) for 25% density		0.79
10	Dry unit weight for 25% density		14.97 N/m^3

Loading Conditions

In this study, two shaking patterns was selected to understand the reliquefaction potential of sand deposits. Repeated shaking events was simulated with varying the acceleration amplitude with a constant frequency of 2 Hz and a 1 min shaking duration for both the patterns.

Test series 1:—First test series was proposed to simulate a series of foreshocks followed by a main-shock event. The acceleration amplitude was applied from 0.1 to 0.4 g with an incremental interval of 0.1 g. The selected shaking sequence represents the 2016 Kumamoto (Japan) earthquake. Liquefaction and subsequent reliquefaction was observed during the repeated loading.

Test series 2:—Second series was proposed for stimulating main-shock event followed by the series of aftershocks. The amplitude acceleration was applied at 0.4–0.1 g with a similar interval. The selected shaking sequence represents the 2011 Tohoku (Japan) earthquake. The main earthquake was preceded by a number of large aftershocks.

Results and Discussion

For each shaking event, the variation in generation and dissipation of EPWP (U_{excess}) time history was monitored with help of piezometers. The EPWP was found to be greater at the bottom, smaller at the top and median in the middle location. This is expected due to the effect of overburden pressure observed at that particular location. The findings of the study are discussed in the subsequent sections.

Table 45.2 Effect of increasing shake pattern on excess pore water pressure

Acceleration type	Acceleration (g)	Maximum excess pore water pressure (kN/m^2)		
		T	*M*	*B*
Increasing	0.10	1.06	1.52	1.75
	0.20	1.18	1.66	1.96
	0.30	1.32	1.83	2.22
	0.40	1.45	1.98	2.40

Increasing Shake Pattern

Incremental shaking was applied to the prepared sand deposit. Table 45.2 shows the generated EPWP for all events shaken through incremental pattern. The generated EPWP was observed increased with increase in acceleration amplitude. From increasing the acceleration amplitude from 0.1 to 0. 4 g successively, the generated EPWP increased at the range of 37%, 30%, and 37% for top, middle and bottom piezometers, respectively. In case of top piezometer, the generated EPWP increased by 11, 12, and 10% for increase in acceleration amplitude after each set of incremental loading. The increment was found to be consistent with all the acceleration loadings. It indicates the increased liquefaction potential and occurrence of reliquefaction, when subjected to successive shaking events.

The occurrence of reliquefaction and the potential effect of preshaking (mainly foreshocks) resulted in the rearrangement of sand particles and increased sand density.

Decreasing Shake Pattern

Followed by incremental pattern, decremental shaking pattern was applied to the freshly prepared sand deposit. Table 45.3 shows the maximum EPWP generated under successive shaking events. The EPWP was found reduced with the decrease in acceleration amplitude. The acceleration amplitude was decreased at an interval of 0.1 g starting from 0.4 and ended with 0.1 g, the generated EPWP was found to be decreased by 24%, 18%, and 17% for top, middle and bottom piezometers, respectively. In case of top piezometer, the EPWP reduced by 9%, 9%, and 7% for the decrease in amplitude, respectively.

When the sand bed liquefied due to the application of strong intense seismic event (0.4 g), high EPWP was developed from the lower part reached the upper part of the ground surface. This resulted in reduced shear stress in the upper part and the shallower sand deposit was made susceptible to reliquefaction even by the earthquakes of lower magnitude. The reason being the occurrence of complete liquefaction during 0.4 g seismic loading, the possible effect of preshaking got nullified and makes the

Table 45.3 Effect of decreasing shake pattern on excess pore water pressure

Acceleration type	Acceleration (g)	Maximum excess pore water pressure (kN/m^2)		
		T	*M*	*B*
Decreasing	0.40	1.83	2.02	2.16
	0.30	1.66	1.89	2.04
	0.20	1.51	1.75	1.92
	0.10	1.40	1.66	1.80

sand deposit more susceptible to loss of shear strength during the successive seismic loading.

Comparison with Past Studies

The effect of preshaking and reliquefaction occurrence subjected to repeated shaking events are mostly reported based on field observations (e.g. 2005 El Mayor-Cucupah earthquakes and 2016 Kumamoto, Japan earthquakes) [18] and are rarely reported through experimental evidences. Experimental evidences obtained from the present study finds good correlation with the findings reported by Padmanabhan and Maheshwari [14]. El-Sekelly et al. [19] reported that factors such as preshaking, earthquake pattern, presence of foreshocks and/or aftershocks and liquefaction history influence the occurrence of reliquefaction and in the development of liquefaction resistance. The experimental observations reported by Ha et al. [20] for successive shaking events with varying the shaking duration shows good agreement with the findings of the present study.

Conclusions

Reliquefaction potential and increased liquefaction resistance under repeated shaking events with different shaking pattern was estimated for Solani sand. Uniaxial shaking table experiments was performed on 25% relatively density saturated sand bed.

1. The sand liquefaction and reliquefaction resistance mainly influenced by occurrence of foreshocks/aftershocks associated with the main-shock. Even the low intense shaking event influence the preshaking history and resulted in soil deposition and densification.
2. Substantial increase in liquefaction resistance was reported, even if preshaking observed for a very short period or caused sometimes by a handful of preshaking events.

3. In incremental shake pattern, the EPWP generation increased due to increase in acceleration amplitude, whereas in decremental shake pattern, the excess pore water pressure decreased to a similar extent due to the reduction in acceleration amplitude. From the observed results, the effect of preshaking history is evident in improving the liquefaction resistance.
4. In both the applied shaking pattern, the effect of preshaking resulted in cementation and/or interlocking of sand grains. This enhances the contact between the sand particles that accumulate with time.

References

1. Ye B, Hu H, Bao X, Lu P (2018) Reliquefaction behavior of sand and its mesoscopic mechanism. Soil Dyn Earthq Eng 114:12–21
2. Ye B, Zhang L, Wang H, Zhang X, Lu P, Ren F (2019) Centrifuge model testing on reliquefaction characteristics of sand. Bull Earthq Eng 17(1):141–157
3. Padmanabhan G, Shanmugam GK (2020) Reliquefaction assessment studies on saturated sand deposits under repeated acceleration loading using 1-g shaking table experiments. J Earthq Eng, pp 1–23
4. Padmanabhan G, Shanmugam GK (2021) Addressing influence of prefabricated vertical drains for liquefaction mitigation under multiple shaking events. In: Sitharam TG et al. (ed) Soil dynamics. Select proceedings of 7th ICRAGEE 2020, Lecture notes in civil engineering, vol. 119, p 203
5. Padmanabhan G, Shanmugam GK (2021) Liquefaction and reliquefaction resistance of saturated sand deposits treated with sand compaction piles. Bull Earthq Eng
6. Hayati H, Andrus RD (2009) Updated liquefaction resistance correction factors for aged sands. J Geotech Geoenviron Eng 135(11):1683–1692
7. Ishihara K, Araki K, Toshiyuki K (2014) Liquefaction in Tokyo Bay and Kanto regions in the 2011 great east Japan earthquake. In: Earthquake geotechnical engineering design. Springer, Cham, pp 93–140
8. Heidari T, Andrus RD (2012) Liquefaction potential assessment of Pleistocene beach sands near Charleston, South Carolina. J of Geotech Geoenviron Eng 138(10):1196–1208
9. Kokusho T, Ito F, Nagao Y, Green AR (2012) Influence of non/low-plastic fines and associated aging effects on liquefaction resistance. J Geotech Geoenviron Eng 138(6):747–756. https://doi.org/10.1061/(ASCE)GT.1943-5606.0000632
10. Andrus RD, Hayati H, Mohanan NP (2009) Correcting liquefaction resistance for aged sands using measured to estimated velocity ratio. J Geotech Geoenviron Eng 135(6):735–744
11. Darby KM, Boulanger RW, DeJong JT, Bronner JD (2018) Progressive changes in liquefaction and cone penetration resistance across multiple shaking events in centrifuge tests. J Geotech Geoenviron Eng 145(3):04018112
12. Dobry R, Thevanayagam S, El-Sekelly W, Abdoun T (2018) Large scale modelling of the effect of preshaking on the liquefaction resistance of clean sand deposits. J Geotech Geoenviron Eng—ASCE
13. El-Sekelly W, Dobry R, Abdoun T, Steidl JH (2017) Two case histories demonstrating the effect of past earthquakes on liquefaction resistance of silty sand. J Geotech Geoenviron Eng 143(6):04017009
14. Padmanabhan G, Maheshwari BK (2021) Case studies on preshaking and reliquefaction potential for different earthquakes in japan. In: Sitharam TG et al. (ed) Soil dynamics. Select proceedings of 7th ICRAGEE 2020, Lecture notes in civil engineering, vol 119

15. Bureau of Indian Standards (1985) IS: 2720 Methods of test for soils—grain size analysis, Part IV
16. Maheshwari BK, Singh HP, Saran S (2012) Effects of reinforcement on liquefaction resistance of Solani sand. J Geotech Geoenviron Eng 138(7):831–840
17. Bureau of Indian Standards (2006) IS: 2720 Methods of test for soils–determination of density index for cohesionless soils, Part XIV
18. Dobry R, Thevanayagam S, El-Sekelly W, Abdoun T, Huang Q (2019) Large-scale modeling of preshaking effect on liquefaction resistance, shear wave velocity, and CPT tip resistance of clean sand. J Geotech Geoenviron Eng 145(10):04019065
19. El-Sekelly W, Dobry R, Abdoun T, Steidl JH (2016) Centrifuge modelling of the effect of preshaking on the liquefaction resistance of silty sand deposits. J Geotech Geoenviron Eng 142(6):04016012
20. Ha IS, Olson SM, Seo MW, Kim MM (2011) Evaluation of reliquefaction resistance using shaking table tests. Soil Dyn Earthq Eng 31(4):682–691

Chapter 46
Failure Mechanisms of Pile-Supported Port Building in Liquefiable Sloping Ground During Earthquake

Monirul Mallick, Kalyan Kumar Mandal, and Ramendu Bikas Sahu

Introduction

The seismic analysis of pile foundations is a complicated soil-structure interaction (SSI) problem. The problem of soil-pile interaction gets further intricated if the piles are founded in liquefiable soils. Soil liquefaction is a great concern for design of pile foundation in high seismicity zone. A significant number of failure and/or damages of pile foundations and pile-supported structures in liquefiable soils are observed during past major earthquakes such as San Francisco (1906), Niigata (1964), Northridge (1994), Kobe (1995), Chi-Chi (1999), Bhuj (2001), and Sumatra (2004) in spite of designing with the latest standard code of practices employing high margin of factor of safety (FOS). So, it is challenging job for geotechnical earthquake engineers to ensure safe and economical design of pile foundation and pile-supported high-rise structures in liquefiable soil.

At present, pile foundation in liquefiable soil is designed based on bending failure mechanism due to inertial loading from superstructure and/or kinematic loading owing to lateral spreading [1–9]. Japanese standard code JRA [10] advises design engineers to design the piles based on bending failure mechanism assuming that pile experienced passive earth pressure by non-liquefiable layer while the pressure on a pile due to liquefiable soil layer is equal to 30% of total overburden pressure. The code also advises design engineers to check the bending failure of piles for kinematic forces and inertial forces individually. Eurocode [11], NEHRP code [12], and Indian standard seismic code [13] also advises to design pile based on bending strength. Eurocode emphasises to consider extra forces on the pile foundation in liquefiable soil due to lateral spreading, specifically in existence of non-liquefiable soil layer

M. Mallick (✉) · K. K. Mandal · R. B. Sahu
Department of Civil Engineering, Jadavpur University, Kolkata 700032, India
e-mail: monirulmallick64@gmail.com

K. Muthukkumaran et al. (eds.), *Soil Dynamics, Earthquake and Computational Geotechnical Engineering*, Lecture Notes in Civil Engineering 300,
https://doi.org/10.1007/978-981-19-6998-0_46

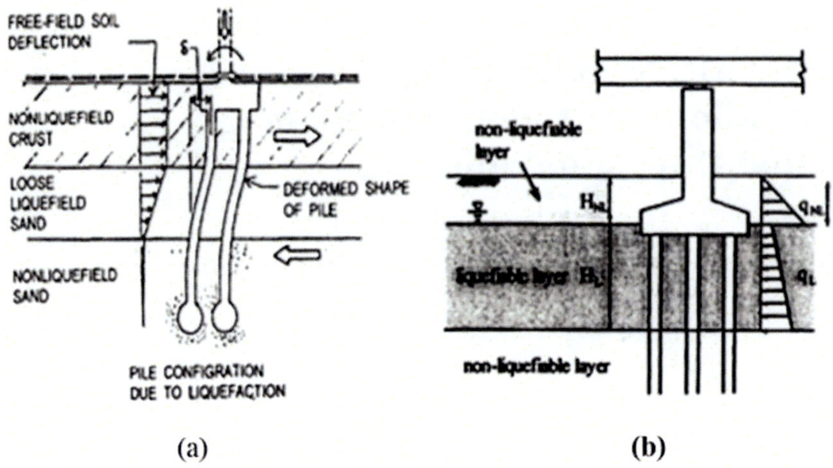

Fig. 46.1 Idealization bending failure of pile by **a** Finn and Thavaraj [8], and **b** JRA code [10]

overlying liquefiable soil layer. Figure 46.1 shows the idealization of bending failure mechanism.

However, recent research suggested that pile foundations in liquefiable soil may be failed due to structural failures (shear, bending, and buckling), geotechnical failure (excessive settlement), or combination thereof depending upon the thickness and position of liquefiable crust [14]. Bending-buckling interaction is also reported as failure mechanism of piles in liquefiable soil by several authors [15–18]. During earthquake-induced soil liquefaction, pile loses its lateral support and acts as an unsupported structural column. This axially loaded column is prone to buckling failure in the direction of least bending stiffness even without lateral spreading loading. So, Euler's buckling criteria must be satisfied for analysis of piles in liquefiable soil. All these failures may happen due to static load. Pile foundations may also be damaged because of dynamic amplification during the earthquake. The frequency of the pile-supported structures significantly altered during earthquake-induced soil liquefaction. If the altered frequency matches with the driving frequency of the earthquake, pile may be failed because of dynamic amplification due to resonance effect. Various possible failure mechanisms of pile-supported structures in liquefiable sloping ground are shown in Fig. 46.2 [19].

Objective of This Study

Bending and buckling are two distinct approaches of structural design. Buckling criteria will not be fulfilled automatically by designing the pile against bending criteria. Bending is stable failure mode which depends on bending strength of material (plastic moment capacity, M_p). Contrariwise, buckling is unstable failure mode which

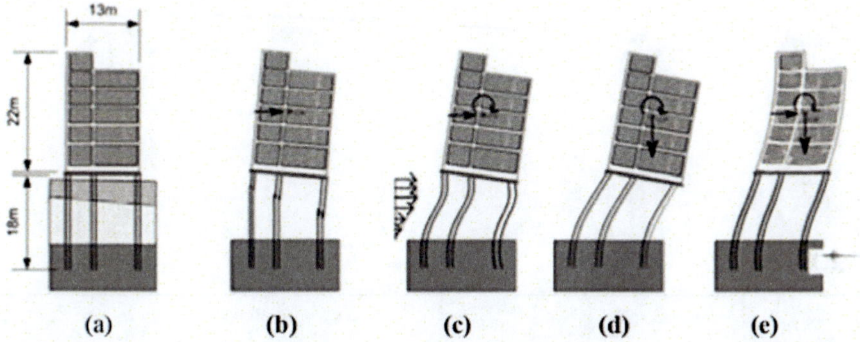

Fig. 46.2 Various failure mechanisms of pile, **a** Kandla Port building with foundation, **b** shear failure, **c** bending failure, **d** buckling failure, and **e** dynamic amplification [19]

depends only on geometric characteristics of the member. There should be a least diameter of pile for safe designing against buckling criteria depending upon the thickness of liquefiable layer. In the present study, a simplified method based on beam on nonlinear Winkler foundation (BNWF) model has been carried out using open-source finite element-based code, OpenSees, to investigate the probable failure mechanism of Kandla Port building during 2001 Bhuj earthquake.

Case Study of Kandla Port Building

The Kandla Port building, 22.0 m high pile-supported six-storeyed RCC frame structure resting on liquefiable sloping ground, was tilted towards seaside after Bhuj earthquake as shown in Fig. 3a [14].

Kandla Port is located in the Kandla Creek and is 50 km from the epicentre of the earthquake. The Kandla Port tower, a 22.0 m high six-storeyed building supported on combined pile-raft foundation, is located proximate to waterfront. The details of foundation and superstructure of Kandla Port building are shown in Fig. 46.3 and tabulated in Tables 46.1 and 46.2 [14, 20].

The typical soil profile of Kandla Port site at Gujarat state of India is consisting of recent unconsolidated layer of clay, silt, and sand. The ground slope is about 5° towards seaside. The groundwater table (GWT) is located at 1.5 m below ground surface [21].

The input motion considered is the 2001 Bhuj earthquake. The maximum bedrock level acceleration (MBRA) and mean period were 0.106 g and 0.598 s, respectively.

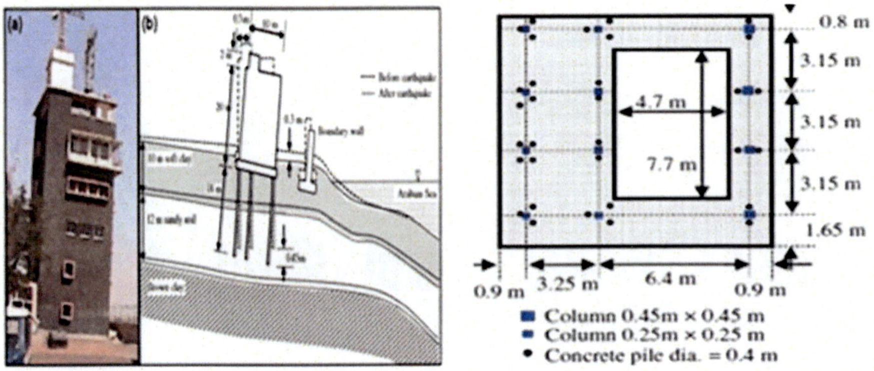

(a) (b)

Fig. 46.3 **a** Tilting and pre- and post-earthquake configuration, and **b** foundation layout plan of Kandla Port building [14]

Table 46.1 Pile properties considered in the present numerical study [14, 20]

Properties	Values
Characteristic compressive strength of concrete (f_{ck})	30 MPa
Young's modulus (E)	30 GPa
Poisson's ratio (μ)	0.30
Unit weight (γ)	25 kN/m^3
Length of pile (L)	18.0 m
Diameter of pile (d)	0.4 m

Table 46.2 Structural details of Kandla Port building [14]

Description	Values
Number of storeys	6 nos
Building height (h)	22.0 m from GL
Building type	RCC frame structure
Building dimensions at plinth level ($l \times b$)	9.6 m × 9.8 m
Foundation type	Combined piled raft
Foundation-raft dimensions	11.45 m × 11.9 m × 0.5 m
Numbers of column	12 nos
Column size	0.45 m × 0.45 m (4 nos.) and 0.25 m × 0.25 m (8 nos.)
Numbers of pile	32 nos
Total weight of building	10,749 kN
Load on each pile	336 kN

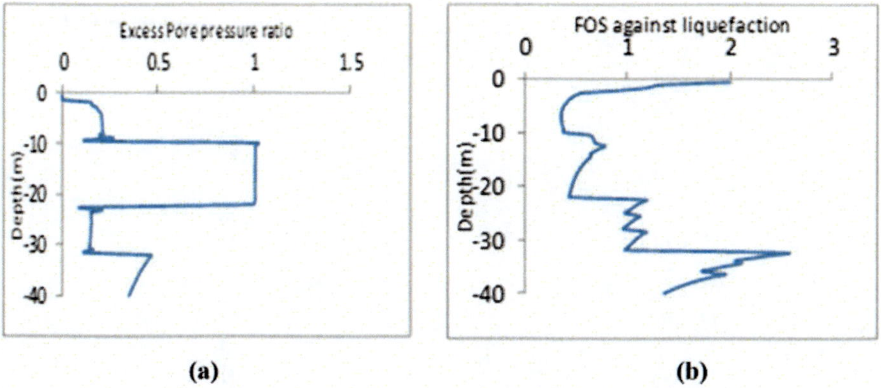

Fig. 46.4 Profile of **a** excess pore pressure ratio, and **b** factor of safety (FOS) against liquefaction potential

Ground Response Analysis (GRA) and Liquefaction Assessment

1D effective stress-based free-field nonlinear ground response analysis of the port site has been performed to assess the liquefaction susceptibility using 2001 Bhuj earthquake motion at bedrock level through computer program Cyclic1D [21]. The profile of excess pore pressure (EPP) ratio and factor of safety against liquefaction (FOS) using simplified deterministic approach of Kandla Port site are shown in Fig. 4a and b, respectively [21].

The EPP ratio is almost 1 for soil layer of 10.0–22.0 m depth under the Bhuj earthquake. So, the intermediate fine to coarse sandy strata is prone to liquefaction. Also, the FOS is less than 1 for top clay (1.5–10.0 m) and intermediate sandy layer (10.0–22.0 m) under Bhuj motion. So, the upper clay stratum experiences ground deformation and cracking due to cyclic failure. Moreover, intermediate sand layer (10.0–22.0 m) is susceptible to liquefaction. Hence, stiffness degradation has been considered for soil layer between 10.0 and 22.0 m for evaluating the pile response.

Factor of Safety Against Kinematic Bending Failure

The seismic behaviour of piles of Kandla Port building under liquefiable condition has been analysed using pseudo-static approach based on BNWF model. The open-source finite element-based code, OpenSees [22], is used to perform the kinematic bending analysis. The bottom 8.0 m of the pile was embedded in liquefiable soil, and top 10.0 m was passing through non-liquefied soft clay layer. The displacement-based beam element and nonlinear spring element are used for simulating pile and soil, respectively. The PGD profile obtained from GRA is employed to the soil end of

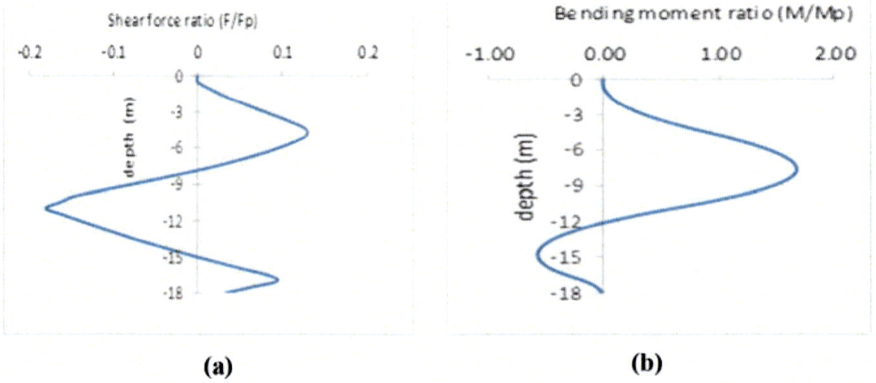

Fig. 46.5 Variation of **a** shear force ratio, and **b** bending moment ratio with depth

the p-y springs as kinematic loading to simulate lateral spreading. Detail modelling and analysis can be found in [21] where it was showed that pile foundation in Kandla port site was prone to settlement failure owing to soil liquefaction.

The shear force and bending moment obtained from pseudo-static analysis are normalized with plastic shear (F_p) and moment (M_p) capacity to obtain shear force and bending moment ratio. The plastic shear capacity and moment capacity (F_p) of M-30 grade RCC pile of diameter 0.4 m are calculated [14, 20] to be 460 kN and 143 kN-m, respectively. The profile of shear force and bending moment ratio obtained from the analysis is shown in Fig. 5a and b, respectively. It is seen that maximum shear force ratio (0.18) is far less than 1 and bending moment ratio (1.68) is more than 1.

So, shear force demand is less than shear capacity of pile and pile is safe against shear failure. However, the pile is unsafe against kinematic bending failure and plastic hinge is expected to form near the boundary between liquefiable and non-liquefiable layer.

Factor of Safety Against Buckling Failure

The maximum axial load on pile during earthquake will be increased due to additional dynamic axial load for inertial action of the superstructure. If P_s is the static axial load acting on each pile head from superstructure and α is the dynamic axial load factor, then the maximum axial load (P_d) can be estimated using Eq. (46.1) as proposed by Bhattacharya [23]. α may vary from 0.1 to 0.5 depending up on the type of superstructure, supporting soil condition, and characteristics of input earthquake motion.

$$P_d = P_s(1 + \alpha) \tag{46.1}$$

The static axial load acting on each pile of Kandla Port building from superstructure (P_s) is 336 kN.

At the beginning of the earthquake before liquefaction

The inertial forces generated due to vibration of the superstructure transmitted to the pile foundations through the pile cap. The inertial force transmitted as lateral force and overturning moment to the pile through pile cap. The overturning moment is further transmitted as asymmetrical axial forces through pile cap. During soil liquefaction, the stiffness of soil-pile system changes significantly and period of the superstructure also changes. The additional axial load on pile foundations due to change of period of structure is calculated before and after soil liquefaction.

Before soil liquefaction, the fundamental period of RCC building is calculated as per IS 1893 [13] as

$$T_b = (0.09 * h)/b^{0.5} \tag{46.2}$$

where h = height of building = 22.0 m and b = base dimension of the building at plinth level = 9.8 m for this case. The period calculated using Eq. (46.2) is 0.63 s. The spectral acceleration coefficient (S_a/g) for time period between 0.10 and 0.67 s is 2.50 for soft soil site Type-III as per IS: 1893.

So, the base shear of Kandla Port building can be estimated as

$$V_b = A_h \times W \tag{46.3}$$

where A_h = seismic coefficient and W = total seismic weight of the building.

$$A_h = Z/2 * I/R * (S_a/g) \tag{46.4}$$

The values of Z = zone factor = 0.36 for zone V, I = importance factor = 1.5, and R = response reduction factor = 3.0 for Kandla Port building as per IS: 1893.

The value of A_h is estimated to be 0.225 for this case, and total seismic weight is calculated as

$$W = \text{DL} + 0.25\,\text{LL} \tag{46.5}$$

where DL = total dead load of the building = 9688 kN and LL = total live load of the building = 983 kN [14].

Hence, total seismic weight of Kandla Port building obtained using Eq. (46.5) is 9933.75 kN.

Therefore, the base shear of Kandla Port building using Eq. (46.3) is evaluated to be 2235 kN.

The pile foundation is to withstand this base shear, and lateral load on each pile is 70 kN. According to Davisson and Robinson [24], laterally load pile foundations may be assumed to be fixed at depth of 1.8 T below ground level, where T is the relative stiffness factor which may be calculated using the following equation:

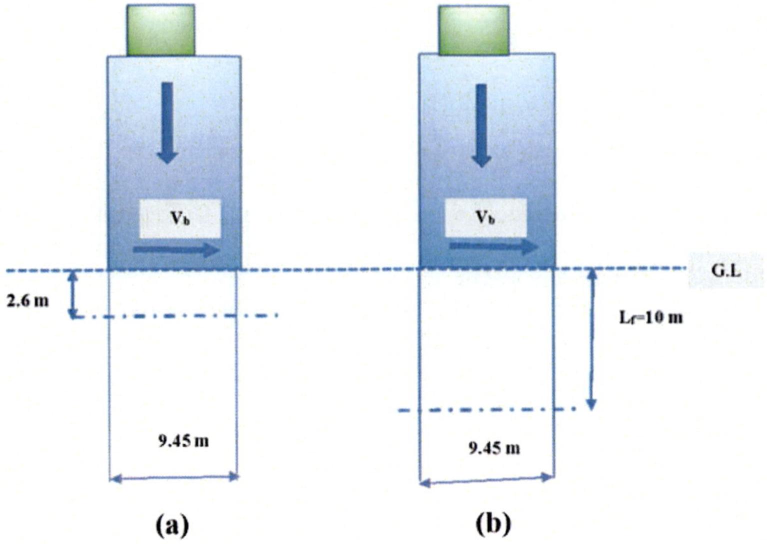

Fig. 46.6 Depth of fixity of pile **a** before, and **b** after full liquefaction

$$T = (\mathrm{EI}/n_h)^{0.2} \tag{46.6}$$

where EI = flexural rigidity of pile and n_h = coefficient of modulus of subgrade reaction of soil.

The value of n_h can be taken as 6000 kN/m^3 for soft clay from IS 2911 [25].

The relative stiffness factor (T) can now be evaluated using Eq. (46.6) as 1.44 m. Thus, the depth of fixity may be assumed at 2.60 m below ground level (Fig. 6a). Now, the rocking moment in pile foundation may be evaluated as

$$M_R = 2235 \times 2.60 = 5811\,\mathrm{kN\,m} \tag{46.7}$$

The extra axial compressive load on each pile may be estimated using following formula:

$$P_{\mathrm{ext}} = (M_R * r_i)/\sum r_i^2 = 74.32\,\mathrm{kN} \tag{46.8}$$

using the values of r_i = 11.9/2 = 5.95 m and $\Sigma r_i{}^2$ = 402.68 m from Fig. 3b of foundation plan of Kandla Port building, where r_i = perpendicular distance from centre line of footing to the particular pile. So, the maximum increase of axial load at corner pile of one side of the pile cap is close to 22.12%.

During liquefaction stage

The period of vibration of Kandla Port building is calculated at liquefied condition. The top 10 m of pile is passing through non-liquefiable crust, and bottom 8 m is passing through in liquefiable soil. Hence, depth of fixity of pile is assumed at the interface of liquefiable and non-liquefiable crust [19].

So, the pile will be fixed at 10.0 m below ground level (Fig. 6b). The stiffness of Kandla Port building during liquefied condition may be estimated using $L_f = 10.0$ m by following equation:

$$k = 32 \times 12\,\mathrm{EI}/L_f^3 = 14469.12\,\mathrm{kN/m} \tag{46.9}$$

Here, the total stiffness of soil-pile system is supposed to be mainly contributed by stiffness of pile foundations as stiffness of liquefiable soil is negligible when compared to RCC piles. The time period of Kandla Port building at liquefied condition is now calculated using following equation:

$$T_{b,\mathrm{liq}} = 2\pi(M/k)^{0.5} = 1.73\,\mathrm{s} \tag{46.10}$$

where M = mass of Kandla Port building = 10,749 kN [14].

So, the period of vibration of the building is increased by 2.75 (1.73/0.63) times and the spectral acceleration coefficient (S_a/g) for time period of 1.73 s would be 1.67/1.73 = 0.965 according to IS:1893.

The value of A_h will be now 0.087 for $S_a/g = 0.965$ using Eq. (46.4). Therefore, the base shear would be 863 kN using Eq. (46.3). The maximum increase of axial load on corner pile is calculated using Eq. (46.8) as

$$P_{\mathrm{ext}} = (863 \times 10.0 \times 5.15)/402.68 = 110.37\,\mathrm{kN} \tag{46.11}$$

The increase of axial load is around 32.85%. Thus, the axial load on pile may be increased up to 33% due to soil liquefaction. Therefore, the dynamic axial load factor (α) may be considered as 0.35 for calculation of the maximum axial load (P_d) during dynamic condition and the maximum axial load (P_d) estimated using Eq. (46.1) is 454 kN.

During liquefied condition, the fixity point of the pile is assumed to be effectively held in position but not restrained against rotation and top is restrained against rotation but not held in position which is feasible as top non-liquefiable layer may move laterally with bottom liquefiable layer in sloping ground. So, effective length (L_e) of pile as per Table 28 of IS: 456-2000 [20] = 2 × 10 = 20 m.

The Euler's buckling load (P_{cr}) for 0.4 m diameter pile is calculated using following expression

$$P_{\mathrm{cr}} = \left(\pi^2 \times \mathrm{EI}\right)/\left(L_e^2\right) \tag{46.12}$$

The value of P_{cr} obtained using Eq. (46.12) is 930 kN. Therefore, the factor of safety (FOS) against buckling failure = 930/454 = 2.05.

Usually, a FOS 3 is preferred by structural engineers for linear elastic buckling taking into account eccentricities, decaying of elastic stiffness because of plastic yielding, and inevitable imperfections of pile section. In this case, FOS is less than

3. So, the pile is also prone to buckling failure during soil liquefaction which is not addressed in standard code of practice.

Failure Due to Dynamic Amplification

During soil liquefaction, the time period of Kandla Port building increases to 1.73 s which is far away from the mean period of 0.598 s of 2001 Bhuj earthquake motion. So, the possibility of pile failure due to dynamic amplification owing to resonance effect is not expected in this case.

Conclusion

In the present study, a simplified methodology has been formulated using a case study of Kandla Port building (Gujarat) supported on pile foundation to demonstrate the possibility of buckling failure and dynamic amplification due to resonance effect for piles in liquefiable soil. The current design codes recommend to design the piles in liquefiable sloping ground based on bending failure mechanism due to inertial or kinematic loading neglecting buckling instability and the dynamic effects of axial load. However, primary requirement of axially loaded pile in liquefiable soil is to check the safety against Euler's buckling criteria. Bending and buckling are two distinct approaches of design. Buckling criteria will not be fulfilled automatically by designing the pile against bending criteria. Thus, there is a need to re-examine the safety of existing pile-supported structures designed using present codal provisions based on bending failure mechanism. The pile in liquefiable soil may be failed due to structural failures (shear, bending, and buckling), geotechnical failure (excessive settlement), dynamic amplification owing to resonance, or combination thereof depending upon the thickness and position of liquefiable crust. However, in reality failure of pile may be happened due to nonlinear combination of the above failure mechanisms.

References

1. Finn WDL, Fujita N (2002) Piles in liquefiable soils: seismic analysis and design issues. Soil Dyn Earthq Eng 22(9–12):731–742
2. Berrill JB, Christensen SA, Keenan RP, Okada W, Pettinga JR (2001) Case studies of lateral spreading forces on a piled foundation. Geotechnique 51(6):501–517
3. Dobry R, Abdoun T (2001) Recent studies on seismic centrifuge modelling of liquefaction and its effect on deep foundation. In: Proceedings of 4th international conference on recent advances in geotechnical earthquake engineering and soil dynamics and symposium, San Diego, California, 26–31 Mar 2001

4. Hamada M (2000) Performances of foundations against liquefaction-induced permanent ground displacements. In: Proceedings of the 12th world conference on earthquake engineering, Auckland, New Zealand
5. Hamada M, O'Rourke TD (1992) Case studies of liquefaction and lifeline performance during past earthquakes. In: Japanese Case Studies, Technical Report NCEER—92-0001, vol 1, State University of New York at Buffalo, Buffalo, USA
6. Ishihara K (1993) Liquefaction and flow failure during earthquakes. Geotechnique 43(3):351–415
7. Ishihara K (1997) Geotechnical aspects of the 1995 Kobe earthquake. In: Proceedings of international conference on soil mechanics and foundation engineering, Hamburg, pp 2047–2073
8. Finn WDL, Thavaraj T (2001) Deep foundations in liquefiable soils: case histories, centrifuge tests and methods of analysis. In: Proceedings of the fourth international conference on recent advances in geotechnical engineering and soil dynamics, San Diego, USA, Paper No. SOAP-1
9. Tokimatsu K, Asaka Y (1998) Effects of liquefaction-induced ground displacements on pile performance in the 1995 Hyogeken-Nambu earthquake. Soils Found, pp 163–177
10. JRA-Part V (1996) Design specifications of highway bridge: seismic design. Japan Road Association, Tokyo, Japan
11. Eurocode 8-Part V (1998) Design provisions for earthquake resistance of structures foundations, retaining structures and geotechnical aspects. European Committee for Standardization, Brussels
12. NEHRP: National Earthquake Hazards Reduction Program (2000) Commentary for Federal Emergency Management Agency (FEMA) seismic regulations for new buildings and other structures. USA
13. IS 1893-Part 1 (2002) Criteria for earthquake resistant design of structure. Bureau of Indian Standards, New Delhi, India
14. Dash SR, Govindaraju L, Bhattacharya S (2009) A case study of damages of the Kandla Port and Customs Office tower supported on a mat–pile foundation in liquefied soils under the 2001 Bhuj earthquake. Soil Dyn Earthq Eng 29(2):333–346
15. Bhattacharya S, Madabhushi SPG, Bolton MD (2004) An alternative mechanism of pile failure in liquefiable deposits during earthquakes. Géotechnique 54(3):203–213
16. Knappett JA, Madabhushi SPG (2006) Modelling of liquefaction-induced instability in pile groups in seismic performance and simulation of pile foundations in liquefied and laterally spreading ground. Boulanger RW, Tokimatsu K (eds). Geotechnical Special Publication, American Society of Civil Engineers, Reston, USA, p 145
17. Dash SR, Bhattacharya S, Blakeborough A (2010) Bending-buckling interaction as a failure mechanism of piles in liquefiable soils. Soil Dyn Earthq Eng 30(1–2):32–39
18. Knappett JA, Madabhushi SPG (2012) Effects of axial load and slope arrangement on pile group response in laterally spreading soils. J Geotech Geoenviron Eng ASCE 138:799–809
19. Bhattacharya S, Dash SR, Adhikari S (2008) On the mechanics of failure of pile-supported structures in liquefiable deposits during earthquakes. Curr Sci 94(5):605–611
20. IS 456 (2000) Indian standard on plain and reinforced concrete-code of practice. Bureau of Indian Standards, New Delhi
21. Mallick M, Mandal KK, Sahu RB (2020) A case study of liquefaction-induced damage to a port building supported on pile foundation. In: Proceedings of Indian geotechnical conference, Andhra University, Visakhapatnam, 17–19 Dec 2020, pp 39–50
22. Mazzoni S, McKenna F, Scott MH, Fenves GL et al. (2006) OpenSees command language manual. In: Pacific earthquake engineering research center, University of California, Berkeley
23. Bhattacharya S (2006) Safety assessment of existing piled foundations in liquefiable soils against buckling instability. ISET J Earthq Technol 43(4):133–147

24. Davisson MT, Robinson KE (1965) Bending and buckling of partially embedded piles. In: Proceedings of the sixth international conference on soil mechanics and foundation engineering, vol 2, Montreal, Canada, pp 243–246
25. IS 2911-Part1 Section 4 (1984) Indian standard code of practice for design and construction of pile foundations. Bureau of Indian Standards, New Delhi

Chapter 47
Development of Shear Wave Velocity Profiles Using Active MASW on Kattangal Region Soil Deposits in Calicut of Kerala

P. G. Abhishek and K. Rangaswamy

Introduction

Shear wave velocity mapping has been employed widely as a parameter for seismic hazard and microzonation studies. For an appropriate characterization of a region, it is necessary to have seismic site classification and ground response studies. In most seismic surface studies, a compressional wave source is used where 66% of the generated seismic energy will be imparted as a Rayleigh wave [10]. For a vertical profile of the near-surface soil, the velocity variation for each frequency component of a surface wave velocity has a different velocity of propagation called phase velocity at each distinctive frequency component. This phenomenon can be referred to as dispersion. The dispersive properties of the waves are used to determine the near-surface elastic behaviour of the soil [12].

The development of shear wave velocity (Vs) profile is attained by the analysis of the surface waves, and the Rayleigh wave in the fundamental mode is the generally accepted method utilizing the dispersive properties of the surficial wave [11]. The near-surface velocity (Vs) offers the relevant particulars regarding the statics during the data processing of body waves [6]. The phase velocity of Rayleigh waves is not only a function of frequency. Still, it is also related to several groups the properties of the earth, particularly the speed of the shear waves of the individual soil layers. The shear wave speed profile for the test site has been obtained by reversing the dispersive phase velocity of recorded Rayleigh waves [14, 15].

The shear wave velocity profile (SWV) can then be employed to assess the stiffness of the highest soil layers. In addition, the speed of the shear wave is a crucial parameter in the design of earthquakes when assessing the soil liquefaction and reinforcement potential and when setting the site-specific seismic design loading [13]. Rayleigh

P. G. Abhishek (✉) · K. Rangaswamy
Department of Civil Engineering, National Institute of Technology, Calicut 673601, India
e-mail: abhishekpga123@gmail.com

K. Muthukkumaran et al. (eds.), *Soil Dynamics, Earthquake and Computational Geotechnical Engineering*, Lecture Notes in Civil Engineering 300,
https://doi.org/10.1007/978-981-19-6998-0_47

wave dispersion curves can be constructed from the calculated surface wave data and utilized to establish the SWV profile of the topmost soil layers by using several surface wave methodologies, including surface wave spectral analysis (SASW) and multi-channel surface wave analysis (MASW). These methods are to be effectively carried out in the natural ground and even in the human-made soil fills [2].

MASW is a practical geophysical approach widely accepted due to its non-intrusive nature, and the abundance in data gaining involves easy procedure taking a short time and no boreholes [8, 9]. MASW is considered a perfect method for predicting the SWV, which is used to determine parameters of engineering purposes. These can be used for estimating the 1D and 2D seismic shear wave velocities of different sections [1].

The MASW interprets the lithology, quality of rock, and geotechnical specifications of the area under consideration, which can be used for civil engineering purposes [15]. The shear wave velocity (Vs) is the essential metric; its values are proportional to the shear modulus, density, and damping ratio. As the Vs value lowers, the shear modulus decreases and the damping ratio increases [5].

National Earthquake Hazard Reduction Program (NEHRP)

Many approaches have been developed for classifying soils and rocks based on their site-dependent amplification features [3]. Average SWV to a depth of one-fourth the wavelength of the governing frequency of interest, for example, can be used to describe site amplification [4]. However, since it is relatively difficult to implement, this technique has not been commonly used.

According to NEHRP guidelines for seismic design of the structure, the top 30 m of the soil profiles is used to classify the site under seismic conditions, as shown in Table 47.1. It shows the manual lists position classifications in grades A, B, C, D, E, and F and average shear wave velocities (Vs_{30}) for each 30 m topsoil depth. E and F are considered readily liquefiable in these groups, so proper evaluation and thorough analysis are needed [7].

The presence of loose unconsolidated sediment over the bedrock modifies seismic waves propagating close to the ground surface, causing changes in the properties of soil movement on the surface. In turn, the intensity of ground tremors near the surface is determined mainly by the thickness and rigidity of the soil layers. So, this becomes the need to classify the soil based on the velocity of the surface waves.

The paper discusses the need to determine average shear wave velocities by creating the vertical shear wave velocity profiles. It also calculates the Vs_{30} for the spatially distributed sample sites across the study area and thus classified according to the NEHRP provisions. The area has variations in rainfall, so the need for the study by varied groundwater depth plays a crucial role. Since the shear wave velocity provides a foundation for ground motion response, it helps to verify possible liquefaction effects and an idea of natural frequency amplifications during an earthquake event.

Table 47.1 Soil profile type classification for seismic loading (NEHRP guidelines)

Soil type	General description	Average shear wave velocity for Top 30 m (m/s)
A	Hard rock	>1500
B	Rock	760–1500
C	Very dense soil and soft rock	360–760
D	Stiff soil $15 \le N \le 50$ or $50\ kPa \le Su \le 100\ kPa$	180–360
E	Soil or any profile with more than 3 m of soft clay defined as soil with, $PI > 20$, $w \ge 40\%$ and $Su < 25\ kPa$	$\le$180
F	Soil requiring site-specific evaluation	

Study Area

Kattangal is a small town near the National Institute of Technology Calicut, located in the Kozhikode district in Kerala, India. Kattangal has coordinates of latitudes 11° 18′ 0″ N and longitude 75° 58′ 30″ E in the coastal community of Kerala. It has situated about 22 km from Kozhikode town. The local administration of the village is the Chathamangalam grama panchayath, and this village is an essential educational hub in Kerala. Figure 47.1 shows the location of the study area.

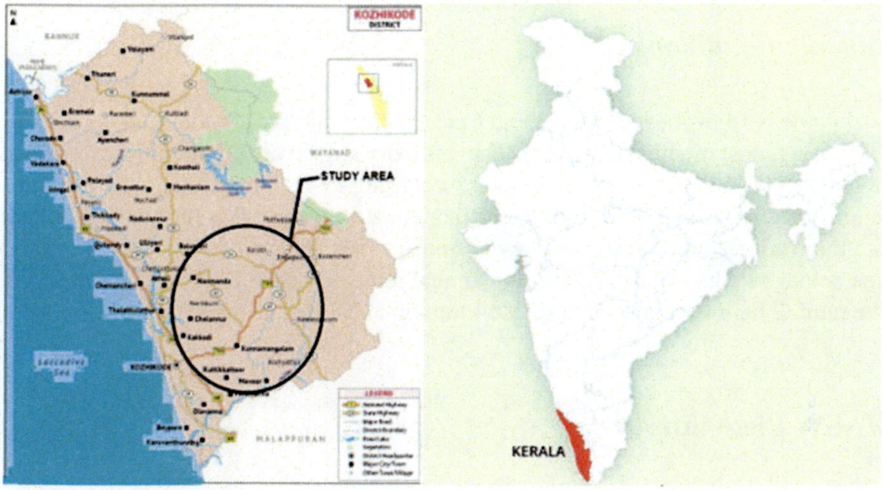

Fig. 47.1 Location of study area

Geology

The geology of the Calicut area can be divided into three belts: (i) a longitudinal NW–SE gneissic belt that runs from north to south, (ii) a narrow coastal belt, and (iii) a charnockite belt that spans the northeast and south, with clusters of charnockite within the gneissic landscape. These areas may have granite, quartz, quartzite, granulite, well-rounded pebbles, and were of Pleistocene origin. Also, the site is bounded by the quaternary deposits consisting of marine and fluvial origin.

Geomorphology

The Kozhikode district has been divided into physiographic areas such as the coastal plains (i) lowlands (<7.6 m above sea level), (ii) medium land (7.6–76 m above sea level) and (iii) highlands—hilly terrain (above 76 m above sea level) from west to East. The land is gently sloping with a slender shape, 5–10 km widespread with marine, fluvio-marine and fluvio-marine origin deposits. The lowland stretches from South Kadalundi to North Mahé as a thin area of land on the coast. The middle zone is between 7.6 and 76 m above sea level and is classified as low hilly terrain and moderately undulating relief. The landforms consisted of rice fields and coconut palms, cash crops such as rubber were grown.

Climatic Conditions

Kozhikode is 14 m above sea level and has a tropical climate. The rainfall is remarkable during most months of the year. There is a dry season of short duration, having a low impact on the general environment. According to the Köppen Climate Classification System, Zone A represents tropical or equatorial zone. It is further subdivided into three zones, viz. Af, Am and Aw. Zone Am has a short dry season. The study area belongs to this zone. The average annual temperature in Kozhikode is 26.2° C. The rainfall here is approximately 2864 mm or 112.8 inches per year.

MASW Field Survey

For the present investigation, MASW tests have been performed at the spatially selected locations in the Kattangal region. There were 15 sample sites considered for the study. The field setting of the instruments is shown in Fig. 47.2. The experiments were carried out using a geode seismic detector with a seismograph with 24 channels. The wavefields created by an actively generated source of 8 kg sledgehammer on steel plate were obtained using a vertical receiver with a natural frequency of 4.5 Hz (24 numbers). The apparatus used for carrying out the survey is shown in Fig. 47.3.

Although the accession of the entire data beneath the ground is an easy task, the arrangements considered for the field need to be optimized based on the requirement. The most important part of the MASW test's field arrangement is the spacing between the geophone and the distance between the initial receiver and source.

The present study used 24 geophones placed in the field as a series and a linear array at an interval of 1.5 m. The geophones are interconnected using a cable and connected with the geode seismograph, which records as shown in Fig. 47.4. The

Fig. 47.2 Experimental setup used in the field survey

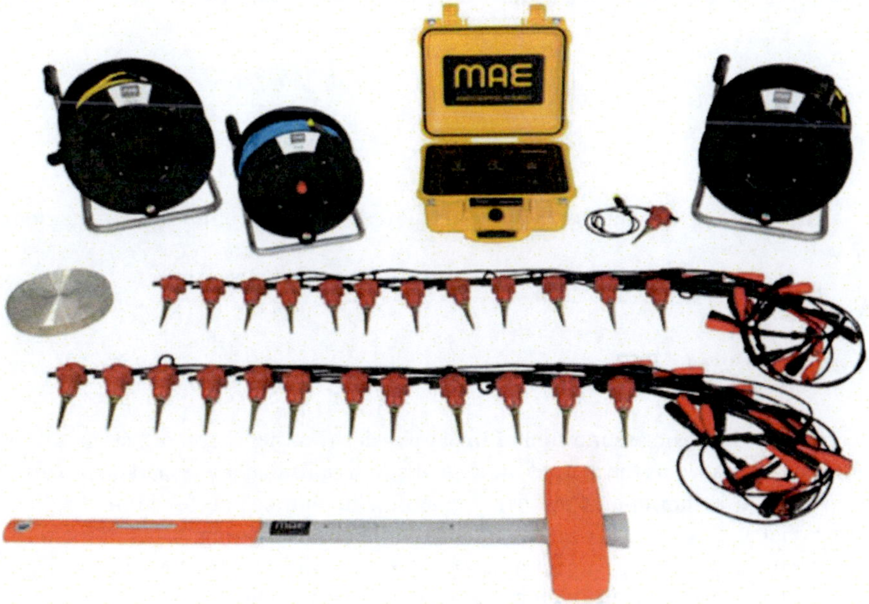

Fig. 47.3 Apparatus used to conduct MASW field survey

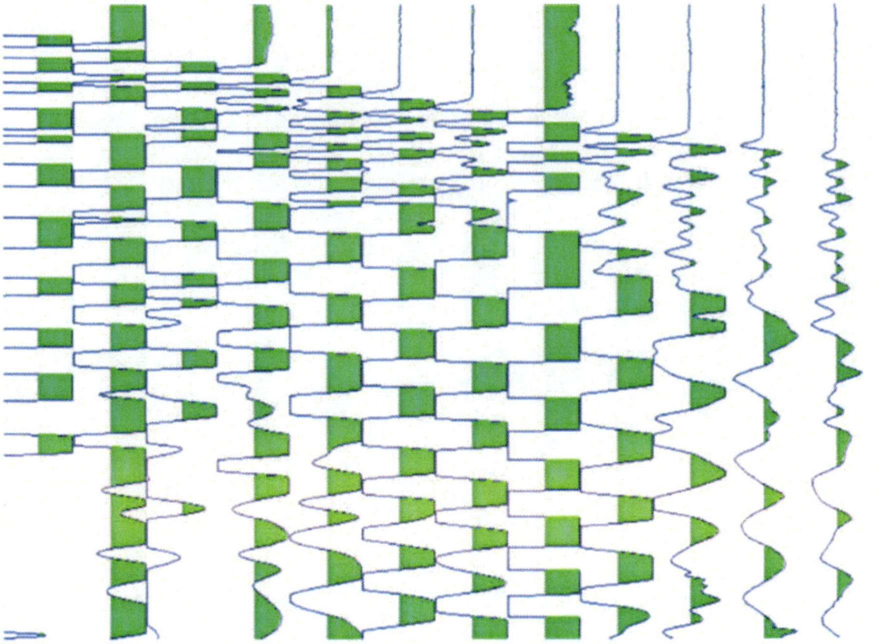

Fig. 47.4 Raw data obtained from seismograph

fieldwork is carried out in 15 spatially distributed sites codes as NC-1, NC-2 up to NC-15.

Analysing Data

The analysis of the raw data to obtain the required velocity profiles involves various steps.

Signal Analysis

According to the basic assumption of linear physics (Fourier theorem), signals have often been portrayed as the sum of individualistic signals that are remarked as signal harmonics. These harmonics are trigonometric sine and cosine functions for one-dimensional analysis and act independently while not interacting.

Harmonic analysis (FFT spectral analysis) is the fundamental method for signal spectral characterization. Rayleigh wave analysis using the MASW technique is conducted with the spectral process of the signal within the transformed domain.

Wherever signal for Rayleigh waves can be determined from different styles of signals, additional perceptive however the Rayleigh waves move with propagating at a speed that's propagating may be a function of frequency. The dispersion continuum describes the relationship between speed and frequency. The experiment dispersion curve is the best-known in the f-k domain, and it represents the spectrum's maximum amplitudes in this domain.

Data Acquisition

Multi-channel surface wave analysis is a non-destructive geophysical method that creates SWV profiles (Vs versus depth) by analysing the shear wave from multi-channel recordings. The main task in this data acquisition is to achieve the shear wave speed of the subsurface below the surface in such a way that it is at least 30 m below it. Therefore, the recording frequency should be lower so that the phase velocity has to be achieved at longer wavelengths. The corresponding tracks from the seismograph will be imported, as shown in Fig. 47.5.

The acquisition setup is interconnected to the seismograph and coupled to a computer running the SYSMATRACK software that records the waves and saves

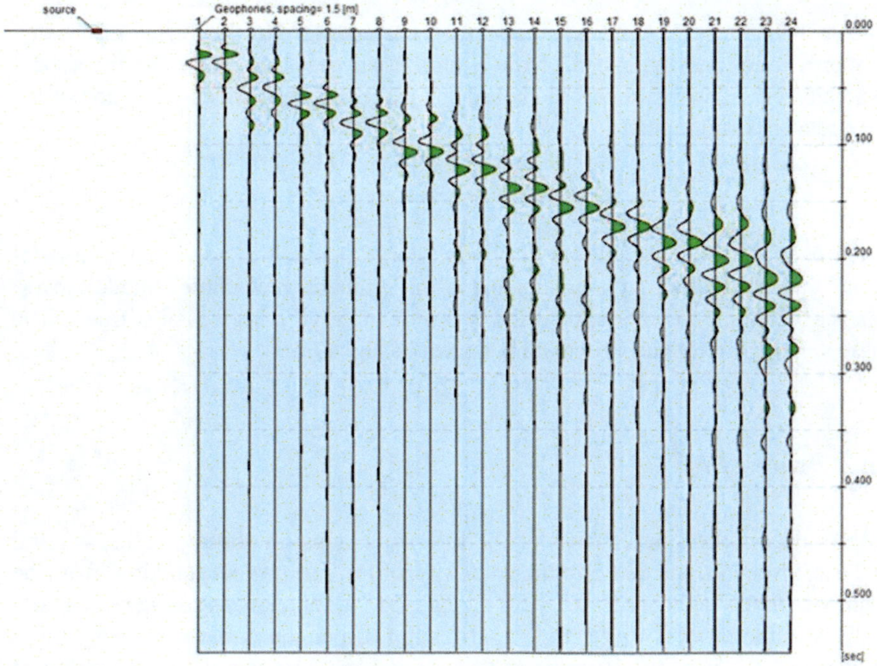

Fig. 47.5 Track records imported from geophone channels

Table 47.2 Data acquisition parameters

Parameters	Range/value
No. of geophones	24
Interspacing of geophones	1.5 m
Rate of samplings	0.5 s
Array length	36 m
Receiver geophones	4.5 Hz (vertical)
Impact source	8 kg sledge hammer
Offset distance from source	5 m

the data in SEG.2 format in the file selected by the user. The acquisition parameters are given in Table 47.2.

A mass of 8 kg was used as a source, triggered on a circular steel plate with a diameter of 200 mm, connected to the seismograph with a trigger geophone. The SYSMATRACK software will provide the primary raw seismic plots obtained from the 24 channels to process the data.

Modelling

A geotechnical model formed synthetically is characterized by the thickness, density, Poisson's ratio, and the speed of the S and P waves. It simulates the theoretical dispersion curve, which shows the speed and length of the wave as a function of the wave relates correlation:

$$v = \lambda \times \nu \tag{47.1}$$

An overlay of the theoretical dispersion curve is obtained using actual experimental data after altering the parameters of the synthetic geotechnical model. Inversion is a technique for creating a velocity profile of varying degrees of stiffness. The phase velocity–frequency spectrum is shown in Fig. 47.6.

Data Processing

MASW data processing consists of the following steps: importing tracks, spectral analysis, inversion, and determining velocity profiles. The raw data obtained from the seismograph has been processed with the EASYMASW analysis software. Spectral analysis is the process to get the experimental dispersion curve by performing a double discrete Fourier transform in time and space. In general, the scattering of

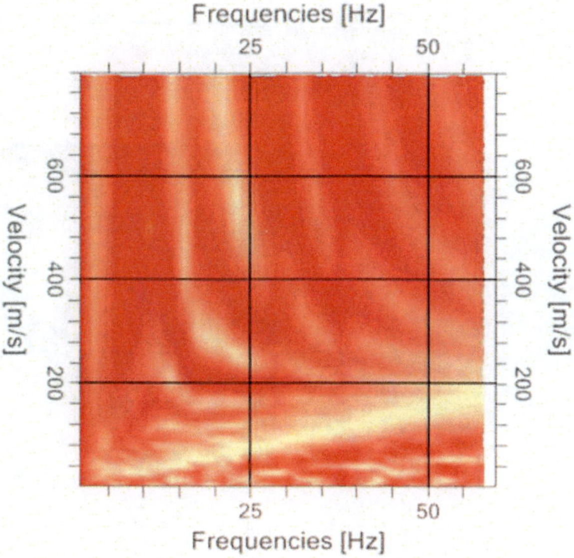

Fig. 47.6 Phase velocity–frequency spectrum

surface waves is related to the properties of the underground soil since different frequencies represent different soil thicknesses and travel at different speeds.

The dispersion curve obtained from the geophones is critical in generating the required SWV profile. It is an iterative process that compares the profile of the starting shear wave to the established theoretical curve. The software applies a heuristic algorithm during the inversion process to determine the soil stratigraphy model, which produces a dispersion curve identical to that acquired from the field survey, as shown in Fig. 47.7. An inversion process has compared the actual SPT data of a typical borehole in the study area with the theoretical dispersion curve. It proves that the correlation between SPT and MASW data is well matched as per NEHRP guidelines. The dispersion curve after inversion gives the corresponding shear wave velocity profile for the specified site. A typical sample of shear wave velocity profile for representative site NC-1 is as shown in Fig. 47.8.

Results and Interpretation

The average shear wave velocities for the first 30 m soil depth are deduced from the dispersion spectrum. Each dispersion curve is inverted to create the 1D shear wave velocity profiles. As a result of the data processing, the vertical SWV profile is attained. The processing is carried out without considering the saturation condition shown in Table 47.3.

The results are also presented by varying the groundwater table depths for 15 soil sample sites. Water table depth has changed in the range of 0–8 m from the

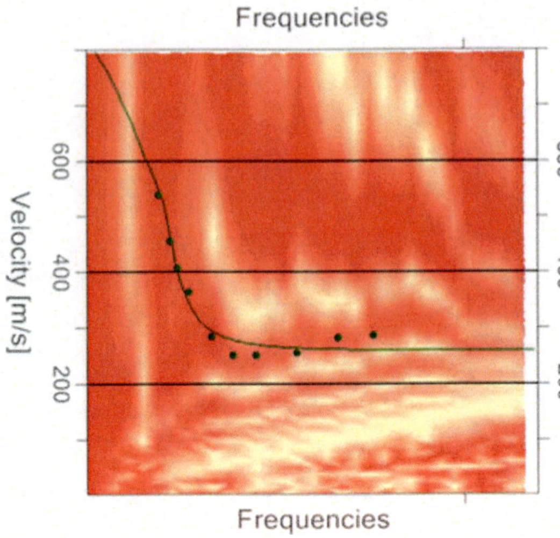

Fig. 47.7 Dispersion curve obtained after inversion

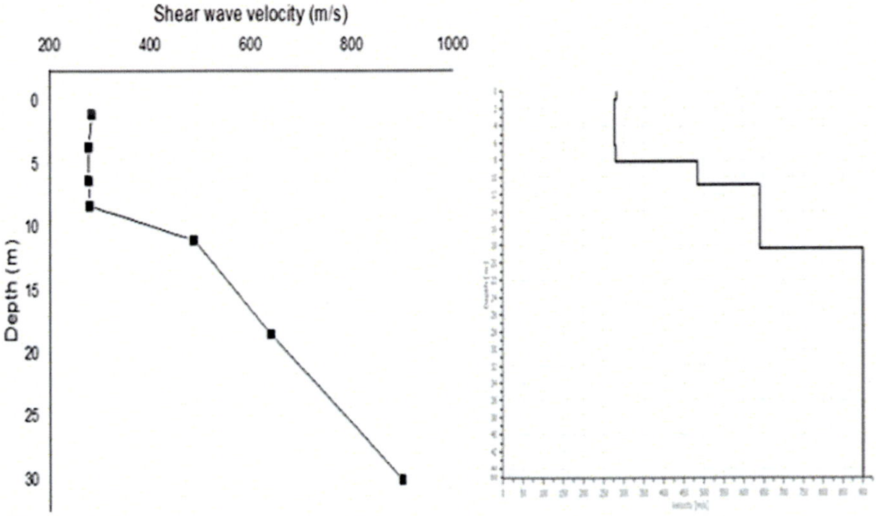

Fig. 47.8 Shear wave velocity profile for sample site NC-1

ground surface, and Fig. 47.9 shows the variation of shear wave velocity profile with different water table levels for sample site NC-1. The comparison of average shear wave velocities with varying water tables is also shown in Table 47.4.

Table 47.3 Average shear wave velocity of top 30 m soil depth at field condition

Code	Site	Lattitude	Longitude	(Vs_{30}) m/s
NC1	Malayamma Road	N 11° 20′ 19.085	E 75° 58′ 09.882″	540.18
NC2	Temple Road	N 11° 18′ 35.93″	E 75° 54′ 49.231″	526.01
NC3	Spring Valley	N 11° 19′ 27.764″	E 75° 56′ 14.558″	538.68
NC4	Guest House	N 11° 19′ 7.158″	E 75° 55′ 54.623	502.73
NC5	Mega Ladies Hostel	N 11° 18′ 57.864″	E 75° 55′ 50.862	483.07
NC6	East Campus	N 11° 19′ 24.564″	E 75° 56′ 16.794″	547.23
NC7	Men's Hostel	N 11° 19′ 12.708″	E 75° 56′ 15.131″	532.20
NC8	Surveying Field	N 11° 19′ 19.934″	E 75° 56′ 2.812″	492.15
NC9	Bhaskara Hall	N 11° 19′ 15.189″	E 75° 56′ 1.03″	521.53
NC10	Architecture Dept	N 11° 19′ 20.34″	E 75° 56′ 12.584″	463.00
NC11	Waste Treatment Plant	N 11° 18′ 54.887″	E 75° 55′ 46.745″	601.25
NC12	Water Resources Lab	N 11° 19′ 23.11″	E 75° 56′ 9.594″	578.59
NC13	Geotechnical Lab	N 11° 19′ 20.759″	E 75° 56′ 6.318″	465.43
NC14	Administrative Block	N 11° 19′ 12.595″	E 75° 55′ 48.645″	502.02
NC15	SBI Junction	N 11° 19′ 31.062″	E 75° 55′ 39.372″	591.08

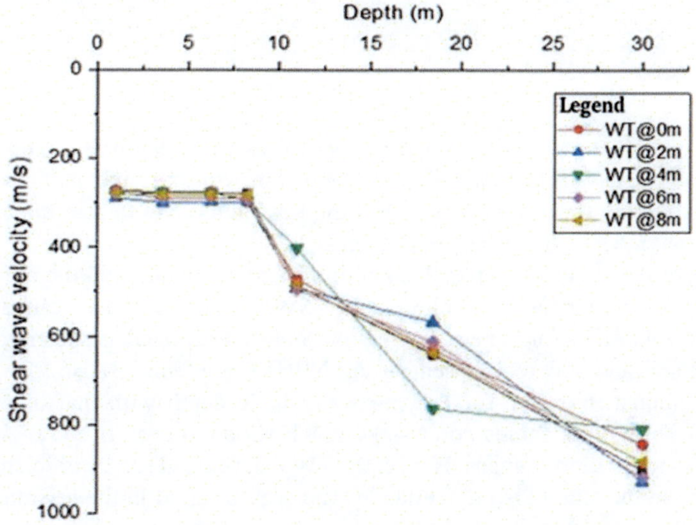

Fig. 47.9 Variation of shear wave velocity for site NC-1

Table 47.4 Average shear wave velocity at different depths of ground water tables

Code	Average V_{S30} (m/s) for depth of water table below ground surface				
	0 m	2 m	4 m	6 m	8 m
NC1	547.71	547.90	543.04	544.67	541.43
NC2	530.14	533.78	532.35	529.28	527.82
NC3	539.22	537.96	539.09	535.23	537.29
NC4	494.00	494.04	498.36	497.66	499.90
NC5	476.42	470.59	475.64	473.42	482.86
NC6	530.90	541.09	543.62	546.32	548.13
NC7	531.91	530.36	528.38	530.24	532.85
NC8	490.64	489.56	490.44	491.78	491.87
NC9	516.12	512.85	514.44	517.30	519.80
NC10	461.25	461.03	462.02	461.29	461.79
NC11	589.71	589.87	589.82	591.41	599.60
NC12	567.70	574.94	575.39	579.01	576.59
NC13	461.20	461.99	461.95	462.20	462.52
NC14	495.76	492.01	496.93	501.10	504.18
NC15	579.62	578.70	579.64	585.40	592.72

Conclusion

Shallow seismic geophysical tests are a powerful tool for identifying underground layers of soil. MASW acts as an efficient tool that provides a quick and inexpensive means of extracting shear wave velocities from Rayleigh waves to determine various properties that meet engineering requirements.

This paper discussed the MASW system of field surveying and the application in the geotechnical means of development of SWV profile of the Kattangal region and their site classification. The study region can be categorized as site class C with very dense soil and soft rock based on the NEHRP guidelines. Also, the average SWV was almost the same for varying water table depth with that of dry state conditions. Thus, V_{S30} values don't show much variation even in the region with significant precipitation changes. Hence, the SWV developed can be efficiently used to characterize the study area dynamically and may be used in the seismic hazard analysis in the near future.

References

1. Anbazhagan P, Sitharam TG (2008) Mapping of average shear wave velocity for Bangalore region: a case study. J Environ Eng Geophys 13(2):69–84
2. Bessason B, Erlingsson S (2011) Shear wave velocity in surface sediments. JÖKULL 61

3. Borcherdt RD (1991) GIS mapping of strong ground shaking for seismic zonation. In: Proceedings Pacific conference on earthquake engineering, Auckland, New Zealand, Nov 1991, pp 20–24
4. Joyner WB, Fumal TE (1984) Use of measured shear wave velocity for predicting geologic site effects on strong ground motion. In: Proceedings of the 8th world conference on earthquake engineering, San Francisco, CA, pp 777–783
5. Karl L, Haegeman W, Degrande G (2006) Determination of the material damping ratio and the shear wave velocity with the seismic cone penetration test. Soil Dyn Earthq Eng 26:1111–1126
6. Mari JL (1984) Estimation of static correction for shear-wave profiling using the dispersion properties of love waves. Geophysics 49:1169–1179
7. NEHRP (2003) NEHRP recommended provisions for seismic regulations for new buildings and other structures (FEMA 368 and 369). 2000. Building Seismic Safety Council, National Institute of Building Sciences, Washington, DC
8. Park CB, Miller RD, Xia J, Ivanov J (2007) Multichannel analysis of surface waves (MASW)—active and passive methods. Lead Edge 26:60–64
9. Park CB, Miller RD, Xia J (1999) Multichannel analysis of surface waves. Geophysics 64(3):800–808
10. Richart FE, Hall JR, Woods RD (1970) Vibrations of soils and foundations. Prentice-Hall, Englewood Cliffs, NJ
11. Sebastiano F (2018) Guidelines for the good practice of surface wave analysis: a product of the Inter PACIFIC project. Bull Earthq Eng 16:2367–2420
12. Stokoe II KH, Wright SG, Bay JA, Roesset JM (1994) Characterization of geotechnical sites by SASW method. In Geophysical characterization of sites, pp 15–25
13. Sairam B, Singh AP, Vandana P, Sumer C, Ravi KM (2019) VS30 mapping and site characterization in the seismically active intraplate region of Western India: implications for risk mitigation. Near Surf Geophys 17:533–546
14. Xia J, Miller RD, Park CB (1999) Estimation of near surface shear-wave velocity by inversion of Rayleigh wave. Geophysics 64(3):691–700
15. Xia J, Miller RD, Xu Y, Luo Y, Chen C, Liu J, Ivanov J, Zeng C (2009) High frequency Rayleigh-wave method. J Earth Sci 20:563–579

Chapter 48
Role of Damping in Response of Hammer Foundation

Partha Gupta, Swagata Bisoi, and B. C. Chattopadhyay

Introduction

Hammer foundation is a common element seen in many mechanical engineering works. Such foundations are used to house an ANVIL on which a RAM is allowed to fall in a controlled fashion from a specified height for stamping or forging works. In earlier days, weights of the ram and height of fall were small and the energy of impact due to falling ram on the anvil was not very high. Older practice for the design of the foundation of such cases was to provide a massive block foundation to house the anvil. Generally, the weight of the block provided was nearly 120 times of the weight of the ram [1]. However, with industrial advancement over the world, the adopted weight of ram and height of fall of the ram are getting increased for industrial reasons and as a result, the impact energy of the falling ram on the anvil are also becoming large. The containment of its effect on resulting vibrations of the foundation to keep the right quality of stamping or forging works and also prevention of the propagation of vibration from the foundation to adjoining structures in the establishment become a serious concern for acceptable design practice. Apart from making such hammer foundation safe from bearing capacity failure during impact, for industrial requirement the amplitude of motion of the foundation is required to be maintained at permissible limit value for the safe working condition in the working enclosure and to keep propagation of vibration to adjoining structure to an allowable limit. To limit the amplitude of vibration due to the impact of falling ram on the anvil, Barkan [1] first introduced the method of design and construction by introducing an

P. Gupta (✉)
Geotechnical Engineering, Meghnad Saha Institute of Technology, Kolkata 700150, India
e-mail: parthagte@gmail.com

S. Bisoi · B. C. Chattopadhyay
Department of Civil Engineering, Meghnad Saha Institute of Technology, Kolkata 700150, India
e-mail: swagata.bisoi@msit.edu.in

K. Muthukkumaran et al. (eds.), *Soil Dynamics, Earthquake and Computational Geotechnical Engineering*, Lecture Notes in Civil Engineering 300,
https://doi.org/10.1007/978-981-19-6998-0_48

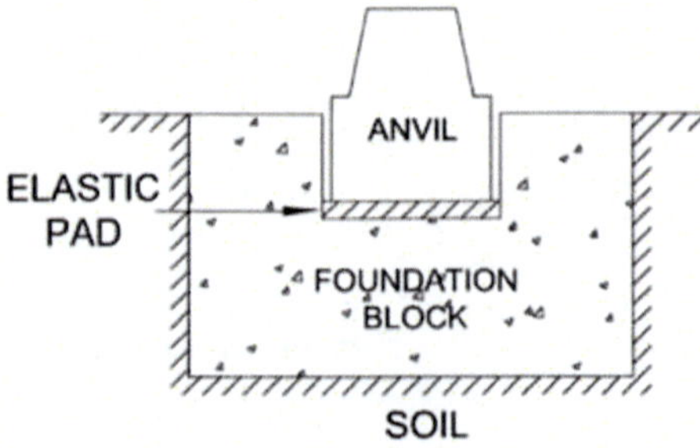

Fig. 48.1 Modified hammer foundation according to Barkan [1]

elastic pad between the anvil and the foundation (Fig. 48.1). The impact of the falling ram was faced by the elastic pad support through its elastic spring effects, and the subsequent reaction of the foundation is resisted by the elastic spring characteristics of the supporting soils. Relevant design details were also provided by the Barkan [1]. Presently, the design of hammer foundation in India is practiced following the above principle accepted by the IS: 2974 (Part-II)—1980. For design, the foundation is assumed to be placed without any embedment into the soil with a horizontal ground surface. Further, the damping property of the supporting soil mass is also neglected.

Literature Review

Prevailing Method of Design of Hammer Foundation

Previously, it was believed that foundation for a hammer should be massive and must be embedded at a considerable depth. Main objective was to provide such dimension so that static elastic settlement would be larger than the amplitude of vertical vibration. Barkan [1] proposed the insertion of an elastic pad between the anvil and the provided foundation to reduce the impact of falling ram to the foundation resulting decrease in amplitude. IS: 2974 (Part-II)—1980 had adopted the principle of design of hammer foundation presented by Barkan [1] (Fig. 48.1).

Indian Standard Method of Design: (IS: 2974 (Part-2)—1980) [3]

Following Barkan's method of design, foundation system is represented in Fig. 48.2, where m_1 = mass of foundation and frame if the latter is mounted on the foundation, m_2 = mass of anvil, $K_1 = C'_u . A_1$ = equivalent spring constant of soil under consideration, $C'_u = \lambda . C_u$ = modified coefficient of uniform elastic compression to take into account impact condition which is different from periodic loading, λ

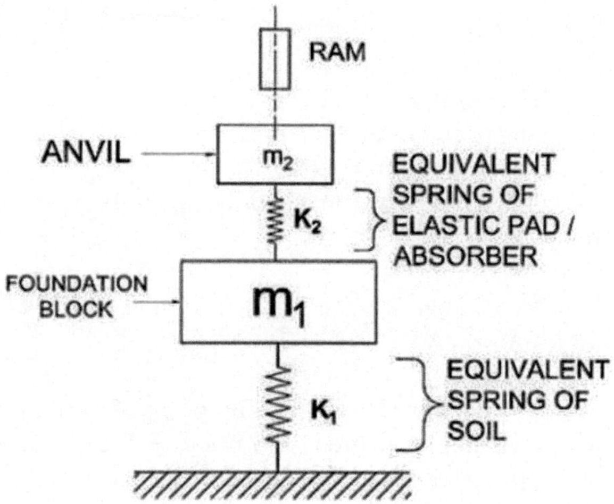

Fig. 48.2 Schematic diagram for hammer foundation as per IS 2974 (Part-2)—1981

= multiplying factor that governs the relationship between C_u and C'_u for impact depending upon soil type, C_u = coefficient of uniform elastic compression, $K_2 = \frac{E}{b} \times A_2$ = equivalent spring constant of the elastic pad under anvil, E = Young's modulus of the pad material, b = thickness of the pad, A_1 = area of foundation in contact with soil, A_2 = area of pad and $z_1,\ z_2$ = displacement vector of foundation and anvil, respectively, from the equilibrium position. The equations of motion in free vibration are:

$$m_1.\ddot{z}_1 + K_1.z_1 + K_2.(z_1 - z_2) = 0 \tag{48.1}$$

$$m_2.\ddot{z}_2 + K_2.(z_2 - z_1) = 0 \tag{48.2}$$

The resulting amplitudes are:

$$z_1 = -\frac{\left(\omega_{na}^2 - \omega_{n1}^2\right)\left(\omega_{na}^2 - \omega_{n2}^2\right)}{\left(\omega_{n1}^2 - \omega_{n2}^2\right)\omega_{na}^2\omega_{n1}}v_0 \tag{48.3}$$

$$z_2 = -\frac{\left(\omega_{na}^2 - \omega_{n1}^2\right)}{\left(\omega_{n1}^2 - \omega_{n2}^2\right)\omega_{n2}}v_0 \tag{48.4}$$

where ω_{n1} and ω_{n2} are the natural frequencies of the system, ω_{na} = circular natural frequency of the anvil on the pad = $\sqrt{\frac{K_2}{m_2}}$ and ω_{nl} = Limiting natural frequency of the foundation and anvil on the soil = $\sqrt{\frac{K_1}{m_{1+}m_2}}$, and v_0 is the velocity of the anvil after impact.

Fulan [5]

Fulan [5] presented a method to analyse hammer foundation considering the effect of damping of supporting soil mass for which schematic presentation of its hammer foundation is shown in Fig. 48.3.

According to Fulan, maximum amplitude of the foundation considering soil damping is

$$A_{Z\max} = \frac{m_0 v_0 (1 + \Psi)}{\sqrt{k_z (m_1 + m_0)}} e^{\left(\frac{-D}{\sqrt{1-D^2}} \frac{\sqrt{1-D^2}}{D}\right)} \tag{48.5}$$

where v_0 = maximum speed of falling part of hammer (m/s), m_1 = mass of hammer frame, anvil and foundation (t-s^2/m), m_0 = mass of falling part of hammer (t-s^2/m), Ψ = influence coefficient of speed resilience in impact, D = damping ratio, C = damping coefficient of natural sub grade (t-s/m), K_z = compression rigidity of natural subgrade (t/m).

According to the analysis by Fulan, the following observations can be drawn:

1. The calculated amplitude of hammer foundation of damped one-degree-freedom decreases with the increase of damping ratio.
2. The circular frequency of free vibration of hammer foundation decreases with the increase of damping ratio, too.

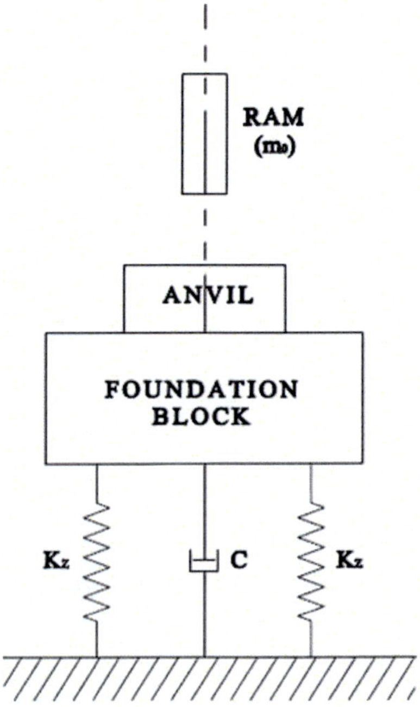

Fig. 48.3 Schematic diagram for hammer foundation after Fulan [5]

3. Time for the hammer foundation reaches its maximum amplitude of vibration independent of damping ratio. The greater is the damping ratio, shorter is the time required.

Chowdhury et al. [2]

Chowdhury et al. carried out an analysis of a Hammer Foundation shown schematically in Fig. 48.4 having non-proportional damping effect of soil incorporated by converting it into an approximate equivalent Rayleigh damping.

The free vibration of motion including damping is given by:

$$[m_1\, 0\, 0\, m_2][\ddot{x}_1 \ddot{x}_2] + [c_1 + c_2 - c_2 - c_2 c_2][\dot{x}_1 \dot{x}_2] + [k_1 + k_2 - k_2 - k_2 k_2][x_1 x_2] = 0 \tag{48.6}$$

where m_1 = mass of the foundation block plus frame resting on it if any, m_2 = mass of the anvil resting on elastic pad, k_1 = soil spring to be calculated from Barkan or Richart's formula, k_2 = spring value for the elastic pad, c_1 = damping of the soil to

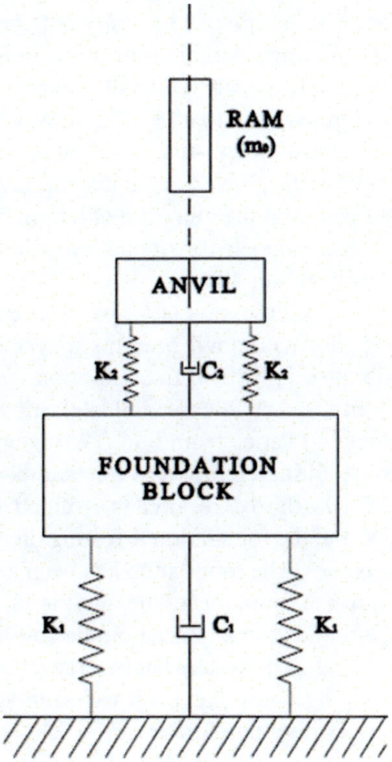

Fig. 48.4 Hammer foundation as defined by Chowdhury et al. [2]

be obtained from Richart's formula, c_2 = damping of the elastic pad, x_2 = amplitude vector of the anvil and x_1 = amplitude vector of the Foundation.

Vertical displacements are:

$$[X] = \sum_{i=1}^{N} \left\{ \frac{[\varphi]^{\mathrm{T}}[M][V_0]}{\omega_{di}} \right\} [\varphi] e^{-\zeta_i \omega_{ni} t} \sin \sin[\omega_i t] \quad (48.7)$$

where $[X]$ = deflection vector; (2×1), $[M]$ = a square mass matrix (2×2), $[V_0]$ = velocity of the anvil after impact, ω_{di} = damped frequencies, ω_{ni} = natural frequencies, $[\varphi]$ = eigen vectors of the system, ζ_i = damping ratio.

Scope and Objective

In the present codal analysis of hammer foundation as per IS: 2974 (Part-2)—1981, damping of foundation soil media has not been taken into account and since such damping has influence on the response of foundation vibration in vertical mode, it is understood that the damping of foundation soil will affect the mass of needed foundation to cause specified vertical vibration amplitude. Furthermore, the damping factor at the site can be easily obtained from same block vibration test suggested for determining coefficient of uniform elastic compression (C_{U}) of the soil at the site [4]. Thus, consideration of damping factor of foundation soil when considered in design will lead to affect the mass of the foundation block and thus a lot of economy of foundation construction may be affected.

With the above aim, a programme was undertaken to study the effect of damping at the site on the vibration of hammer foundation resting on horizontal ground. Four different models for simulation of resulting vertical free vibrations are considered as shown in Fig. 48.5.

In first case, no elastic pad is used between anvil and foundation and foundation together with anvil presents a system with one degree of freedom (Fig. 5a). The vibration is vertical free vibration.

In the second case, soil is elastic but having damping of various order (damping ratio (D) varies from 0 to 0.25) (Fig. 5b). The same foundation has been analysed for resulting vibration as free damped vertical vibration.

In the third case, anvil is assumed to be placed on the foundation through an elastic pad and the foundation is resting on elastic soil as adopted in IS: 2974, Part-2: 1980. (Fig. 5c). The foundation has been analysed in the conventional manner.

In the fourth case, the anvil is placed on the foundation through elastic pad and the foundation is placed on the soil having both elasticity and damping coefficients (Fig. 5d). The system has been analysed for free vibration under impact loading. Here, damping factor has been varied to study the effect of mass of foundation required for permissible vibration.

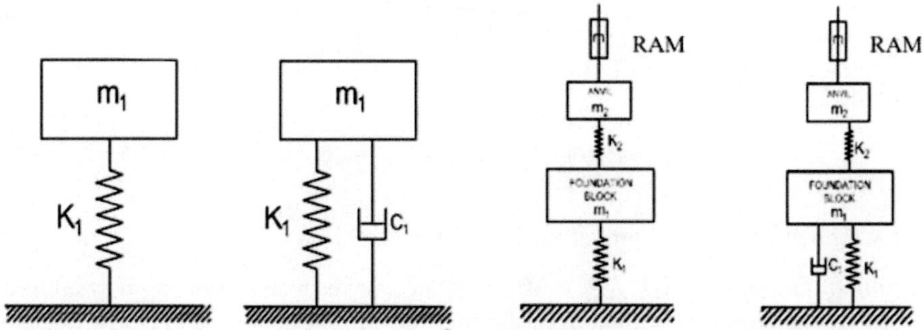

Model – I
Anvil is embedded in foundation which is resting on elastic soil.
(a)

Model –II
Anvil is embedded in foundation which is resting on soil having elastic and damping properties.
(b)

Model – III
Anvil is embedded in foundation through an elastic pad which is resting on elastic soil.
(c)

Model – IV
Anvil is embedded in foundation through an elastic pad which is resting on soil having elastic and damping properties.
(d)

Fig. 48.5 Considered models for analysis of vertical free vibration of the hammer foundation

Proposed Analysis

In this presentation, weight of the hammer foundation will be determined for all the four cases for a ram of weight W_0 falling from a height h on an anvil having weight W_2 and base area A_2. The values of the above parameters are chosen to the similar values used by Barkan [1] for detailed design of a hammer foundation placed on brown clay with some sand and silt. The chosen values are given below:

$W_0 = 5$ tons, $h = 1$ m, $W_2 = 100$ tons, $A_2 = 5.5$ m^2. Coefficient of uniform elastic compression (C_u) of the soil is 5×10^3 (t/m^3). Unit weight (γ_s) and safe bearing capacity (p_0) of soil below foundation are 1.6 (t/m^3) and 20 (t/m^2), respectively. For this ram falling from a height of 1 m, corresponding value of v_0 is calculated as 0.441 m/s.

Model I

This is the simplest condition where the foundation together with the anvil presents a system with one degree of freedom without effect of damping (Fig. 5a).

Equation of motion:

$$m_1.\ddot{z}_1 + k_1.z_1 = 0 \tag{48.8}$$

Equation of amplitude:

$$z_1(t) = A_1.\sin(\omega_{nz}.t + \varphi) \tag{48.9}$$

Applying boundary condition: at t = 0, z = 0 and $\dot{z}_1 = v_0$.

Maximum amplitude: $A_z = \frac{v_0}{\omega_{nz}}$. After putting the respective values, the required weight of the hammer foundation for permissible amplitude of 1 mm is 374 tons.

Model II

In addition, with previous model, damping of the soil has been considered (Fig. 5b).

Equation of motion:

$$m_1.\ddot{z}_1 + c_1.\dot{z}_1 + k_1.z_1 = 0 \tag{48.10}$$

Equation of displacement:

$$z_1 = A.e^{-D.\omega_{nz}.t}\sin(\omega_0.t + \varphi) \tag{48.11}$$

Applying boundary condition: at t = 0, z = 0 and $\dot{z}_1 = v_0$.

Maximum amplitude:

$$A_z = \frac{v_0}{\omega_{nz}.\sqrt{D^2 - 1}} e^{-\left(\frac{\Pi}{2}\right)\left(\frac{D}{\sqrt{1-D^2}}\right)} \tag{48.12}$$

After putting the respective values, the weight of the foundation (W_1) in terms of damping ratio 'D', can be represented as

$$W_1 = \left[\left[\frac{1}{2} \times \left[\left(49 \times \left[\frac{e^{-\pi\left(\frac{D}{\sqrt{1-D^2}}\right)}}{1 - D^2}\right] - 1\right) \pm \sqrt{\left(49 \times \left[\frac{e^{-\pi\left(\frac{D}{\sqrt{1-D^2}}\right)}}{1 - D^2}\right] - 1\right) - 1}\right]\right] \times 9.81\right] - 100 \tag{48.13}$$

Now, the required weight of foundation corresponding to different values of damping ratio (D) are given in Table 48.1 and shown in Fig. 48.6.

Table 48.1 Weight of foundation for different damping ratios (SDOF System)

Damping ratio (*D*) (%)	Weight of the foundation (W_1) tons
5	302
10	244
15	195
20	154
25	118

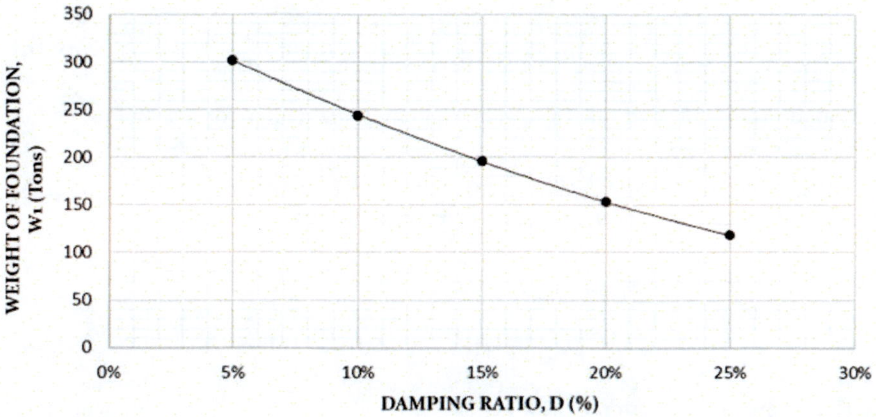

Fig. 48.6 Effect of damping on weight of foundation required to 1 mm maximum vertical vibration (SDOF System)

Model III

The method from IS: 2974, Part-2: 1980, has been used in this case to estimate the weight of foundation to keep vertical vibration near 1 mm (Fig. 5c).

The amplitude of foundation for the case is given as:

$$z_1 = -\frac{\left(\omega_{na}^2 - \omega_{n1}^2\right)\left(\omega_{na}^2 - \omega_{n2}^2\right)}{\left(\omega_{n1}^2 - \omega_{n2}^2\right)\omega_{na}^2\omega_{n1}} v_a \quad (48.14)$$

Substituting the corresponding values of ω_{n1}, ω_{n2}, ω_{na}, ω_{nl} and v_a in Eq. (48.14), the weight of foundation was found to be 270 tons.

Model IV

In this case, the anvil is placed on the foundation through elastic pad and the foundation is placed on the soil having both elasticity and damping coefficient (Fig. 5d).

The equation of motions for free vibration for this case is given by

$$m_1.\ddot{z}_1 + c_1.\dot{z}_1 + k_1.z_1 + k_2(z_1 - z_2) = 0 \quad (48.15)$$

$$m_2.\ddot{z}_2 + k_2(z_2 - z_1) = 0 \quad (48.16)$$

Equations (48.15) and (48.16) can be written in the matrix form as below

$$[m_1\, 0\, 0\, m_2]\{\ddot{z}_1 \ddot{z}_2\} + [c_1\, 0\, 0\, 0]\{\dot{z}_1 \dot{z}_2\} + [k_1 + k_2 - k_2 - k_2 k_2]\{z_1 z_2\} = 0 \quad (48.17)$$

and can be represented as:

$$[M]\{\ddot{Z}\} + [C]\{\dot{Z}\} + [K]\{Z\} = 0 \quad (48.18)$$

The eigen solution $\left[[K] - [M]\omega^2\right].\{\varphi\} = 0$ reduces to $\left|k_1 + k_2 - m_1.\omega^2 - k_2 - k_2 k_2 - m_2.\omega^2\right| = 0.$

which on expansion gives two roots ω_1 and ω_2 of the equation, which are the two coupled natural frequency of the system. For each value of ω_1 and ω_2, one gets corresponding eigen vector $[\varphi]^1$ and $[\varphi]^2$, respectively.

Complete eigen vector for the system, $\{\varphi\} = \left\{\varphi_1^1\, \varphi_1^2\, \varphi_2^1\, \varphi_2^2\right\}$ and $\{\varphi\}^{\mathrm{T}} = \left\{\varphi_1^1 \varphi_2^1 \varphi_1^2 \varphi_2^2\right\}$.

In the whole procedure, damping matrix is non-proportional and do not decouple; hence, a transformation is made to make non-proportional damping to proportional Rayleigh damping. By considering the two modes separately (i.e. φ_1 and φ_2) and comparing the above equations, from mathematical symmetry, it can be written that

$$\sum_{i=1}^{2} 2D_i\omega_i = \{\varphi_i\}^{\mathrm{T}}[C]\{\varphi_i\} \quad (48.19)$$

From Eq. (48.19), two values of damping ratio can be obtained for two modes of vibration (i.e. D_1 and D_2).

An orthogonal transformation of Eq. (48.18) is performed to convert the coupled matrix to uncoupled one:

$$\left[\{\varphi\}^{\mathrm{T}}[M]\{\varphi\}\right]\{\ddot{Z}\} + \left[\{\varphi\}^{\mathrm{T}}[C]\{\varphi\}\right]\{\dot{Z}\} + \left[\{\varphi\}^{\mathrm{T}}[K]\{\varphi\}\right]\{Z\} = 0 \quad (48.20)$$

which decouples to

$$\sum_{i=1}^{n}\left[\{\ddot{z}_i\}+2D_i\omega_i\{\dot{z}_i\}+\omega_i^2\{z_i\}\right]=0 \tag{48.21}$$

Here, $\{Z\}$ is coupled and to uncoupled it; $\{X\}=\{\varphi\}\{Z\}$

where $\{\varphi\}$ = transformation matrix which transform a couple matrix to uncouple one.

Then

$$\{X\}=\sum_{i=1}^{n}\{\varphi\}\{z_i\} \tag{48.22}$$

Equation (48.21) is analogous to the equation of motion for a single degree freedom system of free damped condition. Hence, for under damped condition, the solution is given by:

$$x=e^{-D\omega_n t}[C_1\sin\sin(\omega_d t)+C_2\cos\cos(\omega_d t)] \tag{48.23}$$

where $\omega_d=\omega_n\sqrt{1-D^2}$.

For our condition, in the transformed coordinates, Eq. (48.23) gets decoupled; hence,

$$\{z_i\}=e^{-D_i\omega_{ni}t}[A_i\sin\sin(\omega_{di}t)+B_i\cos\cos(\omega_{di}t)] \tag{48.24}$$

$$\{X\}=\sum_{i=1}^{n}\{\varphi\}e^{-D_i\omega_{ni}t}[A_i\sin\sin(\omega_{di}t)+B_i\cos\cos(\omega_{di}t)] \tag{48.25}$$

Multiplying both side of Eq. (48.25) by the term $\{\varphi\}^{\mathrm{T}}[M]$, one gets

$$\{\varphi\}^{\mathrm{T}}[M]\{X\}=\sum_{i=1}^{n}\{\varphi\}^{\mathrm{T}}[M]\{\varphi\}e^{-D_i\omega_{ni}t}[A_i\sin\sin(\omega_{di}t)+B_i\cos\cos(\omega_{di}t)] \tag{48.26}$$

Now imposing boundary condition at, $t=0$; $\{X\}=0$ and $\{\dot{X}\}=v_0$. From Eq. (48.25), one gets the equation of amplitude:

$$\{X\}=\sum_{i=1}^{n}\left[\frac{\{\varphi\}^{\mathrm{T}}[M]v_0}{\omega_{di}}\right]\{\varphi\}e^{-D_i\omega_{ni}t}\sin\sin(\omega_{di}t) \tag{48.27}$$

Table 48.2 Weight of foundation for different damping ratios (2-DOF System)

Damping ratio (D) (%)	Weight of the foundation (W_1) tons
5	161
10	147
15	135
20	125

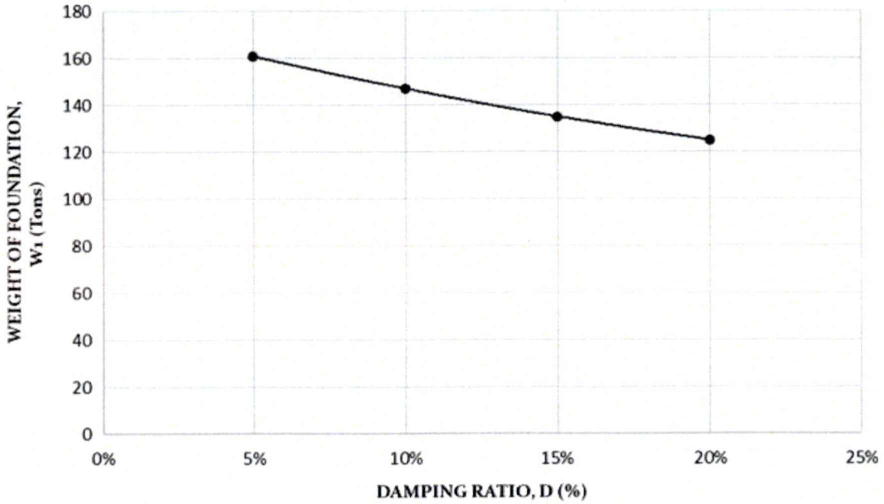

Fig. 48.7 Effect of damping on weight of foundation required to 1 mm maximum vertical vibration (2-DOF System)

For a particular value of damping, maximum amplitude is calculated by varying the weight of the foundation and thereafter the required weight of the foundation corresponding to 1 mm. Amplitude was obtained. The same procedure was repeated for various value of damping ratio, and corresponding weight of foundation is given in Table 48.2 and plotted against damping ratio in Fig. 48.7.

Conclusions

In the present design of hammer foundation, damping of foundation soil is neglected. Four different methods have been presented without and with damping of the foundation soil. The following conclusions are made:

1. If the anvil is embedded rigidly with the foundation, mass of the foundation required is much higher than the mass when damping of soil is consider for a permissible vibration amplitude.

2. If the anvil is placed over the foundation through an elastic pad for same permissible vertical vibration amplitude, the required mass of the foundation become much lesser than when no elastic pad is provided. This is true for the cases when damping is considered or not.
3. For hammer foundation, when anvil is placed over an elastic pad over the foundation for the same permissible vertical vibration amplitude, required mass gradually decreased with increase in available damping ratio.

References

1. Barkan DD (1962) Dynamics of bases and foundation. McGraw-Hill
2. Chowdhury I, Ghosh B, Dasgupta SP (2002) Analysis of hammer foundation considering soil damping. In: ResearchGate, Jan 2002
3. IS 2974-Part 2 (1980) Code of practice for design and construction of machine foundations, Part 2: Foundations for impact type machines (hammer foundations). Bureau of Indian Standards, New Delhi
4. IS 5249-Part 2 (1992) Determination of dynamic properties of soil—method of test. Bureau of Indian Standards, New Delhi
5. Fulan P (1988) Theory and experiment of hammer foundation vibration. In: Proceedings: second international conference on case histories in geotechnical engineering, 1–5 June 1988, St. Louis, Mo., Paper No. 4.05

Chapter 49
Liquefaction Analysis of Earth Embankment Dam Structure

Neetha Prabhakar and K. Rangaswamy

Introduction

The earth embankment dam is an artificial barrier constructed to store water for numerous purposes. Earth embankment must resist the horizontal loads due to earth and water pressures from storage containments or reservoirs. The most criticized case concerning seismic behaviour is earthquake-induced liquefaction of the foundation soil or within the dam. Liquefaction occurs in loose or saturated cohesionless soils and sensitive clays, resulting in large permanent deformations of the ground. Pressures generated in liquefied sand cause excess water to force its way upwards to the ground surface. There have been numerous researches over the past years, mainly after the San Fernando earthquake (1971), which led to the Lower San Fernando dam's disastrous failure and critical displacements of Upper San Fernando dam. The design and the construction of earth embankment dam structures are among the main challenges in geotechnical engineering because of variations in the foundation conditions and available construction materials. The soils that are vulnerable to liquefaction may affect the rate of pore pressure built up. It results in loss of strength of saturated sand due to sudden or cyclic loading, which causes severe damage for slight disturbances of soil fabric.

Some of the past research studies carried out on seismic stability and liquefaction susceptibility of dikes and dams. Melo and Sharma [1] performed a parametric study on a soil embankment of varied soil and seismic parameters. The weighted average seismic coefficients were analysed for possible correlations. It provides a more rational approach for selecting seismic coefficients for pseudo-static analysis.

N. Prabhakar (✉)
National Institute of Technology, Calicut 673601, India
e-mail: Prabhakarneetha@gmail.com

K. Rangaswamy
Faculty, National Institute of Technology, Calicut 673601, India

K. Muthukkumaran et al. (eds.), *Soil Dynamics, Earthquake and Computational Geotechnical Engineering*, Lecture Notes in Civil Engineering 300,
https://doi.org/10.1007/978-981-19-6998-0_49

Petalas et al. [2] validated and verified the UBC3D-PLM model implemented in PLAXIS as a user-defined soil model is presented. It predicted undrained triaxial and simple shear behaviour of loose Syncrude sand and is compared with experimental data.

Daftari and Kudla [3] used the UBCSAND model in PLAXIS software to evaluate liquefaction. It used the actual data of the superstition hills earthquake 1987 in the Imperial Valley. I also simulated using the UBC3D-PLM model.

Dinesh et al. [4] performed numerical analysis using UBCSAND and the PM4SAND model to evaluate soil liquefaction. This paper used these models in predicting the soil behaviour under cyclic loading and comparison.

This paper presents the results of finite element modelling of the seismic stability of the earth embankment dam structure using MIDAS GTS NX software. This study concentrates on evaluating the liquefaction analysis for the determination of vertical and horizontal deformations and excess pore pressures, which are generated as a result of seismic loading using MIDAS GTS NX software. It estimates its dynamic response using the nonlinear UBC 3D sand consecutive model.

Methodology

The seismic analysis of earthen embankment dam structure in liquefiable foundation soil deposit using time history analysis is performed using the finite element software MIDAS GTS NX. A typical cross-section of earth embankment dam structure and a representative ground motion are selected, and a UBC sand model is formed. A general assessment of the model is considered to determine the liquefaction susceptibility and uncertainties.

Geometry of the Model

The geometry of an earth embankment dam structure is selected from the published paper, as shown in Fig. 49.1. It has 30 m high, and the inclination of upstream and downstream slopes is 2H:1V. The top width of the dam crest is taken as 10 m, and the base width is 210 m in length. It is resting on a 10-m-thick sand foundation deposit underlain by 20-m bedrock. The upstream water level in the earth embankment dam is varied between 0 and 28 m depths. Figure 49.2 shows the 3D meshing geometry of the zoned earth embankment dam model created from the GT software.

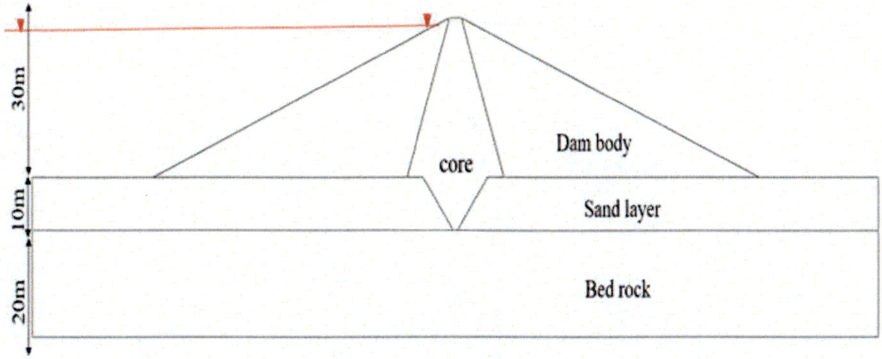

Fig. 49.1 Cross section of earth embankment dam model

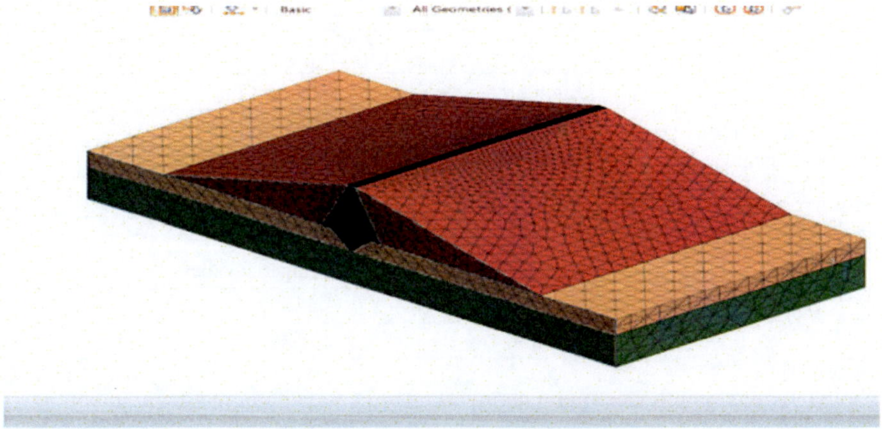

Fig. 49.2 3D meshing of zoned earth embankment dam model

Input Parameters

The core and bedrock are modelled as plane strain elements using a simple Mohr–Coulomb model. The dam body and foundation layer are modelled using the UBC sand model. The input soil parameters used for modelling are presented in Table 49.1. As it is seismic analysis, it requires horizontal motion data, i.e. acceleration versus time data; hence, the El Centro (Imperial Valley) earthquake input motion with a magnitude of 6.9 was provided as a seismic load. The earthquake incident happened on 12th May 2012. The details regarding the earthquake are as follows, shown in Table 49.2 (Fig. 49.3).

Table 49.1 Input model parameters for finite element model

ID	1	2	3	4
Name	Core	Dam body	Foundation layer	Bed rock
Model type	Mohr–Coulomb	Modified UBC	Modified UBC	Mohr–Coulomb
Elastic modulus (KN/m^2)	4.00E+05	7.28E+05	6.00E+05	2.00E+07
Poisson's ratio (ν)	0.35	0.3	0.2	0.2
Unit weight γ (KN/m^3)	20	19.7	20	23
Unit weight (Saturated, KN/m^3)	21	21.8	21	23
Cohesion (c) (KN/m^2)	300	0	0	1000
Int. friction angle, φ, °	35.6	–	–	43
Damping ratio	0.05	–	–	0.05
K_0	0.5	–	–	0.6

Table 49.2 El Centro earthquake details (official site-meteorological department India)

Origin time	04:36:47
Latitude	32° 43′ 59.99″ N
Longitude	−115° 29′ 59.99″ W
Depth (km)	16
Magnitude	6.9
Region	Imperial valley
Record duration	58 s
Direction	N-S

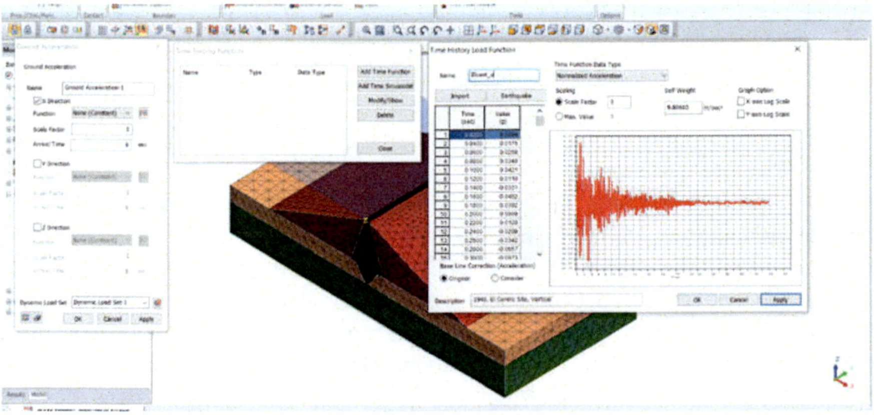

Fig. 49.3 Input time history of El Centro earthquake

Result and Discussions

A detailed numerical study was performed on earthquake-induced permanent deformations, pore pressure ratio, and post-seismic stability analysis. It studied the effect of liquefaction phenomena of an earth embankment dam subjected to seismic loading and various parameters. When a barrier is subjected to seismic excitation, movement of the dam and supporting ground takes place, resulting in deformation.

Earthquake Induced Permanent Deformations

When an earth embankment dam is subjected to seismic excitations, movement of the earth dam and supporting ground occurs, resulting in horizontal and vertical displacements. Figures 49.4 and 49.5 indicate that the maximum vertical displacements happened on the top of the dam crest, up to 15 s; the displacements are observed with small changes over time history. The vertical and horizontal displacement then changes direction to reach equilibrium eventually. In horizontal displacement, the changes are relatively uniform for up to 4 s; after which, the displacements change in a positive order and with a significant slope. The trend of horizontal displacements is towards the downstream side as a result of the water load. This oscillating motion is due to the force of the earthquake.

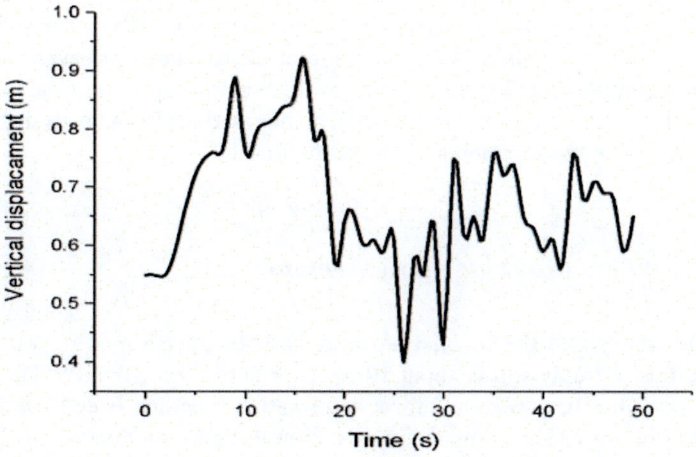

Fig. 49.4 Variation of vertical displacement with acceleration time history

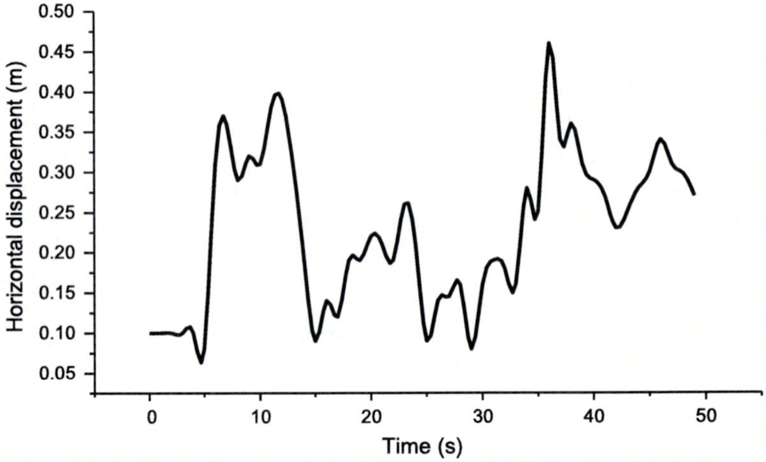

Fig. 49.5 Variation of horizontal displacement with acceleration time history

Effect of PGA on Settlement

The effect of PGA on settlement of earth embankment dam subjected to seismic motion with their pseudo-static coefficients of 0.1, 0.2, 0.3, and 0.4 is taken for this study. From Fig. 49.6, it can be inferred that indicating that the earth dam subjected to seismic input with 0.3-g PGA exhibited higher vertical displacements, and seismic load with 0.2-g PGA showed the least displacements irrespective of pseudo-static coefficients. Results show that the deformation of the earth dam is severe under the earthquake magnitude of 0.3 g, which reduces the bearing capacity of the dam body and leads to dam break. An earth dam may fail at intermediate PGA levels that depend on the dam body's natural period and soil properties.

Effect of Young's Modulus on Settlement

The seismic analysis of the foundation layer has been performed by varying the elastic modulus of sandy soil is about 20×10^3 kPa to 60×10^3 kPa. The highest settlement value has been obtained for a deformation modulus $E = 20 \times 10^3$ kPa, as shown in Fig. 49.7. The compaction operation leads to a reduction of the voids filled with air without expulsion of water. When soil is tight, the relative movements of the grains are reduced or eliminated. The high densities have the same effect.

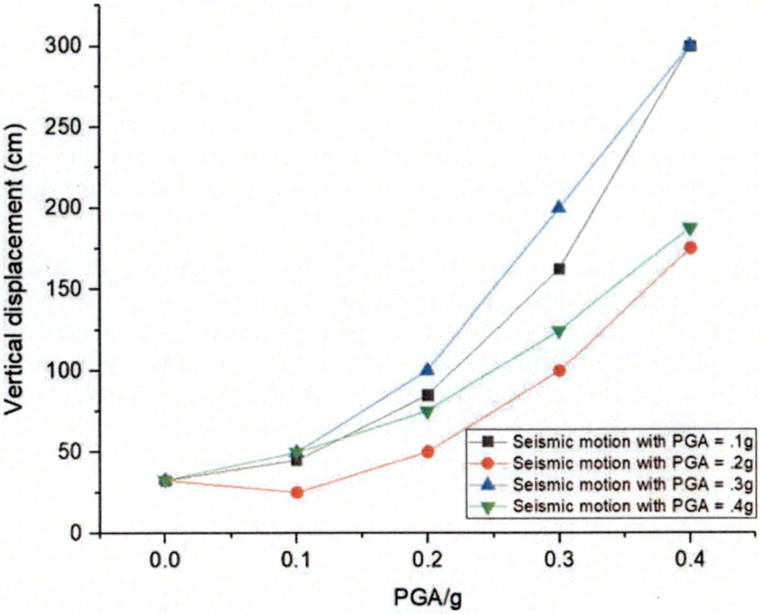

Fig. 49.6 Variation of vertical displacements on earth embankment dam with peak ground accelerations

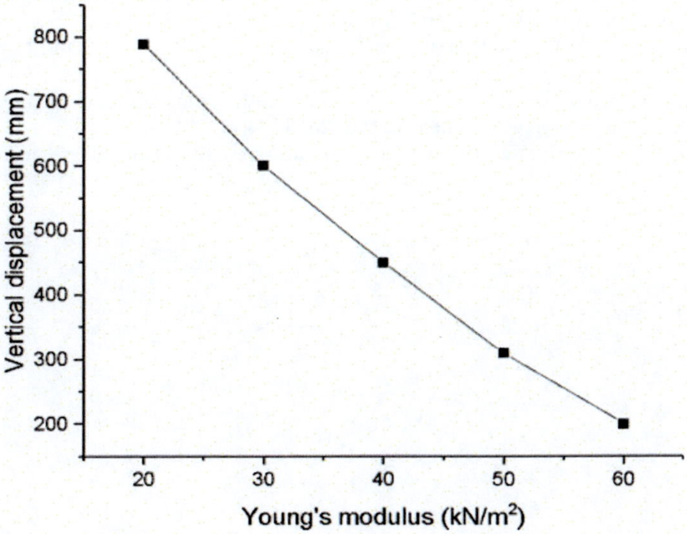

Fig. 49.7 Variation of settlement of the earth dam in relation to the values of the elastic moduli of the foundation layer

Conclusion

In this paper, the seismic behaviour of the earth embankment dam has been evaluated. The dynamic finite element analyses performed by varying the Young's modulus, and PGA resulted in liquefaction potential, permanent deformations, pore pressure ratio, and post-seismic stability. The conclusions are summarized in the following section.

- Settlement of earth dam structure subjected to seismic loads is decreased with the increase of elastic moduli of the foundation soil layer.
- Earth dam subjected to seismic input with 0.3-g PGA exhibiting higher displacements and seismic load with 0.2-g PGA showed least displacements irrespective of pseudo-static coefficients. It also found that vertical and horizontal displacements are increasing with the increase of pseudo-static coefficients.

References

1. Melo C, Sharma S (2004) Seismic coefficients for pseudo static slope analysis. In: 13th world conference on earthquake engineering. Vancouver, B.C., Canada, pp 1–15
2. Petalas A, Galavi V, Bringkreve R (2012) Validation and verification of a practical constitutive model for predicting liquefaction in sands. In: Proceedings of the 22nd European young geotechnical engineers conference 2012, Sweden, pp 167–172
3. Daftari, Kudla W (2014) Prediction of soil liquefaction by using UBC3D-PLM model in PLAXIS. Int J Civil Environ Eng 8(2):1–6
4. Dinesh N, Rajagopal K, Banerjee S (2016) Study of constitutive models for cyclic Liquefaction in sand. In: Indian geotechnical conference IGC 2016, p 4
5. IS:7894—Code of practice for stability analysis of earth dams, Bureau of Indian Standards, New Delhi (1975)